AF595522

Rare Monogenic Diseases: Molecular Pathophysiology and Novel Therapies

Rare Monogenic Diseases: Molecular Pathophysiology and Novel Therapies

Editor

Ivano Condò

MDPI • Basel • Beijing • Wuhan • Barcelona • Belgrade • Manchester • Tokyo • Cluj • Tianjin

Editor
Ivano Condò
Department of Biomedicine and Prevention
University of Rome Tor Vergata
Rome
Italy

Editorial Office
MDPI
St. Alban-Anlage 66
4052 Basel, Switzerland

This is a reprint of articles from the Special Issue published online in the open access journal *International Journal of Molecular Sciences* (ISSN 1422-0067) (available at: www.mdpi.com/journal/ijms/special_issues/rare_monogenic_diseases).

For citation purposes, cite each article independently as indicated on the article page online and as indicated below:

LastName, A.A.; LastName, B.B.; LastName, C.C. Article Title. *Journal Name* **Year**, *Volume Number*, Page Range.

ISBN 978-3-0365-7461-5 (Hbk)
ISBN 978-3-0365-7460-8 (PDF)

Cover image courtesy of Ivano Condò

Contents

International Journal of
Molecular Sciences

MDPI

Editorial

Rare Monogenic Diseases: Molecular Pathophysiology and Novel Therapies

Ivano Condò

Department of Biomedicine and Prevention, University of Rome Tor Vergata, 00133 Rome, Italy; ivano.condo@uniroma2.it

A rare disease is defined by its low prevalence in the general population. Although the fixed threshold slightly varies between different countries, a specific disorder is usually considered to be rare when it affects less than 50–60 in 100,000 individuals. The inherited defects originating from single gene mutations characterize the vast panorama of rare monogenic diseases. More than 4000 monogenic mutations account for at least 80% of all rare diseases, involving a broad spectrum of clinical features and pathophysiological mechanisms. Strikingly, rare monogenic diseases often appear during childhood and remain largely untreatable, underscoring the hard challenges in biology and medicine to identify the underlying molecular mechanisms and to develop effective therapies.

The Special Issue "Rare Monogenic Diseases: Molecular Pathophysiology and Novel Therapies" provides a collection of original research articles and systematic reviews focused on diverse conditions stemming from a pathogenic single gene mutation. However, we are aware of at least 6000 monogenic phenotypes and clearly this cannot be an exhaustive collection of all the rare monogenic diseases. The included articles cover current progress in pathogenesis mechanisms and experimental therapeutics concerning such diverse conditions as neurodegenerative or neurodevelopmental disorders, lysosomal storage diseases, coagulation disorders, hemoglobinopathies, kidney diseases and cancer.

Among rare monogenic diseases, close to 70% of these disorders affects the nervous system. In particular, predominant phenotypic manifestations are represented by neurodevelopmental and neurodegenerative alterations. Loi et al. present preclinical data supporting a new therapeutical strategy for the disease caused by mutations in X-linked cyclin-dependent kinase-like 5 (*CDKL5*) gene [1]. CDKL5 deficiency disorder (CDD) is characterized by defects in neuronal development and survival, a condition with no available therapy. Because single inhibitors able to target GSK-3β kinase and histone deacetylases (HDAC) have been shown to have neuroprotective effects in CDD mice, the authors demonstrate in vitro and in vivo the efficacy of a novel combinatorial therapy using Compound 11 (C11), a synthetic GSK-3β/HDAC dual inhibitor [1]. Trinucleotide repeat disorders, a group of neuropathological conditions caused by abnormal expansion of natural triplet repeats in a specific gene, are the focus of two systematic reviews. The expansion of CAG trinucleotide repeats within the coding sequence of huntingtin gene (*HTT*) generates the elongation of polyglutamine sequence in the mutated HTT protein and triggers Huntington's disease (HD), a neurodegenerative disorder examined by Tomczyk et al. [2]. This review describes the complex network of molecular and cellular changes affecting purine metabolism and signaling in HD. Based on the summarized dysfunctions, the authors analyze and propose potential therapeutic strategies able to target intracellular and extracellular adenosine levels [2]. On the other hand, the pathogenic expansion of CGG triplets occurs in the 5′ non-coding sequence of the Fragile X Mental Retardation 1 (*FMR1*) gene and leads to various early or late-onset Fragile-X-associated syndromes. Valor et al. review the molecular pathogenic mechanisms involved in the adult syndromes, a spectrum of gain-of-function disorders [3]. The authors analyze in detail the mechanistic roles of mutant *FMR1* mRNA (riboCGG), *FMR1*-derived long non-coding RNAs (lncRNAs), RNA-binding

Citation: Condò, I. Rare Monogenic Diseases: Molecular Pathophysiology and Novel Therapies. *Int. J. Mol. Sci.* **2022**, *23*, 6525. https://doi.org/10.3390/ijms23126525

Received: 23 May 2022
Accepted: 7 June 2022
Published: 10 June 2022

proteins and *FMR1*-derived cryptic polyglycine peptide (FMRpolyG). Moreover, an interesting description of established and potential biomarkers for adult Fragile-X-associated syndromes is provided [3].

Lysosomal storage disorders (LSDs) are metabolic diseases that result from genetic enzyme deficiencies. Although more than 50 genes are involved in specific pathogenesis, most storage disorders are systemic because of the ubiquitous functions of lysosomes. Currently, approved drugs for some LSDs include recombinant proteins for enzyme replacement therapy (ERT) and small molecules acting as pharmacological chaperone therapy (PCT). Unfortunately, these approaches present important therapeutic limitations. Total or partial loss of lysosomal α-galactosidase A (GLA) enzyme activity results in Fabry disease (FD), a multisystemic X-linked disorder. Naturally, aggressive disease manifestations develop more frequently in male FD patients. However, depending on the random X chromosome inactivation, female patients may exhibit milder signs and symptoms. Modrego et al. provide a systematic overview of the knowledge stemming from the countless mutations affecting the *GLA* gene in FD patients [4]. By analyzing the structure-function relationships of GLA enzyme, these authors illustrate the potential use of recombinant GLA mutants characterized by increased enzyme activity, improved protein stability or lower immunogenicity. Altogether, the proposed approach may generate novel therapeutics to promptly improve ERT and PCT for the treatment of FD [4]. Mucopolysaccharidosis type II (MPS II) is another X-linked LSD caused by a mutation in the gene encoding iduronate 2-sulphatase (*IDS*). This multisystemic disorder is partially treatable with ERT. However, the recombinant enzyme cannot cross the blood–brain barrier, thus no improvement is possible in central nervous system of MPS II patients with severe forms. Zapolnik and Pyrkosz evaluate the progresses of gene therapy as future alternative to cure MPS II patients [5]. By describing successes and failures of MPS II preclinical and clinical studies based on retroviruses, lentiviruses, adeno-associated viruses and genome editing technology, this review looks at possible achievements of ongoing and upcoming MPS II clinical trials centered on gene therapies [5].

Monogenic blood disorders may be associated with abnormalities in a protein either soluble in blood plasma or present within blood cells. For instance, mutations in the genes encoding coagulation factors critically affect the blood clotting mechanism and lead to bleeding disorders. The review by Rodríguez-Merchán et al. illustrates the promises of gene therapy in the context of hemophilia A, arising from mutations in coagulation factor VIII gene, and hemophilia B, triggered by mutations in coagulation factor IX gene [6]. This paper also inspects the main clinical trials on hemophilia patients to discuss opportunities and limits of current adenoviral vector-mediated gene transfer protocols [6]. Parahemophilia or Owren's disease, is an ultra-rare coagulation disorder associated with mutations in the factor V gene. Bernal et al. describe the molecular study performed to characterize two patients' families and the consequent discovery of a new mutation leading to severe factor Va deficiency [7]. β-thalassemia is instead a blood cells disorder, linked to mutations of the β-globin gene and reduced expression of the encoded adult hemoglobin (HbA) protein in erythrocytes. The original article by Zuccato et al. focuses on the induction of fetal hemoglobin (HbF) as experimental therapy to overcome HbA deficiency in β-thalassemia [8]. This study demonstrates that Cinchonidine and Quinidine, two natural alkaloids from Cinchona plants, are able to induce HbF expression in affected cells derived from β-thalassemia patients and to potentiate the similar effects of sirolimus, a drug currently under clinical evaluation to treat this hemoglobinopathy [8].

The selective expression or the particular role of specific genes in a single tissue explains the appearance of organ-specific inherited diseases. This is the case of genetic disorders of the kidney, which include dominant and recessive forms of cystic diseases, and renal tubulopathies. Mutations in polycystin-1 (*PKD1*) or -2 (*PKD2*) genes lead to autosomal-dominant polycystic kidney disease (ADPKD), whose gender-dependent phenotype was analyzed in the study by Talbi et al. [9]. These results, obtained in mice lacking PKD1 expression, show the involvement of intracellular Ca2+ levels in the more severe phenotype affecting male ADPKD animals. Altogether, identification of the molecular mechanisms underlying enhanced Ca2+ signaling and proliferation in cells from male kidneys may contribute to develop novel therapeutics for ADPKD [9]. The autosomal-recessive form of polycystic kidney disease (ARPKD) mostly arises from defects in the gene named polycystic kidney and hepatic disease 1 (*PKHD1*), whereas a minority of cases is linked to a second causative gene *DZIP1L*. To examine the still unclear molecular pathophysiology of ARPKD, Cordido et al. recapitulate known molecular disease mechanisms and possible therapeutic approaches, from cellular and animal models to clinical trials [10]. The knowledge of ARPKD pathogenic pathways, involving the epidermal growth factor receptor (EGFR) axis, the production of adenylyl cyclase adenosine 3′,5′-cyclic monophosphate (cAMP) and the activation of several protein kinases, begins to stimulate possible pharmacological interventions [10]. Inherited loss of function in various electrolyte transport proteins located along the nephron leads to two types of kidney tubulopathy with overlapping clinical symptoms: Gitelman and Bartter syndromes. The review by Nuñez-Gonzalez et al. aims to explain the different molecular basis of these difficult to diagnose monogenic syndromes. Moreover, the authors provide an overview of current therapeutic approaches and highlight the presence of common and specific options for Gitelman and Bartter patients [11].

Finally, the disparate group of rare monogenic diseases also include well-known genetic mutations that directly cause inherited cancers. These kinds of neoplasia occur by two classic types of germline mutations: inactivating alterations in a tumor suppressor gene or activating variants in a proto-oncogene. Multiple endocrine neoplasia type 1 (MEN1) is triggered by mutations in the tumor suppressor gene *MEN1* encoding for menin, a nuclear protein involved in broad transcriptional regulation. Given the absence of a direct genotype-phenotype correlation in tumorigenesis affecting patients with identical *MEN1* mutations, the review by Marini and Brandi explores the involvement of epigenetic regulations [12]. In light of the molecular effects of microRNA-24 (miR-24) on menin expression, the authors discuss a possible role of this miRNA in MEN1 development and progression. The potential use of RNA therapeutics able to target miR-24 in MEN1 patients is examined as well [12].

Overall, this collection of articles underlines how molecular therapies for a monogenic disorder may target either directly the mutated gene-product or the associated pathway(s). Indeed, as understanding of molecular mechanisms improves, novel therapeutics targeting different disease pathways become conceivable. By providing insights into specific molecular defects and broadly applicable therapies, these contributions may be a valuable resource to stimulate future scientific advances in the field of genetic diseases.

Funding: This research received no external funding.

Conflicts of Interest: The author declares no conflict of interest.

References

1. Loi, M.; Gennaccaro, L.; Fuchs, C.; Trazzi, S.; Medici, G.; Galvani, G.; Mottolese, N.; Tassinari, M.; Giorgini, R.R.; Milelli, A.; et al. Treatment with a GSK-3beta/HDAC Dual Inhibitor Restores Neuronal Survival and Maturation in an In Vitro and In Vivo Model of CDKL5 Deficiency Disorder. *Int. J. Mol. Sci.* **2021**, *22*, 5950. [CrossRef] [PubMed]
2. Tomczyk, M.; Glaser, T.; Slominska, E.M.; Ulrich, H.; Smolenski, R.T. Purine Nucleotides Metabolism and Signaling in Huntington's Disease: Search for a Target for Novel Therapies. *Int. J. Mol. Sci.* **2021**, *22*, 6545. [CrossRef] [PubMed]
3. Valor, L.M.; Morales, J.C.; Hervas-Corpion, I.; Marin, R. Molecular Pathogenesis and Peripheral Monitoring of Adult Fragile X-Associated Syndromes. *Int. J. Mol. Sci.* **2021**, *22*, 8368. [CrossRef] [PubMed]
4. Modrego, A.; Amaranto, M.; Godino, A.; Mendoza, R.; Barra, J.L.; Corchero, J.L. Human alpha-Galactosidase A Mutants: Priceless Tools to Develop Novel Therapies for Fabry Disease. *Int. J. Mol. Sci.* **2021**, *22*, 6518. [CrossRef] [PubMed]
5. Zapolnik, P.; Pyrkosz, A. Gene Therapy for Mucopolysaccharidosis Type II-A Review of the Current Possibilities. *Int. J. Mol. Sci.* **2021**, *22*, 5490. [CrossRef] [PubMed]
6. Rodriguez-Merchan, E.C.; De Pablo-Moreno, J.A.; Liras, A. Gene Therapy in Hemophilia: Recent Advances. *Int. J. Mol. Sci.* **2021**, *22*, 7647. [CrossRef] [PubMed]
7. Bernal, S.; Pelaez, I.; Alias, L.; Baena, M.; De Pablo-Moreno, J.A.; Serrano, L.J.; Camero, M.D.; Tizzano, E.F.; Berrueco, R.; Liras, A. High Mutational Heterogeneity, and New Mutations in the Human Coagulation Factor V Gene. Future Perspectives for Factor V Deficiency Using Recombinant and Advanced Therapies. *Int. J. Mol. Sci.* **2021**, *22*, 9705. [CrossRef] [PubMed]

8. Zuccato, C.; Cosenza, L.C.; Zurlo, M.; Lampronti, I.; Borgatti, M.; Scapoli, C.; Gambari, R.; Finotti, A. Treatment of Erythroid Precursor Cells from beta-Thalassemia Patients with Cinchona Alkaloids: Induction of Fetal Hemoglobin Production. *Int. J. Mol. Sci.* **2021**, *22*, 13433. [CrossRef] [PubMed]
9. Talbi, K.; Cabrita, I.; Schreiber, R.; Kunzelmann, K. Gender-Dependent Phenotype in Polycystic Kidney Disease Is Determined by Differential Intracellular Ca^{2+} Signals. *Int. J. Mol. Sci.* **2021**, *22*, 6019. [CrossRef] [PubMed]
10. Cordido, A.; Vizoso-Gonzalez, M.; Garcia-Gonzalez, M.A. Molecular Pathophysiology of Autosomal Recessive Polycystic Kidney Disease. *Int. J. Mol. Sci.* **2021**, *22*, 6523. [CrossRef] [PubMed]
11. Nunez-Gonzalez, L.; Carrera, N.; Garcia-Gonzalez, M.A. Molecular Basis, Diagnostic Challenges and Therapeutic Approaches of Bartter and Gitelman Syndromes: A Primer for Clinicians. *Int. J. Mol. Sci.* **2021**, *22*, 11414. [CrossRef] [PubMed]
12. Marini, F.; Brandi, M.L. Role of miR-24 in Multiple Endocrine Neoplasia Type 1: A Potential Target for Molecular Therapy. *Int. J. Mol. Sci.* **2021**, *22*, 7352. [CrossRef] [PubMed]

International Journal of
Molecular Sciences

MDPI

Article

Treatment with a GSK-3β/HDAC Dual Inhibitor Restores Neuronal Survival and Maturation in an In Vitro and In Vivo Model of *CDKL5* Deficiency Disorder

Manuela Loi [1,†], Laura Gennaccaro [1,†], Claudia Fuchs [1], Stefania Trazzi [1], Giorgio Medici [1], Giuseppe Galvani [1], Nicola Mottolese [1], Marianna Tassinari [1], Roberto Rimondini Giorgini [2], Andrea Milelli [3] and Elisabetta Ciani [1,*]

1 Department of Biomedical and Neuromotor Sciences, University of Bologna, Piazza di Porta San Donato 2, 40126 Bologna, Italy; manuela.loi3@unibo.it (M.L.); laura.gennaccaro@gmail.com (L.G.); claudiafuchs83@hotmail.com (C.F.); stefania.trazzi3@unibo.it (S.T.); giorgio.medici2@unibo.it (G.M.); giuseppe.galvani2@unibo.it (G.G.); nicola.mottolese@studio.unibo.it (N.M.); marianna.tassinari5@unibo.it (M.T.)
2 Department of Medical and Clinical Sciences, University of Bologna, 40126 Bologna, Italy; roberto.rimondini@unibo.it
3 Department for Life Quality Studies, University of Bologna, 47921 Rimini, Italy; andrea.milelli3@unibo.it
* Correspondence: elisabetta.ciani@unibo.it; Tel.: +39-0512091773; Fax: +39-0512091737
† Authors contributed equally to the work.

Abstract: Mutations in the X-linked cyclin-dependent kinase-like 5 (*CDKL5*) gene cause a rare neurodevelopmental disorder characterized by early-onset seizures and severe cognitive, motor, and visual impairments. To date there are no therapies for *CDKL5* deficiency disorder (CDD). In view of the severity of the neurological phenotype of CDD patients it is widely assumed that *CDKL5* may influence the activity of a variety of cellular pathways, suggesting that an approach aimed at targeting multiple cellular pathways simultaneously might be more effective for CDD. Previous findings showed that a single-target therapy aimed at normalizing impaired GSK-3β or histone deacetylase (HDAC) activity improved neurodevelopmental and cognitive alterations in a mouse model of CDD. Here we tested the ability of a first-in-class GSK-3β/HDAC dual inhibitor, Compound **11** (**C11**), to rescue CDD-related phenotypes. We found that **C11**, through inhibition of GSK-3β and HDAC6 activity, not only restored maturation, but also significantly improved survival of both human *CDKL5*-deficient cells and hippocampal neurons from *Cdkl5* KO mice. Importantly, in vivo treatment with **C11** restored synapse development, neuronal survival, and microglia over-activation, and improved motor and cognitive abilities of *Cdkl5* KO mice, suggesting that dual GSK-3β/HDAC6 inhibitor therapy may have a wider therapeutic benefit in CDD patients.

Keywords: *CDKL5* deficiency disorder; GSK-3β; HDAC6; dual inhibitor; neuronal survival; hippocampal defects; synapse development

Citation: Loi, M.; Gennaccaro, L.; Fuchs, C.; Trazzi, S.; Medici, G.; Galvani, G.; Mottolese, N.; Tassinari, M.; Rimondini Giorgini, R.; Milelli, A.; et al. Treatment with a GSK-3β/HDAC Dual Inhibitor Restores Neuronal Survival and Maturation in an In Vitro and In Vivo Model of *CDKL5* Deficiency Disorder. *Int. J. Mol. Sci.* **2021**, *22*, 5950. https://doi.org/10.3390/ijms22115950

Academic Editor: Ivano Condò

Received: 22 April 2021
Accepted: 28 May 2021
Published: 31 May 2021

1. Introduction

CDKL5 deficiency disorder (CDD) is a complex and severe neurodevelopmental disorder caused by mutations of the *CDKL5* gene [1], for which a cure is not available. Patients with CDD are characterized by early-onset seizures and severe cognitive, motor, visual, and autonomic disturbances [1–4]. Incidence varies from 1:40,000 to 1:60,000 [5], and due to the fact that *CDKL5* is located on the X chromosome, the prevalence of CDD among women is four times higher than in men. Genetic mutations of the *CDKL5* gene cause absence of a functional CDKL5 protein, a serine/threonine kinase that is highly expressed in the brain and, in particular, in neurons [6,7].

In vitro and in vivo models of CDD have helped to provide important insights into the mechanism of CDKL5 functions in neuronal development. CDKL5 has been found to regulate neuronal migration, axon outgrowth, dendritic morphogenesis, and synapse development in cultured rodent neurons as a model system [8–12]. Similarly, *Cdkl5* deficiency in mice impairs spine maturation and dendritic arborization of hippocampal and cortical neurons [13–18], indicating that CDKL5 plays a role in dendritic morphogenesis and synapse development. CDKL5 has been shown to regulate cell survival [19–21]. *CDKL5* deletion in human neuroblastoma cells induces an increase in cell death and in DNA damage-associated biomarkers [19]. During brain aging, loss of *Cdkl5* was shown to decrease neuronal survival in different brain regions such as the hippocampus, cortex, and basal ganglia in *Cdkl5* knockout (KO) mice, paralleled by increased neuronal senescence [22].

In view of the variety of cellular processes regulated by protein kinases and of the severity of the neurological phenotype of CDD patients, it is widely assumed that CDKL5 may have a very complex role in neurons, influencing the activity of a variety of intracellular pathways. Indeed, through a kinome profiling study, Wang and colleagues demonstrated that several signaling transduction pathways involved in neuronal and synaptic plasticity are disrupted in the forebrain of *Cdkl5* KO mice [23]. Among the signaling pathways whose function is altered in the absence of *Cdkl5*, the AKT-GSK-3β signaling pathway is particularly interesting due to its pivotal role in brain development and function [24,25]. Increased GSK-3β activity plays a role in several neurodevelopmental alterations that characterize *Cdkl5*-deficient brains. Importantly, pharmacological inhibition of GSK-3β activity rescues dendritic morphogenesis and synapse development in the *Cdkl5* KO mouse [26,27]. Similar benefits to defective dendritic spine number and dynamics were obtained with the AKT activator IGF-1 in *Cdkl5* KO mice [17]. However, inhibition of GSK-3β activity has been shown to have positive effects in juvenile but not in adult *Cdkl5* KO mice [26], suggesting that pharmacological interventions aimed at normalizing only impaired GSK-3β activity might not be sufficient to restore the defects of a complex disease such as CDD. Interestingly, epigenetic modulators, including histone deacetylase (HDAC) inhibitors, have recently been shown to improve neurodevelopmental and cognitive alterations in *Cdkl5* KO mice [11], suggesting that imbalanced protein acetylation might represent a molecular mechanism that underlies Cdkl5 function.

Combinatorial therapies have recently become one of the most successful drug development strategies for complex diseases. For instance, it was reported that combined inhibition of GSK-3β and HDACs induces synergistic effects compared to the single target drug, with a potential improved therapeutic efficacy [28]. Due to the complexity of CDD, it is worth hypothesizing that the combined inhibition of GSK-3β and HDACs by a multi-target drug might be more efficient then a single-target therapy. Here, we examined the effect of treatment with a recently synthetized first-in-class GSK-3β/HDAC dual inhibitor, Compound **11** (**C11**) [29], to rescue CDD-related phenotypes in in vitro and in vivo models of CDD.

2. Results

2.1. *Treatment with* **C11** *Restores GSK-3β and HDAC6 Activity in a Human Cellular Model of CDKL5 Deficiency*

Compound **11** (**C11**) was selected for its high dual activity against GSK-3β and histone deacetylase 6 (HDAC6) [29], which offer potential nontoxic therapeutic targets for the amelioration of CNS development [30,31]. In order to confirm the dual inhibitory activity of **C11** on GSK-3β and HDAC6, we treated a recently generated human neuronal cell model of *CDKL5* deficiency, the *CDKL5* knockout (KO) SH-SY5Y neuroblastoma cell line (SH-*CDKL5*-KO; [19]), with **C11** (1 and 10 μM), and, as a control for selective GSK-3β inhibition, with NP-12 (Tideglusib; 1 μM).

GSK-3β is a constitutively active serine/threonine kinase that is predominantly modulated by inhibitory serine-9 (Ser9) phosphorylation [32]. Consistently with previous evidence [19], we found that phosphorylation of GSK-3β at Ser-9 was reduced in SH-*CDKL5*-KO compared with parental SH-SY5Y cells (Figure 1A,B). SH-*CDKL5*-KO cells treated with **C11** underwent an increase in Ser-9-phosphorylated GSK-3β levels that, at a dose of 10 μM, became similar to those of parental SH-SY5Y cells (Figure 1A,B). Similarly, NP-12-treated SH-*CDKL5*-KO cells showed an increase in GSK-3β phosphorylation (Figure 1A,B). As a molecular consequence of GSK-3β inhibition in SH-*CDKL5*-KO cells, we evaluated the levels of the downstream GSK-3β target, CRMP2 (collapsing response mediator protein 2), the phosphorylation of which is regulated by GSK-3β [33]. In line with an increased activity of GSK-3β (Figure 1A,B), we found higher phosphorylation levels of CRMP2 (Figure 1C,D) in SH-*CDKL5*-KO cells. Treatment with **C11** recovered CRMP2 phosphorylation in SH-*CDKL5*-KO cells (Figure 1C,D). As expected, we found a complete recovery of CRMP2 phosphorylation in NP12-treated SH-*CDKL5*-KO cells (Figure 1C,D).

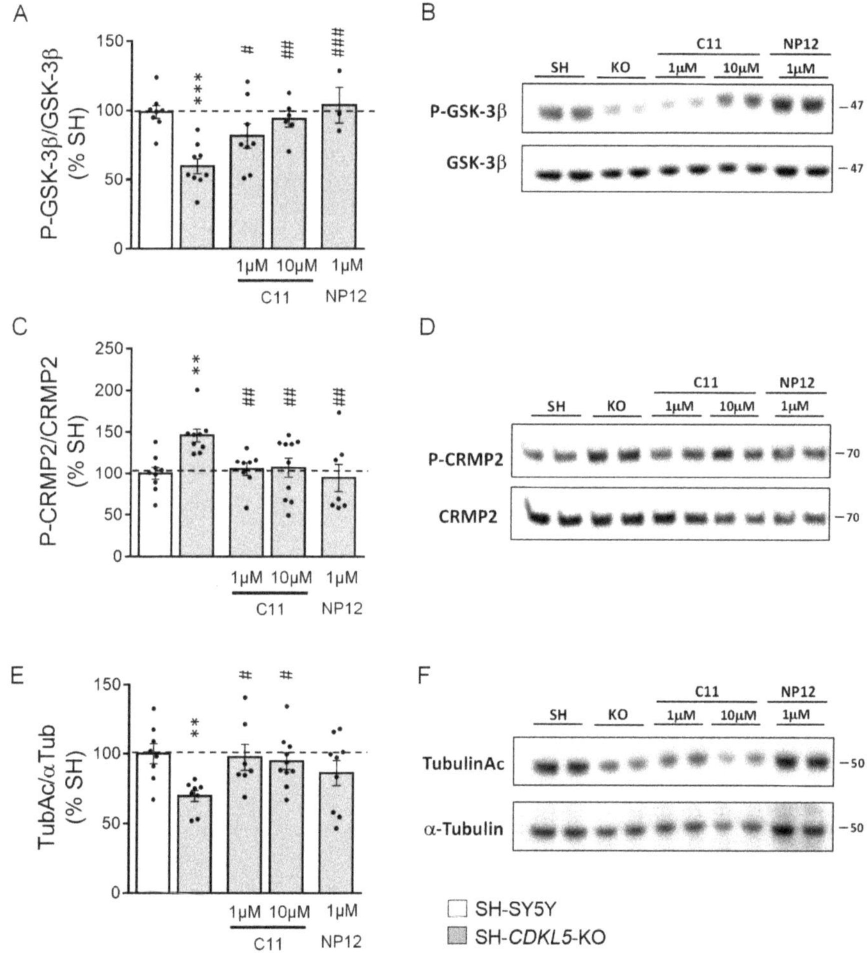

Figure 1. Effect of treatment with **C11** on GSK-3β and HDAC6 signaling in SH-*CDKL5*-KO cells. Western blot analysis of phospho-GSK-3β (Ser9; P-GSK-3β; **A**,**B**), P-collapsin response mediator protein 2-(Thr514P; P-CRMP2; **C**,**D**) and acetylated alpha tubulin (TubulinAc; **E**,**F**) levels in protein extracts from parental cells (SH-SY5Y; n = 8–9), SH-*CDKL5*-KO cells (SH-*CDKL5*-KO; n = 8–9) and SH-*CDKL5*-KO cells treated with **C11** (1 μM and 10 μM; n = 6–9) or NP12 (1 μM; n = 3–8) for 24 h. Immunoblots are examples from two biological replicates of each experimental condition. Histograms on the left show P-GSK-3β, P-CRMP2, and TubulinAc protein levels normalized to corresponding total protein levels. Data are expressed as a percentage of parental cells. Values are represented as means ± SE. ** $p < 0.01$; *** $p < 0.001$ as compared to the vehicle-treated SH-SY5Y condition; # $p < 0.05$; ## $p < 0.01$; ### $p < 0.001$ as compared to the vehicle-treated SH-*CDKL5*-KO condition (Fisher's LSD test after one-way ANOVA).

Since tubulin is a substrate of HDAC6 [34], to confirm the inhibitory activity of **C11** on HDAC6, we investigated acetylated tubulin levels in untreated and treated SH-*CDKL5*-KO cells. Interestingly, we found that SH-*CDKL5*-KO cells showed decreased levels of acetylated tubulin compared with parental SH-SY5Y cells (Figure 1E,F), that were not related to increased HDAC6 levels (Figure S1A,B). Treatment with **C11** recovered acetylated tubulin levels in SH-*CDKL5*-KO cells (Figure 1E,F). Importantly, the effects on tubulin were not observed when SH-*CDKL5*-KO cells were treated with NP12 (Figure 1E,F).

As previously demonstrated [29], the weak in vitro inhibitory activity of **C11** on other HDACs such as HDAC1 was confirmed by the lack of an effect of **C11** treatment on H3 acetylation levels in SH-*CDKL5*-KO cells (Figure S1C,D).

2.2. *Treatment with* **C11** *Restores Neuronal Maturation and Survival of a Human Cellular Model of CDKL5 Deficiency*

When SH-SY5Y cells were treated with retinoic acid (RA) for 5 days, a considerable proportion of cells differentiated to a more neuronal phenotype by extending neuritic processes [19]. As previously reported [19], RA-treated SH-*CDKL5*-KO clones showed reduced neurite outgrowth in comparison with parental cells (Figure 2). Since **C11** and NP12 inhibit GSK-3β at different concentration ranges (**C11**: IC_{50} = 2.7 μM [29]; NP12: IC_{50} = 60 nM), to compare drug efficacy we used **C11** at a concentration that was 10 times higher (10 μM) than that of NP12 (1 μM). We found that treatment with **C11** restored neuritic length in SH-*CDKL5*-KO clones (Figure 2A,D). After treatment with NP-12, neuritic length in SH-*CDKL5*-KO cells increased to levels that were even higher than those of parental cells (Figure 2A,D), confirming the positive effect of GSK-3β inhibition in neuronal maturation [35].

To investigate cell proliferation and viability in SH-*CDKL5*-KO clones, we evaluated the percentage of mitotic and pyknotic nuclei visualized with Hoechst staining. As previously reported [19], SH-*CDKL5*-KO showed a reduced number of mitotic cells (Figure 2B,D) and an increase in the fraction of pyknotic nuclei (Figure 2C,D) compared to parental cells. Interestingly, we found that treatment with **C11** restored cell proliferation (Figure 2B,D) and strongly improved survival (Figure 2C,D) in SH-*CDKL5*-KO cells, while treatment with NP-12 had very limited or no effect (Figure 2B,D).

In parental SH-SY5Y cells, treatment with **C11** had no effect on neuronal maturation and survival (Figure S2A,B).

Increased stress-induced cell death and DNA damage characterizes *CDKL5*-null cells [19]. To test whether **C11** protects SH-*CDKL5*-KO cells from stress, we examined cell viability and induction of DNA damage in SH-*CDKL5*-KO cells after oxidative (H_2O_2) stress. SH-*CDKL5*-KO cells were more sensitive to treatment with H_2O_2 than were parental SH-SY5Y cells, showing an increased number of pyknotic nuclei and γH2AX levels (Figure 2E,F). Treatment with **C11** drastically improved survival and DNA damage induction in SH-*CDKL5*-KO cells (Figure 2E,F). Again, treatment with NP-12 had no effect (Figure 2E,F). In agreement with the increased apoptotic cell death after oxidative stress, SH-*CDKL5*-KO cells showed decreased phosphorylation of AKT (Figure 2G) [36]. Restored levels of AKT phosphorylation in **C11**-treated, but not NP12-treated, SH-*CDKL5*-KO cells (Figure 2G), confirmed the pro-survival effect of **C11**.

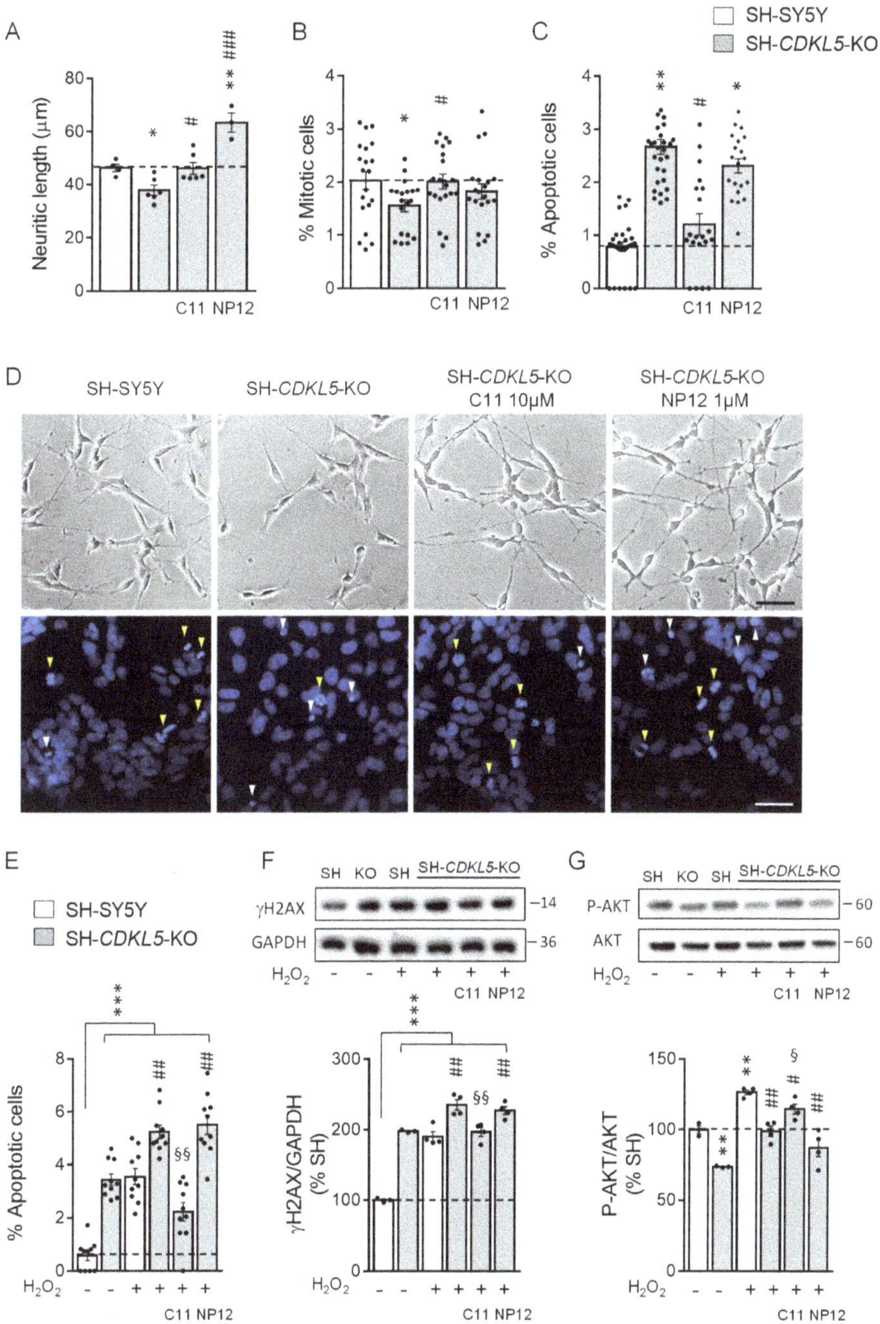

Figure 2. Effect of treatment with **C11** on cell differentiation, proliferation, and survival in SH-*CDKL5*-KO cells. (**A**) Quantification of neurite outgrowth of SH-SY5Y ($n = 4$) and SH-*CDKL5*-KO ($n = 6$) cells treated daily with retinoic acid (RA; 10 μM) for 5 days. SH-*CDKL5*-KO cells were treated with **C11** (10 μM, $n = 6$) or NP12 (1 μM, $n = 6$) every 2 days during retinoic acid differentiation. (**B**,**C**) Percentage of mitotic (**B**) and apoptotic (**C**) cells in proliferating SH-SY5Y ($n = 20$–30)

and SH-*CDKL5* KO cells (n = 20–30); cells were treated with **C11** (10 µM, n = 20) or NP12 (1 µM, n = 20) for 24 h. (**D**) Representative phase-contrast images of neurite outgrowth (upper panel) of cells treated as in (**A**) (scale bar = 50 µm); fluorescence images of Hoechst-stained nuclei (lower panel; white triangles indicate pyknotic nuclei, yellow triangles indicate mitotic nuclei) of SH-SY5Y and SH-*CDKL5* KO cells treated as in (**B**,**C**) (scale bar = 30 µm). (**E**) Percentage of apoptotic cells in untreated (n = 10), and treated with H_2O_2 (200 µM, n = 10) for 24 h, parental SH-SY5Y and SH-*CDKL5*-KO cells, and in SH-*CDKL5*-KO cells co-treated with H_2O_2 (200 µM) and **C11** (10 µM, n = 10) or NP12 (1 µM, n = 10) for 24 h. (**F**,**G**) Western blot analysis of γH2AX (**F**) and phospho-AKT-Ser473 (P-AKT; (**G**)) levels in protein extracts of SH-SY5Y and SH-*CDKL5*-KO cells treated as in E (untreated cells n = 3; treated cells n = 4). Histograms show γH2AX protein levels normalized to GAPDH (**F**), and P-AKT levels normalized to total AKT protein levels (**G**). Examples of γH2AX, GADPH, P-AKT, and AKT immunoblots (upper panels). Values represent mean ± SE of 3 independent experiments. In (**A**–**C**) * $p < 0.05$; ** $p < 0.01$ as compared to the vehicle-treated SH-SY5Y condition; # $p < 0.05$; ### $p < 0.001$ as compared to the vehicle-treated SH-*CDKL5*-KO condition (Fisher's LSD test after one-way ANOVA). In (**E**–**G**) ** $p < 0.01$; *** $p < 0.001$ as compared to the untreated SH-SY5Y cell condition; # $p < 0.05$; ## $p < 0.001$ as compared to the SH-SY5Y + H_2O_2 treated cell condition; § $p < 0.05$; §§ $p < 0.001$ as compared to the SH-*CDKL5*-KO + H_2O_2 treated cell condition (Fisher's LSD test after one-way ANOVA).

2.3. *Treatment with* **C11** *Restores Neuronal Maturation and Survival of Hippocampal Neurons from Cdkl5 KO Mice*

As previously reported [11,12], hippocampal neurons generated from *Cdkl5* −/Y mice had a reduced neuritic length compared to control (+/Y) neurons (Figure 3), as well as a reduced number of dendritic PSD-95 (postsynaptic protein 95) immunoreactive puncta (Figure 3B,C). Treatment with **C11** or NP12 fully restored the reduced dendritic outgrowth of hippocampal neurons from *Cdkl5* −/Y mice (Figure 3A,C), suggesting that GSK-3β signaling plays a major role in hippocampal neuron maturation. In contrast, the number of PSD-95 immunoreactive puncta were not recovered by treatment with NP12 (Figure 3B,C) and only partially recovered by treatment with **C11** (Figure 3B,C).

It has been shown that differentiating hippocampal neurons generated from *Cdkl5* −/Y mice have a reduced survival rate [20], which is evident in the reduced number of *Cdkl5* −/Y neurons present in culture after 10 days of differentiation, in comparison with control (+/Y) cultures (Figure 3D). Treatment with **C11** restored neuronal survival in hippocampal cultures from *Cdkl5* −/Y (Figure 3D). Similarly to SH-*CDKL5*-KO cells, treatment with NP-12 had no effect on *Cdkl5* −/Y hippocampal neuron survival (Figure 3D).

In control (+/Y), neuron treatment with **C11** had no effect on neuronal maturation and survival (Figure S2C,D).

2.4. *Effect of Treatment with* **C11** *on Behaviors of Cdkl5 KO Mice*

Based on the results obtained in vitro we sought to investigate whether dual GSK-3β and HDAC6 inhibition improves the neurodevelopmental alterations that characterize the *Cdkl5* −/Y brain. Since to date there are no studies on the action of **C11** in vivo, a pilot study was performed to examine the effect of an acute treatment with increasing **C11** doses (10, 50, and 100 mg/kg) on GSK-3β inhibition. We found a dose response increase of phosphorylated GSK-3β levels in both the hippocampus and cortex of *Cdkl5* −/Y mice (Figure 4A,B).

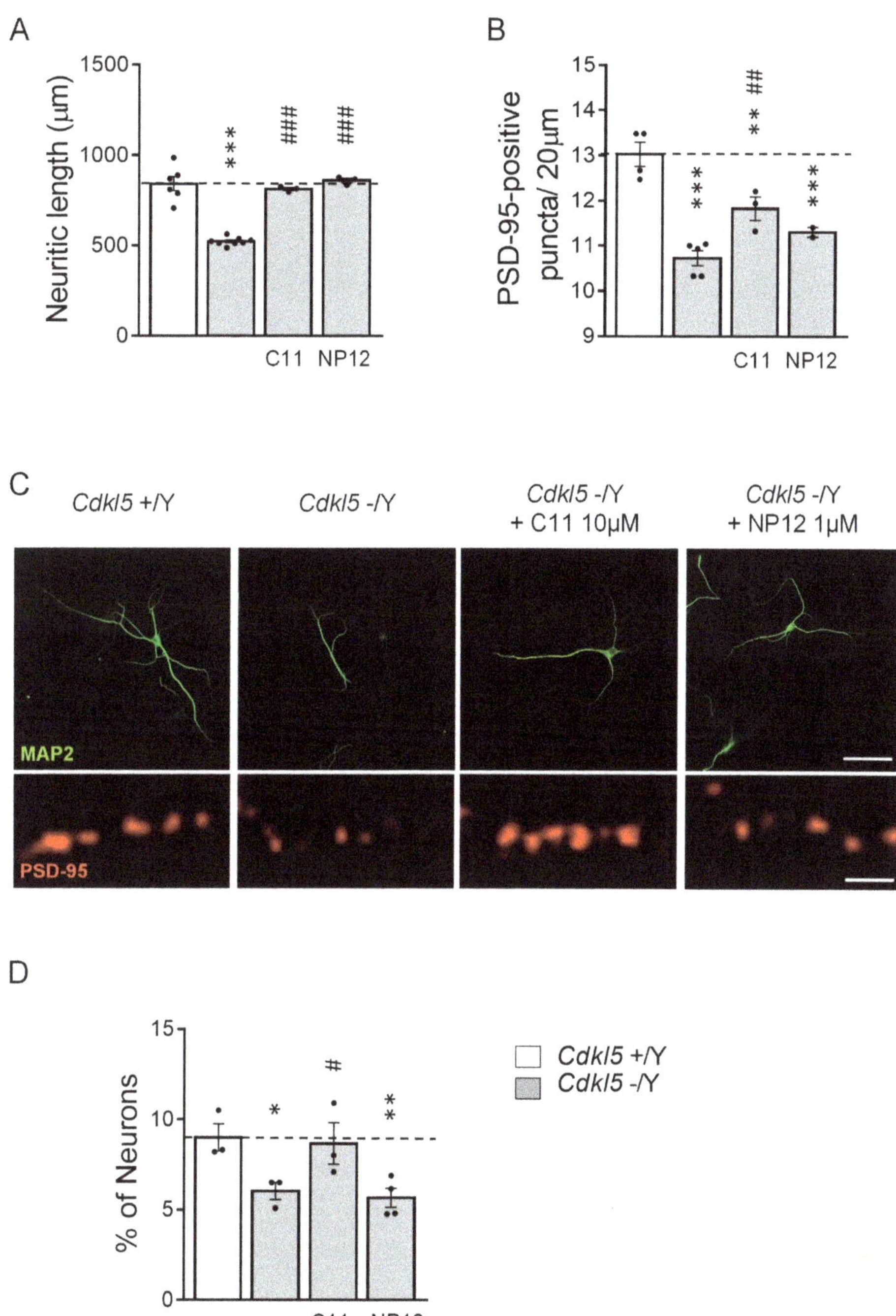

Figure 3. Effect of treatment with **C11** on maturation and survival of hippocampal neurons from *Cdkl5* KO mice. (**A**) Quantification of the total length of MAP2-positive cells from 10-day differentiated (DIV10) hippocampal neurons. On day 2 post-plating (DIV2) hippocampal cultures were treated with vehicle (0.1% DMSO in PBS; *Cdkl5* +/Y; $n = 6$ and *Cdkl5* −/Y;

$n = 8$), **C11** (10 μM; *Cdkl5* −/Y + **C11**; $n = 3$), or NP12 (1 μM; *Cdkl5* −/Y + NP12, $n = 3$), which was then administered on alternate days throughout the entire differentiation period. (**B**) Quantification of the number of PSD-95 immunoreactive puncta per 20 μm in proximal dendrites of hippocampal neurons from vehicle-treated (*Cdkl5* +/Y; $n = 4$ and *Cdkl5* −/Y; $n = 5$), **C11**-treated (*Cdkl5* −/Y + **C11**; $n = 3$), and NP12-treated (*Cdkl5* −/Y + NP12; $n = 2$) hippocampal cultures. (**C**) Upper panels: representative fluorescence images of differentiated MAP2-positive hippocampal neurons from *Cdkl5* +/Y and *dkl5* −/Y mice treated as in (**A**) (scale bar = 40 μm). Lower panels show a magnification of a proximal dendrite immunopositive for PSD-95 (scale bar = 1 μm) of hippocampal neurons from *Cdkl5* +/Y and *Cdkl5* −/Y mice treated as in (**A**). (**D**) Quantitative analysis of the number of MAP2-positive cells in hippocampal cultures from wild-type (*Cdkl5* +/Y; $n = 3$), KO (*Cdkl5* −/Y; $n = 3$), KO treated with **C11** (*Cdkl5* −/Y + **C11**; $n = 3$), and KO treated with NP12 (*Cdkl5* −/Y + NP12; $n = 4$). Values are represented as mean ± SE. Values in D are represented as % of the *Cdkl5* +/Y conditions. * $p < 0.05$; ** $p < 0.01$; *** $p < 0.001$ as compared to the untreated cell *Cdkl5* +/Y condition; # $p < 0.05$; ## $p < 0.01$; ### $p < 0.001$ as compared to the untreated *Cdkl5* −/Y cell condition (Fisher's LSD test after one-way ANOVA).

Based on these results, we treated 1-month-old *Cdkl5* −/Y mice (P30) daily for 15 days with **C11** (50 mg/kg), the lower dose that was shown to improve or even restore phosphorylated GSK-3β (P-GSK-3β) to control levels (Figure 4). Animals were behaviorally tested in the time window shown in Figure 5A and then sacrificed at P60. The effects of treatment on the neuroanatomy of the hippocampal region were examined in mice at P60. We found that treatment had no adverse effect on body weight, indicating that **C11** does not impair animals' well-being (Figure 5B).

Common features of *CDKL5* deficiency disorder include decreased mobility, such as muscular rigidity, and repetitive hand movements (stereotypies), such as clasping and hand-sucking [37].

Catalepsy bar tests are widely used to measure the failure, resulting from muscular rigidity or akinesia, to correct an imposed posture [38]. Using this test we found that *Cdkl5* −/Y mice spent significantly more time hanging onto the bar than did control (+/Y) mice (Figure 5C). Treatment with **C11** restored this motor behavior in *Cdkl5* −/Y mice (Figure 5C), suggesting that treatment improves muscular rigidity due to loss of *Cdkl5*.

Similarly to the human condition, *Cdkl5* KO mice exhibited a hind-limb clasping behavior [13]. In order to examine the effect of treatment with **C11** on motor stereotypies, we evaluated the hind-limb clasping time during the two minutes' testing period (Figure 5D). While control (+/Y) mice exhibited very little hind-limb clasping, *Cdkl5* −/Y mice spent more time in the clasping position (Figure 5D). Importantly, *Cdkl5* −/Y mice treated with **C11** showed a significant decrease in clasping (Figure 5D), indicating a treatment-induced improvement of motor stereotypies.

Animals that had been treated with **C11** were subjected to the Morris Water Maze (MWM) test in order to explore hippocampus-dependent spatial memory. As previously reported [26], we found that, while control (+/Y) mice learned to find the platform by the second day, *Cdkl5* −/Y mice showed a clear learning deficit over the 2–4 days of the trial (Figure 5E). Treatment with **C11** improved the learning ability of *Cdkl5* −/Y mice, which became significantly different from untreated *Cdkl5* −/Y mice on the second and fourth day of testing (Figure 5E). In the probe test, as previously reported [26], *Cdkl5* −/Y mice showed an increased latency to enter the former platform zone (Figure 5F). Treatment with **C11** in *Cdkl5* −/Y mice caused a notable reduction in their latency to enter the former platform zone (Figure 5F), indicating an improvement in spatial memory. Performance in the MWM is influenced by motor functions that can be assessed by analyzing swimming patterns such as swim speed. No difference was observed between untreated and treated *Cdkl5* +/Y and *Cdkl5* −/Y mice as far as swim velocity was concerned (Figure 5G), indicating that the deficit of *Cdkl5* −/Y mice in the hidden platform test was not caused by abnormalities in swimming abilities.

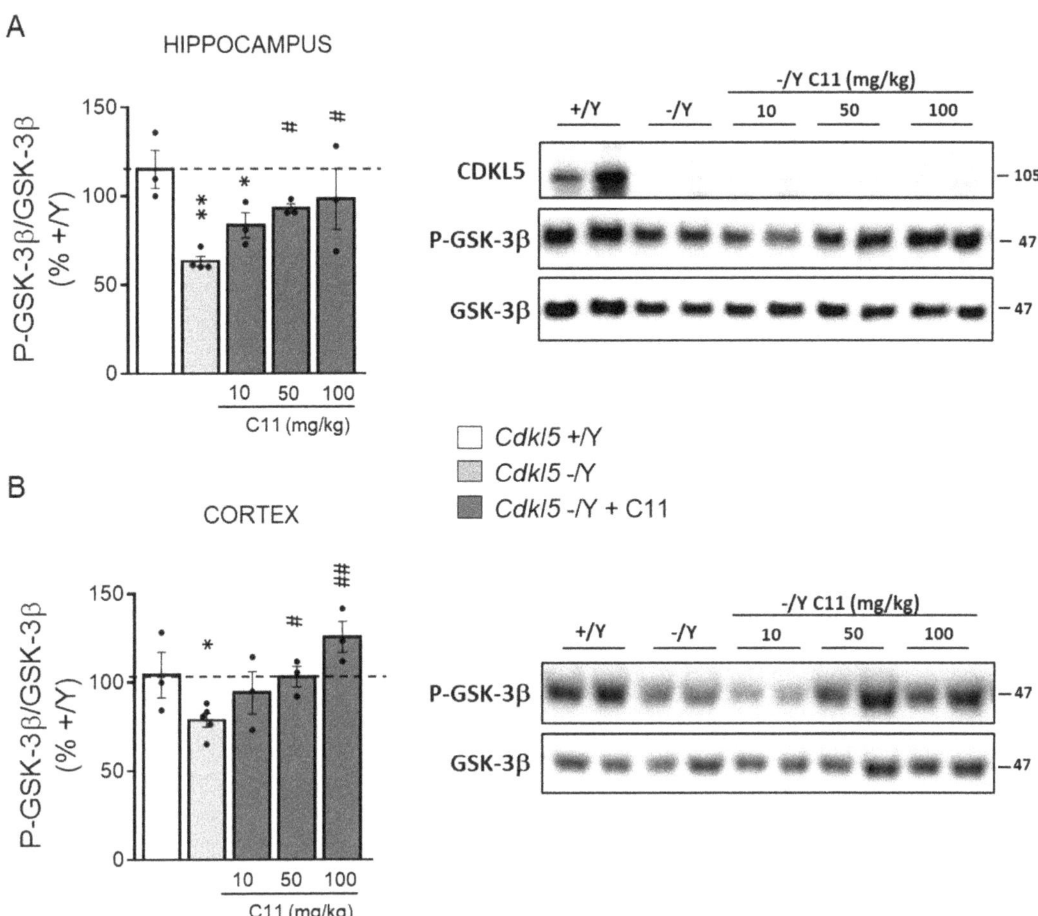

Figure 4. Effect of in vivo treatment with **C11** on GSK-3β activity in the brain of *CDKL5* −/Y mice. Western blot analysis of P-GSK-3β levels normalized to total GSK-3β levels in the hippocampus (**A**) and cortex (**B**) of wild-type (*Cdkl5* +/Y $n = 3$), KO (*Cdkl5* −/Y $n = 4$–5), and KO treated with a single intraperitoneal dose of **C11** (*Cdkl5* −/Y + **C11** 10 mg/kg $n = 3$, 50 mg/kg $n = 3$, 100 mg/kg $n = 3$) mice. Mice were sacrificed 4 h after the treatment. Histograms on the left show P-GSK3β protein levels normalized to corresponding total protein levels. Immunoblots on the right are examples from two animals of each experimental group of CDKL5, P-GSK-3β, and total GSK-3β levels. Values are represented as means ± SE and data are expressed as a percentage of untreated *Cdkl5* +/Y mice. * $p < 0.05$; ** $p < 0.01$ as compared to the untreated *Cdkl5* +/Y condition; # $p < 0.05$; ## $p < 0.01$ as compared to the untreated *Cdkl5* −/Y condition (Fisher's LSD test after one-way ANOVA).

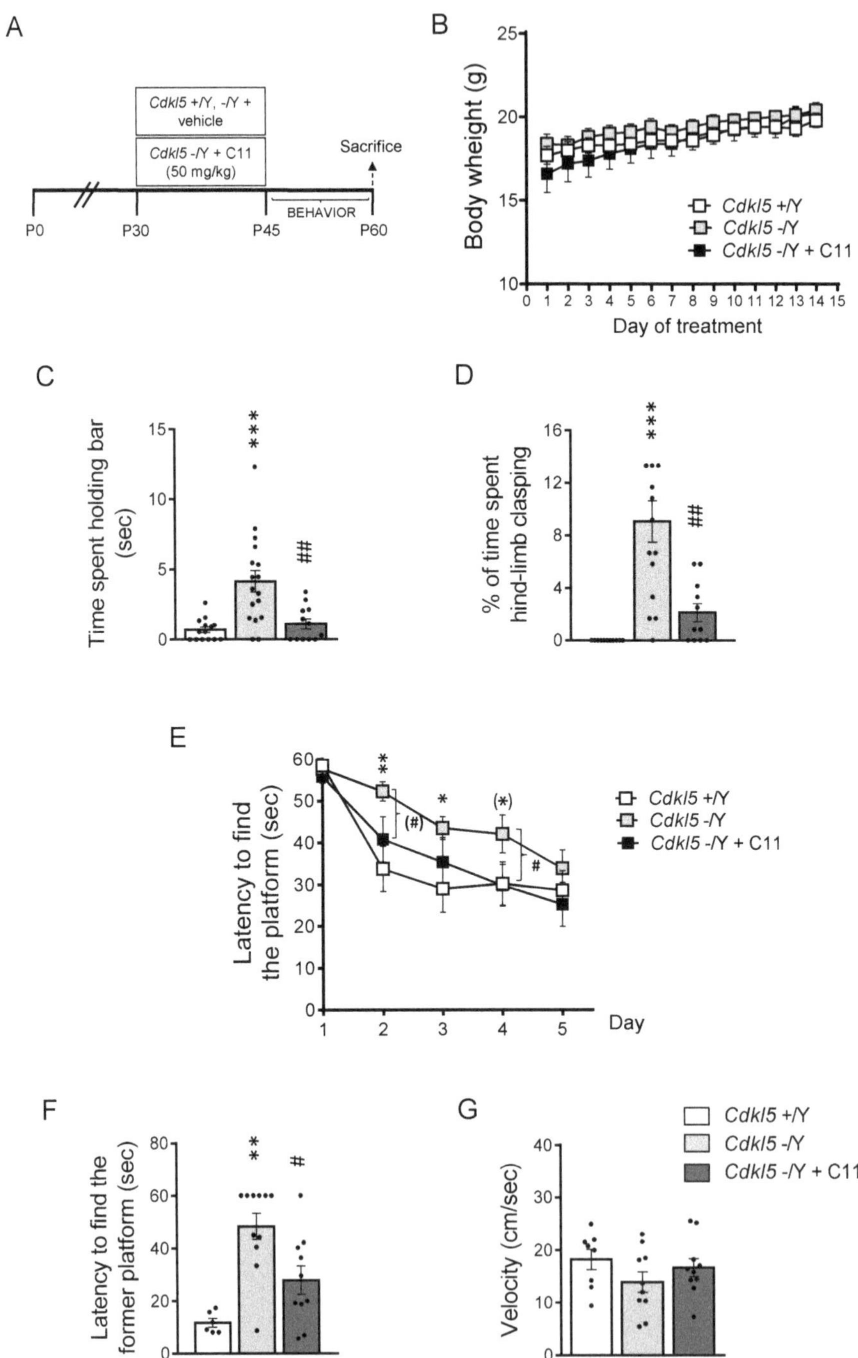

Figure 5. Effect of treatment with **C11** on motor coordination and hippocampus-dependent learning and memory in *Cdkl5* −/Y mice. (**A**) Experimental protocol. Starting from postnatal day 30 (P30), *Cdkl5* +/Y mice were treated with vehicle and *Cdkl5* −/Y male mice were treated with vehicle or **C11** (50 mg/kg), administered via intraperitoneal injection every day for

15 days. Animals from different experimental groups were behaviorally tested. Animals were sacrificed 15 days after the last injection (P60). (**B**) Body weight in grams of vehicle-treated *Cdkl5* +/Y (n = 15) and *Cdkl5* −/Y (n = 17) mice and **C11**-treated *Cdkl5* −/Y (n = 12) mice during the daily treatment period. (**C**) Amount of time taken to remove both paws from the bar in vehicle-treated (*Cdkl5* +/Y n = 14, *Cdkl5* −/Y n = 17) and **C11**-treated (*Cdkl5* −/Y + **C11** n = 12) *Cdkl5* male mice. (**D**) Total amount of time spent hind-limb clasping during a 2 min interval in vehicle-treated (*Cdkl5* +/Y n = 14, *Cdkl5* −/Y n = 17) and **C11**-treated (*Cdkl5* −/Y + **C11** n = 11) *Cdkl5* male mice. (**E**) Spatial learning (5-day learning period) assessed using the Morris Water Maze in vehicle-treated (*Cdkl5* +/Y n = 9, −/Y n = 10) and **C11**-treated (*Cdkl5* −/Y + **C11** n = 10) *Cdkl5* male mice. (**F**) Spatial memory on day 6 (probe test) in mice as in (**D**). Memory was assessed by evaluating the latency to enter the former platform zone. (**G**) Mean velocity during the learning phase in mice as in (**D**). Values represent mean ± SE; (*) p = 0.058; * $p < 0.05$; ** $p < 0.01$; *** $p < 0.001$ as compared to vehicle-treated *Cdkl5* +/Y mice; (#) p = 0.059; # $p < 0.05$; ## $p < 0.01$ as compared to vehicle-treated *Cdkl5* −/Y mice. Dataset in (**B**) Fisher's LSD after two-way ANOVA; dataset in (**E**) Fisher's LSD after three-way mixed ANOVA; datasets in (**C**,**D**,**F**,**G**) Fisher's LSD after one-way ANOVA.

2.5. *Effect of Treatment with* **C11** *on Synapse Development in Cdkl5 KO Mice*

Alteration in synaptogenesis is now considered the main structural brain defect that underlies the behavioral abnormalities in the *Cdkl5* KO mouse [11,17,27,39].

Given the positive effects of treatment with **C11** on behavior in *Cdkl5* −/Y mice, we investigated the effects of **C11** on spine maturation/connectivity. As previously reported [26], spine density of Golgi-stained CA1 hippocampal pyramidal neurons was reduced, as was the percentage of mature vs. immature spines in *Cdkl5* −/Y mice in comparison with control (+/Y) mice (Figure 6A–C). Treatment with **C11** restored the number of dendritic spines (Figure 6A,C) and the balance between immature and mature spines (Figure 6B,C) in *Cdkl5* −/Y mice.

Recruitment of PSD-95 to the postsynaptic compartment is one factor that contributes to the stabilization of synaptic contacts [40]. To confirm that the **C11**-induced restoration of spine maturation was accompanied by a restoration of synaptic connectivity, we evaluated the number of immunoreactive puncta for PSD-95 in the hippocampal of *Cdkl5* −/Y mice. As previously reported [11,41], in *Cdkl5* −/Y mice, the number of postsynaptic terminals was lower in comparison with that of *Cdkl5* +/Y mice (Figure 6D,E). Treatment with **C11** rescued this defect (Figure 6D,E).

2.6. *Effect of Treatment with* **C11** *on Neuronal Survival in Cdkl5 KO Mice*

Recent evidence has shown that, in addition to affecting synaptogenesis, *CDKL5* also impacts on neuronal survival [19–21]. *Cdkl5* KO mice are characterized by decreased survival of hippocampal neurons, which worsens with age [42].

In order to evaluate whether treatment with **C11** affects neuronal survival rate in *Cdkl5* KO mice, we evaluated cell density in the CA1 layer of the hippocampus. We found that *Cdkl5* −/Y mice showed a reduced number of Hoechst-positive nuclei and NeuN-positive pyramidal neurons in the CA1 layer in comparison with control (+/Y) mice (Figure 7A–C), confirming the reduced neuronal survival in the absence of *Cdkl5*. Importantly, treatment with **C11** rescued hippocampal neuronal survival (Figure 7A–C). Otherwise, in vivo treatment with NP12 [26] does not improve survival of pyramidal neurons in the hippocampus in *Cdkl5* −/Y mice (Figure S3), assessed as Hoechst-positive nuclear density in the CA1 layer.

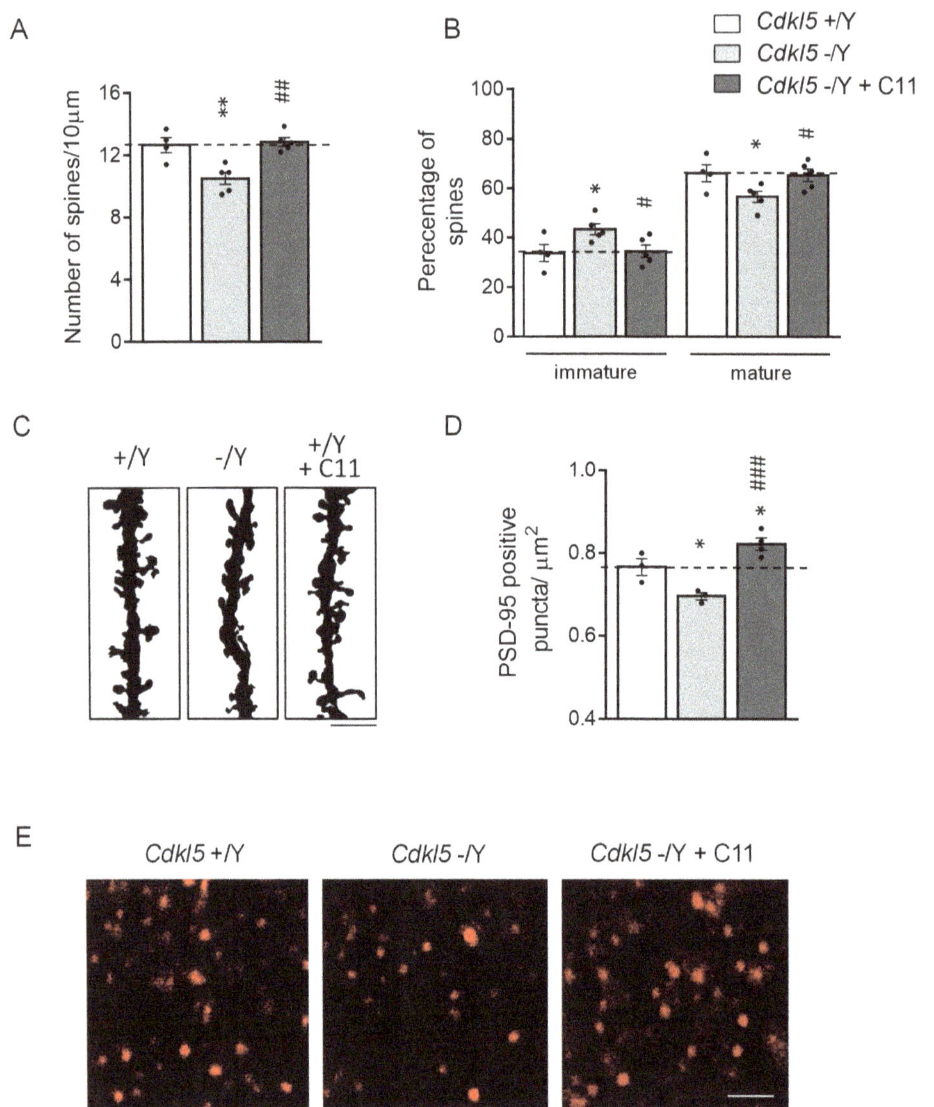

Figure 6. Effect of treatment with **C11** on dendritic spine density and maturation in the hippocampus of *Cdkl5* −/Y mice. (**A**) Dendritic spine density in CA1 pyramidal neurons of intraperitoneal vehicle-treated (*Cdkl5* +/Y $n = 4$, *Cdkl5* −/Y $n = 5$; P60) and **C11**-treated (*Cdkl5* −/Y $n = 4$; P60) *Cdkl5* male mice. Data are expressed as number of spines/10 μm. (**B**) Percentage of immature and mature spines in relation to the total number of protrusions in CA1 pyramidal neurons of mice as in A. (**C**) Examples of Golgi-stained dendritic branches of CA1 pyramidal neurons of one animal from each experimental group. Scale bar = 1 μm. (**D**) Number of fluorescent puncta per μm^2 exhibiting PSD-95 immunoreactivity in the stratum oriens of the hippocampus of intraperitoneal vehicle-treated (*Cdkl5* +/Y $n = 3$; *Cdkl5* −/Y $n = 3$; P60) and **C11**-treated (*Cdkl5* −/Y $n = 4$; P60) *Cdkl5* male mice. (**E**) Representative confocal images of CA1-sections processed for PSD-95 immunohistochemistry of one animal from each experimental group. Scale bar = 2.5 μm. Values are represented as means ± SE. * $p < 0.05$; ** $p < 0.01$ as compared to the vehicle-treated *Cdkl5* +/Y condition; # $p < 0.05$; ## $p < 0.01$; ### $p < 0.001$ as compared to the vehicle-treated *Cdkl5* −/Y condition (datasets in (**A**,**D**), Fisher's LSD after one-way ANOVA; dataset in (**B**), Dunn's test after Kruskall–Wallis).

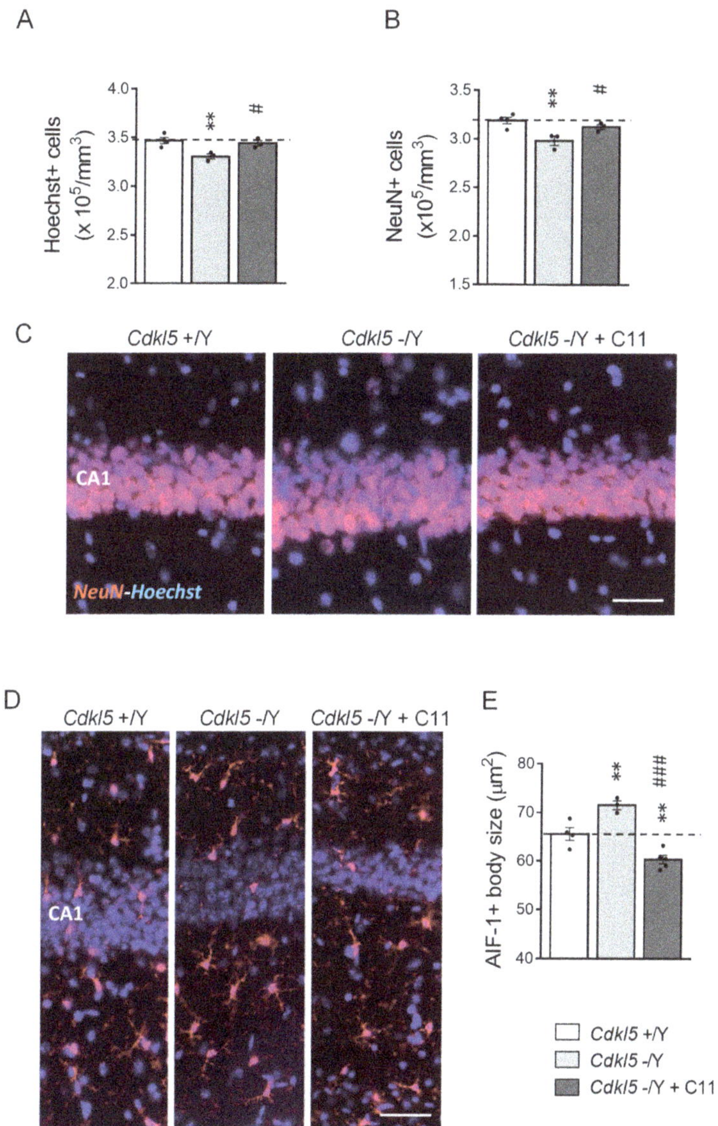

Figure 7. Effect of treatment with **C11** on neuronal survival and microglia activation in the hippocampus of *Cdkl5* −/Y mice. (**A**,**B**) Quantification of Hoechst-positive cells (**A**) and NeuN-positive cells (**B**) in CA1 layer of hippocampal sections from intraperitoneal vehicle-treated (*Cdkl5* +/Y; $n = 4$, *Cdkl5* −/Y; $n = 5$; P60) and **C11**-treated (*Cdkl5* −/Y + **C11**; $n = 4$; P60) mice. (**C**) Representative fluorescence images of immunopositive for NeuN (red) and counterstained with Hoechst (blue) sections in the hippocampal CA1 region of one animal from each group. Scale bar = 50 µm. (**D**) Representative fluorescence images of hippocampal sections processed for AIF-1 immunohistochemistry of one animal from each group. (**E**) Mean microglia cell body size in hippocampal sections from vehicle-treated (*Cdkl5* +/Y; $n = 4$, *Cdkl5* −/Y; $n = 5$; P60) and **C11**-treated (*Cdkl5* −/Y + **C11**; $n = 4$; P60) mice. Scale bar = 50 µm. Values are represented as means ± SE. ** $p < 0.01$, as compared to the vehicle-treated *Cdkl5* +/Y mice; # $p < 0.05$; ### $p < 0.001$ as compared to the vehicle-treated *Cdkl5* −/Y mice (Fisher's LSD test after one-way ANOVA).

2.7. *Effect of Treatment with* **C11** *on Microglia Over-Activation in Cdkl5 KO Mice*

It has been shown that **C11**, in addition to its dual GSK-3β/HDAC6 inhibitory action, decreases neuroinflammation by modulating the microglial phenotypic switch from the M1 (pro-inflammatory) to the M2 (anti-inflammatory) phenotype [29]. We recently found increased microglial activation in the brain of the *Cdkl5* KO mouse [43]; indeed, modulation of microglial activation seems to be a therapeutic approach to be considered for CDD. Similarly to previous findings, we found alterations in microglial cell morphology, with an enlarged body size that is typical of a state of activation [44], in the hippocampus of *Cdkl5* −/Y mice compared to their wild-type counterparts (Figure 7D,E). Interestingly, treatment with **C11** reversed the inflammatory status in *Cdkl5* −/Y mice, bringing microglial soma size to levels that were even lower than those of vehicle-treated control (+/Y) mice (Figure 7D,E).

3. Discussion

It is likely that, due to *CDKL5* failure, alterations in multiple complex molecular networks are involved in the mechanisms that underlie the CDD-related phenotype. The complexity of this disease might limit the efficacy of commonly used single molecular drug therapeutics. Therefore, the severity of the neurological phenotype of CDD patients could be better tackled using a multi-target approach that is able to simultaneously modulate multiple targets involved in the onset of the disease. Our study provides novel evidence that, in in vitro experimental models of CDD, the GSK-3β/HDAC6 dual inhibitor, **C11** [29], is more effective at recovering neuronal survival than treatment with a single inhibitor that is selective for GSK-3β. Importantly, in vivo treatment with **C11** restored synapse development, neuronal survival, and microglia over-activation, and improved motor and cognitive abilities of *Cdkl5* KO mice. Overall, our data suggest that a GSK-3β/HDAC6 dual inhibitor therapy may engender a more effective strategy with which to achieve therapeutic benefits in CDD patients.

To validate the effect of **C11** in CDD we selected, as an in vitro model, a recently generated human cellular model of *CDKL5* deficiency that exhibited alterations in the GSK-3β signal pathway [19] as well as reduced survival and differentiation. Here, we found that *CDKL5* deficiency in these cells causes decreased acetylation of a well-known substrate of HDAC6 deacetylase activity, αtubulin [45]. The decreased αtubulin acetylation, not resulting from deregulated HDAC6 expression (Figure S1A), may be attributed to increased HDAC6 activity which, in turn, could be explained by the complex regulation of HDAC6 activity, including direct or indirect binding partners and kinase-mediated phosphorylation that leads to either increased or reduced HDAC6 activity [46]. Among the kinases that regulate HDAC6 activity, GSK-3β interacts and localizes with HDAC6, and enhances HDAC6 activity; thereby, it may indirectly influence tubulin acetylation [47]. The link between HDAC6 and GSK-3β may explain the partial recovery of tubulin acetylation in *CDKL5* deficient cells obtained with the single GSK-3β inhibitor, NP12.

Abnormal activation of GSK-3β has been associated with several neurological and psychiatric disorders that share developmental abnormalities and altered neurocircuitry maintenance, such as schizophrenia, bipolar disorder, autism, and Alzheimer's disease (AD) [48–50]. The involvement of GSK-3β misregulation in a variety of brain abnormalities strongly supports its pivotal role in controlling basic mechanisms of neuronal function from brain bioenergetics to the establishment of neuronal circuits, modulation of neuronal polarity, migration, and proliferation [24]. In particular, the role of GSK-3β in phosphorylation of cytoskeletal proteins impacts neuronal plasticity, as cytoskeletal constituents are involved in the development and maintenance of neurites, and changes in the rate of stabilization/destabilization of microtubules could influence major cellular compartments of neurons, such as dendrites, spines, axons, and synapses. Confirming previous findings involving GSK-3β inhibitors [26,27], we found that pharmacological inhibition of the GSK-3β activity with **C11** improved or even restored neuronal maturation and connectivity in in vitro and in vivo models of CDD. In both human *CDKL5* deficient cells and hippocampal neurons from *Cdkl5* KO mice a complete restoration of the extension of the neuritic processes was obtained when cells were treated with either **C11** or NP12. Similarly, the effects of in vivo treatment with **C11** or NP12 were comparable in terms of dendritic spines and PSD-95 positive puncta in *CDKL5* KO mice (present findings and those of [41]).

Interestingly, we showed here that **C11** treatment was much more effective at recovering neuronal survival than treatment with NP12 in both in vitro and in vivo experimental models of CDD, suggesting that the **C11** pro-survival effect is conveyed to HDAC6 inhibition. In support of this, our findings are in line with a previous study showing that inhibition of HDAC6 activity in neurons promotes protection against oxidative stress-induced death [30]. There is a growing consensus that HDACs play a role in the regulation of neuronal survival [51–53]. Considerable research activity has focused on histone deacetylase (HDAC) inhibitors as neuroprotective agents for a number of neurodegenerative diseases and CNS injuries [52,54]. HDAC6, a member of Class IIb family of HDACs, regulates trafficking of neurotrophic factor, functions as an αtubulin deacetylase, and modulates mitochondrial transport in hippocampal neurons [55]. Importantly, HDAC6 is involved in several events of the neurodegenerative cascades [56]; specific HDAC6 inhibition exerts neuroprotection by increasing the acetylation levels of αtubulin with subsequent improvement of the axonal transport, which is usually impaired in neurodegenerative disorders [56]. HDAC6 is a specific deacetylase of the homodimeric molecular chaperone Hsp90, which sequesters the heat shock transcription factor 1 (HSF1) [57]. A further mechanism through which the inhibition of HDAC6 could exert its pro-survival action concerns the decreasing of HSP90 deacetylation. A number of reports suggest that HSP90 may be a viable target for neuroprotection, as HSP90 inhibitors have been found to promote HSF1 release and augment the heat-shock response, which protects against neurotoxic insults in a variety of models of neurodegenerative disease [58,59]. While we cannot rule out the possibility that other HDAC family members may also be targets for neuroprotection in *CDKL5* KO cells, our findings showing that treatment with **C11** selectively increases αtubulin acetylation, without altering histone H3 acetylation, strongly confirm the higher specificity of **C11** for HDAC6.

Evidence that microglia-mediated inflammation contributes to the neuropathology of CDD has recently been reported [43]. Chronic activation of microglia may cause reduced neuronal maturation and survival through the release of potentially cytotoxic molecules such as pro-inflammatory cytokines [60,61]. Therefore, suppression of microglia over-activation by **C11** might contribute to restoring neuronal survival and connectivity in the hippocampus of *Cdkl5* KO mice. The anti-inflammatory effect exerted by **C11** may depend on both of its inhibitory activities. GSK-3 inhibitors lead to a reduction in microglia activity [62,63] via increased expression of the inhibitory marker CD200R and diminished inflammatory mediators such as nitric oxide (NO), glutamate, pro-inflammatory cytokines (TNF-α and IL-6) [64]. The mechanism by which HDAC inhibition is anti-inflammatory is not understood, but inhibitors of HDACs reduce the inflammatory response of isolated microglia to stimulants such as lipopolysaccharide [65,66], and in vivo delivery of HDAC inhibitors reduces neuroinflammation in models of brain injury [67,68].

Conclusions

The search for better results in clinical practices, because of the inefficacy of some treatments based on single drugs, has encouraged the adoption of polypharmacology as a new therapeutical strategy. This multi-target approach could be: association of drugs, combination of drugs, or a single drug with multiple ligands. However, the deleterious effects caused by drug–drug interactions and toxicity of combination therapy have reinforced the emergence of novel strategies in drug discovery. The most recent approach is now considering single drugs that concomitantly recognize more than one molecular target. Although GSK-3β and HDAC inhibitors have been previously tested in *Cdkl5* KO mice [11,26,27], this is the first combinatorial approach using a GSK-3β/HDAC6 dual inhibitor to study the beneficial effects of GSK-3β and HDAC6 inhibition in *CDKL5*-null neurons. The importance of the use of an inhibitor with large selectivity for HDAC6 such as **C11** is that it can overcome all of the neuronal toxicity associated with the use of pan-HDAC inhibitors, suggesting that its specific inhibition may eliminate a range of untoward effects seen with the clinical application of pan-HDAC inhibitors in cancer [69]. This is also consistent with the demonstration that mice lacking HDAC6 are viable and develop normally [45]. Our finding that treatment with **C11** improved motor and cognitive behavior in *Cdkl5* KO mice suggest that dual GSK-3β/HDAC6 inhibitors might be an effective treatment option for CDD. Since previous findings showed that a pharmacological intervention aimed at normalizing impaired GSK-3β activity elicits different age-dependent outcomes in *Cdkl5* KO mice [26], a multi-target compound, acting simultaneously on different pathways that are essential for neuronal maturation and survival, might have a better chance of being effective regardless of age. At present we do not know whether the therapeutic effect of the **C11** double inhibitor is more efficient at recovering brain abnormalities in CDD at different developmental ages than a therapy with a selective GSK-3β or HDAC6 inhibitor. Future studies may clarify this issue.

4. Materials and Methods

4.1. *Compound* **11** *Synthesis*

Compound **11** (**C11**) was synthetized as previously reported [29].

4.2. *Cell Lines, Treatments and Measurements*

Human neuroblastoma cell line SH-SY5Y, deriving from The European Collection of Authenticated Cell Cultures (Sigma-Aldrich, St. Louis, MO, USA) and the *CDKL5* knockout (KO) SH-SY5Y neuroblastoma cell line (SH-*CDKL5*-KO; [19]), were maintained in Dulbecco modified Eagle medium (DMEM, Thermo Fisher Scientific, Waltham, MA, USA) supplemented with 10% heat-inactivated FBS, 2 mM of glutamine, and antibiotics (penicillin, 100 U/mL; streptomycin, 100 μg/mL, Thermo Fisher Scientific, Waltham, MA, USA), in a humidified atmosphere of 5% of CO2 at 37 °C. Cell medium was replaced every 3 days and the cells were sub-cultured once they reached 90% confluence.

4.2.1. **C11** and NP-12 Treatments

Cells were plated onto poly-D-lysine-coated slides in a 6-well plate at a density of 2.5×10^5 cells per well in culture medium supplemented with 10% FBS. The day after, cells were exposed to **C11** (1 μM and 10 μM; stock solution 10 mM in 100% DMSO), NP-12 (Tideglusib; 1 μM; stock solution 10 mM in 100% DMSO; Sigma-Aldrich, St. Louis, MO, USA), or vehicle (0.1% DMSO) for 24 h.

4.2.2. Hydrogen Peroxide Treatment

Cells were plated onto poly-D-lysine-coated slides in a 6-well plate at a density of 2.5×10^5 cells per well in culture medium supplemented with 10% FBS. The day after, cells were exposed to H_2O_2 (200 μM; Sigma-Aldrich, St. Louis, MO, USA) and **C11** (10 μM) or NP12 (1 μM) for 24 h. The following day, cells were fixed in a 4% paraformaldehyde 4% glucose solution at 37 °C for 30 min.

4.2.3. Retinoic Acid Induced Differentiation

For differentiation analyses, cells were plated onto poly-D-lysine-coated slides in a 6-well plate at a density of 1×10^5 cells per well in culture medium supplemented with 10% FBS. Twenty-four hours after cell plating, retinoic acid (RA; Sigma-Aldrich, St. Louis, MO, USA) 10 mM in ethanol was added to the medium at 10 μM final concentration each day for 5 days. Cells were co-treated with **C11** (10 μM), NP12 (1 μM), or vehicle (0.1% DMSO) every 2 days.

4.2.4. Analysis of Neurite Outgrowth

Phase contrast images of cells were taken with an Eclipse TE 2000-S microscope (Nikon Instruments Inc., Melville, NY, USAnikon) equipped with a DS-Qi2 digital SLR camera (Nikon Instruments Inc., Melville, NY, USA). Images were taken from random microscopic fields. Neurite outgrowth was measured using the image analysis system Image Pro Plus (Media Cybernetics, Silver Spring, MD, USA). Only cells with neurites longer than one cell body diameter were considered as neurite-bearing cells. In each experiment, a total of 900 cells was analyzed. All experiments were performed at least three times. The total length of neurites was divided by the total number of cells counted in the areas.

4.2.5. Apoptotic and Mitotic Index

Nuclei were stained with Hoechst 33342 (Sigma-Aldrich, St. Louis, MO, USA). Apoptotic cell death was assessed by manually counting the number of pyknotic nuclei and apoptotic bodies and is expressed as a percentage of the total number of cells. The number of mitotic cells was assessed by manually counting the cells in prophase (chromosomes are condensed and visible), metaphase (chromosomes are lined up at the metaphase plate), and anaphase/thelophase (chromosomes are pulled toward and arrive at the opposite poles) and expressed as a percentage of the total number of cells.

4.3. *Colony*

The mice used in this work derive from the *Cdkl5* knockout (KO) strain in the C57BL/6N background developed in [13] and backcrossed in C57BL/6J for three generations. *Cdkl5* KO mice (*Cdkl5* −/Y) and age-matched controls (wild-type; *Cdkl5* +/Y), when possible littermates, were used for all experiments and genotyped as previously described [13]. The day of birth was designated as postnatal day (P) zero and animals with 24 h of age were considered as 1-day-old animals (P1). Mice were housed 3–5 per cage on a 12 h light/dark cycle in a temperature- (23 °C) and humidity-controlled environment with standard mouse chow and water ad libitum. The animals' health and comfort were controlled by the veterinary service. All efforts were made to minimize animal suffering and to keep the number of animals used to a minimum.

4.4. *Primary Hippocampal Cultures, Treatments and Measurements*

Primary hippocampal neuronal cultures were prepared from 1-day-old (P1) wild-type (WT) and *Cdkl5* KO mice as described [70]. Briefly, hippocampi were dissected from mouse brains under a dissection microscope and treated with 2.5% trypsin (Sigma-Aldrich, St. Louis, MO, USA) for 15 min at 37 °C and 1% DNase (Sigma-Aldrich, St. Louis, MO, USA) for 2 min at room temperature before being triturated mechanically with a fire-polished glass pipette to obtain a single-cell suspension. Approximately 1.2×10^5 cells were plated on coverslips coated with poly-L-lysine in 6-well plates and cultured in Neurobasal medium supplemented with B27 (Invitrogen, Thermo Fisher Scientific, Waltham, MA, USA) and glutamine (Invitrogen, Thermo Fisher Scientific, Waltham, MA, USA). Cells were maintained in vitro at 37 °C in a 5% CO2-humified incubator and fixed for immunostaining or western blot analysis on day 10 after plating (DIV10). Treatment was added to the conditioned medium and replenished every second day until DIV10. Cell cultures were fixed in a 4% paraformaldehyde 4% glucose solution at 37 °C for 30 min and processed for immunocytochemistry analysis.

4.4.1. Morphological Analyses

To evaluate neuritic outgrowth and connectivity, fixed cells were stained with the following primary antibodies: a rabbit polyclonal anti-MAP2 (1:100, Merck Millipore, Burlington, MA, USA) and a mouse monoclonal anti-PSD-95 (1:100, Abcam, Cambridge, UK) antibody. Detection was performed using a FITC-conjugated anti-rabbit IgG (1:200, Jackson ImmunoResearch Laboratories, Inc., West Grove, PA, USA) and a Cy3-conjugated anti-mouse IgG (1:200, Jackson ImmunoResearch Laboratories, Inc., West Grove, PA, USA) antibody, respectively. Nuclei were counterstained with Hoechst-33342 (Sigma-Aldrich, St. Louis, MO, USA) and fluorescent images were acquired using a Nikon Eclipse Te600 microscope equipped with a Nikon Digital Camera DXM1200 ATI system (Nikon Instruments, Inc., Melville, NY, USA). The neuritic length of MAP2-positive neurons was measured and quantified by tracing along each neuronal projection using the image analysis system Image Pro Plus (Media Cybernetics, Silver Spring, MD, USA) as previously described [11]. The degree of synaptic innervation was evaluated by counting the number of PSD-95 positive puncta on proximal dendrites and expressed as the number of PSD-95 puncta per 20 μm of neuritic length. Immunofluorescence images were taken with a LEICA TCS SL confocal microscope (Leica Mycrosystems, Wetzlar, Germany). Fifty neurons for each condition were evaluated.

4.4.2. Neuronal Survival

In order to assess survival of differentiated hippocampal neurons in culture, the number of MAP2-positive neurons was counted and expressed as a percentage of the total number of cells in culture evaluated through Hoechst staining.

4.5. *In Vivo Treatment*

C11 was dissolved in a vehicle solution of 5.5% dimethyl sulfoxide (DMSO) in water.

4.5.1. Acute **C11** Treatment

Two-month-old *Cdkl5* +/Y mice were treated with vehicle (5.5% of DMSO in water) and two-month-old *Cdkl5* −/Y mice were treated with vehicle or **C11** (10 mg/kg, 50 mg/kg or 100 mg/kg) administered intraperitoneally. Animals were sacrificed after 4 h.

4.5.2. Chronic **C11** Treatment

Since in the mammalian brain, the dramatic rearrangements in structure and function that characterize childhood/adolescence result in a critical and sensitive period of brain development and suggest that early interventions would provide potential clinical benefits, we decided to chronically treat juvenile *Cdkl5* KO mice [71]. *Cdkl5* +/Y and *Cdkl5* −/Y male mice were randomly assigned to the experimental conditions and, starting from postnatal day 30 (P30), were treated with vehicle (5.5% of DMSO in water) or **C11** (50 mg/kg) administered via a daily intraperitoneal injection for 15 consecutive days. Body weight of the mice was monitored daily.

4.5.3. Chronic NP12 Treatment

Brain sections that were processed for Hoechst staining, and used to evaluate hippocampal neuron survival, derived from animals used in [26]. Briefly, 3-week-old *Cdkl5* +/Y and *Cdkl5* −/Y male mice were treated with vehicle (corn oil; Sigma-Aldrich, St. Louis, MO, USA) or NP-12 (Tideglusib; 20 mg/kg body weight in corn oil; Sigma-Aldrich, St. Louis, MO, USA), administered via subcutaneous injection every other day for 20 days. The dose of NP12 was chosen based on [72].

4.6. *Behavioral Assays*

A total of 44 animals, from 9 litters, were used for behavioral studies. The sequence of the tests was arranged to minimize the effect of one test influencing subsequent evaluation of the next test, and mice were allowed to recover for 1–2 days between different tests. All behavioral studies and analyses were performed blind to genotype. Mice were allowed to habituate to the testing room for at least 1 h before the test, and testing was always performed at the same time of day. Three independent test cohorts were used. The first test cohort consisted of 15 animals (*Cdkl5* +/Y n = 1, *Cdkl5* −/Y n = 8, and *Cdkl5* −/Y + **C11** n = 6) that were tested with the following assays: clasping, catalepsy, and Morris water maze. The second cohort consisted of 18 animals (*Cdkl5* +/Y n = 9, *Cdkl5* −/Y n = 3, and *Cdkl5* −/Y + **C11** n = 6) that were tested with the following assays: clasping, catalepsy, and

Morris Water Maze. The third cohort consisted of 11 animals (*Cdkl5* +/Y $n = 5$ and *Cdkl5* −/Y $n = 6$) that were tested with the following assays: clasping and catalepsy.

4.6.1. Morris Water Maze

Hippocampal-dependent spatial learning and memory was assessed using the Morris Water Maze (MWM) as previously described [26]. Mouse behavior was automatically videotracked (EthoVision 3.1; Noldus Information Technology, Wageningen, the Netherlands). During training, each mouse was subjected to either 1 swimming session of 4 trials (day 1) or 2 sessions of 4 trials per day (days 2–5), with an intersession interval of 1 h (acquisition phase). Mice were allowed to search for the platform for up to 60 s. The latency to find the hidden platform was used as a measure of learning. Twenty-four h after the last acquisition trial, on day 6, the platform was removed and a probe test was run. Animals were allowed to search for the platform for up to 60 s. The latency of the first entrance into the former platform area and the average swim speeds were measured. A visual cue test was conducted after the probe test to assess sensorimotor ability, motivation, and visual ability. The percentage of floating was defined as the percentage of time swimming at a speed slower than 4 cm/s. Animals that failed the visual clue or floating test (1 *Cdkl5* +/Y, 1 *Cdkl5* −/Y and 2 *Cdkl5* −/Y + **C11**) were excluded from the analysis.

4.6.2. Catalepsy Bar Test

The bar was set at a height of 6 cm. Mice were gently positioned, by placing both forelimbs on the bar and their hindlimbs on the floor. The time needed for the mice to remove both paws from the bar was measured using a stopwatch.

4.6.3. Hind-Limb Clasping

Animals were suspended by their tail for 2 min and hind-limb clasping was assessed independently by two operators from video recordings. A clasping event is defined as the retraction of limbs into the body and toward the midline. The time spent hind-limb clasping was expressed as a percentage.

4.7. Western Blot Analysis

Total proteins from SH-SY5Y and SH-*CDKL5*-KO cell lines were lysates in ice-cold RIPA buffer (50 mM Tris–HCl, pH 7.4, 150 mM NaCl, 1% Triton-X100, 0.5% sodium deoxycholate, 0.1% SDS) supplemented with 1mM PMSF, and with 1% protease and phosphatase inhibitor cocktail (Sigma-Aldrich, St. Louis, MO, USA). Total protein from the hippocampus and cortex of 2-month-old *Cdkl5* −/Y and *Cdkl5* +/Y male mice were homogenized in ice-cold RIPA buffer supplemented with 1mM PMSF, and with 1% protease and phosphatase inhibitor cocktail (Sigma-Aldrich, St. Louis, MO, USA). Protein concentration for both cell and tissue extracts was determined using the Bradford method [73]. Equivalent amounts of protein (50 µg) were subjected to electrophoresis on a 4–12% Mini-PROTEAN® TGX™ Gel (Bio-Rad Laboratories, Inc., Hercules, CA, USA) and transferred to a Hybond-ECL nitrocellulose membrane (Amersham—GE Healthcare Life Sciences, Chicago, IL, USA). For acetylated tubulin detection, 10 µg of proteins were loaded on the gel. The primary and secondary antibodies used are listed in Table S1. To quantify post-translational modifications, such as phosphorylation and acetylation, the nitrocellulose membranes were stripped in Restore Western Blot Stripping Buffer (Thermo Fisher Scientific, Waltham, MA, USA) for 20 min and reprobed with the corresponding total protein antibody. Repeated measurements of the same samples were performed by running from two to four different gels. The signal of one sample (internal control) was used to perform a relative analysis of the antigen expression of each sample on the same gel. We considered the control signal as 100 and assigned a value to the other sample as a percentage of the control. Data analysis was performed by averaging the signals obtained in two to four gels for each individual sample. The densitometric analysis of digitized Western blot images was performed using Chemidoc XRS Imaging Systems and Image Lab™ Software (Bio-Rad Laboratories, Inc.,

Hercules, CA, USA). Images acquired with exposition times that generated protein signals out of a linear range were not considered for the quantification.

4.8. Histological and Immunohistochemistry Procedures

For histological and immunohistochemistry analysis, animals were euthanized with isoflurane (2% in pure oxygen) and sacrificed through cervical dislocation. Brains were quickly removed and cut along the midline. Right hemispheres were fixed via immersion in 4% paraformaldehyde in 100 mM phosphate buffer, pH 7.4, stored in fixative for 48 h, kept in 20% sucrose for an additional 24 h, and then frozen with cold ice. Subsequently, hemispheres were cut with a freezing microtome into 30 μm thick coronal sections that were serially collected in anti-freeze solution (30% glycerol; 30% ethylen-glycol; 10% PBS10X; 0.02% sodium azide; MilliQ to volume) and processed for immunohistochemistry procedures as described below. Left hemispheres were Golgi-stained as previously described [27]. All steps of sectioning, imaging, and data analysis were conducted blindly.

4.8.1. NeuN and AIF-1 Immunohistochemistry and Measurements

One out of every 8 serial brain sections from the hippocampal formation was incubated with one of the following primary antibodies: mouse monoclonal anti-NeuN antibody (1:250; Merk Millipore, Burlington, MA, USA) or rabbit polyclonal anti-AIF-1 antibody (1:300; Thermo Fisher Scientific, Waltham, MA, USA). Sections were then incubated for 2 h at room temperature with a Cy3-conjugated anti-mouse secondary antibody (1:200, Jackson ImmunoResearch Laboratories, Inc., West Grove, PA, USA) or with a Cy3-conjugated anti-rabbit secondary antibody (1:200, Jackson ImmunoResearch Laboratories, Inc., West Grove, PA, USA). Nuclei were counterstained with Hoechst-33342 (Sigma-Aldrich, St. Louis, MO, USA). Fluorescent images were acquired using a Nikon Eclipse TE600 microscope equipped with a Nikon Digital Camera DXM1200 ATI System (Nikon Instruments Inc., Melville, NY, USA). The number of Hoechst-positive and NeuN-positive cells were manually counted using the Image Pro Plus software (Media Cybernetics, Silver Spring, MD, USA) and established as cells/mm^3. Starting from 20X magnification images of AIF-1-stained hippocampal slices, AIF-1 positive microglial cell body size was manually drawn using the Image Pro Plus measurement function, and expressed in μm^2.

4.8.2. PSD-95 Immunohistochemistry and Measurements

One out of every 6 free-floating sections from the hippocampal formation was incubated with a rabbit polyclonal anti-PSD-95 antibody (1:1000, Abcam, Cambridge, UK) and with a CY3-conjugated anti-rabbit IgG secondary antibody (1:200, Jackson ImmunoResearch Laboratories, Inc. Laboratories, Inc., West Grove, PA, USA). For quantification of PSD-95 immunoreactive puncta, images from the stratum oriens of the hippocampus were acquired using a LEICA TCS SL confocal microscope (63X oil immersion objective, NA 1.32; zoom factor = 8 Leica Microsystems; Wetzlar, Germany). Three to four sections per animal were analyzed and puncta counts were performed on a single plan, 1024 × 1024 pixel images. Counting was carried out using Image Pro Plus software (Media Cybernetics, Silver Spring, MD, USA).

4.8.3. Spine Density and Morphology

In Golgi-stained 100-μm-thick sections, spines of CA1 pyramidal neurons were counted using a 100× oil immersion objective lens. Dendritic spine density was measured by manually counting the number of dendritic spines on basal dendrites of CA1 pyramidal neurons. In each mouse, 15 dendritic segments (segment length: 10–30 μm) from each zone were analyzed and the linear spine density was calculated by dividing the total number of counted spines by the length of the sampled dendritic segment. The total number of spines was expressed per 10 μm. Based on their morphology, dendritic spines can be divided into five different classes which fall into two categories which also reflect their state of maturation (immature spines: filopodium-like, thin- and stubby-shaped;

mature spines: mushroom- and cup-shaped). The total number of spines was expressed per μm and the number of spines belonging to each class was counted and expressed as a percentage.

4.9. Statistical Analysis

Data from single animals represented the unity of analysis. Results are presented as mean ± standard error (SE). Statistical analysis was performed using GraphPad Prism (version 7, San Diego, CA, USA). All datasets were analyzed using the ROUT method (Q = 1%) to identify significant outliers and the D'Agostino–Pearson omnibus test for normality. Datasets with normal distribution were analyzed for significance using Student's t-test or a one-way analysis of variance (ANOVA) with genotype and treatment as factors. Post hoc multiple comparisons were carried out using the Fisher least significant difference (Fisher LSD) test. For the learning phase of the MWM test, statistical analysis was performed using a repeated three-way mixed ANOVA with genotype and treatment as grouping factors and day as a repeated measure. For categorical data, that is, percentages of spines, we used a chi-squared test. Animals identified as outliers by the use of Grubb's test were excluded from the analyses [74]. A probability level of $p < 0.05$ was considered to be statistically significant.

Supplementary Materials: The following are available online at https://www.mdpi.com/article/10.3390/ijms22115950/s1.

Author Contributions: M.L. and L.G. performed experiments, analyzed data, and contributed to the draft of the paper; C.F. and S.T. performed experiments and analyzed data; G.M., G.G., M.T., N.M. performed experiments; R.R.G. supervised the behavioral study; A.M. synthetized the **C11** compound, E.C. designed the study, analyzed data, and drafted the paper. All authors have read and agreed to the published version of the manuscript.

Funding: This research was funded by the Telethon foundation (grant number GGP19045, awarded to E.C.), and by the Italian parent association "*CDKL5* insieme verso la cura" (to E.C.).

Institutional Review Board Statement: All research and animal care procedures were performed according to protocols approved by the Italian Ministry for Health and by the Bologna University Bioethical Committee (n° 965/2020-PR).

Informed Consent Statement: Not applicable.

Data Availability Statement: The data that support the findings of this study are available upon request.

Conflicts of Interest: The authors declare no conflict of interest.

References

1. Fehr, S.; Wilson, M.; Downs, J.; Williams, S.J.; Murgia, A.; Sartori, S.; Vecchi, M.; Ho, G.; Polli, R.; Psoni, S.; et al. The CDKL5 disorder is an independent clinical entity associated with early-onset encephalopathy. *Eur. J. Hum. Genet.* **2012**, *21*, 266–273. [CrossRef]
2. Demarest, S.; Pestana-Knight, E.M.; Olson, H.E.; Downs, J.; Marsh, E.D.; Kaufmann, W.E.; Partridge, C.-A.; Leonard, H.; Gwadry-Sridhar, F.; Frame, K.E.; et al. Severity Assessment in CDKL5 Deficiency Disorder. *Pediatr. Neurol.* **2019**, *97*, 38–42. [CrossRef]
3. Demarest, S.T.; Olson, H.E.; Moss, A.; Pestana-Knight, E.; Zhang, X.; Parikh, S.; Swanson, L.C.; Riley, K.D.; Bazin, G.A.; Angione, K.; et al. CDKL5 deficiency disorder: Relationship between genotype, epilepsy, cortical visual impairment, and development. *Epilepsia* **2019**, *60*, 1733–1742. [CrossRef]
4. Mangatt, M.; Wong, K.; Anderson, B.; Epstein, A.; Hodgetts, S.; Leonard, H.; Downs, J. Prevalence and onset of comorbidities in the CDKL5 disorder differ from Rett syndrome. *Orphanet J. Rare Dis.* **2016**, *11*, 1–17. [CrossRef]
5. Kothur, K.; Holman, K.; Farnsworth, E.; Ho, G.; Lorentzos, M.; Troedson, C.; Gupta, S.; Webster, R.; Procopis, P.G.; Menezes, M.P.; et al. Diagnostic yield of targeted massively parallel sequencing in children with epileptic encephalopathy. *Seizure* **2018**, *59*, 132–140. [CrossRef]
6. Hector, R.D.; Dando, O.; Landsberger, N.; Kilstrup-Nielsen, C.; Kind, P.C.; Bailey, M.E.S.; Cobb, S.R. Characterisation of CDKL5 Transcript Isoforms in Human and Mouse. *PLoS ONE* **2016**, *11*, e0157758. [CrossRef]

7. Montini, E.; Andolfi, G.; Caruso, A.; Buchner, G.; Walpole, S.M.; Mariani, M.; Consalez, G.G.; Trump, D.; Ballabio, A.; Franco, B. Identification and Characterization of a Novel Serine–Threonine Kinase Gene from the Xp22 Region. *Genomics* **1998**, *51*, 427–433. [CrossRef]
8. Nawaz, M.S.; Giarda, E.; Bedogni, F.; La Montanara, P.; Ricciardi, S.; Ciceri, D.; Alberio, T.; Landsberger, N.; Rusconi, L.; Kilstrup-Nielsen, C. CDKL5 and Shootin1 Interact and Concur in Regulating Neuronal Polarization. *PLoS ONE* **2016**, *11*, e0148634. [CrossRef]
9. Barbiero, I.; Peroni, D.; Tramarin, M.; Chandola, C.; Rusconi, L.; Landsberger, N.; Kilstrup-Nielsen, C. The neurosteroid pregnenolone reverts microtubule derangement induced by the loss of a functional CDKL5-IQGAP1 complex. *Hum. Mol. Genet.* **2017**, *26*, 3520–3530. [CrossRef]
10. Fuchs, C.; Gennaccaro, L.; Ren, E.; Galvani, G.; Trazzi, S.; Medici, G.; Loi, M.; Conway, E.; Devinsky, O.; Rimondini, R.; et al. Pharmacotherapy with sertraline rescues brain development and behavior in a mouse model of CDKL5 deficiency disorder. *Neuropharmacology* **2020**, *167*, 107746. [CrossRef]
11. Trazzi, S.; Fuchs, C.; Viggiano, R.; De Franceschi, M.; Valli, E.; Jedynak, P.; Hansen, F.K.; Perini, G.; Rimondini, R.; Kurz, T.; et al. HDAC4: A key factor underlying brain developmental alterations in CDKL5 disorder. *Hum. Mol. Genet.* **2016**, *25*, 3887–3907. [CrossRef]
12. Trovò, L.; Fuchs, C.; De Rosa, R.; Barbiero, I.; Tramarin, M.; Ciani, E.; Rusconi, L.; Kilstrup-Nielsen, C. The green tea polyphenol epigallocatechin-3-gallate (EGCG) restores CDKL5-dependent synaptic defects in vitro and in vivo. *Neurobiol. Dis.* **2020**, *138*, 104791. [CrossRef]
13. Amendola, E.; Zhan, Y.; Mattucci, C.; Castroflorio, E.; Calcagno, E.; Fuchs, C.; Lonetti, G.; Silingardi, D.; Vyssotski, A.L.; Farley, D.; et al. Mapping Pathological Phenotypes in a Mouse Model of CDKL5 Disorder. *PLoS ONE* **2014**, *9*, e91613. [CrossRef]
14. Fuchs, C.; Gennaccaro, L.; Trazzi, S.; Bastianini, S.; Bettini, S.; Martire, V.L.; Ren, E.; Medici, G.; Zoccoli, G.; Rimondini, R.; et al. Heterozygous CDKL5 Knockout Female Mice Are a Valuable Animal Model for CDKL5 Disorder. *Neural Plast.* **2018**, *2018*, 1–18. [CrossRef]
15. Lupori, L.; Sagona, G.; Fuchs, C.; Mazziotti, R.; Stefanov, A.; Putignano, E.; Napoli, D.; Strettoi, E.; Ciani, E.; Pizzorusso, T. Site-specific abnormalities in the visual system of a mouse model of CDKL5 deficiency disorder. *Hum. Mol. Genet.* **2019**, *28*, 2851–2861. [CrossRef]
16. Ren, E.; Roncacé, V.; Trazzi, S.; Fuchs, C.; Medici, G.; Gennaccaro, L.; Loi, M.; Galvani, G.; Ye, K.; Rimondini, R.; et al. Functional and Structural Impairments in the Perirhinal Cortex of a Mouse Model of CDKL5 Deficiency Disorder Are Rescued by a TrkB Agonist. *Front. Cell. Neurosci.* **2019**, *13*, 169. [CrossRef] [PubMed]
17. Della Sala, G.; Putignano, E.; Chelini, G.; Melani, R.; Calcagno, E.; Ratto, G.M.; Amendola, E.; Gross, C.T.; Giustetto, M.; Pizzorusso, T. Dendritic Spine Instability in a Mouse Model of CDKL5 Disorder Is Rescued by Insulin-like Growth Factor 1. *Biol. Psychiatry* **2016**, *80*, 302–311. [CrossRef]
18. Pizzo, R.; Gurgone, A.; Castroflorio, E.; Amendola, E.; Gross, C.; Sassoè-Pognetto, M.; Giustetto, M. Lack of Cdkl5 Disrupts the Organization of Excitatory and Inhibitory Synapses and Parvalbumin Interneurons in the Primary Visual Cortex. *Front. Cell. Neurosci.* **2016**, *10*, 261. [CrossRef] [PubMed]
19. Loi, M.; Trazzi, S.; Fuchs, C.; Galvani, G.; Medici, G.; Gennaccaro, L.; Tassinari, M.; Ciani, E. Increased DNA Damage and Apoptosis in CDKL5-Deficient Neurons. *Mol. Neurobiol* **2020**, *57*, 2244–2262. [CrossRef] [PubMed]
20. Fuchs, C.; Medici, G.; Trazzi, S.; Gennaccaro, L.; Galvani, G.; Berteotti, C.; Ren, E.; Loi, M.; Ciani, E. CDKL5 deficiency predisposes neurons to cell death through the deregulation of SMAD3 signaling. *Brain Pathol.* **2019**, *29*, 658–674. [CrossRef]
21. Fuchs, C.; Trazzi, S.; Torricella, R.; Viggiano, R.; De Franceschi, M.; Amendola, E.; Gross, C.; Calza, L.; Bartesaghi, R.; Ciani, E. Loss of CDKL5 impairs survival and dendritic growth of newborn neurons by altering AKT/GSK-3beta signaling. *Neurobiol. Dis.* **2014**, *70*, 53–68. [CrossRef]
22. Gennaccaro, L.; Fuchs, C.; Loi, M.; Roncacè, V.; Trazzi, S.; Ait-Bali, Y.; Galvani, G.; Berardi, A.C.; Medici, G.; Tassinari, M.; et al. A GABAB receptor antagonist rescues functional and structural impairments in the perirhinal cortex of a mouse model of CDKL5 deficiency disorder. *Neurobiol. Dis.* **2021**, *153*, 105304. [CrossRef]
23. Wang, I.-T.J.; Allen, M.; Goffin, D.; Zhu, X.; Fairless, A.H.; Brodkin, E.S.; Siegel, S.J.; Marsh, E.D.; Blendy, J.A.; Zhou, Z. Loss of CDKL5 disrupts kinome profile and event-related potentials leading to autistic-like phenotypes in mice. *Proc. Natl. Acad. Sci. USA* **2012**, *109*, 21516–21521. [CrossRef]
24. Salcedo-Tello, P.; Ortiz-Matamoros, A.; Arias, C. GSK3 Function in the Brain during Development, Neuronal Plasticity, and Neurodegeneration. *Int. J. Alzheimer's Dis.* **2011**, *2011*, 1–12. [CrossRef]
25. Dorszewska, J.; Kozubski, W.; Waleszczyk, W.; Zabel, M.; Ong, K. Neuroplasticity in the Pathology of Neurodegenerative Diseases. *Neural Plast.* **2020**, *2020*, 1–2. [CrossRef]
26. Fuchs, C.; Fustini, N.; Trazzi, S.; Gennaccaro, L.; Rimondini, R.; Ciani, E. Treatment with the GSK3-beta inhibitor Tideglusib improves hippocampal development and memory performance in juvenile, but not adult, Cdkl5 knockout mice. *Eur. J. Neurosci.* **2018**, *47*, 1054–1066. [CrossRef]
27. Fuchs, C.; Rimondini, R.; Viggiano, R.; Trazzi, S.; De Franceschi, M.; Bartesaghi, R.; Ciani, E. Inhibition of GSK3beta rescues hippocampal development and learning in a mouse model of CDKL5 disorder. *Neurobiol. Dis.* **2015**, *82*, 298–310. [CrossRef]
28. Sharma, S.; Taliyan, R. Synergistic effects of GSK-3beta and HDAC inhibitors in intracerebroventricular streptozotocin-induced cognitive deficits in rats. *Naunyn Schmiedebergs Arch. Pharmacol.* **2015**, *388*, 337–349. [CrossRef]

29. De Simone, A.; La Pietra, V.; Betari, N.; Petragnani, N.; Conte, M.; Daniele, S.; Pietrobono, D.; Martini, C.; Petralla, S.; Casadei, R.; et al. Discovery of the First-in-Class GSK-3beta/HDAC Dual Inhibitor as Disease-Modifying Agent To Combat Alzheimer's Disease. *ACS Med. Chem. Lett.* **2019**, *10*, 469–474. [CrossRef]
30. Rivieccio, M.A.; Brochier, C.; Willis, D.E.; Walker, B.A.; D'Annibale, M.A.; McLaughlin, K.; Siddiq, A.; Kozikowski, A.P.; Jaffrey, S.R.; Twiss, J.L.; et al. HDAC6 is a target for protection and regeneration following injury in the nervous system. *Proc. Natl. Acad. Sci. USA* **2009**, *106*, 19599–19604. [CrossRef]
31. Luo, J. The role of GSK3beta in the development of the central nervous system. *Front. Biol.* **2012**, *7*, 212–220. [CrossRef] [PubMed]
32. Frame, S.; Cohen, P. GSK3 takes centre stage more than 20 years after its discovery. *Biochem. J.* **2001**, *359 Pt 1*, 1–16. [CrossRef]
33. Yoshimura, T.; Kawano, Y.; Arimura, N.; Kawabata, S.; Kikuchi, A.; Kaibuchi, K. GSK-3beta regulates phosphorylation of CRMP-2 and neuronal polarity. *Cell* **2005**, *120*, 137–149. [CrossRef] [PubMed]
34. Zhang, Y.; Li, N.; Caron, C.; Matthias, G.; Hess, D.; Khochbin, S.; Matthias, P. HDAC-6 interacts with and deacetylates tubulin and microtubules in vivo. *EMBO J.* **2003**, *22*, 1168–1179. [CrossRef]
35. Morales-Garcia, J.A.; Luna-Medina, R.; Alonso-Gil, S.; Sanz-SanCristobal, M.; Palomo, V.; Gil, C.; Santos, A.; Martinez, A.; Perez-Castillo, A. Glycogen Synthase Kinase 3 Inhibition Promotes Adult Hippocampal Neurogenesis in Vitro and in Vivo. *ACS Chem. Neurosci.* **2012**, *3*, 963–971. [CrossRef] [PubMed]
36. Franke, T.F.; Hornik, C.P.; Segev, L.; Shostak, G.A.; Sugimoto, C. PI3K/Akt and apoptosis: Size matters. *Oncogene* **2003**, *22*, 8983–8998. [CrossRef]
37. Bahi-Buisson, N.; Nectoux, J.; Rosas-Vargas, H.; Milh, M.; Boddaert, N.; Girard, B.; Cances, C.; Ville, D.; Afenjar, A.; Rio, M.; et al. Key clinical features to identify girls with CDKL5 mutations. *Brain* **2008**, *131 Pt 10*, 2647–2661. [CrossRef]
38. Banasikowski, T.J.; Beninger, R.J. Haloperidol conditioned catalepsy in rats: A possible role for D1-like receptors. *Int. J. Neuropsychopharmacol.* **2011**, *15*, 1525–1534. [CrossRef] [PubMed]
39. Ricciardi, S.; Ungaro, F.; Hambrock, M.; Rademacher, N.; Stefanelli, G.; Brambilla, D.; Sessa, A.; Magagnotti, C.; Bachi, A.; Giarda, E.; et al. CDKL5 ensures excitatory synapse stability by reinforcing NGL-1-PSD95 interaction in the postsynaptic compartment and is impaired in patient iPSC-derived neurons. *Nat. Cell Biol.* **2012**, *14*, 911–923. [CrossRef]
40. Taft, C.E.; Turrigiano, G.G. PSD-95 promotes the stabilization of young synaptic contacts. *Philos. Trans. R. Soc. B Biol. Sci.* **2014**, *369*, 20130134. [CrossRef] [PubMed]
41. Trazzi, S.; De Franceschi, M.; Fuchs, C.; Bastianini, S.; Viggiano, R.; Lupori, L.; Mazziotti, R.; Medici, G.; Martire, V.L.; Ren, E.; et al. CDKL5 protein substitution therapy rescues neurological phenotypes of a mouse model of CDKL5 disorder. *Hum. Mol. Genet.* **2018**, *27*, 1572–1592. [CrossRef]
42. Gennaccaro, L.; Fuchs, C.; Loi, M.; Pizzo, R.; Alvente, S.; Berteotti, C.; Lupori, L.; Sagona, G.; Galvani, G.; Gurgone, A.; et al. Age-Related Cognitive and Motor Decline in a Mouse Model of CDKL5 Deficiency Disorder is Associated with Increased Neuronal Senescence and Death. *Aging Dis.* **2021**, *12*, 764–785. [CrossRef]
43. Galvani, G.; Mottolese, N.; Gennaccaro, L.; Loi, M.; Medici, G.; Tassinari, M.; Fuchs, C.; Ciani, E.; Trazzi, S. Inhibition of Microglia Over-activation Restores Neuronal Survival in a Mouse Model of CDKL5 Deficent Disorder. *J. Neuroinflamm.* **2021**, preprints.
44. Torres-Platas, S.G.; Comeau, S.; Rachalski, A.; Bo, G.D.; Cruceanu, C.; Turecki, G.; Giros, B.; Mechawar, N. Morphometric characterization of microglial phenotypes in human cerebral cortex. *J. Neuroinflamm.* **2014**, *11*, 12. [CrossRef]
45. Zhang, Y.; Kwon, S.; Yamaguchi, T.; Cubizolles, F.; Rousseaux, S.; Kneissel, M.; Cao, C.; Li, N.; Cheng, H.-L.; Chua, K.; et al. Mice Lacking Histone Deacetylase 6 Have Hyperacetylated Tubulin but Are Viable and Develop Normally. *Mol. Cell. Biol.* **2008**, *28*, 1688–1701. [CrossRef] [PubMed]
46. Li, Y.; Shin, D.; Kwon, S.H. Histone deacetylase 6 plays a role as a distinct regulator of diverse cellular processes. *FEBS J.* **2012**, *280*, 775–793. [CrossRef]
47. Chen, S.; Owens, G.C.; Makarenkova, H.; Edelman, D.B. HDAC6 Regulates Mitochondrial Transport in Hippocampal Neurons. *PLoS ONE* **2010**, *5*, e10848. [CrossRef] [PubMed]
48. Mazanetz, M.P.; Fischer, P. Untangling tau hyperphosphorylation in drug design for neurodegenerative diseases. *Nat. Rev. Drug Discov.* **2007**, *6*, 464–479. [CrossRef]
49. Lovestone, S.; Killick, R.; Di Forti, M.; Murray, R. Schizophrenia as a GSK-3 dysregulation disorder. *Trends Neurosci.* **2007**, *30*, 142–149. [CrossRef] [PubMed]
50. Jope, R.S.; Roh, M.S. Glycogen synthase kinase-3 (GSK3) in psychiatric diseases and therapeutic interventions. *Curr. Drug Targets* **2006**, *7*, 1421–1434. [CrossRef]
51. D'Mello, S.R. Histone deacetylases as targets for the treatment of human neurodegenerative diseases. *Drug News Perspect.* **2009**, *22*, 513–524. [CrossRef]
52. Kazantsev, A.G.; Thompson, L.M. Therapeutic application of histone deacetylase inhibitors for central nervous system disorders. *Nat. Rev. Drug Discov.* **2008**, *7*, 854–868. [CrossRef]
53. Sleiman, S.F.; Basso, M.; Mahishi, L.; Kozikowski, A.P.; Donohoe, M.E.; Langley, B.; Ratan, R.R. Putting the 'HAT' back on survival signalling: The promises and challenges of HDAC inhibition in the treatment of neurological conditions. *Expert Opin. Investig. Drugs* **2009**, *18*, 573–584. [CrossRef] [PubMed]
54. Shukla, S.; Tekwani, B.L. Histone Deacetylases Inhibitors in Neurodegenerative Diseases, Neuroprotection and Neuronal Differentiation. *Front. Pharmacol.* **2020**, *11*, 537. [CrossRef] [PubMed]

55. Ganai, S.A. Small-molecule Modulation of HDAC6 Activity: The Propitious Therapeutic Strategy to Vanquish Neurodegenerative Disorders. *Curr. Med. Chem.* **2017**, *24*, 4104–4120. [CrossRef]
56. Simoes-Pires, C.; Zwick, V.; Nurisso, A.; Schenker, E.; Carrupt, P.A.; Cuendet, M. HDAC6 as a target for neurodegenerative diseases: What makes it different from the other HDACs? *Mol. Neurodegener.* **2013**, *8*, 7. [CrossRef]
57. Scroggins, B.T.; Robzyk, K.; Wang, D.; Marcu, M.G.; Tsutsumi, S.; Beebe, K.; Cotter, R.J.; Felts, S.; Toft, D.; Karnitz, L.; et al. An Acetylation Site in the Middle Domain of Hsp90 Regulates Chaperone Function. *Mol. Cell* **2007**, *25*, 151–159. [CrossRef] [PubMed]
58. Fujimoto, M.; Takaki, E.; Hayashi, T.; Kitaura, Y.; Tanaka, Y.; Inouye, S.; Nakai, A. Active HSF1 Significantly Suppresses Polyglutamine Aggregate Formation in Cellular and Mouse Models. *J. Biol. Chem.* **2005**, *280*, 34908–34916. [CrossRef]
59. Waza, M.; Adachi, H.; Katsuno, M.; Minamiyama, M.; Sang, C.; Tanaka, F.; Inukai, A.; Doyu, M.; Sobue, G. 17-AAG, an Hsp90 inhibitor, ameliorates polyglutamine-mediated motor neuron degeneration. *Nat. Med.* **2005**, *11*, 1088–1095. [CrossRef] [PubMed]
60. Lull, M.E.; Block, M.L. Microglial activation and chronic neurodegeneration. *Neurotherapeutics* **2010**, *7*, 354–365. [CrossRef]
61. Wang, W.-Y.; Tan, M.-S.; Yu, J.T.; Tan, L. Role of pro-inflammatory cytokines released from microglia in Alzheimer's disease. *Ann. Transl. Med.* **2015**, *3*, 136.
62. Yuskaitis, C.J.; Jope, R.S. Glycogen synthase kinase-3 regulates microglial migration, inflammation, and inflammation-induced neurotoxicity. *Cell. Signal.* **2009**, *21*, 264–273. [CrossRef] [PubMed]
63. Green, H.F.; Nolan, Y.M. GSK-3 mediates the release of IL-1beta, TNF-alpha and IL-10 from cortical glia. *Neurochem. Int.* **2012**, *61*, 666–671. [CrossRef]
64. Zain, Z.M.; Vidyadaran, S.; Hassan, M. GSK3 Inhibition Reduces Inflammatory Responses of Microglia and Upregulates Il-10 Production. *Mal. J. Med. Health Sci.* **2017**, *13*, 8.
65. Suuronen, T.; Huuskonen, J.; Pihlaja, R.; Kyrylenko, S.; Salminen, A. Regulation of microglial inflammatory response by histone deacetylase inhibitors. *J. Neurochem.* **2003**, *87*, 407–416. [CrossRef] [PubMed]
66. Kannan, V.; Brouwer, N.; Hanisch, U.-K.; Regen, T.; Eggen, B.J.; Boddeke, H.W. Histone deacetylase inhibitors suppress immune activation in primary mouse microglia. *J. Neurosci. Res.* **2013**, *91*, 1133–1142. [CrossRef] [PubMed]
67. Zhang, Z.-Y.; Schluesener, H.J. Oral Administration of Histone Deacetylase Inhibitor MS-275 Ameliorates Neuroinflammation and Cerebral Amyloidosis and Improves Behavior in a Mouse Model. *J. Neuropathol. Exp. Neurol.* **2013**, *72*, 178–185. [CrossRef]
68. Shein, N.A.; Shohami, E. Histone Deacetylase Inhibitors as Therapeutic Agents for Acute Central Nervous System Injuries. *Mol. Med.* **2011**, *17*, 448–456. [CrossRef]
69. Bruserud, O.; Stapnes, C.; Ersvaer, E.; Gjertsen, B.T.; Ryningen, A. Histone deacetylase inhibitors in cancer treatment: A review of the clinical toxicity and the modulation of gene expression in cancer cell. *Curr. Pharm. Biotechnol.* **2007**, *8*, 388–400. [CrossRef]
70. Beaudoin, G.M., 3rd; Lee, S.H.; Singh, D.; Yuan, Y.; Ng, Y.G.; Reichardt, L.F.; Arikkath, J. Culturing pyramidal neurons from the early postnatal mouse hippocampus and cortex. *Nat. Protoc.* **2012**, *7*, 1741–1754. [CrossRef]
71. Finlay, B.L.; Darlington, R.B. Linked regularities in the development and evolution of mammalian brains. *Science* **1995**, *268*, 1578–1584. [CrossRef]
72. Zhou, A.; Lin, K.; Zhang, S.; Chen, Y.; Zhang, N.; Xue, J.; Wang, Z.; Aldape, K.D.; Xie, K.; Woodgett, J.R.; et al. Nuclear GSK3beta promotes tumorigenesis by phosphorylating KDM1A and inducing its deubiquitylation by USP22. *Nat. Cell Biol.* **2016**, *18*, 954–966. [CrossRef]
73. Bradford, M.M. A rapid and sensitive method for the quantitation of microgram quantities of protein utilizing the principle of protein-dye binding. *Anal. Biochem.* **1976**, *72*, 248–254. [CrossRef]
74. Grubbs, F.E. Procedures for Detecting Outlying Observations in Samples. *Technometrics* **1969**, *11*, 21. [CrossRef]

International Journal of *Molecular Sciences*

MDPI

Review

Purine Nucleotides Metabolism and Signaling in Huntington's Disease: Search for a Target for Novel Therapies

Marta Tomczyk [1,*], Talita Glaser [2], Ewa M. Slominska [1], Henning Ulrich [2] and Ryszard T. Smolenski [1,*]

[1] Department of Biochemistry, Medical University of Gdansk, 80-210 Gdansk, Poland; eslom@gumed.edu.pl
[2] Department of Biochemistry, Institute of Chemistry, University of São Paulo, São Paulo 05508-000, Brazil; talita.glaser@usp.br (T.G.); henning@iq.usp.br (H.U.)
* Correspondence: marta.tomczyk@gumed.edu.pl (M.T.); rt.smolenski@gumed.edu.pl (R.T.S.)

Abstract: Huntington's disease (HD) is a multi-system disorder that is caused by expanded CAG repeats within the exon-1 of the huntingtin (*HTT*) gene that translate to the polyglutamine stretch in the HTT protein. HTT interacts with the proteins involved in gene transcription, endocytosis, and metabolism. HTT may also directly or indirectly affect purine metabolism and signaling. We aimed to review existing data and discuss the modulation of the purinergic system as a new therapeutic target in HD. Impaired intracellular nucleotide metabolism in the HD affected system (CNS, skeletal muscle and heart) may lead to extracellular accumulation of purine metabolites, its unusual catabolism, and modulation of purinergic signaling. The mechanisms of observed changes might be different in affected systems. Based on collected findings, compounds leading to purine and ATP pool reconstruction as well as purinergic receptor activity modulators, i.e., P2X7 receptor antagonists, may be applied for HD treatment.

Keywords: purine metabolism; purinergic signaling; Huntington's disease

Citation: Tomczyk, M.; Glaser, T.; Slominska, E.M.; Ulrich, H.; Smolenski, R.T. Purine Nucleotides Metabolism and Signaling in Huntington's Disease: Search for a Target for Novel Therapies. *Int. J. Mol. Sci.* **2021**, *22*, 6545. https://doi.org/10.3390/ijms22126545

Academic Editor: Ivano Condò

Received: 31 May 2021
Accepted: 14 June 2021
Published: 18 June 2021

Publisher's Note: MDPI stays neutral with regard to jurisdictional claims in published maps and institutional affiliations.

1. Introduction

1.1. Huntington's Disease Pathophysiology

Huntington's disease (HD) is a rare neurodegenerative disease that extensively affects the central nervous system. The disorder is inherited in an autosomal dominant manner. Clinically, HD is manifested by the occurrence of cognitive, mental, and motor disorders [1]. One of the earliest signs of the motor disorder in Huntington's disease is chorea, i.e., involuntary dance-like movements. Patients with HD are also characterized by bradykinesia (motor slowness) and dystonia (the occurrence of unnaturally slow, prolonged muscle spasms that cause repetitive torsional movements affecting various parts of the body) [2]. Thus, the motor disorders in Huntington's disease visibly affect the attitude, balance, and gait of HD patients. Furthermore, with the development of the disease, the speech of a patient with HD becomes unclear. Moreover, swallowing difficulties may also occur, which may lead to weight loss [3]. In addition to movement disorders, HD causes prominent changes in personality and mood. Most often HD patients suffer from depression, apathy, anxiety, irritability, outbursts of anger, impulsiveness, obsessive-compulsive syndromes, sleep disorders, and withdrawal from social life [4]. A characteristic feature of Huntington's disease is also cognitive impairment, which affects the understanding, reasoning, and memory. It includes slower thinking, problems with concentration, organization, planning, decision-making, answering questions, short-term memory disorders, as well as limited problem-solving skills and understanding of new information [2]. The incidence of HD in Europe is estimated at 5 to 10 cases per 100,000 people. In adults, the first symptoms appear between 30 and 50 years of age, after which the disease relentlessly progresses over the next 15–20 years.

The genetic cause of HD is the occurrence of multiple repeats of the CAG nucleotide sequence within the huntingtin gene (*HTT*) localized on chromosome 4, which results in the elongation of the polyglutamine stretch in the HTT protein. The number of CAG nucleotide sequence repeats in the healthy population varies from 6 to 35, while the presence of over 36 repeats defines the pathogenic HD allele. In cases with 36 to 39 CAG repeats within the *HTT* gene, the symptoms of the disease may be reduced or completely unnoticeable [1]. Moreover, the expansion length of CAG repeats correlates with the onset of the disease [5]. Huntingtin (HTT) is a multi-domain protein with a size of 348 kDa, with the highest level of HTT, demonstrated in the brain [6,7]. It has also been demonstrated outside the nervous system, in organs such as skeletal muscles or the heart [3]. The elongation of the polyglutamine stretch in exon 1 HTT leads to the formation of insoluble huntingtin aggregates, which are observed in both the early and advanced stages of the disease [8]. Aggregates of the mutated form of HTT (mHTT) have been identified in the brain as well as the outside central nervous system, e.g., in skeletal muscle [9]. In the CNS, mHTT mainly affects the basal ganglia region of the encephalon; this is the main region for voluntary and involuntary motor control, as well as cognition. This mutant protein sensitizes GABAergic neurons, making them vulnerable to NDMA induced excitotoxicity, leading to cell death. On the cellular level, HTT was found in the nucleus, endoplasmic reticulum, Golgi apparatus, and endosomes [10–12]. It has been shown that HTT interacts with proteins involved in gene transcription (e.g., CREB-binding transcription factor (CBP)), intracellular signaling (e.g., HIP14 protein), intracellular transport (e.g., HIP1 protein, HAP1), endocytosis, and metabolism (e.g., PACSIN1 phosphoprotein, vitamin D-binding receptor, hepatic X-receptor) [13,14]. Furthermore, HTT is essential during early embryogenesis and brain development. The inactivation of the *HTT* gene by targeting exon 1 or 5 is lethal in mice on embryonic day 7.5 (E7.5) of mouse development [15]. Biochemical and molecular pathways by which mutant huntingtin affects cellular dysfunction and death remain unclear; however, these might be caused not only by cellular mHTT accumulation but also the loss of HTT function leading to metabolic and signaling cascades impairment. Thus, in this work, we aimed to summarize the knowledge about the dysfunction of intra- and extracellular metabolism related to purines in the most affected by Huntington's disease systems (central nervous system, heart, skeletal muscle), its role in HD pathophysiology, and possible applications in HD treatment.

1.2. Purine Nucleotides Metabolism and Signaling

Purines play an important role as metabolic signals, controlling cellular growth and providing energy to the cell. In the central nervous system (CNS), the balance of nucleotides depends on a continuous supply of preformed purine and pyrimidine rings, mainly in the form of nucleosides. These nucleosides can enter the brain through the blood–brain barrier, or locally supplied by the conversion of extracellular phosphorylated forms (nucleotides) by extracellular nucleotidases located in the neuronal plasma membrane. The ectonucleotidases are divided into four families that differ in the specificity of the substrate and cellular location: nucleoside triphosphate diphosphohydrolases (NTPDases), nucleotide pyrophosphatase/ phosphodiesterases (NPPs), alkaline and acid phosphatases (ALP and ACP, respectively), and ecto-5′-nucleotidase [16–19]. The NTPDase comprises NTPDase1–8; however, just NTPDase1, -2, -3, and -8 can efficiently hydrolyze all nucleotides. The NPP family includes seven members (NPP1–7) but as NTPDASE, only NPP1, NPP2, and NPP3 can hydrolyze nucleotides [17]. The ALP and ACP families comprise many ectoenzymes that dephosphorylate nucleotides (ATP, ADP, and AMP) and diverse substrates. The human 5′-nucleotides family has seven enzymes, although just one is anchored to the plasma membrane, known as CD73 [19,20]. Its main function is the production of extracellular adenosine. Later in the extracellular cascade, this adenosine can be converted to inosine through ecto-adenosine deaminase (eADA), and later to hypoxanthine by purine nucleoside phosphorylase (PNP) [21]. Then, after the transport of nucleosides and inosine/hypoxanthine into the cell, they are converted to AMP, ADP, and ATP by the basic cellular processes similar to those taking place in muscles.

In skeletal muscles and the heart, high energy phosphate produced in oxidative phosphorylation is transported from mitochondria to the contractile apparatus via phosphocreatine (PCr) shuttle. In the mitochondrial inter-membrane space, the energy of the high-energy phosphate bond of ATP can be transferred to creatine by mitochondrial creatine kinase (CK) resulting in the formation of PCr. In the cytosol, PCr can be used to resynthesize ATP from ADP by cytosolic CK. An important aspect of ATP involvement in energy metabolism is ATP degradation to adenosine-5′-diphosphate (ADP) by ATPases (e.g., CK, sodium–potassium, or calcium myosin ATPase). There is also a possibility of further conversion of ADP to AMP that is mediated by adenylate kinase (AK). AMP is a substrate for two alternative pathways and enzymes: (1) 5′-nucleotidase (5NT) dephosphorylating AMP to adenosine that occurs in multiple isoforms, and (2) AMP deaminase (AMPD) converting AMP to inosine monophosphate (IMP). A unique aspect of purine nucleotide metabolism in the skeletal muscle is the function of the purine nucleotide cycle that besides AMPD, involves also adenylosuccinate synthetase, and adenylosuccinate lyase. This cycle plays an important role in energy balance through the maintenance of a high ATP/ADP ratio. Higher levels of intracellular AMP may also activate the AMP-activated protein kinase, an important protein involved in the regulation of cellular energy metabolism at both protein expression and activity levels. IMP is also the final product of purine de novo synthesis as well as purine salvage pathway (formation of IMP from hypoxanthine). The adenosine can be degraded to inosine by adenosine deaminase (ADA). Afterward, inosine can be converted to hypoxanthine by a purine nucleoside phosphorylase (PNP). Hypoxanthine can be converted by xanthine oxidoreductase activity to xanthine and uric acid. Nucleotide breakdown is highly organ and cell-type specific.

In the extracellular space, ATP can act as a signaling molecule by interacting with purinergic P2X and P2Y receptors. While ADP, UTP, UDP, and UDP-glucose interact with P2Y receptor subtypes, P2X receptors are ligand-gated cation channels comprised of seven subtypes (P2X1-7) that assemble in a trimeric structure and upon stimulation allow cations inflow (such as Na^+, K^+, and Ca^{2+}). P2Y receptors consist of eight subtypes (P2Y1, 2, 4, 6, 11, 12, 13, 14), in which P2Y1,2,4,6 and 11 receptors couple to Gq proteins, activating phospholipase C (PLC)-β. Resulting in IP3 and diacylglycerol production, releasing Ca^{2+} from intracellular stores [22]. Instead, P2Y12,13, and 14 receptors couple to Gi proteins, inhibiting cAMP production [22]. While adenosine is a ligand for a category of receptors named P1. This category is composed of A1, A2A, A2B, and A3 subtypes. Within, A1 and A3 receptors are coupled to Gi/Go proteins, inhibit adenylate cyclase and reduce cAMP levels upon activation; A2 receptors are coupled to Gs protein activating adenylate cyclase, increasing intracellular cAMP levels [23].

2. Purines in Huntington's Disease

2.1. Cellular Changes Related to Affected Huntingtin Expression

In 2014, Ismailoglu et al. investigated the metabolic profile of three syngeneic mouse embryonic stem cell (mESC) lines: HTT knock-out (KO), extended poly-Q (Htt-Q140/7), and wildtype mESCs (Htt-Q7/7). They found that HTT KO cells exhibited a 50% decrease in ATP levels, concomitant with 2-fold increases in both ADP and AMP levels, which demonstrated that HTT protein activity is critical for the maintenance of high energy phosphates in the cell. Moreover, HTT KO exhibited unique expression of several molecules involved in purine synthesis, such as 5-aminoimidazole ribonucleotide (AIR), phosphor-ribosyl-formyl-glycineamidine (FGAM), formamidoimidazole-4-carboxamide ribotide (FAICAR), and 5′-phosphoribosyl-4-(N-succinocarboxamide)-5-aminoimidazole (SAICAR). It was accompanied by increased levels of the 5-aminoimidazole-4-carboxamide ribotide (AICAR) and the final pathway product of purine biosynthesis, IMP in those cells. Overall, it reveals a substantial acceleration of purine synthesis and turnover in HTT KO mESCs and suggesting the HTT importance in maintaining its mutual balance [24]. Following these findings, in 2020, we differentiated HTT KO mESCs to neurons as well as to cardiomyocytes and established that besides HTT absence, differentiation was successful [25,26]. It suggested that HTT is not a fundamental protein in cardiomyocyte development. The previous study in human pluripotent stem cells suggested that loss of HTT can reduce the induction of neural and neuronal genes during differentiation [27]. In the case of investigation purine metabolism, there were no changes in extracellular ATP concentration between HTT KO and WT neurons [25]. While HTT KO cardiomyocytes exhibited diminished intracellular ATP pool, which is in the line with data obtained from the mHTT overexpression cellular model [26,28]. HEK 293T cell line transfected with plasmids expressing the mutant exon 1 of the *HTT* gene was characterized also by increased ADA activity, which suggested deteriorations in intracellular purine metabolism. Increased intracellular levels of metabolites such as inosine, hypoxanthine, and adenosine were found in HTT KO mESC [24]. Besides changes in intracellular purine metabolism, HD cells exhibited reduced activities of all extracellular enzymes, including eNTPD, e5NT, and eADA relative to the control [28]. In the case of purinergic signaling, the Ca^{2+} response of P2 receptors is impaired in mouse HTT KO neurons derived from mESC [25], while the P2Y2 receptor response is only impaired in the presence of mHTT. Lack of P2Y2 receptor input induces the cells to upregulate their expression, as a feedback compensation effort [25].

2.2. Purine Nucleotides Metabolism and Signaling in the Central Nervous System in Huntington's Disease

One of the pathological hallmarks of the HD-affected brain is the gradual atrophy of the striatum (caudate nucleus and putamen) [29]. On gross examination, 80% of HD brains show atrophy of the frontal lobes. Thus, bilateral, symmetric atrophy of the striatum is observed in 95% of the HD brains [29]. The mean brain weight in HD patients is approximately 30% lower than in normal individuals. Striatal degeneration may lead to energy metabolism changes. It is clear that neurons are highly dependent on mitochondria ATP and Ca^{2+} buffering to maintain synaptic communication [30]. Moreover, neuronal mitochondria levels require to renew or adapt by efficient biogenesis and mitophagy during their lifespan [31]. There are undisputed data that highlighted the intensive deficits in energy metabolism in the human HD-affected brain. The striatum mitochondrial oxidative metabolism investigation underlined the selective defect of glycolysis in early and clinical symptoms in HD patients [31].

Interestingly, further analysis showed a significant correlation between impaired basal ganglia metabolism and functional capacity of HD patients [32]. Studies concentrated on lactate metabolism in HD-affected brains are ambiguous. Increased lactate levels were observed in the striatum of HD patients, which is discussed as inefficient oxidative phosphorylation leading to lactate accumulation from pyruvate via lactate dehydrogenase [32]. In contrast, reduced levels of lactate and citrate were shown in cerebrospinal fluid which may indicate impairment of glycolysis and TCA cycle function in HD subjects [32].

At the molecular level, brain energy metabolism deterioration included mitochondria dysfunction and trafficking interruption resulted in changes in the activities of molecules involved in energy balance [33]. In few independent studies of the striatum of mHTT knock-in mice, HD patients' postmortem brains, and lymphoblasts, the ATP/ADP ratio was reduced as a consequence of mHTT aggregation [34–36]. Significant reduction in mitochondrial spare respiratory capacity was reported in human HD fibroblasts and immortalized mHTT expressing mouse striatal cells when compared to wild-type cells indicating that mitochondrial bioenergetics is compromised by mHTT and supporting a toxic role of mHTT on mitochondrial bioenergetics [37].

Moreover, expression of full-length mHTT in immortalized striatal progenitor cells, derived from HD mice model *Hdh*Q111, diminished the activity of mitochondrial respiratory chain complex II associated with intense sensitivity to Ca^{2+}-induced decrease in oxygen consumption and mitochondrial membrane potential [33]. Mitochondrial Ca^{2+} transport is powered by the mitochondrial proton gradient, and increased neuronal Ca^{2+} modifies mitochondrial ATP production by uncoupling oxidative phosphorylation. In healthy conditions, Ca^{2+} can promote ATP synthesis by assisting pyruvate dehydrogenase, isocitrate dehydrogenase, α-ketoglutarate dehydrogenase, and ATP synthase complex [31].

The reduced mitochondrial ATP levels and decreased ATP/ADP ratio found in mHTT-containing striatal cells is linked to increased Ca^{2+} influx through N-methyl-D-aspartate (NMDA) receptors, and cell ATP/ADP ratio is normalized by blocking Ca^{2+} influx [36]. Mitochondria incubated with mHTT had increased sensitivity to Ca^{2+}-induced opening of the MPT and release of cytochrome c and apoptosis induction [38], consistent with a direct effect of mHTT on mitochondrial Ca^{2+} handling [33]. As demonstrated by mitochondria isolated from lymphoblast cells of HD patients and HD mouse brain that have a reduced membrane potential and depolarize at lower Ca^{2+} concentrations than control mitochondria [39]. Both wild HTT and mHTT bind to the outer mitochondrial membrane in human neuroblastoma cells and cultured striatal cells [40]. It is established that full-length mHTT may impair mitochondrial motility in neurons through a toxic gain of loss of function from the polyglutamine tract [40]. Moreover, in vitro and in vivo models of HD are characterized by the altered mitochondrial trafficking that precedes neuronal dysfunction [41].

Roles of mitochondria in HD go far beyond ATP production and Ca^{2+} homeostasis; they can also regulate the metabolism of the reactive oxygen species (ROS) and apoptosis [42]. In addition, mHTT can affect the production of neurotrophins, such as brain-derived neurotrophic factor (BDNF), impairing neuronal survival [43]. Noteworthy, BDNF promotes ATP synthesis and mitochondrial efficiency [44,45], which correlates with the fact that neurons are high-demand energy cells [46].

The reduction of mitochondrial bioenergetics in HD could be also a result of impairment of mitochondrial enzymes. Postmortem studies of the striatum of HD patients as well as cultured striatal neurons transfected with N-terminus mHTT showed selective depletion of succinate dehydrogenase associated with decreased complex II enzymatic activity [47,48]. The mHTT can interfere with the TATA box binding protein (TBP)-associated factor 4 (TAF4)/ cyclic adenosine monophosphate (cAMP) response element-binding protein (CREB) signaling pathway [39]. Data are underlining a reduction in CREB phosphorylation and CRE signaling, which may contribute to the down-regulation of molecules involved in energy metabolism, such as peroxisome proliferator-activated receptor-gamma coactivator 1 alpha (PGC-1α) in mHTT expressing striatal cells [34]. Moreover, the reduction in PGC-1α correlates with a diminished number of mitochondria in HD postmortem brain tissue [49,50]. The mHTT can impair the mitochondria biogenesis through PGC-1α transcription inhibition as well. Post-mortem HD brain analysis shows lower levels of PGC-1α [38,51]. In this way, cells increase the anaerobic metabolism in the basal ganglia of the HD patients, leading to improved lactate generation and its accumulation, thus promoting local inflammation [32].

AMP-activated protein kinase (AMPK), the main sensor for cellular energy content is also targeted by mHTT. It is activated by increased AMP/ATP ratios and induces PGC-1α expression. AMPK is located in the nucleus in the HD striatum, downregulating the Bcl-2 family which leads to apoptosis [52,53].

As previously mentioned, brain capacity for *de novo* production of purines and pyrimidine rings is limited. In this regard, the brain demands constant provision of nucleosides produced in the liver and that cross the blood–brain barrier [30]. Adenosine and ATP in the extracellular milieu of the encephalon stimulate P1 and P2 receptors, respectively, acting as co-neurotransmitters. ATP and UTP can be released to the extracellular environment by microglia, astrocytes, and neurons themselves. They are also present in vesicles, where concentrations can be as high as 100 mM (ATP) and 8 mM (UTP) [30]. Thus, the imbalance of intra/extracellular nucleotides in neurons is related to various neurological disorders, such as Parkinson's, Alzheimer's, and Huntington's disease [54]. Figure 1 summarizes deteriorations in purine nucleotides metabolism and signaling in CNS affected by Huntington's disease.

HD is characterized by GABAergic loss due to excessive calcium response to glutamate stimuli. However, some evidence points to the impairment of some metabolic pathways link with purines metabolism [55]. Patassini and coworkers, using human post-mortem brains, underwent metabolomics analyses of many brain areas typically affected in patients of Huntington's disease [56]. They found a decreased level of hypoxanthine but an increased level of inosine, indicating higher activity of ADA and low PNP. On the other hand, another study with a transgenic mouse model (R6/1) of HD highlighted an increased adenosine level in the cerebrospinal fluid in the middle stage of HD [57,58]. Furthermore, microarray analysis of the prefrontal cortex of HD patients showed a significant reduction in the transcript of CD73, suggesting that the synaptic adenosine level converted from AMP might be low. Moreover, the transcript levels of ENTs and AKs in HD patients are higher than in non-HD subjects [59]. Thus, it seems that regulation of the adenosine modulating enzymes in HD is still unclear and requires further investigation.

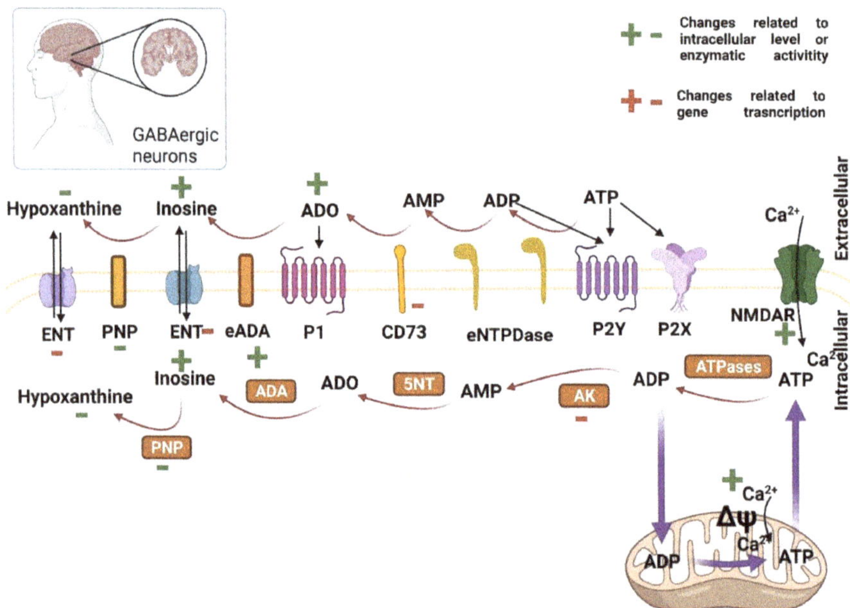

Figure 1. Purine metabolism and signaling in healthy and Huntington's disease striatal GABAergic neurons. In the neurons of the striatum, ATP is heavily produced by mitochondria which also play important role in buffering cytosolic calcium signaling. In the cytosol, many enzymes (orange boxes) can convert ATP to AMP, ADO, IMP, inosine, and hypoxanthine. The latter ones can be transported to the extracellular milieu by ENT. Extracellularly, ATP can also be converted to inosine and hypoxanthine through a similar cascade by extracellular-faced enzymes anchored to the cell membrane. CD73 is the only extracellular 5NT. Pathophysiology of Huntington's disease exerts alterations in the activity of enzymes (marked in green +/−) or by genomic expression (marked in red +/−). In Huntington's disease condition, NMDA-glutamate receptors permit uncontrolled Ca^{2+} influx into cells. The excessive Ca^{2+} is transported to mitochondria, where it disrupts the membrane potential, releasing cytochrome c and promoting cell death. **P1**—Adenosine receptor; **P2X**—ATP receptor channel; **P2Y**—ATP/UTP receptor; **ATPases**—adenosine-5′-triphosphatases; **AK**—adenylate kinase; **eNTPD**—ecto-nucleoside triphosphate diphosphohydrolase; **5NT**—5′ nucleotidase; **CD73**—ecto-5′ nucleotidase; **eADA**—ecto-adenosine deaminase; **ADA**—adenosine deaminase; **PNP**—purine nucleoside phosphorylase; **ENT**—ecto nucleoside transporter. Created with BioRender.com (accessed on 5 May 20021).

So far, few studies have investigated the role of purinergic signaling in Huntington's disease. Regarding the adenosine receptors, A1, which is expressed in neurons, microglia, astrocytes, and oligodendrocytes is impaired in Huntington's disease, this receptor usually protects against degeneration by inhibiting excitatory neurotransmission. A2A receptor is the most studied in HD; however, this is still controversial. Some studies detected down-regulated A2A receptor expression in HD rodent models [60–64], while others improved the motor symptoms by antagonizing the A2A receptor [65,66]. Regarding ATP-sensitive receptors, the P2X7 receptor has been described as a major player, since its antagonism by

Brilliant Blue G or selective P2X7 receptor inhibitors mitigates dyskinesia and body weight loss while preventing neuronal loss in the Tet/HD94 and R6/1 models [67].

Besides the P2X7 receptor, ATP and UTP-sensitive P2Y2 receptor plays important roles in HD. The intracellular signaling triggered by this receptor is impaired in neural precursor cells and neurons of HD human and mouse in vitro models. Moreover, the activity of the P2Y2 receptor favors the differentiation of neural stem cells towards a GABAergic neuronal fate [25]. Reestablishment of the activity of the P2Y2 receptor, promoting BDNF release, may prevent cell death.

2.3. *Purine Nucleotides Metabolism and Signaling in Skeletal Muscle and Heart in Huntington's Disease*

It has been shown that HD patients, except for the central nervous system disorders, are also characterized by a reduced (by about 50%) muscular strength compared to healthy subjects [68]. Moreover, HD mice models were characterized by skeletal muscle atrophy [69]. It is also noted that R6/2 mice have altered the ultrastructure of transverse tubules in skeletal muscle fibers [70]. At the cellular level, aggregation of mHTT, the inclusion of poly-ubiquitinated proteins were found in myofibers and myonuclei in R6/2 mice [9,71]. The mHTT formation in skeletal muscle leads to defects, such as myofiber size reduction, type switching, and denervation [69,72–75]. In addition to changes in myofiber structure, transcriptional deregulation, and deteriorations in energy metabolism occur linked to impaired adenine nucleotide metabolism [76]. It has been noted that the skeletal muscles of HD patients are characterized by dysfunction of oxidative metabolism [77]. Studies in experimental mouse models have also shown that mitochondria isolated from the quadriceps muscle of the R6/2 mice model were characterized by reduced activity of the respiratory chain complexes [78]. Increased production of energy substrates such as lactate and acetate were also shown which confirms the presence of oxidative metabolism disorders [79]. Furthermore, in vitro myocyte cultures revealed disturbances of the mitochondrial membrane potential and cytochrome c release [80]. Moreover, increased levels of the mitochondrial pro-fission factor DRP1 and its phosphorylated active form, and decreased levels of the pro-fusion factor MFN2 in quadriceps of the R6/2 mice model were detected [78]. Interestingly, experimental studies have also shown a reduction in the level of PGC-1α, one of the proteins activated by peroxisomal proliferator gamma (PPAR γ) in skeletal muscles of HD mice models as well as HD patients [50]. Additionally, the pharmacological activation of this co-activator led to increased expression of the skeletal muscle fiber proteins that suggested an important role of energy metabolism abnormalities in development of HD related myopathy [81]. Recently, Miller et al. report that the decrease of adenine nucleotides strictly linked with energy metabolism of the cell may lead to increased nucleotide degradation and contribute to the general pathophysiology of skeletal muscle atrophy in numerous disease states and conditions [82].

Besides skeletal muscle pathology, multiple epidemiological studies have shown that heart failure is the second cause of death in HD patients [83,84]. Reduced cortical and subcortical blood flow and heart rate have been reported as examples of pathological abnormalities in HD hearts [85–88]. HD patients' heart rate variability pattern is consistent with a higher sympathetic prevalence [89]. Further studies with HD animal models reaffirmed cardiac pathological events, such as variations in the heart rate and cardiac remodeling [90–92]. HD mouse models also revealed heart contractile dysfunctions, which might be a part of dilated cardiomyopathy. These changes were accompanied by an increased expression of fetal genes (the same is observed during the pathological remodeling of the heart) or the presence of interstitial fibrosis [90]. Interestingly, hearts of the HD mouse model R6/2 did not react to the same extent upon long-term treatment with isoproterenol (a compound that causes hypertrophy of the heart), as wild-type mouse hearts, suggesting the presence of signaling dysfunction that stimulates heart remodeling [93]. Hearts of HD mice models contained an increased number of apoptotic cells and degree of fibrosis [94]. Cardiomyocytes of the BACHD HD mouse model also showed electromechanical abnormalities, including prolonged action potential or arrhythmic contractions. Cellular arrhythmia was accompanied by increased activity of Ca^{2+}/calmodulin-dependent protein kinase II, suggesting disturbed calcium metabolism in the cell [95]. Dridi et al. show that intracellular calcium (Ca^{2+}) leaks via post-translationally modified ryanodine receptor/intracellular calcium release (RyR) channels may play an important role in HD pathology [96]. Furthermore, abnormalities of superoxide dismutase activity and glutathione peroxidase in the mitochondria of cardiomyocytes were observed. Cardiomyocytes from R6/2 mice also showed abnormalities in mitochondrial structure (loss of longitudinal shape and changes in mitochondrial density) that may lead to deteriorations in energy metabolism [97]. The studies performed on this model have also shown that the hearts of these mice are characterized by decreased activity of the mammalian target of rapamycin kinase complex 1 (mTORC1) that could be the cause of the reduced heart weight which is observed in this strain as well as the lack of resistance to severe and chronic stress [98]. Nevertheless, Kojer et al. highlighted no changes in mitochondrial oxidative chain complexes in R6/2 mice hearts [78]. Mechanistically, pathological huntingtin was found to be responsible for cellular death via inhibition of the proteasome in the cytosol and the nucleus of cardiomyocytes of R6/2 mice [97]. On the other hand, some studies indicated the mHTT absence may be the underlying mechanism, as observed in HD mouse hearts [78,90]. Taken together, cardiac dysfunction in HD might derive not only from autonomic nervous system dysfunction but also from several cellular and tissue defects, such as energy metabolism impairment, associated with dysfunctional cardiac purine metabolism and signaling [99]. Figures 2 and 3 illustrate changes in purine nucleotides metabolism and signaling in Huntington's disease-related myopathies.

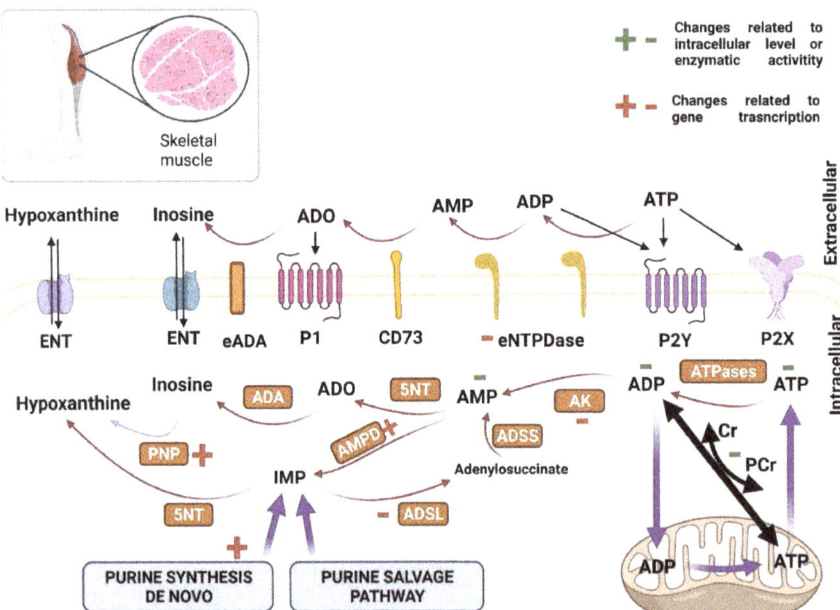

Figure 2. Purine metabolism and signaling in healthy and Huntington's disease skeletal muscles. In the skeletal muscles, ATP is heavily produced by mitochondria and transported to the cytosol through the phosphocreatine shuttle. In the cytosol, many enzymes (orange boxes) can convert ATP to AMP, ADO, IMP, inosine, and hypoxanthine. The later ones can be transported to the extracellular milieu by ENT. Extracellularly, ATP can be converted to inosine through a similar cascade by extracellular-faced enzymes anchored to the cell membrane. CD73 is the only extracellular 5NT. Pathophysiology of Huntington's disease exerts alterations in the activity of enzymes (marked in green +/−) or by genomic expression (marked in red +/−). **P1**—Adenosine receptor; **P2X**—ATP receptor channel; **P2Y**—ATP/UTP receptor; **ATPases**—adenosine-5′-triphosphatases; **AK**—adenylate kinase; **eNTPD**—ecto-nucleoside triphosphate diphosphohydrolase; **AMPD**—AMP deaminase; **5NT**—5′ nucleotidase; **CD73**—ecto-5′ nucleotidase; **eADA**—ecto-adenosine deaminase; **ADA**—adenosine deaminase; **PNP**—purine nucleoside phosphorylase; **ADSL**—adenylosuccinate lyase; **ADSS**—adenylosuccinate synthetase; **ENT**—ecto nucleoside transporter. Created with BioRender.com (accessed on 5 May 20021).

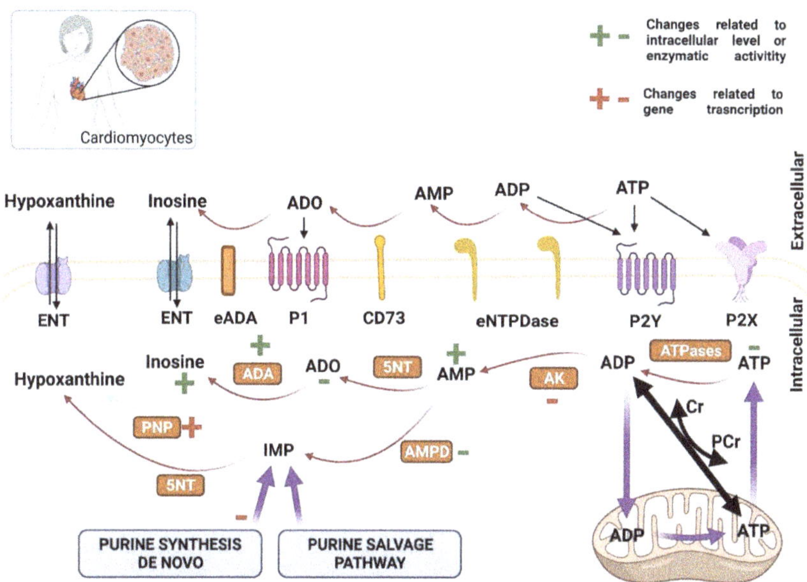

Figure 3. Purine metabolism and signaling in healthy and Huntington's disease cardiomyocytes. In cardiomyocytes, ATP is heavily produced by mitochondria and transported to the cytosol through the phosphocreatine shuttle. In the cytosol, many enzymes (orange boxes) can convert ATP to AMP, ADO, IMP, inosine, and hypoxanthine. The later ones can be transported to the extracellular milieu by ENT. Extracellularly, ATP can also be converted to inosine through a similar cascade by extracellular-faced enzymes anchored to the cell membrane. CD73 is the only extracellular 5NT. Pathophysiology of Huntington's disease exerts alterations in the activity of enzymes (marked in green +/−) or by genomic expression (marked in red +/−). **P1**—Adenosine receptor; **P2X**—ATP receptor channel; **P2Y**—ATP/UTP receptor; **ATPases**—adenosine-5′-triphosphatases; **AK**—adenylate kinase; **eNTPD**—ecto-nucleoside triphosphate diphosphohydrolase; **AMPD**—AMP deaminase; **5NT**—5′ nucleotidase; **CD73**—ecto-5′ nucleotidase; **eADA**—ecto-adenosine deaminase; **ADA**—adenosine deaminase; **PNP**—purine nucleoside phosphorylase; **ADSL**—adenylosuccinate lyase; **ADSS**—adenylosuccinate synthetase; **ENT**—ecto nucleoside transporter. Created with BioRender.com (accessed on 5 May 20021).

Research with HD patients detected reduced phosphocreatine to inorganic phosphate ratio in skeletal muscle of the symptomatic HD patients at rest (analyzed with a non-invasive 31P-MRS method). Moreover, muscle ATP/phosphocreatine and inorganic phosphate levels were significantly reduced in both symptomatic and presymptomatic HD subjects [100]. Furthermore, the maximum rate of mitochondrial ATP production during recovery from exercise was reduced by 44% in symptomatic HD patients and by 35% in presymptomatic HD carriers [77]. In the case of experimental models of HD, we have established that R6/2, as well as *Hdh*Q150 HD mice, exhibited decreased ATP, ADP, and AMP concentrations in *Extensor digitorum longus*, *Tibialis anterior*, and *Soleus*. Moreover, a significant reduction of phosphocreatine (PCr) and creatine (Cr) levels and the PCr/Cr ratio was observed in the examined skeletal muscles [75]. Similar changes were observed in HD mouse model hearts. We observed decreased concentrations of ATP and phosphocreatine as well as diminished ATP/ADP ratios. Interestingly, in contrast to skeletal muscles, this was accompanied by increased AMP levels [101]. Elevated concentration of cardiac AMP may activate AMP-regulated protein kinase (AMPK), which was shown as an increased AMPK phosphorylation status and enhanced AMPK mRNA transcript (observed also in HD-affected skeletal muscle) [102].

The functionality of the ATP-phosphocreatine shuttle, the transcriptional signature of genes involved in purine metabolism in HD-affected skeletal muscle and hearts were also assessed [101,102]. In the case of genes involved in purine nucleotide catabolism, skeletal muscle mRNA levels of Adenylate kinase 1 exhibited significant down-regulation while the mRNA levels of ecto-5′-nucleotidase remained unchanged in skeletal muscle (*Extensor digitorum longus, Tibialis anterior,* and *Soleus*). The same parameters were also studied in R6/2 mice hearts and demonstrated similar trends, such as reduced *Ak1* (Adenylate kinase 1) and no changes in *Nt5e* transcript levels [101]. Nevertheless, cardiac e5NT activity was significantly reduced [103]. Similarly, cardiac *Ampd3* (Adenosine monophosphate deaminase 3) transcript levels remained unchanged but AMPD activity was significantly reduced in R6/2 mice hearts, which could be caused by increased functional protein turnover or activity modulation in HD [101,103]. On the other hand, HD mouse skeletal muscle exhibited a significant up-regulation of the *Ampd3*, which could be caused by its alteration by denervation which is exclusively linked to skeletal muscle atrophy [104]. Moreover, HD mouse model skeletal muscles and hearts exhibited significant expression down-regulation of two genes involved in the purine nucleotide cycle-adenylosuccinate lyase (*Adsl,*) and adenylosuccinate lyase 1. Expression profiles of selected genes involved in the intracellular adenosine metabolism such as *Ada* (Adenosine deaminase) and *Adk* (Adenosine kinase) were also measured. Significant up-regulation for the cardiac *Ada* transcript (no changes in skeletal muscle *Ada* transcript) and in the cardiac and skeletal muscle transcription levels of adenosine kinase were noted. This agrees with observed increased ADA activity, inosine concertation, and reduced adenosine levels in HD mouse model hearts [103]. Furthermore, mRNA levels of *Pnp* (Purine nucleoside phosphorylase) as well as PNP activity, which degrades inosine, remain unchanged in HD mouse hearts, while *Pnp* transcript levels were significantly and uniformly up-regulated in each examined HD mouse skeletal muscle. This observation suggests intensive inosine degradation to hypoxanthine and its release from the skeletal muscle cell. Extracellular hypoxanthine influx might activate its degradation by other cells to xanthine and then uric acid by xanthine oxidase or xanthine dehydrogenase. Moreover, inhibition of xanthine oxidase should protect against muscle mass loss, thus its activation may implicate opposite effects [105]. Interestingly, the skeletal muscle transcript of *Xdh*-(Xanthine dehydrogenase) was significantly up-regulated while in heart was unaltered, which confirms our earlier thesis. Furthermore, up-regulation of the mRNA levels of one of the enzymes involved in purine synthesis de novo, such as phosphoribosylglycinamide formyltransferase (*Gart*), was noted in HD skeletal muscle. A different observation was obtained in *Hdh*Q150 mouse hearts, in whose *Gart* transcription was reduced. Nevertheless, cardiac and skeletal muscle transcript levels of other genes such as adenine phosphoribosyltransferase (*Aprt*) and amidophosphoribosyltransferase (*Ppat)* mRNA remained unchanged. Acceleration of purine de novo synthesis in HD skeletal muscle may indicate a greater demand for purine pool reconstruction in the cell. Probably, in HD mouse hearts, the signaling pathways aiming stabilized normal purine synthesis might be interrupted. Transcript rate analysis of genes involved in extracellular metabolism of purine nucleotides revealed significant down-regulation of *Entpd2* (Ectonucleoside triphosphate diphosphohydrolase 2) in investigated HD-affected skeletal muscle and heart. Nevertheless, eNTPD activity analysis did not reveal any changes in its cardiac activity in R6/2 mice.

In the case of purinergic signaling, there are still no literature data available regarding its role in skeletal muscle and heart dysfunction related to HD. Nevertheless, it is well known that enhanced expression of specific purinergic system elements for example in dystrophic muscles are important for dystrophic pathophysiology and could increase its severity considerably [106]. Dystrophin mutations in Duchene muscular dystrophy (DMD) coincide with significant P2X7 upregulation in muscle and alter receptor signaling in mouse dystrophic myoblasts and myofibers. Thus, P2X7 overexpression combined with the extracellular ATP-rich environment leads to cell dysfunction and death and ultimately to ineffective skeletal muscle regeneration [107]. Moreover, Ryten et al. demonstrated sequential expression of other purinergic receptors, such as P2X5, P2Y1, and P2X2 subtypes during the process of muscle regeneration [108]. Increased expression of the P2X1 receptor was reported in the aria of patients with dilated cardiomyopathy [109]. The protective role of P2X receptor activation (reduction of hypertrophy and increasing a life span) was highlighted in the calsequestrin overexpression model of cardiomyopathy [110,111].

Furthermore, roles for P2X7 receptors in dilated cardiomyopathy have been reported similarly to skeletal muscle pathologies [112]. Based on this data, the role of purinergic signaling in HD affected skeletal muscle and heart might be a new and interesting topic to delve into.

3. Purine Nucleotides Metabolism and Signaling as a Target for Huntington's Disease Therapies

As presented in this review, there are many changes related to the purine metabolism in the central nervous system, skeletal muscle, and heart of HD patients, animals as well as cellular models. Nevertheless, these changes in those systems might involve different mechanisms. As described earlier, some studies indicated mHTT absence in HD mouse model hearts. Thus, the cardiac purine metabolism derangement could be caused only by HTT protein signaling dysfunction not by mutant protein accumulation, while in CNS and skeletal muscle purine metabolism impairment may be caused by loss of HTT protein function together with cellular mHTT accumulation. Deterioration in intracellular purine metabolism leads to the accumulation of purine metabolites that might be released from the cell via transporters. Indeed, we observed a significant accumulation of purine catabolites in HD mouse models and patients' serum, which was strongly correlated with HD progression [101]. Nucleosides, as well as nucleotides released from the cell, might be degraded in the extracellular purine metabolism and activate purinergic signaling pathways. In summary, purine metabolism and signaling impairment might affect the development and progression of energy metabolism, signaling, and hence function derangements in HD-affected systems. Therefore, therapies leading to their improvement may have promising clinical implications in HD treatment.

Based on collected findings, therapeutic strategies may include compounds that directly correct disrupted ATP levels and lead to adenine nucleotide pool reconstitution in HD. Compounds, such as the coenzyme Q_{10} or creatinine, were widely tested and even investigated in clinical trials, but results were not promising [113]. An alternative might be the application of PPAR agonists, which have already undergone preclinical studies for the treatment of CNS, cardiovascular as well as skeletal muscle diseases. In 2016, the PPAR delta receptor agonist KD3010 was tested in the HD N171-82Q mouse model. Studies revealed improved motor function, reducing the progression of the neurodegenerative process and longer survival of treated animals [114]. Interestingly, studies performed by our team showed that simultaneous administration of adenosine metabolism inhibitors and substrates for adenine nucleotide synthesis improved mechanical functions of the heart, energy metabolism, and normalized adenine nucleotide pools [115,116], pointing at possible applications in HD treatment.

AMPD makes part of another altered purine pathway in HD. Skeletal muscles of the HD mouse model showed enhanced AMPD activity, while its activity in the heart was reduced. Pharmacological inhibition, as well as AMPD expression deletion in mice, led to a substantial enhancement of skeletal muscle contraction, induced mainly by AMP accumulation [117]. Moreover, AMPD inhibition seems to be also protective in cardiovascular diseases [118,119]. AMPD activity reduction noted in HD mouse model heart led to increased cardiac AMP levels and AMPK activation, but due to increased ADA activity, it was not capable of preventing intracellular adenosine pool depletion. Thus, in HD affected hearts adenosine levels augment, with suggested cardioprotective properties [120]. Interestingly, adenosine has been used to treat also epileptic diseases and seizures that are commonly observed in the juvenile form of HD. Thus, adenosine might be also a target for HD-affected CNS [121,122].

Furthermore, drugs increasing not only the intracellular but also the extracellular adenosine levels in HD-affected brain and heart might be protective. As described earlier, the most important enzyme that controls extracellular adenosine metabolism balance is eADA. It is well known that extracellular adenosine pool depletion is an important factor in the development of cardiovascular pathologies. Recently, we highlighted the therapeutic perspectives of eADA inhibition in the treatment of cardiovascular diseases such as atherosclerosis, myocardial ischemia-reperfusion injury, or hypertension [123]. Moreover, a study underlined that intrastriatal administration of ecto-nucleoside transporter (ENT) inhibitors increased the extracellular level of adenosine in the striatum of R6/2 mice to a much higher level, compared to controls, and improved HD mouse survival [124].

Impaired purinergic signaling in HD in CNS concerns mainly P2X7 and P2Y2 receptors. P2X7 antagonism in HD prevents neuronal death [67]. Nevertheless, only the P2X7 receptor seems to be important in skeletal muscle as well as heart failure. It is known that antagonizing or knocking out P2X7 or its downstream effectors, caspase-1 or NLRP3, in animal models decreased infarct size, improved cardiac function, and enhanced survival post myocardial infraction via reduced interelukin1β and intereukin18 levels in the heart [125–128]. Furthermore, genetic ablation and pharmacological inhibition of the P2X7 axis alleviated dystrophic phenotypes in mouse models of dystrophinopathy and sarcoglycanopathy [107]. Therefore, its antagonism may be a more suitable approach to treat not only HD-affected CNS system but also skeletal muscle and heart.

Author Contributions: M.T. conceived the idea; M.T. and T.G. wrote the manuscript and prepared the graphs; E.M.S., H.U., and R.T.S. supervised the work. All authors have read and agreed to the published version of the manuscript.

Funding: This work was supported by the National Science Centre of Poland (grant number 2016/22/M/NZ4/00678). Furthermore, H.U. acknowledges funding for his research on Huntington's disease from the Sâo Paulo Research Foundation FAPESP (grant No. 2018/07366-4).

Institutional Review Board Statement: Not applicable.

Informed Consent Statement: Not applicable.

Data Availability Statement: No new data were created or analyzed in this study. Data sharing does not apply to this article.

Conflicts of Interest: The authors declare no conflict of interest.

References

1. Walker, F. Huntington's disease. *Lancet* **2007**, *369*, 218–228. [CrossRef]
2. Sturrock, A.; Leavitt, B.R. The Clinical and Genetic Features of Huntington Disease. *J. Geriatr. Psychiatry Neurol.* **2010**, *23*, 243–259. [CrossRef]
3. McColgan, P.; Tabrizi, S.J. Huntington's disease: A clinical review. *Eur. J. Neurol.* **2018**, *25*, 24–34. [CrossRef]
4. Epping, E.A.; Paulsen, J.S. Depression in the early stages of Huntington disease. *Neurodegener. Dis. Manag.* **2011**, *1*, 407–414. [CrossRef]
5. Ross, C.A.; Tabrizi, S. Huntington's disease: From molecular pathogenesis to clinical treatment. *Lancet Neurol.* **2011**, *10*, 83–98. [CrossRef]
6. Li, S.H.; Schilling, G.; Young, W.S., 3rd; Li, X.J.; Margolis, R.L.; Stine, O.C.; Wagster, M.V.; Abbott, M.H.; Franz, M.L.; Ranen, N.G.; et al. Huntington's Disease Gene (IT15) Is Widely Expressed in Human and Rat Tissues. *Neuron* **1993**, *11*, 985–993. [CrossRef]
7. Li, S.-H.; Li, X.-J. Huntingtin–protein interactions and the pathogenesis of Huntington's disease. *Trends Genet.* **2004**, *20*, 146–154. [CrossRef] [PubMed]
8. Davies, S.W.; Turmaine, M.; Cozens, B.A.; DiFiglia, M.; Sharp, A.H.; Ross, C.A.; Scherzinger, E.; Wanker, E.E.; Mangiarini, L.; Bates, G. Formation of Neuronal Intranuclear Inclusions Underlies the Neurological Dysfunction in Mice Transgenic for the HD Mutation. *Cell* **1997**, *90*, 537–548. [CrossRef]
9. Moffitt, H.; McPhail, G.D.; Woodman, B.; Hobbs, C.; Bates, G.P. Formation of Polyglutamine Inclusions in a Wide Range of Non-CNS Tissues in the HdhQ150 Knock-In Mouse Model of Huntington's Disease. *PLoS ONE* **2009**, *4*, e8025. [CrossRef]
10. Hoffner, G.; Kahlem, P.; Djian, P. Perinuclear localization of huntingtin as a consequence of its binding to microtubules through an interaction with beta-tubulin: Relevance to Huntington's disease. *J. Cell Sci.* **2002**, *115 Pt 5*, 941–948. [CrossRef]
11. Godin, J.; Colombo, K.; Molina-Calavita, M.; Keryer, G.; Zala, D.; Charrin, B.C.; Dietrich, P.; Volvert, M.-L.; Guillemot, F.; Dragatsis, I.; et al. Huntingtin Is Required for Mitotic Spindle Orientation and Mammalian Neurogenesis. *Neuron* **2010**, *67*, 392–406. [CrossRef] [PubMed]
12. Veliera, J.; Kima, M.; Schwarza, C.; Kim, T.W.; Sappa, E.; Chaseb, K.; Aroninb, N.; Di Figliaa, M. Wild-Type and Mutant Huntingtins Function in Vesicle Trafficking in the Secretory and Endocytic Pathways. *Exp. Neurol.* **1998**, *152*, 34–40. [CrossRef] [PubMed]
13. Harjes, P.; Wanker, E.E. The hunt for huntingtin function: Interaction partners tell many different stories. *Trends Biochem. Sci.* **2003**, *28*, 425–433. [CrossRef]

14. Schulte, J.; Littleton, J.T. The biological function of the Huntingtin protein and its relevance to Huntington's Disease pathology. *Curr. Trends Neurol.* **2011**, *5*, 65–78. [PubMed]
15. Duyao, M.P.; Auerbach, A.B.; Ryan, A.; Persichetti, F.; Barnes, G.T.; McNeil, S.M.; Ge, P.; Vonsattel, J.P.; Gusella, J.F.; Joyner, A.L.; et al. Inactivation of the mouse Huntington's disease gene homolog Hdh. *Science* **1995**, *269*, 407–410. [CrossRef] [PubMed]
16. Robson, S.C.; Sévigny, J.; Zimmermann, H. The E-NTPDase family of ectonucleotidases: Structure function relationships and pathophysiological significance. *Purinergic Signal.* **2006**, *2*, 409–430. [CrossRef] [PubMed]
17. Yegutkin, G.G. Enzymes involved in metabolism of extracellular nucleotides and nucleosides: Functional implications and measurement of activities. *Crit. Rev. Biochem. Mol. Biol.* **2014**, *49*, 473–497. [CrossRef]
18. Yegutkin, G.G. Nucleotide- and nucleoside-converting ectoenzymes: Important modulators of purinergic signalling cascade. *Biochim. Biophys. Acta* **2008**, *1783*, 673–694. [CrossRef]
19. Kukulski, F.; Lévesque, S.A.; Sévigny, J. Impact of Ectoenzymes on P2 and P1 Receptor Signaling. *Adv. Pharmacol.* **2011**, *61*, 263–299. [CrossRef]
20. Lin, J.H.-C.; Takano, T.; Arcuino, G.; Wang, X.; Hu, F.; Darzynkiewicz, Z.; Nunes, M.; Goldman, S.; Nedergaard, M. Purinergic signaling regulates neural progenitor cell expansion and neurogenesis. *Dev. Biol.* **2007**, *302*, 356–366. [CrossRef]
21. Shukla, V.; Zimmermann, H.; Wang, L.; Kettenmann, H.; Raab, S.; Hammer, K.; Sévigny, J.; Robson, S.C.; Braun, N. Functional expression of the ecto-ATPase NTPDase2 and of nucleotide receptors by neuronal progenitor cells in the adult murine hippocampus. *J. Neurosci. Res.* **2005**, *80*, 600–610. [CrossRef] [PubMed]
22. von Kügelgen, I. Pharmacology of P2Y receptors. *Brain Res. Bull.* **2019**, *151*, 12–24. [CrossRef] [PubMed]
23. Knight, G. Purinergic Receptors. In *Encyclopedia of Neuroscience*; Elsevier BV: Amsterdam, The Netherlands, 2009; pp. 1245–1252.
24. Ismailoglu, I.; Chen, Q.; Popowski, M.; Yang, L.; Gross, S.S.; Brivanlou, A.H. Huntingtin protein is essential for mitochondrial metabolism, bioenergetics and structure in murine embryonic stem cells. *Dev. Biol.* **2014**, *391*, 230–240. [CrossRef]
25. Glaser, T.; Shimojo, H.; Ribeiro, D.E.; Martins, P.P.L.; Beco, R.P.; Kosinski, M.; Sampaio, V.F.A.; Corrêa-Velloso, J.; Oliveira-Giacomelli, Á.; Lameu, C.; et al. ATP and spontaneous calcium oscillations control neural stem cell fate determination in Huntington's disease: A novel approach for cell clock research. *Mol. Psychiatry* **2020**, 1–18. [CrossRef] [PubMed]
26. Tomczyk, M.; Glaser, T.; Ulrich, H.; Slominska, E.M.; Smolenski, R.T. Huntingtin protein maintains balanced energetics in mouse cardiomyocytes. *Nucleosides Nucleotides Nucleic Acids* **2020**, 1–8. [CrossRef] [PubMed]
27. Irmak, D.; Fatima, A.; Gutiérrez-Garcia, R.; Rinschen, M.M.; Wagle, P.; Altmüller, J.; Arrigoni, L.; Hummel, B.; Klein, C.; Frese, C.K.; et al. Mechanism suppressing H3K9 trimethylation in pluripotent stem cells and its demise by polyQ-expanded huntingtin mutations. *Hum. Mol. Genet.* **2018**, *27*, 4117–4134. [CrossRef]
28. Toczek, M.; Pierzynowska, K.; Kutryb-Zajac, B.-; Gaffke, L.; Slominska, E.M.; Wegrzyn, G.; Smolenski, R. Characterization of adenine nucleotide metabolism in the cellular model of Huntington's disease. *Nucleosides Nucleotides Nucleic Acids* **2018**, *37*, 630–638. [CrossRef] [PubMed]
29. Coppen, E.M.; Van Der Grond, J.; Roos, R.A.C. Atrophy of the putamen at time of clinical motor onset in Huntington's disease: A 6-year follow-up study. *J. Clin. Mov. Disord.* **2018**, *5*, 2. [CrossRef]
30. Ipata, P.L. Origin, utilization, and recycling of nucleosides in the central nervous system. *Adv. Physiol. Educ.* **2011**, *35*, 342–346. [CrossRef] [PubMed]
31. Jodeiri Farshbaf, M.; Ghaedi, K. Huntington's Disease and Mitochondria. *Neurotox. Res.* **2017**, *32*, 518–529. [CrossRef]
32. Herben-Dekker, M.; Van Oostrom, J.C.H.; Roos, R.A.C.; Jurgens, C.K.; Witjes-Ané, M.-N.W.; Kremer, H.P.H.; Leenders, K.L.; Spikman, J.M. Striatal metabolism and psychomotor speed as predictors of motor onset in Huntington's disease. *J. Neurol.* **2014**, *261*, 1387–1397. [CrossRef]
33. Intihar, T.A.; Martinez, E.A.; Gomez-Pastor, R. Mitochondrial Dysfunction in Huntington's Disease; Interplay Between HSF1, p53 and PGC-1α Transcription Factors. *Front. Cell. Neurosci.* **2019**, *13*, 103. [CrossRef] [PubMed]
34. Gines, S.; Seong, I.S.; Fossale, E.; Ivanova, E.; Trettel, F.; Gusella, J.F.; Wheeler, V.C.; Persichetti, F.; MacDonald, M.E. Specific progressive cAMP reduction implicates energy deficit in presymptomatic Huntington's disease knock-in mice. *Hum. Mol. Genet.* **2003**, *12*, 497–508. [CrossRef] [PubMed]
35. Mochel, F.; Durant, B.; Meng, X.; O'Callaghan, J.; Yu, H.; Brouillet, E.; Wheeler, V.C.; Humbert, S.; Schiffmann, R.; Durr, A. Early Alterations of Brain Cellular Energy Homeostasis in Huntington Disease Models. *J. Biol. Chem.* **2012**, *287*, 1361–1370. [CrossRef] [PubMed]
36. Seong, I.S.; Ivanova, E.; Lee, J.-M.; Choo, Y.S.; Fossale, E.; Anderson, M.; Gusella, J.F.; Laramie, J.M.; Myers, R.H.; Lesort, M.; et al. HD CAG repeat implicates a dominant property of huntingtin in mitochondrial energy metabolism. *Hum. Mol. Genet.* **2005**, *14*, 2871–2880. [CrossRef] [PubMed]
37. Siddiqui, A.; Rivera-Sánchez, S.; Castro, M.D.R.; Acevedo-Torres, K.; Rane, A.; Torres-Ramos, C.A.; Nicholls, D.G.; Andersen, J.K.; Ayala-Torres, S. Mitochondrial DNA damage Is associated with reduced mitochondrial bioenergetics in Huntington's disease. *Free. Radic. Biol. Med.* **2012**, *53*, 1478–1488. [CrossRef] [PubMed]
38. Fuster-Matanzo, A.; Llorens-Martín, M.; De Barreda, E.G.; Avila, J.; Hernandez, F. Different Susceptibility to Neurodegeneration of Dorsal and Ventral Hippocampal Dentate Gyrus: A Study with Transgenic Mice Overexpressing GSK3β. *PLoS ONE* **2011**, *6*, e27262. [CrossRef]
39. Cui, L.; Jeong, H.; Borovecki, F.; Parkhurst, C.N.; Tanese, N.; Krainc, D. Transcriptional Repression of PGC-1α by Mutant Huntingtin Leads to Mitochondrial Dysfunction and Neurodegeneration. *Cell* **2006**, *127*, 59–69. [CrossRef]

40. Yablonska, S.; Ganesan, V.; Ferrando, L.M.; Kim, J.; Pyzel, A.; Baranova, O.V.; Khattar, N.K.; Larkin, T.M.; Baranov, S.V.; Chen, N.; et al. Mutant huntingtin disrupts mitochondrial proteostasis by interacting with TIM. *Proc. Natl. Acad. Sci. USA* **2019**, *116*, 16593–16602. [CrossRef]
41. Orr, A.L.; Li, S.; Wang, C.-E.; Li, H.; Wang, J.; Rong, J.; Xu, X.; Mastroberardino, P.G.; Greenamyre, J.T.; Li, X.-J. N-Terminal Mutant Huntingtin Associates with Mitochondria and Impairs Mitochondrial Trafficking. *J. Neurosci.* **2008**, *28*, 2783–2792. [CrossRef]
42. Mattson, M.P.; Gleichmann, M.; Cheng, A. Mitochondria in Neuroplasticity and Neurological Disorders. *Neuron* **2008**, *60*, 748–766. [CrossRef] [PubMed]
43. Greenberg, M.E.; Xu, B.; Lu, B.; Hempstead, B.L. New Insights in the Biology of BDNF Synthesis and Release: Implications in CNS Function. *J. Neurosci.* **2009**, *29*, 12764–12767. [CrossRef] [PubMed]
44. Markham, A.; Cameron, I.; Franklin, P.; Spedding, M. BDNF increases rat brain mitochondrial respiratory coupling at complex I, but not complex II. *Eur. J. Neurosci.* **2004**, *20*, 1189–1196. [CrossRef]
45. Zuccato, C.; Ciammola, A.; Rigamonti, D.; Leavitt, B.R.; Goffredo, D.; Conti, L.; MacDonald, M.E.; Friedlander, R.M.; Silani, V.; Hayden, M.; et al. Loss of Huntingtin-Mediated BDNF Gene Transcription in Huntington's Disease. *Science* **2001**, *293*, 493–498. [CrossRef]
46. Fontán-Lozano, Á.; López-Lluch, G.; Delgado-García, J.M.; Navas, P.; Carrión, Á.M. Molecular Bases of Caloric Restriction Regulation of Neuronal Synaptic Plasticity. *Mol. Neurobiol.* **2008**, *38*, 167–177. [CrossRef]
47. Browne, S.E.; Bowling, A.C.; MacGarvey, U.; Baik, M.J.; Berger, S.C.; Muquit, M.M.K.; Bird, E.D.; Beal, M.F. Oxidative damage and metabolic dysfunction in Huntington's disease: Selective vulnerability of the basal ganglia. *Ann. Neurol.* **1997**, *41*, 646–653. [CrossRef] [PubMed]
48. Benchoua, A.; Trioulier, Y.; Zala, D.; Gaillard, M.-C.; Lefort, N.; Dufour, N.; Saudou, F.; Elalouf, J.-M.; Hirsch, E.; Hantraye, P.; et al. Involvement of Mitochondrial Complex II Defects in Neuronal Death Produced by N-Terminus Fragment of Mutated Huntingtin. *Mol. Biol. Cell* **2006**, *17*, 1652–1663. [CrossRef] [PubMed]
49. Chaturvedi, R.K.; Calingasan, N.Y.; Yang, L.; Hennessey, T.; Johri, A.; Beal, M.F. Impairment of PGC-1alpha expression, neuropathology and hepatic steatosis in a transgenic mouse model of Huntington's disease following chronic energy deprivation. *Hum. Mol. Genet.* **2010**, *19*, 3190–3205. [CrossRef]
50. Chaturvedi, R.K.; Adhihetty, P.; Shukla, S.; Hennessy, T.; Calingasan, N.; Yang, L.; Starkov, A.; Kiaei, M.; Cannella, M.; Sassone, J.; et al. Impaired PGC-1α function in muscle in Huntington's disease. *Hum. Mol. Genet.* **2009**, *18*, 3048–3065. [CrossRef]
51. Lin, J.; Wu, P.-H.; Tarr, P.T.; Lindenberg, K.S.; St-Pierre, J.; Zhang, C.-Y.; Mootha, V.K.; Jager, S.; Vianna, C.R.; Reznick, R.M.; et al. Defects in Adaptive Energy Metabolism with CNS-Linked Hyperactivity in PGC-1α Null Mice. *Cell* **2004**, *119*, 121–135. [CrossRef]
52. Chou, S.-Y.; Lee, Y.-C.; Chen, H.-M.; Chiang, M.-C.; Lai, H.-L.; Chang, H.-H.; Wu, Y.-C.; Sun, C.-N.; Chien, C.-L.; Lin, Y.-S.; et al. CGS21680 attenuates symptoms of Huntington's disease in a transgenic mouse model. *J. Neurochem.* **2005**, *93*, 310–320. [CrossRef] [PubMed]
53. Ju, T.; Chen, H.-M.; Lin, J.-T.; Chang, C.-P.; Chang, W.-C.; Kang, J.-J.; Sun, C.-P.; Tao, M.-H.; Tu, P.-H.; Chang, C.; et al. Nuclear translocation of AMPK-α1 potentiates striatal neurodegeneration in Huntington's disease. *J. Cell Biol.* **2011**, *194*, 209–227. [CrossRef]
54. Glaser, T.; Andrejew, R.; Oliveira-Giacomelli, Á.; Ribeiro, D.E.; Marques, L.B.; Ye, Q.; Ren, W.-J.; Semyanov, A.; Illes, P.; Tang, Y.; et al. Purinergic Receptors in Basal Ganglia Diseases: Shared Molecular Mechanisms between Huntington's and Parkinson's Disease. *Neurosci. Bull.* **2020**, *36*, 1299–1314. [CrossRef] [PubMed]
55. Dayalu, P.; Albin, R.L. Huntington Disease: Pathogenesis and Treatment. *Neurol. Clin.* **2015**, *33*, 101–114. [CrossRef] [PubMed]
56. Patassini, S.; Begley, P.; Xu, J.; Church, S.J.; Reid, S.J.; Kim, E.H.; Curtis, M.; Dragunow, M.; Waldvogel, H.J.; Snell, R.G.; et al. Metabolite mapping reveals severe widespread perturbation of multiple metabolic processes in Huntington's disease human brain. *Biochim. Biophys. Acta* **2016**, *1862*, 1650–1662. [CrossRef]
57. Nambron, R.; Silajdzic, E.; Kalliolia, E.; Ottolenghi, C.; Hindmarsh, P.; Hill, N.R.; Costelloe, S.J.; Martin, N.G.; Positano, V.; Watt, H.C.; et al. A Metabolic Study of Huntington's Disease. *PLoS ONE* **2016**, *11*, e014648. [CrossRef]
58. Gianfriddo, M.; Melani, A.; Turchi, D.; Giovannini, M.; Pedata, F. Adenosine and glutamate extracellular concentrations and mitogen-activated protein kinases in the striatum of Huntington transgenic mice. Selective antagonism of adenosine A2A receptors reduces transmitter outflow. *Neurobiol. Dis.* **2004**, *17*, 77–88. [CrossRef]
59. Lee, C.-F.; Chern, Y. Adenosine Receptors and Huntington's Disease. *Int. Rev. Neurobiol.* **2014**, *119*, 195–232. [CrossRef]
60. Glass, M.; Dragunow, M.; Faull, R. The pattern of neurodegeneration in Huntington's disease: A comparative study of cannabinoid, dopamine, adenosine and GABAA receptor alterations in the human basal ganglia in Huntington's disease. *Neuroscience* **2000**, *97*, 505–519. [CrossRef]
61. Bauer, A.; Zilles, K.; Matusch, A.; Holzmann, C.; Riess, O.; Von Hörsten, S. Regional and subtype selective changes of neurotransmitter receptor density in a rat transgenic for the Huntington's disease mutation. *J. Neurochem.* **2005**, *94*, 639–650. [CrossRef]
62. Cha, J.-H.J.; Frey, A.S.; Alsdorf, S.A.; Kerner, J.A.; Kosinski, C.M.; Mangiarini, L.; Penney, J.B.; Davies, S.W.; Bates, G.; Young, A.B. Altered neurotransmitter receptor expression in transgenic mouse models of Huntington's disease. *Philos. Trans. R. Soc. B Biol. Sci.* **1999**, *354*, 981–989. [CrossRef]

63. Mievis, S.; Blum, D.; Ledent, C. A2A receptor knockout worsens survival and motor behaviour in a transgenic mouse model of Huntington's disease. *Neurobiol. Dis.* **2011**, *41*, 570–576. [CrossRef]
64. Villar-Menéndez, I.; Blanch, M.; Tyebji, S.; Pereira-Veiga, T.; Albasanz, J.L.; Martín, M.; Ferrer, I.; Pérez-Navarro, E.; Barrachina, M. Increased 5-Methylcytosine and Decreased 5-Hydroxymethylcytosine Levels are Associated with Reduced Striatal A2AR Levels in Huntington's Disease. *NeuroMolecular Med.* **2013**, *15*, 295–309. [CrossRef]
65. Domenici, M.; Scattoni, M.L.; Martire, A.; Lastoria, G.; Potenza, R.; Borioni, A.; Venerosi, A.; Calamandrei, G.; Popoli, P. Behavioral and electrophysiological effects of the adenosine A2A receptor antagonist SCH 58261 in R6/2 Huntington's disease mice. *Neurobiol. Dis.* **2007**, *28*, 197–205. [CrossRef]
66. Li, W.; Silva, H.; Real, J.I.; Wang, Y.-M.; Rial, D.; Li, P.; Payen, M.-P.; Zhou, Y.; Muller, C.E.; Tomé, A.R.; et al. Inactivation of adenosine A2A receptors reverses working memory deficits at early stages of Huntington's disease models. *Neurobiol. Dis.* **2015**, *79*, 70–80. [CrossRef]
67. Díaz-Hernández, M.; Díez-Zaera, M.; Sánchez-Nogueiro, J.; Gómez-Villafuertes, R.; Canals, J.M.; Alberch, J.; Miras-Portugal, M.T.; Lucas, J.J. Altered P2X7-receptor level and function in mouse models of Huntington's disease and therapeutic efficacy of antagonist administration. *FASEB J.* **2009**, *23*, 1893–1906. [CrossRef] [PubMed]
68. Busse, M.E.; Hughes, G.; Wiles, C.M.; Rosser, A.E. Use of hand-held dynamometry in the evaluation of lower limb muscle strength in people with Huntington's disease. *J. Neurol.* **2008**, *255*, 1534–1540. [CrossRef]
69. Ribchester, R.R.; Thomson, D.; Wood, N.I.; Hinks, T.; Gillingwater, T.H.; Wishart, T.M.; Court, F.A.; Morton, A.J. Progressive abnormalities in skeletal muscle and neuromuscular junctions of transgenic mice expressing the Huntington's disease mutation. *Eur. J. Neurosci.* **2004**, *20*, 3092–3114. [CrossRef] [PubMed]
70. Romer, S.H.; Metzger, S.; Peraza, K.; Wright, M.C.; Jobe, D.S.; Song, L.-S.; Rich, M.M.; Foy, B.D.; Talmadge, R.J.; Voss, A.A. A mouse model of Huntington's disease shows altered ultrastructure of transverse tubules in skeletal muscle fibers. *J. Gen. Physiol.* **2021**, *153*, 153. [CrossRef]
71. Orth, M.; Cooper, J.M.; Bates, G.; Schapira, A.H.V. Inclusion formation in Huntington's disease R6/2 mouse muscle cultures. *J. Neurochem.* **2003**, *87*, 1–6. [CrossRef] [PubMed]
72. Valadão, P.A.C.; De Aragão, B.C.; Andrade, J.N.; Magalhães-Gomes, M.P.S.; Foureaux, G.; Joviano-Santos, J.V.; Nogueira, J.C.; Machado, T.; De Jesus, I.C.G.; Nogueira, J.M.; et al. Abnormalities in the Motor Unit of a Fast-Twitch Lower Limb Skeletal Muscle in Huntington's Disease. *ASN Neuro* **2019**, *11*, 11. [CrossRef]
73. Valadão, P.A.C.; De Aragão, B.C.; Andrade, J.N.; Magalhães-Gomes, M.P.S.; Foureaux, G.; Joviano-Santos, J.V.; Nogueira, J.C.; Ribeiro, F.; Tapia, J.C.; Guatimosim, C. Muscle atrophy is associated with cervical spinal motoneuron loss in BACHD mouse model for Huntington's disease. *Eur. J. Neurosci.* **2016**, *45*, 785–796. [CrossRef]
74. Strand, A.D.; Aragaki, A.K.; Shaw, D.; Bird, T.; Holton, J.; Turner, C.; Tapscott, S.J.; Tabrizi, S.; Schapira, A.H.; Kooperberg, C.; et al. Gene expression in Huntington's disease skeletal muscle: A potential biomarker. *Hum. Mol. Genet.* **2005**, *14*, 1863–1876. [CrossRef] [PubMed]
75. Mielcarek, M.; Toczek, M.; Smeets, C.J.L.M.; Franklin, S.A.; Bondulich, M.K.; Jolinon, N.; Muller, T.; Ahmed, M.; Dick, J.R.T.; Piotrowska, I.; et al. HDAC4-Myogenin Axis As an Important Marker of HD-Related Skeletal Muscle Atrophy. *PLoS Genet.* **2015**, *11*, e1005021. [CrossRef]
76. Ezielonka, D.; Epiotrowska, I.; Marcinkowski, J.T.; Emielcarek, M. Skeletal muscle pathology in Huntington's disease. *Front. Physiol.* **2014**, *5*, 380. [CrossRef]
77. Saft, C.; Zange, J.; Andrich, J.; Müller, K.; Lindenberg, K.; Landwehrmeyer, B.; Vorgerd, M.; Kraus, P.H.; Przuntek, H.; Schöls, L.; et al. Mitochondrial impairment in patients and asymptomatic mutation carriers of Huntington's disease. *Mov. Disord.* **2005**, *20*, 674–679. [CrossRef]
78. Kojer, K.; Hering, T.; Bazenet, C.; Weiss, A.; Herrmann, F.; Taanman, J.-W.; Orth, M. Huntingtin Aggregates and Mitochondrial Pathology in Skeletal Muscle but not Heart of Late-Stage R6/2 Mice. *J. Huntington's Dis.* **2019**, *8*, 145–159. [CrossRef]
79. Tsang, T.M.; Woodman, B.; McLoughlin, G.A.; Griffin, J.L.; Tabrizi, S.J.; Bates, G.P.; Holmes, E. Metabolic Characterization of the R6/2 Transgenic Mouse Model of Huntington's Disease by High-Resolution MAS1H NMR Spectroscopy. *J. Proteome Res.* **2006**, *5*, 483–492. [CrossRef]
80. Ciammola, A.; Sassone, J.; Alberti, L.; Meola, G.; Mancinelli, E.; Russo, M.A.; Squitieri, F.; Silani, V. Increased apoptosis, huntingtin inclusions and altered differentiation in muscle cell cultures from Huntington's disease subjects. *Cell Death Differ.* **2006**, *13*, 2068–2078. [CrossRef] [PubMed]
81. Johri, A.; Calingasan, N.Y.; Hennessey, T.M.; Sharma, A.; Yang, L.; Wille, E.; Chandra, A.; Beal, M.F. Pharmacologic activation of mitochondrial biogenesis exerts widespread beneficial effects in a transgenic mouse model of Huntington's disease. *Hum. Mol. Genet.* **2011**, *21*, 1124–1137. [CrossRef]
82. Miller, S.G.; Hafen, P.S.; Brault, J.J. Increased Adenine Nucleotide Degradation in Skeletal Muscle Atrophy. *Int. J. Mol. Sci.* **2019**, *21*, 88. [CrossRef]
83. Lanska, D.J.; LaVine, L.; Schoenberg, B.S. Huntington's disease mortality in the United States. *Neurology* **1988**, *38*, 769. [CrossRef] [PubMed]
84. Chiu, E.; Alexander, L. Causes of death in Huntington's Disease. *Med. J. Aust.* **1982**, *1*, 153. [CrossRef] [PubMed]
85. Andrich, J.; Schmitz, T.; Saft, C.; Postert, T.; Kraus, P.; Epplen, J.T.; Przuntek, H.; Agelink, M.W. Autonomic nervous system function in Huntington's disease. *J. Neurol. Neurosurg. Psychiatry* **2002**, *72*, 726–731. [CrossRef] [PubMed]

86. Hasselbalch, S.G.; Oberg, G.; Sorensen, S.A.; Andersen, A.R.; Waldemar, G.; Schmidt, J.F.; Fenger, K.; Paulson, O.B. Reduced regional cerebral blood flow in Huntington's disease studied by SPECT. *J. Neurol. Neurosurg. Psychiatry* **1992**, *55*, 1018–1023. [CrossRef]
87. Melik, Z.; Kobal, J.; Cankar, K.; Strucl, M. Microcirculation response to local cooling in patients with Huntington's disease. *J. Neurol.* **2012**, *259*, 921–928. [CrossRef]
88. Sharma, K.R.; Romano, J.G.; Ayyar, D.R.; Rotta, F.T.; Facca, A.; Sanchez-Ramos, J. Sympathetic Skin Response and Heart Rate Variability in Patients With Huntington Disease. *Arch. Neurol.* **1999**, *56*, 1248–1252. [CrossRef]
89. Terroba-Chambi, C.; Bruno, V.; Vigo, D.E.; Merello, M. Heart rate variability and falls in Huntington's disease. *Clin. Auton. Res.* **2021**, *31*, 281–292. [CrossRef]
90. Mielcarek, M.; Inuabasi, L.; Bondulich, M.K.; Muller, T.; Osborne, G.; Franklin, S.A.; Smith, D.L.; Neueder, A.; Rosinski, J.; Rattray, I.; et al. Dysfunction of the CNS-Heart Axis in Mouse Models of Huntington's Disease. *PLoS Genet.* **2014**, *10*, e1004550. [CrossRef] [PubMed]
91. Wood, N.I.; Sawiak, S.J.; Buonincontri, G.; Niu, Y.; Kane, A.D.; Carpenter, T.A.; Giussani, D.; Morton, A.J. Direct Evidence of Progressive Cardiac Dysfunction in a Transgenic Mouse Model of Huntington's Disease. *J. Huntington's Dis.* **2012**, *1*, 57–64. [CrossRef]
92. Kiriazis, H.; Jennings, N.L.; Davern, P.; Lambert, G.; Su, Y.; Pang, T.; Du, X.; La Greca, L.; Head, G.; Hannan, A.J.; et al. Neurocardiac dysregulation and neurogenic arrhythmias in a transgenic mouse model of Huntington's disease. *J. Physiol.* **2012**, *590*, 5845–5860. [CrossRef]
93. Mielcarek, M.; Bondulich, M.K.; Inuabasi, L.; Franklin, S.A.; Muller, T.; Bates, G.P. The Huntington's Disease-Related Cardiomyopathy Prevents a Hypertrophic Response in the R6/2 Mouse Model. *PLoS ONE* **2014**, *9*, e108961. [CrossRef] [PubMed]
94. Wu, B.-T.; Chiang, M.-C.; Tasi, C.-Y.; Kuo, C.-H.; Shyu, W.-C.; Kao, C.-L.; Huang, C.-Y.; Lee, S.-D. Cardiac Fas-Dependent and Mitochondria-Dependent Apoptotic Pathways in a Transgenic Mouse Model of Huntington's Disease. *Cardiovasc. Toxicol.* **2016**, *16*, 111–121. [CrossRef] [PubMed]
95. Joviano-Santos, J.V.; Santos-Miranda, A.; Botelho, A.F.M.; De Jesus, I.C.G.; Andrade, J.N.; Barreto, T.D.O.; Magalhães-Gomes, M.P.S.; Valadão, P.A.C.; Cruz, J.; Melo, M.M.; et al. Increased oxidative stress and CaMKIIactivity contribute to electro-mechanical defects in cardiomyocytes from a murine model of Huntington's disease. *FEBS J.* **2019**, *286*, 110–123. [CrossRef]
96. Dridi, H.; Liu, X.; Yuan, Q.; Reiken, S.; Yehia, M.; Sittenfeld, L.; Apostolou, P.; Buron, J.; Sicard, P.; Matecki, S.; et al. Role of defective calcium regulation in cardiorespiratory dysfunction in Huntington's disease. *JCI Insight* **2020**, *5*. [CrossRef]
97. Mihm, M.J.; Amann, D.M.; Schanbacher, B.L.; Altschuld, R.A.; Bauer, J.A.; Hoyt, K.R. Cardiac dysfunction in the R6/2 mouse model of Huntington's disease. *Neurobiol. Dis.* **2007**, *25*, 297–308. [CrossRef]
98. Child, D.; Lee, J.H.; Pascua, C.J.; Chen, Y.H.; Monteys, A.M.; Davidson, B.L. Cardiac mTORC1 Dysregulation Impacts Stress Adaptation and Survival in Huntington's Disease. *Cell Rep.* **2018**, *23*, 1020–1033. [CrossRef] [PubMed]
99. Critchley, B.J.; Isalan, M.; Mielcarek, M. Neuro-Cardio Mechanisms in Huntington's Disease and Other Neurodegenerative Disorders. *Front. Physiol.* **2018**, *9*, 559. [CrossRef]
100. Lodi, R.; Schapira, A.H.V.; Manners, D.; Styles, P.; Wood, N.W.; Taylor, D.J.; Warner, T.T. Abnormal in Vivo Skeletal Muscle Energy Metabolism in Huntington's Disease and Dentatorubropallidoluysian Atrophy. *Ann. Neurol.* **2000**. [CrossRef]
101. Toczek, M.; Zielonka, D.; Zukowska, P.; Marcinkowski, J.T.; Slominska, E.; Isalan, M.; Smolenski, R.T.; Mielcarek, M. An impaired metabolism of nucleotides underpins a novel mechanism of cardiac remodeling leading to Huntington's disease related cardiomyopathy. *Biochim. Biophys. Acta* **2016**, *1862*, 2147–2157. [CrossRef] [PubMed]
102. Mielcarek, M.; Smolenski, R.; Isalan, M. Transcriptional Signature of an Altered Purine Metabolism in the Skeletal Muscle of a Huntington's Disease Mouse Model. *Front. Physiol.* **2017**, *8*. [CrossRef]
103. Toczek, M.; Kutryb-Zajac, B.; Zukowska, P.; Slominska, E.M.; Isalan, M.; Mielcarek, M.; Smolenski, R. Changes in cardiac nucleotide metabolism in Huntington's disease. *Nucleosides Nucleotides Nucleic Acids* **2016**, *35*, 707–712. [CrossRef]
104. Fortuin, F.D.; Morisaki, T.; Holmes, E.W. Subunit composition of AMPD varies in response to changes in AMPD1 and AMPD3 gene expression in skeletal muscle. *Proc. Assoc. Am. Physicians* **1996**, *108*, 329–333. [PubMed]
105. Ferrando, B.; Gomez-Cabrera, M.C.; Salvador-Pascual, A.; Puchades, C.; Derbré, F.; Gratas-Delamarche, A.; Laparre, L.; Olaso-González, G.; Cerda, M.; Viosca, E.; et al. Allopurinol partially prevents disuse muscle atrophy in mice and humans. *Sci. Rep.* **2018**, *8*, 1–12. [CrossRef]
106. Krasowska, E.; Róg, J.; Sinadinos, A.; Young, C.N.J.; Górecki, D.C.; Zabłocki, K. Purinergic receptors in skeletal muscles in health and in muscular dystrophy. *Postępy Biochemii* **2014**, *60*, 483–489. [PubMed]
107. Górecki, D.C. P2X7 purinoceptor as a therapeutic target in muscular dystrophies. *Curr. Opin. Pharmacol.* **2019**, *47*, 40–45. [CrossRef]
108. Ryten, M.; Yang, S.Y.; Dunn, P.M.; Goldspink, G.; Burnstock, G. Purinoceptor expression in regenerating skeletal muscle in the mdx mouse model of muscular dystrophy and in satellite cell cultures. *FASEB J.* **2004**, *18*, 1404–1406. [CrossRef] [PubMed]
109. Berry, D.A.; Barden, J.A.; Balcar, V.J.; Keogh, A.; Dos Remedios, C.G. Increase in expression of P2X1 receptors in the atria of patients suffering from dilated cardiomyopathy. *Electrophoresis* **1999**, *20*, 2059–2064. [CrossRef]
110. Yang, A.; Sonin, D.; Jones, L.; Barry, W.H.; Liang, B.T. A beneficial role of cardiac P2X4 receptors in heart failure: Rescue of the calsequestrin overexpression model of cardiomyopathy. *Am. J. Physiol. Circ. Physiol.* **2004**, *287*, H1096–H1103. [CrossRef] [PubMed]

111. Shen, J.-B.; Cronin, C.; Sonin, D.; Joshi, B.V.; Nieto, M.G.; Harrison, D.; Jacobson, K.A.; Liang, B.T. P2X purinergic receptor-mediated ionic current in cardiac myocytes of calsequestrin model of cardiomyopathy: Implications for the treatment of heart failure. *Am. J. Physiol. Circ. Physiol.* **2007**, *292*, H1077–H1084. [CrossRef] [PubMed]
112. Martinez, C.G.; Zamith-Miranda, D.; Da Silva, M.G.; Ribeiro, K.C.; Brandão, I.T.; Silva, C.L.; Diaz, B.L.; Bellio, M.; Persechini, P.M.; Kurtenbach, E. P2×7 purinergic signaling in dilated cardiomyopathy induced by auto-immunity against muscarinic M2 receptors: Autoantibody levels, heart functionality and cytokine expression. *Sci. Rep.* **2015**, *5*, 16940. [CrossRef] [PubMed]
113. Duan, W.; Jiang, M.; Jin, J. Metabolism in HD: Still a relevant mechanism? *Mov. Disord.* **2014**, *29*, 1366–1374. [CrossRef] [PubMed]
114. Dickey, A.S.; Pineda, V.V.; Tsunemi, T.; Liu, P.P.; Miranda, H.C.; Gilmore-Hall, S.K.; Lomas, N.; Sampat, K.R.; Buttgereit, A.; Torres, M.-J.M.; et al. PPAR-δ is repressed in Huntington's disease, is required for normal neuronal function and can be targeted therapeutically. *Nat. Med.* **2016**, *22*, 37–45. [CrossRef]
115. Smolenski, R.T.; Raisky, O.; Slominska, E.M.; Abunasra, H.; Kalsi, K.K.; Jayakumar, J.; Suzuki, K.; Yacoub, M.H. Protection from reperfusion injury after cardiac transplantation by inhibition of adenosine metabolism and nucleotide precursor supply. *Circulation* **2001**, *104*, 246–252. [CrossRef] [PubMed]
116. Smolenski, R.; Kalsi, K.K.; Zych, M.; Kochan, Z.; Yacoub, M.H. Adenine/Ribose Supply Increases Adenosine Production and Protects ATP Pool in Adenosine Kinase-inhibited Cardiac Cells. *J. Mol. Cell. Cardiol.* **1998**, *30*, 673–683. [CrossRef]
117. Plaideau, C.; Lai, Y.-C.; Kviklyte, S.; Zanou, N.; Löfgren, L.; Andersén, H.; Vertommen, D.; Gailly, P.; Hue, L.; Bohlooly, Y.M.; et al. Effects of Pharmacological AMP Deaminase Inhibition and Ampd1 Deletion on Nucleotide Levels and AMPK Activation in Contracting Skeletal Muscle. *Chem. Biol.* **2014**, *21*, 1497–1510. [CrossRef] [PubMed]
118. Smolenski, R.T.; Rybakowska, I.; Turyn, J.; Romaszko, P.; Zabielska, M.; Taegtmeyer, A.; Słomińska, E.M.; Kaletha, K.K.; Barton, P.J.R. AMP deaminase 1 gene polymorphism and heart disease-a genetic association that highlights new treatment. *Cardiovasc. Drugs Ther.* **2014**, *28*, 183–189. [CrossRef]
119. Zabielska, M.A.; Borkowski, T.; Slominska, E.M.; Smolenski, R.T. Inhibition of AMP deaminase as therapeutic target in cardiovascular pathology. *Pharmacol. Rep.* **2015**, *67*, 682–688. [CrossRef]
120. Guieu, R.; Deharo, J.-C.; Maille, B.; Crotti, L.; Torresani, E.; Brignole, M.; Parati, G. Adenosine and the Cardiovascular System: The Good and the Bad. *J. Clin. Med.* **2020**, *9*, 1366. [CrossRef]
121. Boison, D. Role of adenosine in status epilepticus: A potential new target? *Epilepsia* **2013**, *54 (Suppl. 6)* (Suppl. 6), 20–22. [CrossRef]
122. Boison, D. Adenosine dysfunction in epilepsy. *Glia* **2012**, *60*, 1234–1243. [CrossRef]
123. Kutryb-Zajac, B.; Mierzejewska, P.; Slominska, E.M.; Smolenski, R.T. Therapeutic Perspectives of Adenosine Deaminase Inhibition in Cardiovascular Diseases. *Molecules* **2020**, *25*, 4652. [CrossRef] [PubMed]
124. Kao, Y.-H.; Lin, M.-S.; Chen, C.-M.; Wu, Y.-R.; Chen, H.-M.; Lai, H.-L.; Chern, Y.; Lin, C.-J. Targeting ENT1 and adenosine tone for the treatment of Huntington's disease. *Hum. Mol. Genet.* **2016**, *26*, 467–478. [CrossRef]
125. Mezzaroma, E.; Toldo, S.; Farkas, D.; Seropian, I.M.; Van Tassell, B.W.; Salloum, F.; Kannan, H.R.; Menna, A.C.; Voelkel, N.F.; Abbate, A. The inflammasome promotes adverse cardiac remodeling following acute myocardial infarction in the mouse. *Proc. Natl. Acad. Sci. USA* **2011**, *108*, 19725–19730. [CrossRef] [PubMed]
126. Gao, H.; Yin, J.; Shi, Y.; Hu, H.; Li, X.; Xue, M.; Cheng, W.; Wang, Y.; Li, X.; Li, Y.; et al. Targeted P2X7R shRNA delivery attenuates sympathetic nerve sprouting and ameliorates cardiac dysfunction in rats with myocardial infarction. *Cardiovasc. Ther.* **2016**, *35*, e12245. [CrossRef] [PubMed]
127. Bracey, N.A.; Beck, P.L.; Muruve, D.A.; Hirota, S.A.; Guo, J.; Jabagi, H.; Jr, J.R.W.; Macdonald, J.A.; Lees-Miller, J.P.; Roach, D.; et al. The Nlrp3 inflammasome promotes myocardial dysfunction in structural cardiomyopathy through interleukin-1β. *Exp. Physiol.* **2013**, *98*, 462–472. [CrossRef]
128. Frantz, S.; Ducharme, A.; Sawyer, D.; Rohde, L.E.; Kobzik, L.; Fukazawa, R.; Tracey, D.; Allen, H.; Lee, R.T.; Kelly, R.A. Targeted deletion of caspase-1 reduces early mortality and left ventricular dilatation following myocardial infarction. *J. Mol. Cell. Cardiol.* **2003**, *35*, 685–694. [CrossRef]

International Journal of
Molecular Sciences

MDPI

Review

Molecular Pathogenesis and Peripheral Monitoring of Adult Fragile X-Associated Syndromes

Luis M. Valor [1,2,3,4,*], Jorge C. Morales [3,4], Irati Hervás-Corpión [3,4] and Rosario Marín [3,5]

1 Instituto de Investigación Sanitaria y Biomédica de Alicante (ISABIAL), 03010 Alicante, Spain
2 Laboratorio de Apoyo a la Investigación, Hospital General Universitario de Alicante, Av. Pintor Baeza 12, 03010 Alicante, Spain
3 Instituto de Investigación e Innovación Biomédica de Cádiz (INiBICA), 11009 Cádiz, Spain; jorgcmorales@hotmail.com (J.C.M.); ihervas91@gmail.com (I.H.-C.); charoma29i@yahoo.com (R.M.)
4 Unidad de Investigación, Hospital Universitario Puerta del Mar, Av. Ana de Viya 21, 11009 Cádiz, Spain
5 Unidad de Genética, Hospital Universitario Puerta del Mar, Av. Ana de Viya 21, 11009 Cádiz, Spain
* Correspondence: valor_lui@externos.gva.es; Tel.: +34-965-913-988

Abstract: Abnormal trinucleotide expansions cause rare disorders that compromise quality of life and, in some cases, lifespan. In particular, the expansions of the CGG-repeats stretch at the 5′-UTR of the Fragile X Mental Retardation 1 (*FMR1*) gene have pleiotropic effects that lead to a variety of Fragile X-associated syndromes: the neurodevelopmental Fragile X syndrome (FXS) in children, the late-onset neurodegenerative disorder Fragile X-associated tremor-ataxia syndrome (FXTAS) that mainly affects adult men, the Fragile X-associated primary ovarian insufficiency (FXPOI) in adult women, and a variety of psychiatric and affective disorders that are under the term of Fragile X-associated neuropsychiatric disorders (FXAND). In this review, we will describe the pathological mechanisms of the adult "gain-of-function" syndromes that are mainly caused by the toxic actions of CGG RNA and FMRpolyG peptide. There have been intensive attempts to identify reliable peripheral biomarkers to assess disease progression and onset of specific pathological traits. Mitochondrial dysfunction, altered miRNA expression, endocrine system failure, and impairment of the GABAergic transmission are some of the affectations that are susceptible to be tracked using peripheral blood for monitoring of the motor, cognitive, psychiatric and reproductive impairment of the CGG-expansion carriers. We provided some illustrative examples from our own cohort. Understanding the association between molecular pathogenesis and biomarkers dynamics will improve effective prognosis and clinical management of CGG-expansion carriers.

Keywords: FXTAS; FXPOI; FXAND; premutation; blood; biomarker; FMR1; FMRP; endocrine; mitochondria; miRNA; transcription; GABA; telomere

Citation: Valor, L.M.; Morales, J.C.; Hervás-Corpión, I.; Marín, R. Molecular Pathogenesis and Peripheral Monitoring of Adult Fragile X-Associated Syndromes. *Int. J. Mol. Sci.* **2021**, *22*, 8368. https://doi.org/10.3390/ijms22168368

Academic Editor: Kurt A. Jellinger

Received: 13 July 2021
Accepted: 30 July 2021
Published: 4 August 2021

1. The Multiple Faces of FMR1 in Pathology

1.1. Brief Description of the Fragile X-Associated Syndromes

The Fragile X Mental Retardation 1 (*FMR1*) gene is located on the X chromosome and encodes the polysome-associated RNA-binding protein FMRP, which plays key roles in neuronal development and synaptic plasticity through the regulation of mRNAs at the level of traffic, stability, splicing and both somatic and presynaptic translation [1]. The 5′-UTR of the *FMR1* gene contains a stretch of CGG repeats that causes a series of pathological conditions when exceeding 54 repeats, although there is increasing evidence of an association between the "gray zone", ranging from 45 up to 54 triplet copies (which forms an unstable *FMR1* allele that can be expanded in successive generations [2]) with atypical parkinsonism [3]. The silencing of the *FMR1* gene occurs over 200 CGG repeats (the so-called full mutation, FM), resulting in severe diminished FMRP expression during development and in consequence the onset of the Fragile X syndrome (FXS, OMIM #300624). FXS is the most common cause of monogenic intellectual disability, with a high incidence of autistic features, hyperactivity behaviour and seizures, and usually accompanied by macroorchidism and distinct facial features [4].

The so-called premutation (PM), whose carriers can exhibit diverse symptomatology comprising both neural and non-neural pathological traits with different degrees of severity and compromised well-being, is defined in the range of 55 and 200 repeats. The worldwide prevalence of the PM is approximately 1:300 in females and 1:850 in males [5]. The risk to develop FXTAS (OMIM #300623) is higher in men than in women: from 17% to 75% of the PM male carriers as age increases, and only from 8 to 16% in PM carrier women [6,7] who exhibit a diverse phenotype much milder than those observed in men, possibly due to the female inactivation of chromosome X [8]. FXTAS is a late onset neurodegenerative disorder characterized by intention tremor usually accompanied by slow movement and parkinsonism, which develop into cerebellar gait ataxia and dystonia as the disease progresses, and by cognitive deficits (e.g., short-term memory loss, executive functions impairment, language capacity affectation), mood disorders and neuropsychiatric alterations, peripheral neuropathy, sleep disturbance and other signs [6,9,10]. The main radiological signs consist of an increased signal in brain white matter, especially in the middle cerebellar peduncles (MCP) in men [11]. MCP are the main afferent pathway to the cerebellum and are primarily composed by white fibers projected from the contralateral pontine nuclei as part of the cortico-ponto-cerebellar pathway that control motor tasks, planning and initiation of movements [12]. In addition, FXTAS patients show smaller volume in overall brain and particularly cerebellum related to PM carriers without diagnosis [13]. The presence of eosinophilic intranuclear ubiquitin-positive inclusions in brain, spinal cord and peripheral tissues is a hallmark of this disorder [14–16].

More prevalent in women is the development of the Fragile X-associated primary ovarian insufficiency (FXPOI, OMIM #311360): 24% of the PM women whereas the prevalence of POI in the general population is 1% [17]. This form is the most common inheritable ovarian dysfunction, which symptoms include irregular menstruation cycles, reduced fertility and early menopause onset [18]. More precisely, FXPOI involves ovarian hormonal dysfunction alongside with follicle depletion before the age of 40 years, having a direct impact on menstrual cycle regularity (that can lead to amenorrhea) and the ability to conceive, together with a variety of indirect consequences derived from chronic hypoestrogenism, such as early onset of osteoporosis and bone fracture, impaired vascular endothelial function, earlier onset of coronary heart disease and increased cardiovascular mortality, and higher risk of psychiatric symptomatology (e.g., anxiety, depression, etc.) than women with normal ovarian functionality [19–24].

This review is focused on the adult Fragile X-associated syndromes, starting with a brief description of the molecular pathogenic mechanisms caused by the CGG expansion, followed by a discussion about how some molecular and cellular alterations can be monitored at the peripheral level, mainly through the assessment of blood and derivatives: serum, plasma and peripheral blood mononuclear cells (PBMCs).

1.2. The Molecular Pathogenesis of Adult Fragile X-Associated Syndromes

In FXS, the methylation of the FM together with the neighbouring CpG island of the *FMR1* promoter triggers the silencing of the *FMR1* gene, leading to the absence or a drastic reduction of FMRP during development [25]. In contrast, FXTAS and FXPOI are primarily gain-of-function disorders, as FM carriers do not develop these syndromes. As we will see later in the section "Is there a Fragile X spectrum disorder?", PM carriers may exhibit some forms of FXS signs. The main actions of the PM are summarized below (Figure 1).

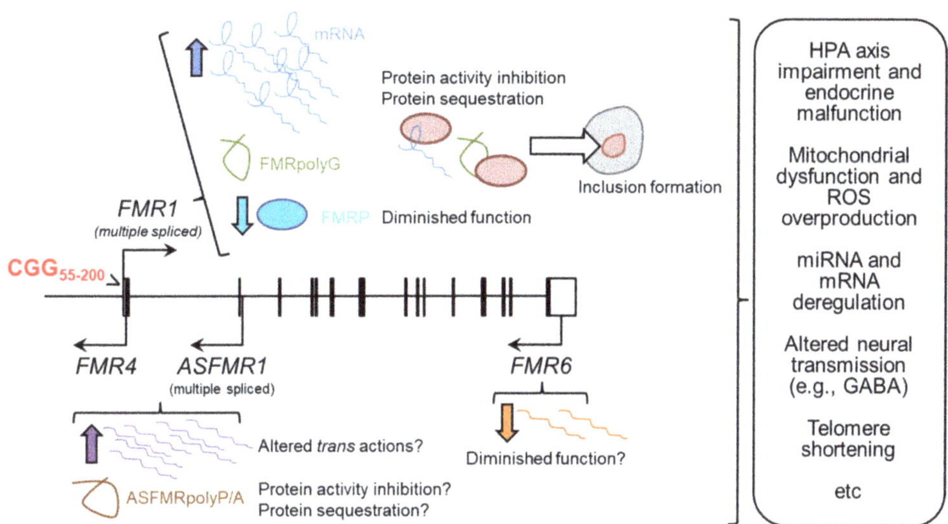

Figure 1. Summary of the molecular mechanisms proposed for the PM of the *FMR1* gene. Arrows indicate increases/decreases of RNA (*FMR1* and the long non-coding RNAs *ASFMR1*/*FMR4* and *FMR6*) or protein (FMRP) in PM carriers compared to controls. Pink ovals represent free proteins that interact/are sequestered/are aggregated by the CGG hairpin of the *FMR1* mRNA and/or FMRpolyG peptide.

1.2.1. RNA Toxicity

Adult Fragile X-associated syndromes were firstly proposed to be the result of RNA toxicity, based on the elevated levels of *FMR1* mRNA and its presence in the ubiquitin-positive inclusions across different cell types in PM carriers [26–28]. The mechanism for *FMR1* upregulation in PM carriers is not well understood, and only correlative evidence for an increase in the binding of epigenetic marks at the *FMR1* gene promoter has been found, as in the case of H3K9ac and H3K9me2 in PM infertile women [29]. As reported in locus-directed stable cell lines, CGG repeats in the PM size alone is not sufficient to increase the mRNA levels of a reporter gene [30], therefore additional regulatory mechanisms may exist. The preferential usage of particular sites for transcription start and 3′-polydenylation in the brains of PM carriers compared to controls [31] may be indicative of differences in the regulation of *FMR1* gene expression. Another possibility is that preceding downregulation of specific miRNAs (see section "miRNAs as potential biomarkers in Fragile X-associated syndromes") may influence the decay of *FMR1* mRNA, as in the case of miR-221 in synaptosomal preparations from KI mouse model, and miR-139-5p in lymphocyte-derived cell lines from patients, both with demonstrated capability to interact with the 3′UTR of *FMR1* mRNA [32,33].

Besides *FMR1* mRNA, the intranuclear inclusions also contain several proteins, suggesting that their sequestration by the PM RNA (riboCGG) caused a titration effect that impaired the biological processes in which these proteins are involved. For a detailed discussion of the role of the sequestered proteins in the FXTAS, see [34,35]. Mass spectrometry analyses, validated in silico predictions and screens for modulators of neurodegeneration eye phenotype in Drosophila PM models have provided a comprehensive list of potential interacting partners with riboCGG that, not surprisingly, included several RNA-binding proteins [36–41]. The first reported proteomics approach determined the composition of inclusions isolated from frontal cortex of FXTAS patients by fluorescent-activated cell sorting (FACS) techniques, taking advantage of the particle characteristics of inclusions and ubiquitin labelling. This analysis revealed the presence of RNA-binding proteins (e.g., the heterogeneous nuclear ribonucleoprotein A2/B1 (hnRNP A2/B1) and the myotonic dystrophy-related muscleblind-like splicing regulator 1 (MBNL1)), but also histones (e.g., H2A, H2B, H4), intermediate filaments and microtubules (e.g., NEFL, laminA/C, tubulins), and myelin associated proteins (e.g., MBP, CNPase) [36]. The interaction of hnRNP A2/B1 was later confirmed using a riboCGG in the PM range as a bait for mouse cerebellar proteins, which appeared besides another RNA-binding protein, Pur-α [37]. However, a subsequent study claimed that most of these proteins bound preferentially to non-pathogenic short sizes of riboCGG, whereas other proteins that included the DiGeorge Syndrome Critical Region Gene 8 (DGCR8) seemed to bind exclusively to longer CGG repeats within the pathogenic range [39] (see the section "miRNAs as potential biomarkers in Fragile X-associated syndromes"). Lastly, analysis of frontal cortex inclusions from FXTAS patients (isolated thanks to their autofluorescent properties, avoiding any bias from immunolabelling) resulted in the identification of > 200 proteins although no one showed any predominant abundance: histones, ubiquitin, proteins involved in RNA-binding (e.g., hnRNA A1 and A3), RNA splicing (e.g., U2AF, SFPQ), and protein turnover (e.g., chaperones and proteasomal-ubiquitin members such as SUMO 2 and p62), DNA damage repair mediators (e.g., RAD50, RPA1, XRCC6), etc. Of these, only five proteins (SUMO2, p62, ubiquitin, Myeloid leukemia factor 2 (MLF2) and MBP) were especially enriched in FXTAS aggregates related to total protein and related to the proteomics profile of control samples [40]. Based on this observation, the authors of this work postulated that inclusions are accumulations of proteins ready for removal but instead they aggregate once the capacity of proteasomal degradation is exceeded [40].

1.2.2. Production of FMRpolyG Peptides

In addition to RNA, a pathogenic role has been proposed for a cryptic polyglycine peptide, FMRpolyG, that is produced by a repeat associated non-AUG (RAN) translation process occurring at the 5′-expanded CGG region of *FMR1* [42,43]. First discovered in the *ATXN8* gene which is involved in the Spinocerebellar ataxia 8 (SCA8), the products of RAN translation have been detected in vivo for other genes containing repetitive nucleotides such as Huntington's disease (HD), myotronic dystrophy type 1 and 2 (DM1, DM2), amyotrophic lateral sclerosis (ALS) and frontotemporal dementia (FTD) [44]. In the *FMR1* locus, RAN translation can generate three types of homopeptides depending on the translation initiation (FMRpolyR, FMRpolyG and FMRpolyA), being the most efficient produced and the most well characterized the FMRpolyG (reviewed in [35]). In addition, FMRpolyG is the only species that have been found in intranuclear aggregates in neural and non-neural tissues (e.g., kidney, heart, thyroid, adrenal gland) [45], and in the cytoplasm of mural granulosa cells [46]. More recently, a vascular phenotype has been observed in the brain of PM carriers linked to the detection of FMRpolyG/p62+ inclusions [47]. Evidence for FMRpolyG toxicity has been collected from Drosophila and mouse models expressing different variants of engineered FMRpolyG aimed at enhancing or precluding such toxicity [42,48]. However, a recent study suggested that the appearance of FMRpolyG in neuronal intranuclear inclusions is not sufficient per se to trigger a clear phenotype in an inducible mouse model [49]. This is in contrast with the results obtained in a cellular model specifically designed to express the FMRpolyG in the absence of a CGG mRNA-dependent phenotype, in which cell viability was compromised and the nuclear lamina architecture was disrupted [50]. The C-terminus of the FMRpolyG (consisted on 42 aas after the polyG stretch) seems to be fundamental to the peptide toxicity as binds to cellular proteins such as LAP2β, a lamina associated protein, leading to the disorganization of the nuclear lamina architecture [48] (see the section "Transcriptional dysregulation in Fragile X-associated syndromes"). Another proposed mechanism for FMRpolyG-mediated toxicity is the acceleration of the ubiquitin-proteasome system impairment, as observed in Drosophila and cellular preparations [51]. This is in apparent disagreement with the proteomics study described above indicating that the proportion of FMRpolyG in intranuclear inclusions was too low in comparison with the total amount of proteins, putting into question its relevance in FXTAS pathogenesis [40].

1.2.3. The Potential Role of Long Non-Coding FMR1 Isoforms

Long non-coding RNAs (lncRNAs) are mainly transcribed by RNA polymerase II, and are often subjected to 5′-end capping, 3′-end polyadenylation and splicing as protein-coding mRNAs, and can play multiple roles in the regulation of transcription and splicing, translation, DNA replication and response to DNA damage through their interaction with RNA-binding proteins, transcription factors and regulatory protein complexes such as the polycomb repressive complex 2 (PRC2) [52]. This type of transcripts have been also described for the *FMR1* locus, starting with the identification of the antisense FMR1 transcript (*ASFMR1*) between 2.5 kb upstream and 10 kb downstream of the transcription start site (TSS) of the *FMR1* gene, encompassing exons 1 and 2, intron 1, and upstream regions [53]. This region also includes *FMR4*, a non-coding antisense RNA of 2.4 kb which expression is initiated at the 5′UTR of the *FMR1* gene [54] and can be considered one of the multiple spliced isoforms of *ASFMR1*. Two additional *FMR1*-derived lncRNAs have been detected by using a combination of rapid amplification of cDNA ends (RACE) and next-generation sequencing (NGS) [55]: *FMR5*, a sense-oriented and unspliced 800 nt transcript in which 5′-end starts approximately 1 kb upstream of the TSS of *FMR1*, and *FMR6*, a 600 nt antisense transcript that originates in the 3′UTR of *FMR1* gene and ends in exon 15, sharing the same splice junction with *FMR1* mRNA.

The potential relevance of *FMR1*-derived lncRNAs in Fragile X-related syndromes is based upon the deregulation observed in PM carriers that can be associated with pathological traits, as further discussed in the section "The molecular outputs of the *FMR1* locus as biomarkers in pathology". Expression of *ASFMR1* and *FMR4* is increased in the PM condition compared to controls, but negligible in FM condition in brain, peripheral blood leukocytes and lymphoblastoid cell lines [53–55]. *FMR5* and *FMR6* are also widely expressed across brain areas and peripheral blood samples, but whereas *FMR6* expression was reduced in PM and FM carriers related to controls in post-mortem brain tissue, *FMR5* levels did not show differences between the three groups of individuals [55]. In granulosa cells of PM women, the RNA levels of *FMR6* (but not those of *FMR4*) showed a significant negative linear correlation with the number of retrieved oocytes and a significant non-linear association with the number of CGG repeats (with the highest levels in women with mid-size CGG repeats) [56], indicating the *FMR6* may be involved in ovarian dysfunction. However, the pathogenic mechanisms underlying their contribution in disease are largely unknown, although there are clues indicating a *trans* action that is independent on *FMR1* regulation. RNA interference and over-expression assays in transient transfection experiments revealed that the expression of *FMR1* and *FMR4* is not influenced by each other [54]. *FMR4* has anti-apoptotic and cell proliferative actions in human cell lines and neural precursors [54,57] and may have a physiological function during neural development [58]. Manipulation of the *FMR4* levels in HEK293T cells leads to genome-wide changes in gene expression and DNA occupancy of H3K4me3 and K27me3, two epigenetic marks associated with active and silent genes respectively [57,58], indicating that *FMR4* is a *trans* regulator of gene expression and chromatin modulation. Based on these results, it is expected that the actions of *FMR1*-lncRNAs can be altered in PM cells.

In addition, *ASFMR1* and *FMR4* contains CCG repeats in the complementary strand that may be deleterious as well: in this regard, ASFMRpolyP and ASFMRpolyA peptides as a result of RAN translation of the *ASFMR1* transcript can be detected in FXTAS brains [59].

1.2.4. Reduction of FMRP Activity

Elevation of *FMR1* mRNA has been detected concomitant to a modest reduction of FMRP in some (although not all) PM individuals [60,61], possibly as a homeostatic mechanism to minimize the deleterious actions of the mRNA [62]. Such reduction may act in combination with the gain-of-function component of the PM to exacerbate some aspects of the complex symptomatology exhibited by the carriers. More specifically, this contribution has been proposed to potentially explain mental traits as we will discuss below, due to the role of FMRP in synapse regulation [63]. Noticeably, PM carriers showing both

FMR1 mRNA increase and FMRP decrease may develop psychotic and bipolar disorder features that are extremely rare in FXTAS and FXS patients, probably due to a synergistic effect of the gain- and loss-of-function components of CGG expansion [64].

1.3. Is There a Fragile X Spectrum Disorder?

Some authors have suggested that all clinical manifestations associated with the PM of the *FMR1* gene constitute a spectrum of varying degrees of penetrance and severity of these clinical signs, in which the final diagnosis of FXTAS or FXPOI may represent extreme forms of cognitive and endocrine impairment that may be present in other PM carriers in a milder manner. Therefore, the motor, cognitive and reproductive impairments generally ascribed to FXTAS and FXPOI diagnoses do not follow a strict boundary between sexes. Meanwhile PM women who are asymptomatic for classical FXTAS may develop subtle motor and cognitive impairments that are suspicious to be cerebellar-dependent [65], PM men can develop intranuclear inclusions in brain as well as in testicular tissues, prominently in the Leydig cells that are responsible for secreting testosterone to maintain spermatogenesis [15,16,66]. A potential association of such inclusions with cases of infertility [66] resembles the situation of granulosa cells and POI in PM women. Moreover, there is also a major risk of macroorchidism in PM men, a clinical feature usually linked to FXS that has been correlated with lower verbal and intelligence quotient (IQ) in the examined individuals [67].

Thus, manifestation of specific traits can be largely dependent on unknown genetic interactions, and/or the degree of mosaicism between the gain- and loss-of-function components of the CGG expansion that may differ throughout neural and non-neural tissues. For instance, patients who presented both *FMR1* mRNA increase and FMRP decrease manifested a symptomatology resembling combined FXS and FXTAS [64], although the mosaicism of methylated and unmethylated FM alleles can also explain the coexistence of FXS and FXTAS-related diagnosis in the same individuals [64,68]. It has been documented that pediatric PM carriers may develop attention deficit hyperactivity disorder (ADHD), anxiety, autistic features, seizures and other psychiatric symptoms that are reminiscent of FXS but to a much lesser extent and prevalence compared to FXS patients [69]. Therefore, PM boys have higher risk to be diagnosed of autism spectrum disorder (ASD) or ADHD than age-matched PM girls, although in both groups the risk was higher than in controls, leading to the hypothesis that reduced protein levels of *FMR1* are responsible for the mild signs of the developmental and cognitive impairments observed in PM carriers that are characteristic of FXS patients [70,71].

However, the majority of PM children do not have psychiatric conditions but they are at higher risk than the general population to develop them in the mid-adulthood, independently of a FXTAS diagnosis. The term Fragile X-Associated Neuropsychiatric Disorders (FXAND) has been coined to group the diverse neuropsychiatric problems observed in PM carriers that precede the onset of characteristic FXTAS but recognizes a series of mental disturbances in those PM carriers that do not fall into the diagnosis of FXTAS and can be more common than expected in these individuals [72]: anxiety, high sensitivity to external stimuli, depression, ADHD, obsessive-compulsive disorder, chronic pain and fatigue, sleep disturbance, and drug abuse (probably as part of a self-medicating behaviour) [72,73]. This term also considers non-neurological symptoms, as in the case of autoimmune conditions that mainly appear in PM females (hypothyroidism, fibromyalgia, irritable bowel disease, etc.) [19].

2. Peripheral Blood as Source of Biomarkers for Adult Fragile X-Associated Syndromes

Due to the pleotropic actions of the *FMR1* PM which complex manifestations of symptoms is being increasingly recognized in the affected population [9], it is indispensable to monitor the health status of their carriers through the molecular assessment of biomarkers that should follow essential requisites of being cost-effective, easily quantifiable in accessible tissues or fluids, and provide a linear correlation with disease progression (or reversal in case of treatment) [74]. In this sense, blood-based biomarkers offer a low-invasive alternative to screen and monitor different organ functions and associated diseases in clinical practice. Accomplishment of such requisites can be challenging in the case of FXTAS, in which the affected nervous tissues are not accessible and the potential biomarkers depend on either the non-neural component of the disease or in the extravasation of metabolites and other molecules to the periphery that should take place in early stages. Thus, interrogation of the heterogeneous cellular fraction (focused on PBMCs) relies on the assumption that the molecular events occurring in the affected organs that trigger the manifested symptomatology in PM carriers can be shared, or at least have a correlative impact, on these cells. In this context, plasma and serum can be relatively homogeneous sources of biomarkers, through the measurement circulating cell-secreted proteins, metabolites, extracellular vesicles and other components for patient screening. Omics approaches can discover novel and unforeseen biomarkers through the description of altered patterns of blood cells RNAs [75,76] and plasma metabolites in PM carriers compared to controls [77–79], as we will discuss later.

However, interpretation of the results obtained in most of the associative studies conducted in PM carriers should be taken as preliminary in the absence of a validation in independent cohorts, due to the small number of recruited volunteers (usually consisted on few individuals to few dozens) that is characteristic of studies in rare disorders. This can be an important reason for the lack of significant differences between PM carriers with or without a diagnosis of FXTAS or FXPOI, unless we consider that molecular alterations precede the onset of the symptomatology required for a conclusive diagnosis.

2.1. The Molecular Outputs of the FMR1 Locus as Biomarkers in Fragile X-Associated Pathology

The most examined peripheral parameter as a putative biomarker of the vast array of Fragile X-related symptoms and alterations is the number of CGG repeats in blood cells. In this sense, the most well-established association using the number of CGG repeats is the non-linear correlation with FXPOI diagnosis in which PM carriers with medium size CGG tracks have higher risk for ovarian insufficiency, altered cycles and dizygotic twinning [17,80]. Quantification of the number of repeats assumes that CGG size is fairly stable in somatic cells across lifespan [81,82] leading to negligible (or at least correlative) differences between peripheral blood cells and the affected organ. However, instability cannot be entirely discarded [83] as this is commonplace in other trinucleotide repeats disorders [84–86]. In fact, somatic expansion rates in PM alleles have been found to be different across human and mouse organs, being high in testis and certain brain areas (e.g., amygdala and striatum) but relatively stable in lymphocytes [87]. Such difference may introduce discrepancies in the number of repeats analyzed from blood and tissues of the same individuals.

Since the realization that *FMR1* mRNA was upregulated in blood samples bearing the PM range compared to controls, concomitant with a reduction in the number of FMRP$^+$ lymphocytes [60], and fueled by early indications of an association of these changes with cognitive deficits in certain PM carriers [88], it has been usual to implement the examination of mRNA and/or protein levels besides the number of CGG repeats as potential correlates of the development of specific symptoms and the severity of the general pathological phenotype. This is particularly interesting, because such associations allow inferring whether the gain- and/or loss-of-function components of the *FMR1* PM (corresponding to RNA toxicity and FMRP activity reduction, respectively) can explain the onset of specific signs. In this direction, there have been attempts to correlate CGG size and *FMR1* mRNA levels with IQ [89] and psychiatric symptoms [90,91]. Of special interest, some studies have tried to correlate these peripheral parameters with alterations in brain structure and brain activation patterns as measured using non-invasive imaging techniques. These approaches also permit the identification of the neural substrates responsible for the altered brain functions of clinical relevance for the PM carriers. A summary of these studies is shown in Table 1.

Table 1. Studies conducted in PM carriers searching for correlation between brain imaging measurements and *FMR1*-derived parameters: number of CGG repeats, *FMR1* mRNA levels and FMRP expression.

PM Carriers	Correlated Signs	*FMR1* Output	Reference
Males	Less voxel density in grey and white matter of brain areas, including cerebellum, brainstem and others Diminished grey matter of amygdala and hippocampal complex, left thalamus and brainstem	CGG repeats (no correlation with mRNA) % FMRP⁺-lymphocytes (no correlation with mRNA)	[92]
Males	Reduced left hippocampal activation during recall task, correlated with psychiatric assessment in the absence of hippocampal volume change	mRNA (no correlation with CGG repeats)	[93]
Males without FXTAS	Reduced activity of right ventral inferior frontal cortex in verbal working memory in both carriers	mRNA	[94]
Males	Decreased parahippocampal activation in working memory task	Blood FMRP levels	[95]
Females without FXTAS	Activity of right dorsolateral prefrontal cortex in correct encoded trials Decreased fronto-parietal activity in a magnitude estimation task	mRNA CGG repeats (no correlation with mRNA)	[96]
Males with/without FXTAS	White matter structural connectivity of thesuperior cerebellar peduncles in both carriers	CGG repeats and mRNA	[97]
Males	Cerebellar volume Defective anticipatory postural responses during stepping	CGG repeats mRNA and CGG repeats	[98]
Males without FXTAS	Motor dysfunction: tremor, balance and brain activation during random finger tapping	(no correlation with mRNA)	[99]

Apart from the canonical mRNA for *FMR1*, the non-coding transcripts generated from the *FMR1* locus have been also screened for clinical associations, based on previous observations showing increased expression in peripheral leukocytes [53,54]. For instance, small CGG expansion *FMR1* alleles (i.e., gray zone and lower PM sizes) play a significant role in the development of the parkinsonian-like phenotype that has been associated with elevated *FMR1* and *ASFMR1*/*FMR4* levels in PBMCs, in parallel to deregulation of mitochondrial genes [100]. Moreover, a spliced variant of *ASFMR1* with a 84 nt-deletion near the TSS was more elevated than the unspliced transcript in PM carriers compared to controls, and have been tentatively associated with FXTAS diagnosis [101]. Another highly similar splicing isoform that lacks 131 bp near the TSS (*ASFMR1*-Iso131bp) was also higher expressed in PM carriers independently of a FXTAS diagnosis compared to controls, with a trend to be correlated with tremor intensity as measured in a series of tasks involving postural and intentional tremor, postural sway, manual coordination, and reaction time [102]. Despite not showing differences between FXTAS and non-FXTAS individuals, the increase in the levels of *ASFMR1*-Iso131bp has been later ascribed to the acquisition FXTAS symptomatology during a longitudinal study [103].

To add more complexity to the potential contribution of *FMR1* RNAs in the etiology of neural and non-neural symptomatology in PM carriers, up to 49 *FMR1* alternative splicing isoforms have been detected in blood and different tissues (brain, muscle, testis and heart) thanks to single-molecule long-read sequencing that enabled the mapping of all possible splicing combinations in a single *FMR1* mRNA molecule [104,105]. An important fraction of these isoforms were increased, and in some cases exclusive, in the PM group [104,105], thus providing new candidates for peripheral exploration. According to the authors, the most notable isoforms were those named as Iso10 and Iso10b that, together with isoforms Iso4 and Iso4b, lack the nuclear export signal (NES) and the glycine-arginine-rich (RGG) motif involved with RNA binding [106]. As happened with *ASFMR1*-Iso131bp, these isoforms increased during the transition from disease-free stage to declaration of FXTAS in PM men [103], suggesting that they may be candidates as biomarkers of disease progression. In agreement with this conclusion, there was a significant association between increased expression of Iso4/4b with the decreased width of MCP only in those individuals that become diagnosed of FXTAS during this longitudinal study [103].

Finally, the *FMR1* locus contains additional features that have been examined as potential peripheral biomarkers. The absence of AGG interruptions in the CGG tracts, leading to uninterrupted CGG repeats that may predispose to germline expansion of low and medium size PM to produce FM during offspring transmission [107], have shown some correlation with neurological signs that deserves further confirmation [101]. Moreover, cortical white matter thickness and executive dysfunction have been associated with changes in the methylation patterns of CpG dinucleotides located in the first exon/intron boundary in the peripheral blood in PM women who were asymptomatic for FXTAS [108,109].

2.2. Endocrine Biomarkers in Premutation Carriers

In POI, measurements of serum levels of follicular stimulating (FSH) and anti-Müllerian (AMH) hormones, estradiol (E2) and inhibin B can be performed using standard protocols to assess the ovarian reserve [110]. A proteomic approach enlarged the list of possible candidates as biomarkers of POI with the identification of proteins with potential actions over reproductive functions (ceruloplasmin, complement C3, fibrinogen α, fibrinogen β, and sex hormone-binding globin SHBG) [111]. Although measurements of circulating endocrine hormones and other compounds can monitor the POI linked to CGG PM (FXPOI), they hardly discriminate between POI of different origin in the absence of a genetic test for *FMR1* CGG repeats. In other words, the hormone profiles found in FXPOI patients are not specific of the PM but, instead, they relate to reduced ovarian reserve.

The most well-established change in PM women is the elevation of circulating FSH levels, as it is one of the requisites for POI diagnosis, that has been shown to be independent on the menstrual cycle phase, regularity of menstrual cycles and oral contraception treatment [112–115], and can explain the decrease in the number of follicles and the reduction of the follicle phase and menstrual cycle length [112]. This phenomenon was reproduced in mouse models in which there was a loss of all types of follicles despite normal development of the ovary and normal establishment of the primordial follicle pool [116–118]. In these mice, elevation of FSH led to follicle depletion due to the activation of the primordial follicle pool and accelerated atresia of follicles, whereas diminished levels of luteinizing hormone (LH) levels reduced ovulation; thus, murine ovaries showed the downregulation of genes related to LH-induced ovulation including the LH receptor, together with a reduction in the phosphorylation of Akt and mTOR [117] that can be relevant molecular events, given the relevant roles of PI3K-Akt and mTOR signalling pathways in granulosa cells differentiation and survival, and growth and maturation of the ovary which dysfunctions can provoke POI [119,120]. In addition, it has been suggested that the follicular atresia can be due to the imbalance of granulosa cells cycle towards apoptosis provoked by the PM [121]. FSH is primarily secreted by the anterior pituitary gland that can contain FMR1-inclusions [66,122], indicating that the deleterious effects of the PM may perturb the hypothalamic-pituitary-adrenal axis; of interest, studies in a CGG-PM mouse model revealed that the formation of inclusions in the pituitary and adrenal gland can be an early event that precedes aggregation in most parts of the brain [123].

In PM women, other hormone alterations have also been detected, such as the reduction of AMH [118,124], inhibin A and B and progesterone P4 [112]. Of note, inhibin B exerts an inhibitory action over FSH secretion [125]. However, measurements of ovarian dysfunction (FSH levels) and pituitary-adrenal dysfunction (prolactin, cortisol and ACTH levels) were not correlated with scores in the FXTAS motor rating scale in PM women [91], suggesting that endocrine and motor alterations in PM carriers do not have a simple relationship. Finally, no association has been observed between CGG repeats in the normal range and reproductive parameters (including FSH and AMH levels) [126], but there are some indications regarding high levels of FSH in cases bearing intermediate CGG sizes (35–54 CGG repeats) and fecundity problems and amenorrhea [115], an observation that deserves further confirmation.

2.3. Mitochondrial Dysfunction and Overproduction of Reactive Oxygen Species in Premutation Carriers

The mitochondria are responsible for the cellular energy production through the oxidative phosphorylation process, play key roles in apoptosis induction and homeostasis maintenance. During ATP synthesis by oxidative phosphorylation, mitochondria produce reactive oxygen species (ROS) that are essential for different biological functions as cell cycle progression, immune response and homeostasis, as well as signalling pathways mediators during developmental processes like maturation and differentiation [127–129]. However, increases in the production of oxidant agents are related to DNA damage and cell death, and are associated with human pathologies such as cancer, cardiovascular disease and neurological disorders [78,130–132]. In the case of *FMR1*, the ablation of the gene in a knockout mouse model produced a transient increase in ROS production, NADH-oxidase activity and altered glutathione homeostasis concomitant to enhanced lipid peroxidation and protein carbonylation in brain extracts [133]. Primary hippocampal neurons from CGG-KI mice (with >150 CGG) also showed abnormal mitochondrial trafficking and bioenergetics [134]. Nonetheless, a more detailed description of alterations has been provided in cultured dermal fibroblasts from PM carriers, including decreased NAD- and FAD-linked oxygen uptake rates, uncoupling between electron transport and synthesis of ATP, decreased mitochondrial protein expression, increased oxidative/nitrative stress, higher NADH-dependent hydrogen peroxide production, Zn and Fe transport impairment, abnormal mitochondrial morphology and increased ROS production that usually worsen in FXTAS-declared patients and can be correlated with the number of CGG repeats [78,135–139]. Isolated peripheral lymphocytes also present aberrant mitochondria which can be correlated with clinical hallmarks of FXTAS [140,141] and postmortem brains of FXTAS patients mirrored the decrease in the levels of mature proteins such as Sodium/potassium-transporting ATPase subunit β (ATPB), Manganese superoxide dismutase (MnSOD) and frataxin found in skin fibroblasts [135,137]. Moreover, the integrity and copy number of mitochondrial DNA (mtDNA) can be influenced by ROS production in pathological processes such as cancer, cardiac conditions and neurodegeneration [142–144]. Meanwhile the deleterious effects of ROS on mitochondrial energy production may cause an increment of the mitochondria number per cell in an attempt to preserve homeostasis and the release of cell-free circulating mtDNA, ROS can also promote its degradation depending on the pathological condition [145,146]. In the case of PM carriers, mitochondrial perturbations were not apparently caused by a reduction in mitochondrial numbers as measured by citrate synthase activity or mtDNA copy number per cell [135,137,138,140]. However, other studies reported a decrease in mtDNA copy number and an increase in the rate of deletions in PM carriers compared to controls regardless of FXTAS manifestation [78,141]. It has been suggested that the decrease in mtDNA copy number per cell is more evident in tissues or cells directly relevant with the clinical manifestations of the disease. Thus, lower mtDNA copy number per cell was reported in specific brain areas related to disease progression (cerebellar vermis, dentate nucleus, parietal and temporal cortex) from asymptomatic PM carriers that were exacerbated in FXTAS-declared patients but failed to be significantly different in cultured dermal fibroblasts and in the brains of CGG-KI mice that did not exhibit tremors [147].

Mitochondrial dysfunction in *FMR1*-dependent ovarian insufficiency has been addressed by using female mice of the 130R strain bearing an insertion of expanded CGG repeats in the *Fmr1* gene. Whereas ROS production at the physiological range is necessary for oocyte maturation, ovarian stereoidogenesis, corpus luteal regulation and luteolysis [148], the 130R strain showed decreased primordial follicles in adult stages, histomorphometric evidence of follicle atresia, reduction of granulosa cell number in both growing and adult follicles, presence of ovarian cysts, and diminished fertility among other abnormalities [116,149]. Transmission electron microscopy analysis also revealed a population of abnormal mitochondria in appearance and structure, with evidence of mitochondrial gene expression downregulation, together with reduced mitochondrial mass [149]. Finally, mtDNA copy number was decreased in granulosa cells and metaphase II eggs of this strain, concomitant with reduced mitochondrial mass measured by flow cytometry [149].

This vast catalogue of mitochondrial alterations has been proposed to be caused by the interaction between FMRpolyG-containing aggregates and the mitochondrial membrane, which was concomitant to negative effects on membrane potential, ATP synthesis, assembly of respiratory chain supercomplexes composed by complexes I, III and IV, and gene expression of mitochondrial mRNAs, in the absence of altered mtDNA content in in vitro preparations [150]. This mechanism is plausible to occur in the ovary as well because FMRpolyG has been detected in granulosa cells with expanded CGG repeats [46]. In addition, it is also feasible that the supply of nuclear-encoded mitochondrial genes is compromised [151]. More recently, it has been described that expression of FMRpolyG in the HEK293 cell line can altered the content of nuclear-encoded miRNAs in whole cellular preparations but also in mitochondrial fractions [152], indicating that the altered biogenesis of miRNAs in CGG PM conditions (as we will discuss in brief) also affect mitochondrial functions. However, some miRNAs were enriched in the mitochondria but depleted in whole cell lysates, as in the case of miR-320a, which translocation into the HEK293 mitochondria was apparently increased in response to CGG PM conditions in an attempt to restore mitochondrial functions but failing to form the functional RISC complexes with Argonaute 2 (Ago2) required for regulation of mitochondrial transcripts [152].

In support of mitochondrial dysfunction as an early event in *FMR1*-CGG repeats disorders, it has been reported that the decreased expression of mitochondrial proteins in the brains of FXTAS patients required a lesser number of CGG repeats to be manifested than inclusion formation [137]. Morever, liver steatosis and linked mitochondrial dysfunction appeared in an inducible transgenic mouse model expressing expanded CGG in a timing when inclusions were not detected [153]. Based on these observations, together with the mitochondrial dysfunction observed in carriers without overt symptomatology, we can hypothesize that the effects of mitochondrial impairment and overproduction of ROS could be easily quantified at the peripheral level with sufficient anticipation to the onset of symptoms, providing reliable biomarkers in disease prognosis and therapeutic response. Noticeably, ROS production and mitochondrial function (e.g., citrate synthase activity, ATP production, etc.) in lymphocytes showed significant correlations with clinical outcomes in a cohort of PM women, such as diagnosis of FXTAS and FXTAS stage, diagnosis of FXPOI, number of CGG repeats, anxiety and full-scale IQ [79], suggesting that peripheral bioenergetics is directly linked to Fragile X-associated symptomatology. Metabolomics can go further in tracing the mitochondrial and oxidative perturbations in the plasma of PM carriers by identifying altered mitochondrial metabolic intermediates of the Krebs' cycle [154], in addition to other metabolites that can be explained by oxidative stress-dependent damage to proteins and carbohydrates, indicative of interference with multiple metabolic pathways (e.g., lower catabolism of carnosine, hyperactivation of the polyol metabolic pathway, increased activity of urea cycle) and activation of NADH-dependent antioxidant mechanisms [155]. Another metabolomics study confirmed the increase in metabolites derived from oxidation of carbohydrates and proteins and the hyperactivation of the polyol pathway; in addition, cells from FXTAS-declared carriers (but not those derived from asymptomatic donors) showed significantly higher Tyr nitration of cytoskeletal proteins and increased malondialdehyde (MDA), a marker of lipid peroxidation [78]. Following these observations, protein carbonylation in plasma/serum can be an adequate proxy of ROS overproduction in PM carriers because of ROS release into the bloodstream (Figure A1 Appendix A). As this phenomenon is tightly linked to lipid peroxidation [156], a spectrophotometry method may be preferred to measure protein carbonylation that can be more easily standardised in clinical laboratories [157] than the more sensitive but less routinary high performance liquid chromatography (HPLC) for MDA measurements [158]. A proteomics approach was also applied to PBMC of PM carriers, reflecting alterations in components related with glycolysis and carbohydrate and pyruvate metabolism that were consistent with the Kreb's cycle alterations found in plasma [79].

2.4. Inflammation in the Premutation Condition

Neuroinflammation, defined as the inflammatory response within the brain or spinal cord, is regarded as a relevant contributor of neurodegenerative processes. FXTAS seems not to be an exception as post-mortem putamen of patients showed signs of microglia activation, reflected by either the presence of dystrophic/senescent microglia or increased number of microglia (also detected in other brain areas) related to controls [159]. Moreover, post-mortem cerebellum contained increased levels of the proinflammatory IL-12 and TNFα, with a non-significant trend for IL-2, IL-8 and IL-10 [160]. At the peripheral level, it has been reported an increase in IL-10, an immunosuppressive cytokine that modulate glial activation to exert both protective and deleterious effects in neurodegenerative disorders [161], in the supernatants of PBMCs obtained from PM males that correlated with the number of CGG repeats but not with motor clinical scores [162]. In contrast, no changes were observed for IL-6 and IL-8 [162]. Overall, these few results suggest that exploration of circulating interleukins and cytokines profiles in Fragile X-associated syndromes can be worthy in the search of valuable biomarkers of disease progression.

2.5. miRNAs as Potential Biomarkers in Fragile X-Associated Syndromes

The binding of DGCR8 and its partner DROSHA with expanded CGG repeats and their presence in CGG RNA-containing aggregates indicated that miRNA processing was affected in PM carriers [39]. DGCR8, a double-stranded RNA binding protein, and DROSHA, a double-stranded RNA-specific ribonuclease, are part of the microprocessor complex that process the primary miRNAs (pri-miRNAs) into precursor miRNAs (pre-miRNAs) that are then exported to the cytoplasm for further processing by Dicer into mature miRNAs [163]. According to these roles, the partial sequestration of DGCR8 and DROSHA by CGG repeats led to reduced processing of pri-miRNA into pre-miRNA in cellular preparations [39], a phenomenon that explained the predominant downregulation of mature miRNAs observed in a microarray analysis of cerebellar samples from patients with FXTAS compared to age-matched controls [39].

In contrast to the cerebellar transcriptomics, the pair-wise comparison of peripheral blood gene profiles between FXTAS patients and control donors revealed the predominance of upregulation (12 of 14 miRNAs consistently changed in microarray and next-generation sequencing platforms) [75]. Of these, only miR-27a was also upregulated among the miRNAs derived from extracellular vesicles or lipoprotein complexes [164] of cell-free plasma of our own cohort (see Table A1 Appendix A). This miR-27a was also detected in the serum of individuals affected by age-related macular degeneration [165] and can serve as a potential candidate for further studies as Fragile X-associated biomarkers.

2.6. Transcriptional Dysregulation in Fragile X-Associated Syndromes

Despite the intranuclear inclusions not containing *bona fide* transcription factors, the presence of histones and laminA/C that indicate general chromatin disorganization, RNA-binding proteins that may regulate gene expression (as in the case of hnRNPs [166]) and splicing factors that may cause the deregulation of specific variants (as in the case of TRA2A [41]) suggest that transcriptional dysregulation may be important in PM cells. Furthermore, miRNA deregulation may modulate the transcript decay of downstream targets [167]. As we already introduced in the section entitled "Production of FMRpolyG peptides", there are reports indicating the disruption of the nuclear lamina architecture, a dense fibrillar network that participates in chromatin organization among other functions, which may have a profound impact in transcriptional regulation [168,169]. Such disruption is caused by altered nuclear distribution of laminA/C, observed in immunocytochemistry assays as a loss of the ring-like pattern in the inner nuclear membrane in cultured skin fibroblasts from PM carriers, both symptomatic and asymptomatic for FXTAS [170]. This observation was concomitant with the upregulation of the stress-related components CRYAB, HSP27 and HSP70 (that, together with laminA/C, were also present in FXTAS inclusions [36]). Of note, the inducible expression of a reporter gene bearing the 5′UTR of the *FMR1* bearing long CGG repeats in neuroblastoma-derived cell lines is sufficient to compromise cell viability at the same time disrupting the laminA/C architecture [171,172]. Another observation in favour of the importance of altered patterns of gene expression in PM carriers comes from the global reduction of 5-hydroxymethylcytosine (5hmC), considered as an intermediate state towards DNA demethylation [173,174], in the cerebellum of mouse models for FXTAS that partially affected genes involved in neuronal function and brain development [175].

A microarray analysis revealed that differential gene expression is observable between FXTAS patients and controls at the peripheral blood level. The results included the deregulation of the following genes: those encoding for respiratory chain subunits and superoxide dismutase 1 (SOD1), mTOR signalling pathway components, chromatin remodelling and epigenetic modulators, related to cell death and, interestingly, those involved in other neurodegenerative conditions, in the case of *ATXN3*, *ATXN7*, *APP* and *TARDP* [76]. Some of these genes were also altered in the brains of mouse models, suggesting that an undetermined component of the PM-associated transcriptional dysregulation is shared between central and peripheral cells. Of note, such direct correlation is difficult to observe in other neurodegenerative disorders and requires additional analytical approaches [176]. Importantly, FXTAS patients showed the downregulation of Interferon Regulatory Factor 2 Binding Protein-like (*IRF2BPL*)/Early at Puberty 1 (*EAP1*), firstly described as a neuronal transcriptional regulator of female reproductive function [177] and later associated with the neurodevelopmental epileptic encephalopathy NEDAMSS (neurodevelopmental disorder with regression, abnormal movements, loss of speech, and seizures, OMIM #618088) [178,179] that was also exacerbated in FXPOI patients compared to asymptomatic PM women [76]. This gene provides an example of a putative biomarker that can be useful for FXTAS, FXPOI and other forms of PM-associated disorders. However, downregulation of this gene was not statistically significant in our cohort (Figure A2 Appendix A), exemplifying the difficulties in the validation of peripheral biomarkers.

The same laboratory also reported the gene expression profile of peripheral blood from PM women with or without FXPOI [180], where differential expression compared to control donors were apparent inconsistent with the previous transcriptomics analysis in FXTAS patients [76], probably indicating that tissue-specific phenomena and normal X-chromosome inactivation may imprint distinct transcriptional signatures in blood cells. Actually, the authors were not able to identify any significant differentially expressed gene (e.g., *IRF2BPL*) between conditions after *P*-value adjustment, although they proposed that pathways related to cell cycle, apoptosis, programmed cell death and survival may be altered in PM carriers with FXPOI when considering the functions associated with the genes showing trends for alteration [180]. Examination of larger cohorts will determine the transcriptional signatures linked to PM and whether these signatures share common components among Fragile X-associated syndromes.

2.7. The GABAergic Dysfuntion in Premutation Carriers

GABA is the main inhibitory neurotransmitter in the brain, which deficits have been associated with the cognitive dysfunctions dealing with attention, learning, memory, planning and others that are frequent in neurological disorders including FXS [70,181–183]. In the PM condition, there is increasing evidence of impairment in $GABA_A$ receptor-mediated transmission. A study using non-invasive brain stimulation protocols in a small cohort of PM females indicated reduced $GABA_A$-mediated activity inhibition [184]. In mutant mice, mRNA expression was altered for several subunits encoding $GABA_A$ receptors and other components associated with the GABA signalling pathway: downregulation in the cortex and to a lesser extent in the cerebellum of *Fmr1*-null mutant mice [185–187] and upregulation in the cerebellum but not in the cortex in mice carrying the expanded CGG repeats in the PM range, despite *Fmr1* mRNA upregulation in both brain areas [187]. Cultured hippocampal neurons from PM mice exhibited altered patterns of electrical spontaneous activity as measured in multielectrode arrays (MEA) that consisted of clustered burst firing interspaced with brief periods of inactivity, resulting in higher spike frequency and longer mean burst duration than wild-type neurons; the patterns of Ca2+ oscillations were also altered in a similar fashion [188]. This behaviour was linked to an imbalance in the excitatory mGluR1/5 and inhibitory $GABA_A$-dependent signalling pathways, paralleled by a decrease in the levels of vesicular GABA and Glu transporters. Either inhibition of mGluR1/5 receptors by selective antagonists or enhancement of the $GABA_A$ receptor activity by the allosteric modulator allopregnanolone restored the spontaneous firing activity of PM neurons, meanwhile opposite pharmacological manipulations reproduced the clustered burst firing in wild-type neurons [188].

In the case of *FMR1* PM carriers, individuals of both sexes exhibited a differential metabolomic profile in plasma compared to controls that indicated a slow tricarboxylic acid cycle activity, incipient neuronal degeneration and degradation of neuromodulatory fatty acid amides that may be interrelated [154]. Regarding GABAergic transmission, this profile included the aforementioned metabolic intermediates of the Krebs' cycle (e.g., citrate, aconitate and isocitrate) that might link mitochondrial dysfunction with altered biogenesis of Krebs' cycle-associated neurotransmitters Glu, Asp and GABA, and the bioactive fatty acid oleamide that can modulate multiple neuronal receptors including GABAergic ones [189,190]. Restoration of GABA function by allopregnallone treatment can improve cognition and executive function, working and episodic memories and anxiety in six FXTAS patients [191], leaving a differential metabolomics signature between pre- and post-treatment that included metabolites related to oxidative stress, mitochondrial function and GABA receptor activity [192], demonstrating the feasibility to find biomarkers related to Fragile X-associated dysfunctions that can be followed up in response to specific treatments.

2.8. Telomere Shortening in Premutation Carriers

Telomere lengths consist of TTAGGG repeats in tandem that are reduced by multiple conditions including cellular aging, cellular senescence and apoptosis, cancer processes, heart and lung conditions, Alzheimer's disease, etc. [193–200]. Shorter telomeres have been described in PBMCs of PM males compared to controls or FM carriers, regardless of FXTAS or dementia diagnosis [201–203]. In PM female carriers, this shortening was also present, although the difference related to control donors was more evident with a FXPOI diagnosis in young individuals [204]. This observation is tentative to be generalized to any type of POI, as another study revealed that peripheral leukocytes from women suffering idiopathic POI (i.e., showing amenorrhea and high levels of FSH before 40 years of age, but without any known genetic and environmental cause of chromosome instability) were significantly shorter than controls [205]. Currently, no mechanistic explanation has been offered to explain telomere shortening in peripheral leukocytes except for a generic acceleration of cellular senescence, although other factors can contribute such as the postmenopausal treatment of POI patients with estrogen [206].

3. Conclusions

Intensive studies aimed at identifying appropriate biomarkers to predict and/or monitor the complex diversity of symptomatology outcomes of the adult Fragile X-associated syndromes have identified interesting correlations between the most affected organs, mainly focused on neural and reproductive tissues. The expanding catalogue of clinical signs that also indicate affectation in other organs (e.g., heart) demands an additional effort to establish new correlations. We already have a large variety of peripheral blood measurements (e.g., levels for *FMR1* mRNA and *FMR1*-associated lncRNAs, circulating metabolites and ROS, endocrine hormones, etc.) that can be extended to novel candidates thanks to the application of proteomics, metabolomics and transcriptomics. However, it seems that we need to further assess these biomarkers in tailored longitudinal studies to precisely establish consistent correlations with the development and progression of specific outcomes in PM carriers. Furthermore, few studies have gone beyond the correlative evidence by addressing the connection between the organ dysfunction and the peripheral alterations from a mechanistic point of view. Additionally, we should keep in mind that we implicitly assume that the PM and its effects are uniform across the organism and the peripheral cells behave as nearly carbon-copies of the inaccessible cells of interest. Enhancing our comprehension regarding the molecular mechanisms of PM toxicity is key to understand the alterations occurring at the level of both individual cells and cellular networks, and to provide improved tools for the clinical management of the PM carriers in a personalized manner.

Author Contributions: Conceptualization, writing and bioinformatics analysis, L.M.V.: experimental research of Appendix A, J.C.M., I.H.-C. and R.M.; review and editing, I.H.-C. All authors have read and agreed to the published version of the manuscript.

Funding: L.M.V. is supported by the Programa Estatal de Generación de Conocimiento, financed by the Instituto de Salud Carlos III and Fondo Europeo de Desarrollo Regional 2014-2020 (Grants PI16/00722 and PI19/00125). L.M.V. is the recipient of a Miguel Servet I contract (CP15/00180) followed by a Miguel Servet II contract (CPII20/00025), and J.C.M. is the recipient of a Río Hortega contract (CM18/00043), all financed by the Instituto de Salud Carlos III and Fondo Social Europeo 2014-2020, Programa Estatal de Promoción del Talento y su empleabilidad en I+D+i.

Institutional Review Board Statement: The study was conducted according to the guidelines of the Declaration of Helsinki, and to national and regional law regulations concerning biomedical research in human samples and personal data protection. This study was approved by the Comité de Ética de la Investigación de Cádiz of the Hospital Universitario Puerta del Mar.

Informed Consent Statement: Informed consent was obtained from all subjects involved in the study.

Data Availability Statement: De novo datasets generated and analyzed during the current study are available in the GEO repository under the Accession Number GSE181018.

Conflicts of Interest: The authors declare no conflict of interest.

Appendix A

Volunteers consisted of a total of six controls (five females and one male) and 11 PM carriers (eight females and three males) with ages expressed as median (interquartile range) of 42.5 (5.5) and 44 (21.8), respectively (P-value = 0.88, Mann Whitney U-test). PM carriers were recruited from family pedigrees with diagnosed cases of FXS, with number of CGG repeats of 103 (11.5).

Peripheral blood from these donors was collected in Tempus Blood RNA tubes (Applied Biosystems, Thermo Fisher, Madrid, Spain) and in EDTA-K2 Blood tubes (BD Vacutainer, Madrid, Spain) in parallel. From the Tempus tubes, total blood RNA was purified with the Preserved Blood RNA Purification Kit I (Norgen Biotek, Thorold, ON, Canada) (including on-column DNAse I digestion) and reverse transcribed into cDNA using the RevertAid First-Strand cDNA Synthesis kit (Fermentas, Thermo Fisher, Madrid, Spain), following manufacturers' instructions. qPCR was performed in the Rotor-Gene 6000 Detection System (Corbett) using PyroTaq EvaGreen qPCR Mix Plus (Cmb-Bioline, Madrid, Spain). The PCR cycling conditions were as follows: 95 °C for 15 min and 40 cycles of 95 °C for 15 s, 60 °C for 20 s, 72 °C for 20 s. Each independent sample was normalized using *GAPDH* and *ACTB* and the relative quantitative values were calculated according to the $2^{-\Delta\Delta CT}$ method.

From the EDTA-K2 tubes, plasma was obtained after centrifugation at 3000× g 4 °C for 20 min and was kept at −80 °C until experimentation. Thirty µL of plasma was used to determine protein carbonylation as marker of oxidative stress according to the method described previously [207]. Results were normalized with total protein concentration in plasma as determined using the Pierce BCA Protein Assay Kit protocol (Thermo Fisher, Madrid, Spain). Carbonyl content was expressed as nanomoles / mg plasma protein using a molar extinction coefficient of 0.022 $mM^{-1}cm^{-1}$. Two hundred µL of plasma was used to isolate circulating small RNA using the miRNeasy Serum/Plasma Kit (Qiagen, Hilden, Germany) following manufacturer's instructions.

Small RNA was sent to the Genomics Unit of Centro Pfizer-Universidad de Granada-Junta de Andalucía de Genómica e Investigación Oncológica (GENYO) for library preparation (TruSeq Small RNA Library Preparation Kit, Illumina, San Diego, CA, USA) and single-end sequencing (NextSeq, Illumina), following Illumina's recommendations. After adapter removal using "Cutadapt" [208], reads were aligned to the human genome (hg38 build) using "Bowtie2" (http://bowtie-bio.sourceforge.net/bowtie2/index.shtml, accessed on 02-08-2021) at the Bioinformatics Unit of GENYO. The following Bioconductor-based software packages were used for conversion of BAM files ("RSamtools") and gene annotation ("GenomicFeatures" and "GenomicAlignments") using the Ensembl annotation package "Db.Hsapiens.v86" after filtering the "miRNA" gene biotype. Transcriptomics differential expression was determined between control donors and PM carriers by using DESeq2 software [209]. Results (adjusted p-value < 0.1) are shown in Table A1 of this Appendix and were compared to the microarray and RNA-seq data obtained from whole blood profiling reported in Table S3 of the original publication [75].

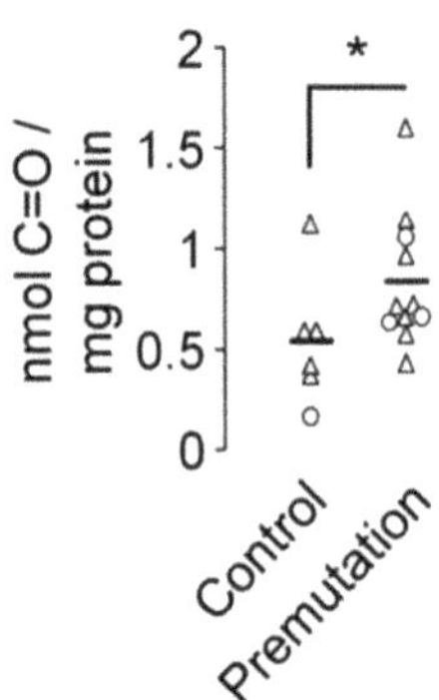

Figure A1. Significant increase of protein carbonylation in the plasma of premutation carriers. Triangles (Δ), female samples; circles (O), male samples. The mean is indicated as a horizontal bar. *, $p < 0.05$, Mann-Whitney U-test.

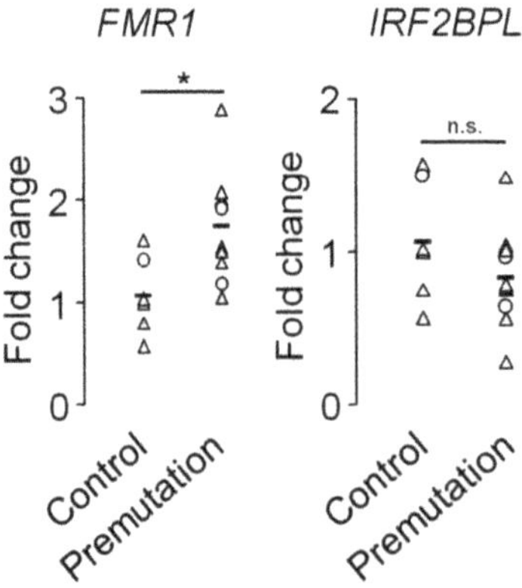

Figure A2. Relative expression of *FMR1* and *IRF2BPL* genes in whole blood of controls and premutation donors. Triangles (Δ), female samples; circles (O), male samples. Of the four males of our cohort, only three were available for these assays. The mean is indicated (horizontal bar). *, $p < 0.05$, Mann-Whitney U-test. n.s., not significant.

Table A1. Differential expression of miRNAs in the plasma of premutation vs. control donors. Potential common genes with a previous report comparing the whole blood profiles from FXTAS patients vs. control donors [75] are indicated as * (upregulation) or ** (downregulation).

miRNA	Ensembl ID	Fold Change (log_2)	Adj p-Value
MIR375	ENSG00000198973	−2.58	5.04×10^{-5}
MIR483**	ENSG00000207805	−2.72	1.06×10^{-2}
MIR4664	ENSG00000265660	−4.30	2.58×10^{-2}
MIR10B	ENSG00000207744	−1.47	3.72×10^{-2}
MIR7159	ENSG00000276824	−2.95	3.72×10^{-2}
MIR125B2	ENSG00000207863	−1.67	4.82×10^{-2}
MIR6730	ENSG00000276830	−4.16	6.60×10^{-2}
	ENSG00000280773	−5.51	8.17×10^{-2}
MIR501	ENSG00000211538	−2.10	9.14×10^{-2}
MIR885**	ENSG00000216135	−2.63	9.99×10^{-2}
MIR27A*	ENSG00000207808	1.29	9.98×10^{-2}
MIR26A1	ENSG00000199075	0.75	9.72×10^{-2}
MIR381	ENSG00000199020	4.10	8.49×10^{-2}
MIR655	ENSG00000207646	5.02	8.49×10^{-2}
MIR377	ENSG00000199015	4.64	6.60×10^{-2}
MIR654	ENSG00000207934	5.00	6.60×10^{-2}
MIR376C	ENSG00000283279	4.77	6.60×10^{-2}
MIR494	ENSG00000194717	2.52	5.56×10^{-2}
MIR889	ENSG00000216099	5.12	6.72×10^{-3}
MIR134	ENSG00000207993	5.72	5.14×10^{-3}
MIR331	ENSG00000199172	5.19	3.57×10^{-3}
MIR766	ENSG00000211578	5.53	8.50×10^{-4}
MIR877	ENSG00000216101	5.11	1.26×10^{-4}
MIR495	ENSG00000207743	5.18	5.74×10^{-5}

References

1. Bagni, C.; Oostra, B.A. Fragile X syndrome: From protein function to therapy. *Am. J. Med. Genet. A* **2013**, *161A*, 2809–2821. [CrossRef]
2. Madrigal, I.; Xuncla, M.; Tejada, M.I.; Martinez, F.; Fernandez-Carvajal, I.; Perez-Jurado, L.A.; Rodriguez-Revenga, L.; Mila, M. Intermediate FMR1 alleles and cognitive and/or behavioural phenotypes. *Eur. J. Hum. Genet.* **2011**, *19*, 921–923. [CrossRef]
3. Hall, D.A.; Nag, S.; Ouyang, B.; Bennett, D.A.; Liu, Y.; Ali, A.; Zhou, L.; Berry-Kravis, E. Fragile X gray zone alleles are associated with signs of parkinsonism and earlier death. *Mov. Disord.* **2020**, *35*, 1448–1456. [CrossRef]
4. Hagerman, R.J.; Hagerman, P. (Eds.) *Fragile-X Syndrome: Diagnosis, Treatment and Research*, 3rd ed.; Johns Hopkins University Press: Baltimore, MD, USA, 2002.
5. Hunter, J.; Rivero-Arias, O.; Angelov, A.; Kim, E.; Fotheringham, I.; Leal, J. Epidemiology of fragile X syndrome: A systematic review and meta-analysis. *Am. J. Med. Genet. A* **2014**, *164A*, 1648–1658. [CrossRef]
6. Jacquemont, S.; Farzin, F.; Hall, D.; Leehey, M.; Tassone, F.; Gane, L.; Zhang, L.; Grigsby, J.; Jardini, T.; Lewin, F.; et al. Aging in individuals with the FMR1 mutation. *Am. J. Ment. Retard.* **2004**, *109*, 154–164. [CrossRef]
7. Seltzer, M.M.; Baker, M.W.; Hong, J.; Maenner, M.; Greenberg, J.; Mandel, D. Prevalence of CGG expansions of the FMR1 gene in a US population-based sample. *Am. J. Med. Genet. B Neuropsychiatr. Genet.* **2012**, *159B*, 589–597. [CrossRef] [PubMed]
8. Schneider, A.; Summers, S.; Tassone, F.; Seritan, A.; Hessl, D.; Hagerman, P.; Hagerman, R. Women with fragile X-associated tremor/ataxia syndrome. *Mov. Disord. Clin. Pract.* **2020**, *7*, 910–919. [CrossRef]
9. Mila, M.; Alvarez-Mora, M.I.; Madrigal, I.; Rodriguez-Revenga, L. Fragile X syndrome: An overview and update of the FMR1 gene. *Clin. Genet.* **2018**, *93*, 197–205. [CrossRef]
10. Hagerman, P. Fragile X-associated tremor/ataxia syndrome (FXTAS): Pathology and mechanisms. *Acta Neuropathol.* **2013**, *126*, 1–19. [CrossRef] [PubMed]
11. Brunberg, J.A.; Jacquemont, S.; Hagerman, R.J.; Berry-Kravis, E.M.; Grigsby, J.; Leehey, M.A.; Tassone, F.; Brown, W.T.; Greco, C.M.; Hagerman, P.J. Fragile X premutation carriers: Characteristic MR imaging findings of adult male patients with progressive cerebellar and cognitive dysfunction. *AJNR Am. J. Neuroradiol.* **2002**, *23*, 1757–1766.
12. Morales, H.; Tomsick, T. Middle cerebellar peduncles: Magnetic resonance imaging and pathophysiologic correlate. *World J. Radiol.* **2015**, *7*, 438–447. [CrossRef]
13. Wang, J.Y.; Hessl, D.; Hagerman, R.J.; Simon, T.J.; Tassone, F.; Ferrer, E.; Rivera, S.M. Abnormal trajectories in cerebellum and brainstem volumes in carriers of the fragile X premutation. *Neurobiol. Aging* **2017**, *55*, 11–19. [CrossRef] [PubMed]

14. Greco, C.M.; Hagerman, R.J.; Tassone, F.; Chudley, A.E.; Del Bigio, M.R.; Jacquemont, S.; Leehey, M.; Hagerman, P.J. Neuronal intranuclear inclusions in a new cerebellar tremor/ataxia syndrome among fragile X carriers. *Brain* **2002**, *125*, 1760–1771. [CrossRef]
15. Gokden, M.; Al-Hinti, J.T.; Harik, S.I. Peripheral nervous system pathology in fragile X tremor/ataxia syndrome (FXTAS). *Neuropathology* **2009**, *29*, 280–284. [CrossRef]
16. Hunsaker, M.R.; Greco, C.M.; Spath, M.A.; Smits, A.P.; Navarro, C.S.; Tassone, F.; Kros, J.M.; Severijnen, L.A.; Berry-Kravis, E.M.; Berman, R.F.; et al. Widespread non-central nervous system organ pathology in fragile X premutation carriers with fragile X-associated tremor/ataxia syndrome and CGG knock-in mice. *Acta Neuropathol.* **2011**, *122*, 467–479. [CrossRef]
17. Wittenberger, M.D.; Hagerman, R.J.; Sherman, S.L.; McConkie-Rosell, A.; Welt, C.K.; Rebar, R.W.; Corrigan, E.C.; Simpson, J.L.; Nelson, L.M. The FMR1 premutation and reproduction. *Fertil. Steril.* **2007**, *87*, 456–465. [CrossRef] [PubMed]
18. Fink, D.A.; Nelson, L.M.; Pyeritz, R.; Johnson, J.; Sherman, S.L.; Cohen, Y.; Elizur, S.E. Fragile X associated primary ovarian insufficiency (FXPOI): Case report and literature review. *Front. Genet.* **2018**, *9*, 529. [CrossRef]
19. Hunter, J.E.; Rohr, J.K.; Sherman, S.L. Co-occurring diagnoses among FMR1 premutation allele carriers. *Clin. Genet.* **2010**, *77*, 374–381. [CrossRef]
20. Van der Stege, J.G.; Groen, H.; van Zadelhoff, S.J.; Lambalk, C.B.; Braat, D.D.; van Kasteren, Y.M.; van Santbrink, E.J.; Apperloo, M.J.; Weijmar Schultz, W.C.; Hoek, A. Decreased androgen concentrations and diminished general and sexual well-being in women with premature ovarian failure. *Menopause* **2008**, *15*, 23–31. [CrossRef] [PubMed]
21. Gallagher, J.C. Effect of early menopause on bone mineral density and fractures. *Menopause* **2007**, *14*, 567–571. [CrossRef]
22. Atsma, F.; Bartelink, M.L.; Grobbee, D.E.; van der Schouw, Y.T. Postmenopausal status and early menopause as independent risk factors for cardiovascular disease: A meta-analysis. *Menopause* **2006**, *13*, 265–279. [CrossRef]
23. Mondul, A.M.; Rodriguez, C.; Jacobs, E.J.; Calle, E.E. Age at natural menopause and cause-specific mortality. *Am. J. Epidemiol.* **2005**, *162*, 1089–1097. [CrossRef] [PubMed]
24. Kalantaridou, S.N.; Naka, K.K.; Papanikolaou, E.; Kazakos, N.; Kravariti, M.; Calis, K.A.; Paraskevaidis, E.A.; Sideris, D.A.; Tsatsoulis, A.; Chrousos, G.P.; et al. Impaired endothelial function in young women with premature ovarian failure: Normalization with hormone therapy. *J. Clin. Endocrinol. Metab.* **2004**, *89*, 3907–3913. [CrossRef] [PubMed]
25. Nobile, V.; Pucci, C.; Chiurazzi, P.; Neri, G.; Tabolacci, E. DNA methylation, mechanisms of FMR1 inactivation and therapeutic perspectives for fragile X syndrome. *Biomolecules* **2021**, *11*, 296. [CrossRef]
26. Swinnen, B.; Robberecht, W.; Van Den Bosch, L. RNA toxicity in non-coding repeat expansion disorders. *EMBO J.* **2020**, *39*, e101112. [CrossRef]
27. Tassone, F.; Iwahashi, C.; Hagerman, P.J. FMR1 RNA within the intranuclear inclusions of fragile X-associated tremor/ataxia syndrome (FXTAS). *RNA Biol.* **2004**, *1*, 103–105. [CrossRef]
28. Elizur, S.E.; Lebovitz, O.; Derech-Haim, S.; Dratviman-Storobinsky, O.; Feldman, B.; Dor, J.; Orvieto, R.; Cohen, Y. Elevated levels of FMR1 mRNA in granulosa cells are associated with low ovarian reserve in FMR1 premutation carriers. *PLoS ONE* **2014**, *9*, e105121. [CrossRef]
29. Eslami, H.; Eslami, A.; Favaedi, R.; Asadpour, U.; Zari Moradi, S.; Eftekhari-Yazdi, P.; Madani, T.; Shahhoseini, M.; Mohseni Meybodi, A. Epigenetic aberration of FMR1 gene in infertile women with diminished ovarian reserve. *Cell J.* **2018**, *20*, 78–83. [CrossRef]
30. Solvsten, C.; Nielsen, A.L. FMR1 CGG repeat lengths mediate different regulation of reporter gene expression in comparative transient and locus specific integration assays. *Gene* **2011**, *486*, 15–22. [CrossRef]
31. Tassone, F.; De Rubeis, S.; Carosi, C.; La Fata, G.; Serpa, G.; Raske, C.; Willemsen, R.; Hagerman, P.J.; Bagni, C. Differential usage of transcriptional start sites and polyadenylation sites in FMR1 premutation alleles. *Nucleic Acids Res.* **2011**, *39*, 6172–6185. [CrossRef]
32. Zongaro, S.; Hukema, R.; D'Antoni, S.; Davidovic, L.; Barbry, P.; Catania, M.V.; Willemsen, R.; Mari, B.; Bardoni, B. The 3′ UTR of FMR1 mRNA is a target of miR-101, miR-129-5p and miR-221: Implications for the molecular pathology of FXTAS at the synapse. *Hum. Mol. Genet.* **2013**, *22*, 1971–1982. [CrossRef] [PubMed]
33. Dolskiy, A.A.; Yarushkin, A.A.; Grishchenko, I.V.; Lemskaya, N.A.; Pindyurin, A.V.; Boldyreva, L.V.; Pustylnyak, V.O.; Yudkin, D.V. miRNA expression and interaction with the 3′UTR of FMR1 in FRAXopathy pathogenesis. *Noncoding RNA Res.* **2021**, *6*, 1–7. [CrossRef] [PubMed]
34. Xu, K.; Li, Y.; Allen, E.G.; Jin, P. Therapeutic development for CGG repeat expansion-associated neurodegeneration. *Front. Cell. Neurosci.* **2021**, *15*, 655568. [CrossRef]
35. Glineburg, M.R.; Todd, P.K.; Charlet-Berguerand, N.; Sellier, C. Repeat-associated non-AUG (RAN) translation and other molecular mechanisms in Fragile X tremor ataxia syndrome. *Brain Res.* **2018**, *1693*, 43–54. [CrossRef]
36. Iwahashi, C.K.; Yasui, D.H.; An, H.J.; Greco, C.M.; Tassone, F.; Nannen, K.; Babineau, B.; Lebrilla, C.B.; Hagerman, R.J.; Hagerman, P.J. Protein composition of the intranuclear inclusions of FXTAS. *Brain* **2006**, *129*, 256–271. [CrossRef]
37. Jin, P.; Duan, R.; Qurashi, A.; Qin, Y.; Tian, D.; Rosser, T.C.; Liu, H.; Feng, Y.; Warren, S.T. Pur alpha binds to rCGG repeats and modulates repeat-mediated neurodegeneration in a Drosophila model of fragile X tremor/ataxia syndrome. *Neuron* **2007**, *55*, 556–564. [CrossRef]

38. Sofola, O.A.; Jin, P.; Qin, Y.; Duan, R.; Liu, H.; de Haro, M.; Nelson, D.L.; Botas, J. RNA-binding proteins hnRNP A2/B1 and CUGBP1 suppress fragile X CGG premutation repeat-induced neurodegeneration in a Drosophila model of FXTAS. *Neuron* **2007**, *55*, 565–571. [CrossRef]
39. Sellier, C.; Freyermuth, F.; Tabet, R.; Tran, T.; He, F.; Ruffenach, F.; Alunni, V.; Moine, H.; Thibault, C.; Page, A.; et al. Sequestration of DROSHA and DGCR8 by expanded CGG RNA repeats alters microRNA processing in fragile X-associated tremor/ataxia syndrome. *Cell Rep.* **2013**, *3*, 869–880. [CrossRef]
40. Ma, L.; Herren, A.W.; Espinal, G.; Randol, J.; McLaughlin, B.; Martinez-Cerdeno, V.; Pessah, I.N.; Hagerman, R.J.; Hagerman, P.J. Composition of the intranuclear inclusions of fragile X-associated tremor/ataxia syndrome. *Acta Neuropathol. Commun.* **2019**, *7*, 143. [CrossRef] [PubMed]
41. Cid-Samper, F.; Gelabert-Baldrich, M.; Lang, B.; Lorenzo-Gotor, N.; Ponti, R.D.; Severijnen, L.; Bolognesi, B.; Gelpi, E.; Hukema, R.K.; Botta-Orfila, T.; et al. An integrative study of protein-RNA condensates identifies scaffolding RNAs and reveals players in fragile X-associated tremor/ataxia syndrome. *Cell Rep.* **2018**, *25*, 3422–3434. [CrossRef]
42. Todd, P.K.; Oh, S.Y.; Krans, A.; He, F.; Sellier, C.; Frazer, M.; Renoux, A.J.; Chen, K.C.; Scaglione, K.M.; Basrur, V.; et al. CGG repeat-associated translation mediates neurodegeneration in fragile X tremor ataxia syndrome. *Neuron* **2013**, *78*, 440–455. [CrossRef] [PubMed]
43. Bonapace, G.; Gullace, R.; Concolino, D.; Iannello, G.; Procopio, R.; Gagliardi, M.; Arabia, G.; Barbagallo, G.; Lupo, A.; Manfredini, L.I.; et al. Intracellular FMRpolyG-Hsp70 complex in fibroblast cells from a patient affected by fragile X tremor ataxia syndrome. *Heliyon* **2019**, *5*, e01954. [CrossRef]
44. Castelli, L.M.; Huang, W.P.; Lin, Y.H.; Chang, K.Y.; Hautbergue, G.M. Mechanisms of repeat-associated non-AUG translation in neurological microsatellite expansion disorders. *Biochem. Soc. Trans.* **2021**, *49*, 775–792. [CrossRef] [PubMed]
45. Buijsen, R.A.; Sellier, C.; Severijnen, L.A.; Oulad-Abdelghani, M.; Verhagen, R.F.; Berman, R.F.; Charlet-Berguerand, N.; Willemsen, R.; Hukema, R.K. FMRpolyG-positive inclusions in CNS and non-CNS organs of a fragile X premutation carrier with fragile X-associated tremor/ataxia syndrome. *Acta Neuropathol. Commun.* **2014**, *2*, 162. [CrossRef]
46. Friedman-Gohas, M.; Elizur, S.E.; Dratviman-Storobinsky, O.; Aizer, A.; Haas, J.; Raanani, H.; Orvieto, R.; Cohen, Y. FMRpolyG accumulates in FMR1 premutation granulosa cells. *J. Ovarian Res.* **2020**, *13*, 22. [CrossRef]
47. Dijkstra, A.A.; Haify, S.N.; Verwey, N.A.; Prins, N.D.; van der Toorn, E.C.; Rozemuller, A.J.M.; Bugiani, M.; den Dunnen, W.F.A.; Todd, P.K.; Charlet-Berguerand, N.; et al. Neuropathology of FMR1-premutation carriers presenting with dementia and neuropsychiatric symptoms. *Brain Commun.* **2021**, *3*, fcab007. [CrossRef]
48. Sellier, C.; Buijsen, R.A.M.; He, F.; Natla, S.; Jung, L.; Tropel, P.; Gaucherot, A.; Jacobs, H.; Meziane, H.; Vincent, A.; et al. Translation of expanded CGG repeats into FMRpolyG is pathogenic and may contribute to fragile X tremor ataxia syndrome. *Neuron* **2017**, *93*, 331–347. [CrossRef] [PubMed]
49. Haify, S.N.; Mankoe, R.S.D.; Boumeester, V.; van der Toorn, E.C.; Verhagen, R.F.M.; Willemsen, R.; Hukema, R.K.; Bosman, L.W.J. Lack of a clear behavioral phenotype in an inducible FXTAS mouse model despite the presence of neuronal FMRpolyG-positive aggregates. *Front. Mol. Biosci.* **2020**, *7*, 599101. [CrossRef]
50. Hoem, G.; Bowitz Larsen, K.; Overvatn, A.; Brech, A.; Lamark, T.; Sjottem, E.; Johansen, T. The FMRpolyGlycine protein mediates aggregate formation and toxicity independent of the CGG mRNA hairpin in a cellular model for FXTAS. *Front. Genet.* **2019**, *10*, 249. [CrossRef]
51. Oh, S.Y.; He, F.; Krans, A.; Frazer, M.; Taylor, J.P.; Paulson, H.L.; Todd, P.K. RAN translation at CGG repeats induces ubiquitin proteasome system impairment in models of fragile X-associated tremor ataxia syndrome. *Hum. Mol. Genet.* **2015**, *24*, 4317–4326. [CrossRef]
52. Statello, L.; Guo, C.J.; Chen, L.L.; Huarte, M. Gene regulation by long non-coding RNAs and its biological functions. *Nat. Rev. Mol. Cell Biol.* **2021**, *22*, 96–118. [CrossRef] [PubMed]
53. Ladd, P.D.; Smith, L.E.; Rabaia, N.A.; Moore, J.M.; Georges, S.A.; Hansen, R.S.; Hagerman, R.J.; Tassone, F.; Tapscott, S.J.; Filippova, G.N. An antisense transcript spanning the CGG repeat region of FMR1 is upregulated in premutation carriers but silenced in full mutation individuals. *Hum. Mol. Genet.* **2007**, *16*, 3174–3187. [CrossRef] [PubMed]
54. Khalil, A.M.; Faghihi, M.A.; Modarresi, F.; Brothers, S.P.; Wahlestedt, C. A novel RNA transcript with antiapoptotic function is silenced in fragile X syndrome. *PLoS ONE* **2008**, *3*, e1486. [CrossRef] [PubMed]
55. Pastori, C.; Peschansky, V.J.; Barbouth, D.; Mehta, A.; Silva, J.P.; Wahlestedt, C. Comprehensive analysis of the transcriptional landscape of the human FMR1 gene reveals two new long noncoding RNAs differentially expressed in Fragile X syndrome and Fragile X-associated tremor/ataxia syndrome. *Hum. Genet.* **2014**, *133*, 59–67. [CrossRef]
56. Elizur, S.E.; Dratviman-Storobinsky, O.; Derech-Haim, S.; Lebovitz, O.; Dor, J.; Orvieto, R.; Cohen, Y. FMR6 may play a role in the pathogenesis of fragile X-associated premature ovarian insufficiency. *Gynecol. Endocrinol.* **2016**, *32*, 334–337. [CrossRef]
57. Peschansky, V.J.; Pastori, C.; Zeier, Z.; Wentzel, K.; Velmeshev, D.; Magistri, M.; Silva, J.P.; Wahlestedt, C. The long non-coding RNA FMR4 promotes proliferation of human neural precursor cells and epigenetic regulation of gene expression in trans. *Mol. Cell. Neurosci.* **2016**, *74*, 49–57. [CrossRef]
58. Peschansky, V.J.; Pastori, C.; Zeier, Z.; Motti, D.; Wentzel, K.; Velmeshev, D.; Magistri, M.; Bixby, J.L.; Lemmon, V.P.; Silva, J.P.; et al. Changes in expression of the long non-coding RNA FMR4 associate with altered gene expression during differentiation of human neural precursor cells. *Front. Genet.* **2015**, *6*, 263. [CrossRef]

59. Krans, A.; Kearse, M.G.; Todd, P.K. Repeat-associated non-AUG translation from antisense CCG repeats in fragile X tremor/ataxia syndrome. *Ann. Neurol.* **2016**, *80*, 871–881. [CrossRef]
60. Tassone, F.; Hagerman, R.J.; Taylor, A.K.; Gane, L.W.; Godfrey, T.E.; Hagerman, P.J. Elevated levels of FMR1 mRNA in carrier males: A new mechanism of involvement in the fragile-X syndrome. *Am. J. Hum. Genet.* **2000**, *66*, 6–15. [CrossRef] [PubMed]
61. Kenneson, A.; Zhang, F.; Hagedorn, C.H.; Warren, S.T. Reduced FMRP and increased FMR1 transcription is proportionally associated with CGG repeat number in intermediate-length and premutation carriers. *Hum. Mol. Genet.* **2001**, *10*, 1449–1454. [CrossRef]
62. Oostra, B.A.; Willemsen, R. A fragile balance: FMR1 expression levels. *Hum. Mol. Genet.* **2003**, *12*, R249–R257. [CrossRef]
63. Bagni, C.; Zukin, R.S. A synaptic perspective of fragile X syndrome and autism spectrum disorders. *Neuron* **2019**, *101*, 1070–1088. [CrossRef]
64. Schneider, A.; Winarni, T.I.; Cabal-Herrera, A.M.; Bacalman, S.; Gane, L.; Hagerman, P.; Tassone, F.; Hagerman, R. Elevated FMR1-mRNA and lowered FMRP—A double-hit mechanism for psychiatric features in men with FMR1 premutations. *Transl. Psychiatry* **2020**, *10*, 205. [CrossRef]
65. Storey, E.; Bui, M.Q.; Stimpson, P.; Tassone, F.; Atkinson, A.; Loesch, D.Z. Relationships between motor scores and cognitive functioning in FMR1 female premutation X carriers indicate early involvement of cerebello-cerebral pathways. *Cerebellum Ataxias* **2021**, *8*, 15. [CrossRef]
66. Greco, C.M.; Soontrapornchai, K.; Wirojanan, J.; Gould, J.E.; Hagerman, P.J.; Hagerman, R.J. Testicular and pituitary inclusion formation in fragile X associated tremor/ataxia syndrome. *J. Urol.* **2007**, *177*, 1434–1437. [CrossRef]
67. Lozano, R.; Summers, S.; Lozano, C.; Mu, Y.; Hessl, D.; Nguyen, D.; Tassone, F.; Hagerman, R. Association between macroorchidism and intelligence in FMR1 premutation carriers. *Am. J. Med. Genet. A* **2014**, *164A*, 2206–2211. [CrossRef] [PubMed]
68. Hwang, Y.T.; Aliaga, S.M.; Arpone, M.; Francis, D.; Li, X.; Chong, B.; Slater, H.R.; Rogers, C.; Bretherton, L.; Hunter, M.; et al. Partially methylated alleles, microdeletion, and tissue mosaicism in a fragile X male with tremor and ataxia at 30 years of age: A case report. *Am. J. Med. Genet. A* **2016**, *170*, 3327–3332. [CrossRef]
69. Bailey, D.B., Jr.; Raspa, M.; Olmsted, M.; Holiday, D.B. Co-occurring conditions associated with FMR1 gene variations: Findings from a national parent survey. *Am. J. Med. Genet. A* **2008**, *146A*, 2060–2069. [CrossRef]
70. Hagerman, R.; Hoem, G.; Hagerman, P. Fragile X and autism: Intertwined at the molecular level leading to targeted treatments. *Mol. Autism* **2010**, *1*, 12. [CrossRef] [PubMed]
71. Hagerman, R.; Au, J.; Hagerman, P. FMR1 premutation and full mutation molecular mechanisms related to autism. *J. Neurodev. Disord.* **2011**, *3*, 211–224. [CrossRef] [PubMed]
72. Hagerman, R.J.; Protic, D.; Rajaratnam, A.; Salcedo-Arellano, M.J.; Aydin, E.Y.; Schneider, A. Fragile X-Associated Neuropsychiatric Disorders (FXAND). *Front. Psychiatry* **2018**, *9*, 564. [CrossRef]
73. Schneider, A.; Johnston, C.; Tassone, F.; Sansone, S.; Hagerman, R.J.; Ferrer, E.; Rivera, S.M.; Hessl, D. Broad autism spectrum and obsessive-compulsive symptoms in adults with the fragile X premutation. *Clin. Neuropsychol.* **2016**, *30*, 929–943. [CrossRef]
74. Strimbu, K.; Tavel, J.A. What are biomarkers? *Curr. Opin. HIV AIDS* **2010**, *5*, 463–466. [CrossRef]
75. Alvarez-Mora, M.I.; Rodriguez-Revenga, L.; Madrigal, I.; Torres-Silva, F.; Mateu-Huertas, E.; Lizano, E.; Friedlander, M.R.; Marti, E.; Estivill, X.; Mila, M. MicroRNA expression profiling in blood from fragile X-associated tremor/ataxia syndrome patients. *Genes Brain Behav.* **2013**, *12*, 595–603. [CrossRef]
76. Mateu-Huertas, E.; Rodriguez-Revenga, L.; Alvarez-Mora, M.I.; Madrigal, I.; Willemsen, R.; Mila, M.; Marti, E.; Estivill, X. Blood expression profiles of fragile X premutation carriers identify candidate genes involved in neurodegenerative and infertility phenotypes. *Neurobiol. Dis.* **2014**, *65*, 43–54. [CrossRef]
77. Cao, Y.; Peng, Y.; Kong, H.E.; Allen, E.G.; Jin, P. Metabolic alterations in FMR1 premutation carriers. *Front. Mol. Biosci.* **2020**, *7*, 571092. [CrossRef] [PubMed]
78. Song, G.; Napoli, E.; Wong, S.; Hagerman, R.; Liu, S.; Tassone, F.; Giulivi, C. Altered redox mitochondrial biology in the neurodegenerative disorder fragile X-tremor/ataxia syndrome: Use of antioxidants in precision medicine. *Mol. Med.* **2016**, *22*, 548–559. [CrossRef] [PubMed]
79. Napoli, E.; McLennan, Y.A.; Schneider, A.; Tassone, F.; Hagerman, R.J.; Giulivi, C. Characterization of the metabolic, clinical and neuropsychological phenotype of female carriers of the premutation in the X-Linked FMR1 gene. *Front. Mol. Biosci.* **2020**, *7*, 578640. [CrossRef]
80. Allen, E.G.; Sullivan, A.K.; Marcus, M.; Small, C.; Dominguez, C.; Epstein, M.P.; Charen, K.; He, W.; Taylor, K.C.; Sherman, S.L. Examination of reproductive aging milestones among women who carry the FMR1 premutation. *Hum. Reprod.* **2007**, *22*, 2142–2152. [CrossRef] [PubMed]
81. Brouwer, J.R.; Huizer, K.; Severijnen, L.A.; Hukema, R.K.; Berman, R.F.; Oostra, B.A.; Willemsen, R. CGG-repeat length and neuropathological and molecular correlates in a mouse model for fragile X-associated tremor/ataxia syndrome. *J. Neurochem.* **2008**, *107*, 1671–1682. [CrossRef]
82. White, P.J.; Borts, R.H.; Hirst, M.C. Stability of the human fragile X (CGG)(n) triplet repeat array in *Saccharomyces cerevisiae* deficient in aspects of DNA metabolism. *Mol. Cell. Biol.* **1999**, *19*, 5675–5684. [CrossRef]
83. Kononenko, A.V.; Ebersole, T.; Vasquez, K.M.; Mirkin, S.M. Mechanisms of genetic instability caused by (CGG)n repeats in an experimental mammalian system. *Nat. Struct. Mol. Biol.* **2018**, *25*, 669–676. [CrossRef] [PubMed]
84. Jones, L.; Hughes, A. Pathogenic mechanisms in Huntington's disease. *Int. Rev. Neurobiol.* **2011**, *98*, 373–418. [CrossRef]

85. Savic Pavicevic, D.; Miladinovic, J.; Brkusanin, M.; Svikovic, S.; Djurica, S.; Brajuskovic, G.; Romac, S. Molecular genetics and genetic testing in myotonic dystrophy type 1. *Biomed. Res. Int.* **2013**, *2013*, 391821. [CrossRef]
86. Schmidt, M.H.M.; Pearson, C.E. Disease-associated repeat instability and mismatch repair. *DNA Repair* **2016**, *38*, 117–126. [CrossRef]
87. Lokanga, R.A.; Entezam, A.; Kumari, D.; Yudkin, D.; Qin, M.; Smith, C.B.; Usdin, K. Somatic expansion in mouse and human carriers of fragile X premutation alleles. *Hum. Mutat.* **2013**, *34*, 157–166. [CrossRef] [PubMed]
88. Tassone, F.; Hagerman, R.J.; Taylor, A.K.; Mills, J.B.; Harris, S.W.; Gane, L.W.; Hagerman, P.J. Clinical involvement and protein expression in individuals with the FMR1 premutation. *Am. J. Med. Genet.* **2000**, *91*, 144–152. [CrossRef]
89. Allen, E.G.; Sherman, S.; Abramowitz, A.; Leslie, M.; Novak, G.; Rusin, M.; Scott, E.; Letz, R. Examination of the effect of the polymorphic CGG repeat in the FMR1 gene on cognitive performance. *Behav. Genet.* **2005**, *35*, 435–445. [CrossRef] [PubMed]
90. Hessl, D.; Tassone, F.; Loesch, D.Z.; Berry-Kravis, E.; Leehey, M.A.; Gane, L.W.; Barbato, I.; Rice, C.; Gould, E.; Hall, D.A.; et al. Abnormal elevation of FMR1 mRNA is associated with psychological symptoms in individuals with the fragile X premutation. *Am. J. Med. Genet. B Neuropsychiatr. Genet.* **2005**, *139B*, 115–121. [CrossRef] [PubMed]
91. Hall, D.; Todorova-Koteva, K.; Pandya, S.; Bernard, B.; Ouyang, B.; Walsh, M.; Pounardjian, T.; Deburghraeve, C.; Zhou, L.; Losh, M.; et al. Neurological and endocrine phenotypes of fragile X carrier women. *Clin. Genet.* **2016**, *89*, 60–67. [CrossRef]
92. Moore, C.J.; Daly, E.M.; Tassone, F.; Tysoe, C.; Schmitz, N.; Ng, V.; Chitnis, X.; McGuire, P.; Suckling, J.; Davies, K.E.; et al. The effect of pre-mutation of X chromosome CGG trinucleotide repeats on brain anatomy. *Brain* **2004**, *127*, 2672–2681. [CrossRef] [PubMed]
93. Koldewyn, K.; Hessl, D.; Adams, J.; Tassone, F.; Hagerman, P.J.; Hagerman, R.J.; Rivera, S.M. Reduced hippocampal activation during recall is associated with elevated FMR1 mRNA and psychiatric symptoms in men with the fragile X premutation. *Brain Imaging Behav.* **2008**, *2*, 105–116. [CrossRef]
94. Hashimoto, R.; Backer, K.C.; Tassone, F.; Hagerman, R.J.; Rivera, S.M. An fMRI study of the prefrontal activity during the performance of a working memory task in premutation carriers of the fragile X mental retardation 1 gene with and without fragile X-associated tremor/ataxia syndrome (FXTAS). *J. Psychiatr. Res.* **2011**, *45*, 36–43. [CrossRef]
95. Wang, J.M.; Koldewyn, K.; Hashimoto, R.; Schneider, A.; Le, L.; Tassone, F.; Cheung, K.; Hagerman, P.; Hessl, D.; Rivera, S.M. Male carriers of the FMR1 premutation show altered hippocampal-prefrontal function during memory encoding. *Front. Hum. Neurosci.* **2012**, *6*, 297. [CrossRef] [PubMed]
96. Kim, S.Y.; Hashimoto, R.; Tassone, F.; Simon, T.J.; Rivera, S.M. Altered neural activity of magnitude estimation processing in adults with the fragile X premutation. *J. Psychiatr. Res.* **2013**, *47*, 1909–1916. [CrossRef]
97. Wang, J.Y.; Hessl, D.; Schneider, A.; Tassone, F.; Hagerman, R.J.; Rivera, S.M. Fragile X-associated tremor/ataxia syndrome: Influence of the FMR1 gene on motor fiber tracts in males with normal and premutation alleles. *JAMA Neurol.* **2013**, *70*, 1022–1029. [CrossRef] [PubMed]
98. Hocking, D.R.; Birch, R.C.; Bui, Q.M.; Menant, J.C.; Lord, S.R.; Georgiou-Karistianis, N.; Godler, D.E.; Wen, W.; Hackett, A.; Rogers, C.; et al. Cerebellar volume mediates the relationship between FMR1 mRNA levels and voluntary step initiation in males with the premutation. *Neurobiol. Aging* **2017**, *50*, 5–12. [CrossRef]
99. Brown, S.S.G.; Basu, S.; Whalley, H.C.; Kind, P.C.; Stanfield, A.C. Age-related functional brain changes in FMR1 premutation carriers. *Neuroimage Clin.* **2018**, *17*, 761–767. [CrossRef] [PubMed]
100. Loesch, D.Z.; Godler, D.E.; Evans, A.; Bui, Q.M.; Gehling, F.; Kotschet, K.E.; Trost, N.; Storey, E.; Stimpson, P.; Kinsella, G.; et al. Evidence for the toxicity of bidirectional transcripts and mitochondrial dysfunction in blood associated with small CGG expansions in the FMR1 gene in patients with parkinsonism. *Genet. Med.* **2011**, *13*, 392–399. [CrossRef] [PubMed]
101. Vittal, P.; Pandya, S.; Sharp, K.; Berry-Kravis, E.; Zhou, L.; Ouyang, B.; Jackson, J.; Hall, D.A. ASFMR1 splice variant: A predictor of fragile X-associated tremor/ataxia syndrome. *Neurol. Genet.* **2018**, *4*, e246. [CrossRef] [PubMed]
102. Al Olaby, R.R.; Tang, H.T.; Durbin-Johnson, B.; Schneider, A.; Hessl, D.; Rivera, S.M.; Tassone, F. Assessment of molecular measures in non-FXTAS male premutation carriers. *Front. Genet.* **2018**, *9*, 302. [CrossRef]
103. Zafarullah, M.; Tang, H.T.; Durbin-Johnson, B.; Fourie, E.; Hessl, D.; Rivera, S.M.; Tassone, F. FMR1 locus isoforms: Potential biomarker candidates in fragile X-associated tremor/ataxia syndrome (FXTAS). *Sci. Rep.* **2020**, *10*, 11099. [CrossRef]
104. Pretto, D.I.; Eid, J.S.; Yrigollen, C.M.; Tang, H.T.; Loomis, E.W.; Raske, C.; Durbin-Johnson, B.; Hagerman, P.J.; Tassone, F. Differential increases of specific FMR1 mRNA isoforms in premutation carriers. *J. Med. Genet.* **2015**, *52*, 42–52. [CrossRef] [PubMed]
105. Tseng, E.; Tang, H.T.; AlOlaby, R.R.; Hickey, L.; Tassone, F. Altered expression of the FMR1 splicing variants landscape in premutation carriers. *Biochim. Biophys. Acta Gene Regul. Mech.* **2017**, *1860*, 1117–1126. [CrossRef] [PubMed]
106. Vasilyev, N.; Polonskaia, A.; Darnell, J.C.; Darnell, R.B.; Patel, D.J.; Serganov, A. Crystal structure reveals specific recognition of a G-quadruplex RNA by a beta-turn in the RGG motif of FMRP. *Proc. Natl. Acad. Sci. USA* **2015**, *112*, E5391–E5400. [CrossRef] [PubMed]
107. Yrigollen, C.M.; Durbin-Johnson, B.; Gane, L.; Nelson, D.L.; Hagerman, R.; Hagerman, P.J.; Tassone, F. AGG interruptions within the maternal FMR1 gene reduce the risk of offspring with fragile X syndrome. *Genet. Med.* **2012**, *14*, 729–736. [CrossRef]
108. Cornish, K.M.; Kraan, C.M.; Bui, Q.M.; Bellgrove, M.A.; Metcalfe, S.A.; Trollor, J.N.; Hocking, D.R.; Slater, H.R.; Inaba, Y.; Li, X.; et al. Novel methylation markers of the dysexecutive-psychiatric phenotype in FMR1 premutation women. *Neurology* **2015**, *84*, 1631–1638. [CrossRef]

109. Shelton, A.L.; Cornish, K.M.; Kolbe, S.; Clough, M.; Slater, H.R.; Li, X.; Kraan, C.M.; Bui, Q.M.; Godler, D.E.; Fielding, J. Brain structure and intragenic DNA methylation are correlated, and predict executive dysfunction in fragile X premutation females. *Transl. Psychiatry* **2016**, *6*, e984. [CrossRef]
110. Amer Abed, F.; Ezzat Maroof, R.; Al-Nakkash, U.M.A. Comparing the diagnostic accuracy of anti-mullerian hormone and follicle stimulating hormone in detecting premature ovarian failure in Iraqi women by ROC analysis. *Rep. Biochem. Mol. Biol.* **2019**, *8*, 126–131.
111. Lee, D.H.; Pei, C.Z.; Song, J.Y.; Lee, K.J.; Yun, B.S.; Kwack, K.B.; Lee, E.I.; Baek, K.H. Identification of serum biomarkers for premature ovarian failure. *Biochim. Biophys. Acta Proteins Proteom.* **2019**, *1867*, 219–226. [CrossRef]
112. Welt, C.K.; Smith, P.C.; Taylor, A.E. Evidence of early ovarian aging in fragile X premutation carriers. *J. Clin. Endocrinol. Metab.* **2004**, *89*, 4569–4574. [CrossRef] [PubMed]
113. Murray, A.; Webb, J.; MacSwiney, F.; Shipley, E.L.; Morton, N.E.; Conway, G.S. Serum concentrations of follicle stimulating hormone may predict premature ovarian failure in FRAXA premutation women. *Hum. Reprod.* **1999**, *14*, 1217–1218. [CrossRef] [PubMed]
114. Hundscheid, R.D.; Braat, D.D.; Kiemeney, L.A.; Smits, A.P.; Thomas, C.M. Increased serum FSH in female fragile X premutation carriers with either regular menstrual cycles or on oral contraceptives. *Hum. Reprod.* **2001**, *16*, 457–462. [CrossRef]
115. Barasoain, M.; Barrenetxea, G.; Huerta, I.; Telez, M.; Carrillo, A.; Perez, C.; Criado, B.; Arrieta, I. Study of FMR1 gene association with ovarian dysfunction in a sample from the Basque Country. *Gene* **2013**, *521*, 145–149. [CrossRef]
116. Hoffman, G.E.; Le, W.W.; Entezam, A.; Otsuka, N.; Tong, Z.B.; Nelson, L.; Flaws, J.A.; McDonald, J.H.; Jafar, S.; Usdin, K. Ovarian abnormalities in a mouse model of fragile X primary ovarian insufficiency. *J. Histochem. Cytochem.* **2012**, *60*, 439–456. [CrossRef]
117. Lu, C.; Lin, L.; Tan, H.; Wu, H.; Sherman, S.L.; Gao, F.; Jin, P.; Chen, D. Fragile X premutation RNA is sufficient to cause primary ovarian insufficiency in mice. *Hum. Mol. Genet.* **2012**, *21*, 5039–5047. [CrossRef]
118. Rohr, J.; Allen, E.G.; Charen, K.; Giles, J.; He, W.; Dominguez, C.; Sherman, S.L. Anti-Mullerian hormone indicates early ovarian decline in fragile X mental retardation (FMR1) premutation carriers: A preliminary study. *Hum. Reprod.* **2008**, *23*, 1220–1225. [CrossRef]
119. Fan, H.Y.; Liu, Z.; Cahill, N.; Richards, J.S. Targeted disruption of Pten in ovarian granulosa cells enhances ovulation and extends the life span of luteal cells. *Mol. Endocrinol.* **2008**, *22*, 2128–2140. [CrossRef]
120. Makker, A.; Goel, M.M.; Mahdi, A.A. PI3K/PTEN/Akt and TSC/mTOR signaling pathways, ovarian dysfunction, and infertility: An update. *J. Mol. Endocrinol.* **2014**, *53*, R103–R118. [CrossRef]
121. Rose, B.I.; Brown, S.E. An explanation of the mechanisms underlying fragile X-associated premature ovarian insufficiency. *J. Assist. Reprod. Genet.* **2020**, *37*, 1313–1322. [CrossRef] [PubMed]
122. Louis, E.; Moskowitz, C.; Friez, M.; Amaya, M.; Vonsattel, J.P. Parkinsonism, dysautonomia, and intranuclear inclusions in a fragile X carrier: A clinical-pathological study. *Mov. Disord.* **2006**, *21*, 420–425. [CrossRef]
123. Brouwer, J.R.; Severijnen, E.; de Jong, F.H.; Hessl, D.; Hagerman, R.J.; Oostra, B.A.; Willemsen, R. Altered hypothalamus-pituitary-adrenal gland axis regulation in the expanded CGG-repeat mouse model for fragile X-associated tremor/ataxia syndrome. *Psychoneuroendocrinology* **2008**, *33*, 863–873. [CrossRef] [PubMed]
124. Spath, M.A.; Feuth, T.B.; Allen, E.G.; Smits, A.P.; Yntema, H.G.; van Kessel, A.G.; Braat, D.D.; Sherman, S.L.; Thomas, C.M. Intra-individual stability over time of standardized anti-Mullerian hormone in FMR1 premutation carriers. *Hum. Reprod.* **2011**, *26*, 2185–2191. [CrossRef] [PubMed]
125. Creus, M.; Penarrubia, J.; Fabregues, F.; Vidal, E.; Carmona, F.; Casamitjana, R.; Vanrell, J.A.; Balasch, J. Day 3 serum inhibin B and FSH and age as predictors of assisted reproduction treatment outcome. *Hum. Reprod.* **2000**, *15*, 2341–2346. [CrossRef]
126. Maslow, B.S.; Davis, S.; Engmann, L.; Nulsen, J.C.; Benadiva, C.A. Correlation of normal-range FMR1 repeat length or genotypes and reproductive parameters. *J. Assist. Reprod. Genet.* **2016**, *33*, 1149–1155. [CrossRef] [PubMed]
127. Yang, Y.; Bazhin, A.V.; Werner, J.; Karakhanova, S. Reactive oxygen species in the immune system. *Int. Rev. Immunol.* **2013**, *32*, 249–270. [CrossRef] [PubMed]
128. Terzi, A.; Suter, D.M. The role of NADPH oxidases in neuronal development. *Free Radic. Biol. Med.* **2020**, *154*, 33–47. [CrossRef]
129. Yun, H.R.; Jo, Y.H.; Kim, J.; Shin, Y.; Kim, S.S.; Choi, T.G. Roles of autophagy in oxidative stress. *Int. J. Mol. Sci.* **2020**, *21*, 3289. [CrossRef]
130. Srinivas, U.S.; Tan, B.W.Q.; Vellayappan, B.A.; Jeyasekharan, A.D. ROS and the DNA damage response in cancer. *Redox Biol.* **2019**, *25*, 101084. [CrossRef]
131. Singh, A.; Kukreti, R.; Saso, L.; Kukreti, S. Oxidative stress: A key modulator in neurodegenerative diseases. *Molecules* **2019**, *24*, 1583. [CrossRef]
132. Panth, N.; Paudel, K.R.; Parajuli, K. Reactive oxygen species: A key hallmark of cardiovascular disease. *Adv. Med.* **2016**, *2016*, 9152732. [CrossRef] [PubMed]
133. El Bekay, R.; Romero-Zerbo, Y.; Decara, J.; Sanchez-Salido, L.; Del Arco-Herrera, I.; Rodriguez-de Fonseca, F.; de Diego-Otero, Y. Enhanced markers of oxidative stress, altered antioxidants and NADPH-oxidase activation in brains from Fragile X mental retardation 1-deficient mice, a pathological model for Fragile X syndrome. *Eur. J. Neurosci.* **2007**, *26*, 3169–3180. [CrossRef] [PubMed]
134. Kaplan, E.S.; Cao, Z.; Hulsizer, S.; Tassone, F.; Berman, R.F.; Hagerman, P.J.; Pessah, I.N. Early mitochondrial abnormalities in hippocampal neurons cultured from Fmr1 pre-mutation mouse model. *J. Neurochem.* **2012**, *123*, 613–621. [CrossRef] [PubMed]

135. Napoli, E.; Ross-Inta, C.; Wong, S.; Omanska-Klusek, A.; Barrow, C.; Iwahashi, C.; Garcia-Arocena, D.; Sakaguchi, D.; Berry-Kravis, E.; Hagerman, R.; et al. Altered zinc transport disrupts mitochondrial protein processing/import in fragile X-associated tremor/ataxia syndrome. *Hum. Mol. Genet.* **2011**, *20*, 3079–3092. [CrossRef] [PubMed]
136. Napoli, E.; Song, G.; Wong, S.; Hagerman, R.; Giulivi, C. Altered bioenergetics in primary dermal fibroblasts from adult carriers of the FMR1 premutation before the onset of the neurodegenerative disease fragile X-associated tremor/ataxia syndrome. *Cerebellum* **2016**, *15*, 552–564. [CrossRef] [PubMed]
137. Ross-Inta, C.; Omanska-Klusek, A.; Wong, S.; Barrow, C.; Garcia-Arocena, D.; Iwahashi, C.; Berry-Kravis, E.; Hagerman, R.J.; Hagerman, P.J.; Giulivi, C. Evidence of mitochondrial dysfunction in fragile X-associated tremor/ataxia syndrome. *Biochem. J.* **2010**, *429*, 545–552. [CrossRef]
138. Alvarez-Mora, M.I.; Rodriguez-Revenga, L.; Madrigal, I.; Guitart-Mampel, M.; Garrabou, G.; Mila, M. Impaired mitochondrial function and dynamics in the pathogenesis of FXTAS. *Mol. Neurobiol.* **2017**, *54*, 6896–6902. [CrossRef]
139. Nobile, V.; Palumbo, F.; Lanni, S.; Ghisio, V.; Vitali, A.; Castagnola, M.; Marzano, V.; Maulucci, G.; De Angelis, C.; De Spirito, M.; et al. Altered mitochondrial function in cells carrying a premutation or unmethylated full mutation of the FMR1 gene. *Hum. Genet.* **2020**, *139*, 227–245. [CrossRef] [PubMed]
140. Loesch, D.Z.; Annesley, S.J.; Trost, N.; Bui, M.Q.; Lay, S.T.; Storey, E.; De Piazza, S.W.; Sanislav, O.; Francione, L.M.; Hammersley, E.M.; et al. Novel blood biomarkers are associated with white matter lesions in fragile X- associated tremor/ataxia syndrome. *Neurodegener. Dis.* **2017**, *17*, 22–30. [CrossRef]
141. Napoli, E.; Schneider, A.; Hagerman, R.; Song, G.; Wong, S.; Tassone, F.; Giulivi, C. Impact of FMR1 premutation on neurobehavior and bioenergetics in young monozygotic twins. *Front. Genet.* **2018**, *9*, 338. [CrossRef] [PubMed]
142. Alfatni, A.; Riou, M.; Charles, A.L.; Meyer, A.; Barnig, C.; Andres, E.; Lejay, A.; Talha, S.; Geny, B. Peripheral blood mononuclear cells and platelets mitochondrial dysfunction, oxidative stress, and circulating mtDNA in cardiovascular diseases. *J. Clin. Med.* **2020**, *9*, 311. [CrossRef]
143. Afrifa, J.; Zhao, T.; Yu, J. Circulating mitochondria DNA, a non-invasive cancer diagnostic biomarker candidate. *Mitochondrion* **2019**, *47*, 238–243. [CrossRef]
144. Gambardella, S.; Limanaqi, F.; Ferese, R.; Biagioni, F.; Campopiano, R.; Centonze, D.; Fornai, F. ccf-mtDNA as a potential link between the brain and immune system in neuro-immunological disorders. *Front. Immunol.* **2019**, *10*, 1064. [CrossRef]
145. Lebedeva, M.A.; Eaton, J.S.; Shadel, G.S. Loss of p53 causes mitochondrial DNA depletion and altered mitochondrial reactive oxygen species homeostasis. *Biochim. Biophys. Acta* **2009**, *1787*, 328–334. [CrossRef] [PubMed]
146. Chen, J.; Zhang, L.; Yu, X.; Zhou, H.; Luo, Y.; Wang, W.; Wang, L. Clinical application of plasma mitochondrial DNA content in patients with lung cancer. *Oncol. Lett.* **2018**, *16*, 7074–7081. [CrossRef]
147. Alvarez-Mora, M.I.; Podlesniy, P.; Gelpi, E.; Hukema, R.; Madrigal, I.; Pagonabarraga, J.; Trullas, R.; Mila, M.; Rodriguez-Revenga, L. Fragile X-associated tremor/ataxia syndrome: Regional decrease of mitochondrial DNA copy number relates to clinical manifestations. *Genes Brain Behav.* **2019**, *18*, e12565. [CrossRef] [PubMed]
148. Kala, M.; Shaikh, M.V.; Nivsarkar, M. Equilibrium between anti-oxidants and reactive oxygen species: A requisite for oocyte development and maturation. *Reprod. Med. Biol.* **2017**, *16*, 28–35. [CrossRef] [PubMed]
149. Conca Dioguardi, C.; Uslu, B.; Haynes, M.; Kurus, M.; Gul, M.; Miao, D.Q.; De Santis, L.; Ferrari, M.; Bellone, S.; Santin, A.; et al. Granulosa cell and oocyte mitochondrial abnormalities in a mouse model of fragile X primary ovarian insufficiency. *Mol. Hum. Reprod.* **2016**, *22*, 384–396. [CrossRef] [PubMed]
150. Gohel, D.; Sripada, L.; Prajapati, P.; Singh, K.; Roy, M.; Kotadia, D.; Tassone, F.; Charlet-Berguerand, N.; Singh, R. FMRpolyG alters mitochondrial transcripts level and respiratory chain complex assembly in Fragile X associated tremor/ataxia syndrome [FXTAS]. *Biochim. Biophys. Acta Mol. Basis Dis.* **2019**, *1865*, 1379–1388. [CrossRef] [PubMed]
151. Gohel, D.; Berguerand, N.C.; Tassone, F.; Singh, R. The emerging molecular mechanisms for mitochondrial dysfunctions in FXTAS. *Biochim. Biophys. Acta Mol. Basis Dis.* **2020**, *1866*, 165918. [CrossRef]
152. Gohel, D.; Sripada, L.; Prajapati, P.; Currim, F.; Roy, M.; Singh, K.; Shinde, A.; Mane, M.; Kotadia, D.; Tassone, F.; et al. Expression of expanded FMR1-CGG repeats alters mitochondrial miRNAs and modulates mitochondrial functions and cell death in cellular model of FXTAS. *Free Radic. Biol. Med.* **2021**, *165*, 100–110. [CrossRef] [PubMed]
153. Hukema, R.K.; Buijsen, R.A.; Raske, C.; Severijnen, L.A.; Nieuwenhuizen-Bakker, I.; Minneboo, M.; Maas, A.; de Crom, R.; Kros, J.M.; Hagerman, P.J.; et al. Induced expression of expanded CGG RNA causes mitochondrial dysfunction in vivo. *Cell Cycle* **2014**, *13*, 2600–2608. [CrossRef] [PubMed]
154. Giulivi, C.; Napoli, E.; Tassone, F.; Halmai, J.; Hagerman, R. Plasma biomarkers for monitoring brain pathophysiology in FMR1 premutation carriers. *Front. Mol. Neurosci.* **2016**, *9*, 71. [CrossRef] [PubMed]
155. Giulivi, C.; Napoli, E.; Tassone, F.; Halmai, J.; Hagerman, R. Plasma metabolic profile delineates roles for neurodegeneration, pro-inflammatory damage and mitochondrial dysfunction in the FMR1 premutation. *Biochem. J.* **2016**, *473*, 3871–3888. [CrossRef]
156. Suzuki, Y.J.; Carini, M.; Butterfield, D.A. Protein carbonylation. *Antioxid. Redox Signal.* **2010**, *12*, 323–325. [CrossRef]
157. Marrocco, I.; Altieri, F.; Peluso, I. Measurement and clinical significance of biomarkers of oxidative stress in humans. *Oxid. Med. Cell. Longev.* **2017**, *2017*, 6501046. [CrossRef]
158. Moselhy, H.F.; Reid, R.G.; Yousef, S.; Boyle, S.P. A specific, accurate, and sensitive measure of total plasma malondialdehyde by HPLC. *J. Lipid Res.* **2013**, *54*, 852–858. [CrossRef] [PubMed]

159. Martinez Cerdeno, V.; Hong, T.; Amina, S.; Lechpammer, M.; Ariza, J.; Tassone, F.; Noctor, S.C.; Hagerman, P.; Hagerman, R. Microglial cell activation and senescence are characteristic of the pathology FXTAS. *Mov. Disord.* **2018**, *33*, 1887–1894. [CrossRef]
160. Dufour, B.D.; Amina, S.; Martinez-Cerdeno, V. FXTAS presents with upregulation of the cytokines IL12 and TNFalpha. *Parkinsonism Relat. Disord.* **2021**, *82*, 117–120. [CrossRef]
161. Porro, C.; Cianciulli, A.; Panaro, M.A. The regulatory role of IL-10 in neurodegenerative diseases. *Biomolecules* **2020**, *10*, 1017. [CrossRef]
162. Marek, D.; Papin, S.; Ellefsen, K.; Niederhauser, J.; Isidor, N.; Ransijn, A.; Poupon, L.; Spertini, F.; Pantaleo, G.; Bergmann, S.; et al. Carriers of the fragile X mental retardation 1 (FMR1) premutation allele present with increased levels of cytokine IL-10. *J. Neuroinflammation* **2012**, *9*, 238. [CrossRef] [PubMed]
163. Michlewski, G.; Caceres, J.F. Post-transcriptional control of miRNA biogenesis. *RNA* **2019**, *25*, 1–16. [CrossRef] [PubMed]
164. Salloum-Asfar, S.; Satheesh, N.J.; Abdulla, S.A. Circulating miRNAs, small but promising biomarkers for autism spectrum disorder. *Front. Mol. Neurosci.* **2019**, *12*, 253. [CrossRef] [PubMed]
165. Romano, G.L.; Platania, C.B.M.; Drago, F.; Salomone, S.; Ragusa, M.; Barbagallo, C.; Di Pietro, C.; Purrello, M.; Reibaldi, M.; Avitabile, T.; et al. Retinal and circulating miRNAs in age-related macular degeneration: An in vivo animal and human study. *Front. Pharmacol.* **2017**, *8*, 168. [CrossRef]
166. Low, Y.H.; Asi, Y.; Foti, S.C.; Lashley, T. Heterogeneous nuclear ribonucleoproteins: Implications in neurological diseases. *Mol. Neurobiol.* **2021**, *58*, 631–646. [CrossRef] [PubMed]
167. Tan, H.; Poidevin, M.; Li, H.; Chen, D.; Jin, P. MicroRNA-277 modulates the neurodegeneration caused by Fragile X premutation rCGG repeats. *PLoS Genet.* **2012**, *8*, e1002681. [CrossRef] [PubMed]
168. Schermelleh, L.; Carlton, P.M.; Haase, S.; Shao, L.; Winoto, L.; Kner, P.; Burke, B.; Cardoso, M.C.; Agard, D.A.; Gustafsson, M.G.; et al. Subdiffraction multicolor imaging of the nuclear periphery with 3D structured illumination microscopy. *Science* **2008**, *320*, 1332–1336. [CrossRef]
169. Smith, C.L.; Poleshko, A.; Epstein, J.A. The nuclear periphery is a scaffold for tissue-specific enhancers. *Nucleic Acids Res.* **2021**, *49*, 6181–6195. [CrossRef]
170. Garcia-Arocena, D.; Yang, J.E.; Brouwer, J.R.; Tassone, F.; Iwahashi, C.; Berry-Kravis, E.M.; Goetz, C.G.; Sumis, A.M.; Zhou, L.; Nguyen, D.V.; et al. Fibroblast phenotype in male carriers of FMR1 premutation alleles. *Hum. Mol. Genet.* **2010**, *19*, 299–312. [CrossRef]
171. Hoem, G.; Raske, C.R.; Garcia-Arocena, D.; Tassone, F.; Sanchez, E.; Ludwig, A.L.; Iwahashi, C.K.; Kumar, M.; Yang, J.E.; Hagerman, P.J. CGG-repeat length threshold for FMR1 RNA pathogenesis in a cellular model for FXTAS. *Hum. Mol. Genet.* **2011**, *20*, 2161–2170. [CrossRef]
172. Arocena, D.G.; Iwahashi, C.K.; Won, N.; Beilina, A.; Ludwig, A.L.; Tassone, F.; Schwartz, P.H.; Hagerman, P.J. Induction of inclusion formation and disruption of lamin A/C structure by premutation CGG-repeat RNA in human cultured neural cells. *Hum. Mol. Genet.* **2005**, *14*, 3661–3671. [CrossRef] [PubMed]
173. Ito, S.; Shen, L.; Dai, Q.; Wu, S.C.; Collins, L.B.; Swenberg, J.A.; He, C.; Zhang, Y. Tet proteins can convert 5-methylcytosine to 5-formylcytosine and 5-carboxylcytosine. *Science* **2011**, *333*, 1300–1303. [CrossRef] [PubMed]
174. Bachman, M.; Uribe-Lewis, S.; Yang, X.; Williams, M.; Murrell, A.; Balasubramanian, S. 5-Hydroxymethylcytosine is a predominantly stable DNA modification. *Nat. Chem.* **2014**, *6*, 1049–1055. [CrossRef] [PubMed]
175. Yao, B.; Lin, L.; Street, R.C.; Zalewski, Z.A.; Galloway, J.N.; Wu, H.; Nelson, D.L.; Jin, P. Genome-wide alteration of 5-hydroxymethylcytosine in a mouse model of fragile X-associated tremor/ataxia syndrome. *Hum. Mol. Genet.* **2014**, *23*, 1095–1107. [CrossRef] [PubMed]
176. Andrade-Navarro, M.A.; Muhlenberg, K.; Spruth, E.J.; Mah, N.; Gonzalez-Lopez, A.; Andreani, T.; Russ, J.; Huska, M.R.; Muro, E.M.; Fontaine, J.F.; et al. RNA Sequencing of human peripheral blood cells indicates upregulation of immune-related genes in Huntington's disease. *Front. Neurol.* **2020**, *11*, 573560. [CrossRef] [PubMed]
177. Heger, S.; Mastronardi, C.; Dissen, G.A.; Lomniczi, A.; Cabrera, R.; Roth, C.L.; Jung, H.; Galimi, F.; Sippell, W.; Ojeda, S.R. Enhanced at puberty 1 (EAP1) is a new transcriptional regulator of the female neuroendocrine reproductive axis. *J. Clin. Investig.* **2007**, *117*, 2145–2154. [CrossRef]
178. Marcogliese, P.C.; Shashi, V.; Spillmann, R.C.; Stong, N.; Rosenfeld, J.A.; Koenig, M.K.; Martinez-Agosto, J.A.; Herzog, M.; Chen, A.H.; Dickson, P.I.; et al. IRF2BPL is associated with neurological phenotypes. *Am. J. Hum. Genet.* **2018**, *103*, 245–260. [CrossRef]
179. Tran Mau-Them, F.; Guibaud, L.; Duplomb, L.; Keren, B.; Lindstrom, K.; Marey, I.; Mochel, F.; van den Boogaard, M.J.; Oegema, R.; Nava, C.; et al. De novo truncating variants in the intronless IRF2BPL are responsible for developmental epileptic encephalopathy. *Genet. Med.* **2019**, *21*, 1008–1014. [CrossRef]
180. Alvarez-Mora, M.I.; Rodriguez-Revenga, L.; Madrigal, I.; Garcia-Garcia, F.; Duran, M.; Dopazo, J.; Estivill, X.; Mila, M. Deregulation of key signaling pathways involved in oocyte maturation in FMR1 premutation carriers with Fragile X-associated primary ovarian insufficiency. *Gene* **2015**, *571*, 52–57. [CrossRef]
181. Schur, R.R.; Draisma, L.W.; Wijnen, J.P.; Boks, M.P.; Koevoets, M.G.; Joels, M.; Klomp, D.W.; Kahn, R.S.; Vinkers, C.H. Brain GABA levels across psychiatric disorders: A systematic literature review and meta-analysis of (1) H-MRS studies. *Hum. Brain Mapp.* **2016**, *37*, 3337–3352. [CrossRef]
182. Chiapponi, C.; Piras, F.; Caltagirone, C.; Spalletta, G. GABA system in schizophrenia and mood disorders: A mini review on third-generation imaging studies. *Front. Psychiatry* **2016**, *7*, 61. [CrossRef]

183. Prevot, T.; Sibille, E. Altered GABA-mediated information processing and cognitive dysfunctions in depression and other brain disorders. *Mol. Psychiatry* **2021**, *26*, 151–167. [CrossRef] [PubMed]
184. Conde, V.; Palomar, F.J.; Lama, M.J.; Martinez, R.; Carrillo, F.; Pintado, E.; Mir, P. Abnormal GABA-mediated and cerebellar inhibition in women with the fragile X premutation. *J. Neurophysiol.* **2013**, *109*, 1315–1322. [CrossRef]
185. Gantois, I.; Vandesompele, J.; Speleman, F.; Reyniers, E.; D'Hooge, R.; Severijnen, L.A.; Willemsen, R.; Tassone, F.; Kooy, R.F. Expression profiling suggests underexpression of the GABA(A) receptor subunit delta in the fragile X knockout mouse model. *Neurobiol. Dis.* **2006**, *21*, 346–357. [CrossRef] [PubMed]
186. D'Hulst, C.; De Geest, N.; Reeve, S.P.; Van Dam, D.; De Deyn, P.P.; Hassan, B.A.; Kooy, R.F. Decreased expression of the GABAA receptor in fragile X syndrome. *Brain Res.* **2006**, *1121*, 238–245. [CrossRef]
187. D'Hulst, C.; Heulens, I.; Brouwer, J.R.; Willemsen, R.; De Geest, N.; Reeve, S.P.; De Deyn, P.P.; Hassan, B.A.; Kooy, R.F. Expression of the GABAergic system in animal models for fragile X syndrome and fragile X associated tremor/ataxia syndrome (FXTAS). *Brain Res.* **2009**, *1253*, 176–183. [CrossRef] [PubMed]
188. Cao, Z.; Hulsizer, S.; Tassone, F.; Tang, H.T.; Hagerman, R.J.; Rogawski, M.A.; Hagerman, P.J.; Pessah, I.N. Clustered burst firing in FMR1 premutation hippocampal neurons: Amelioration with allopregnanolone. *Hum. Mol. Genet.* **2012**, *21*, 2923–2935. [CrossRef]
189. Coyne, L.; Lees, G.; Nicholson, R.A.; Zheng, J.; Neufield, K.D. The sleep hormone oleamide modulates inhibitory ionotropic receptors in mammalian CNS in vitro. *Br. J. Pharmacol.* **2002**, *135*, 1977–1987. [CrossRef]
190. Akanmu, M.A.; Adeosun, S.O.; Ilesanmi, O.R. Neuropharmacological effects of oleamide in male and female mice. *Behav. Brain Res.* **2007**, *182*, 88–94. [CrossRef]
191. Wang, J.Y.; Trivedi, A.M.; Carrillo, N.R.; Yang, J.; Schneider, A.; Giulivi, C.; Adams, P.; Tassone, F.; Kim, K.; Rivera, S.M.; et al. Open-label allopregnanolone treatment of men with fragile X-associated tremor/ataxia syndrome. *Neurotherapeutics* **2017**, *14*, 1073–1083. [CrossRef] [PubMed]
192. Napoli, E.; Schneider, A.; Wang, J.Y.; Trivedi, A.; Carrillo, N.R.; Tassone, F.; Rogawski, M.; Hagerman, R.J.; Giulivi, C. Allopregnanolone treatment improves plasma metabolomic profile associated with GABA metabolism in fragile X-associated tremor/ataxia syndrome: A pilot study. *Mol. Neurobiol.* **2019**, *56*, 3702–3713. [CrossRef]
193. Blackburn, E.H.; Epel, E.S.; Lin, J. Human telomere biology: A contributory and interactive factor in aging, disease risks, and protection. *Science* **2015**, *350*, 1193–1198. [CrossRef] [PubMed]
194. Hayashi, M.T. Telomere biology in aging and cancer: Early history and perspectives. *Genes Genet. Syst.* **2018**, *92*, 107–118. [CrossRef]
195. Martinez, P.; Blasco, M.A. Heart-breaking telomeres. *Circ. Res.* **2018**, *123*, 787–802. [CrossRef]
196. Chen, M.S.; Lee, R.T.; Garbern, J.C. Senescence mechanisms and targets in the heart. *Cardiovasc. Res.* **2021**, in press. [CrossRef]
197. Vodicka, P.; Andera, L.; Opattova, A.; Vodickova, L. The interactions of DNA repair, telomere homeostasis, and p53 mutational status in solid cancers: Risk, prognosis, and prediction. *Cancers* **2021**, *13*, 479. [CrossRef] [PubMed]
198. Vaiserman, A.; Krasnienkov, D. Telomere length as a marker of biological age: State-of-the-art, open issues, and future perspectives. *Front. Genet.* **2020**, *11*, 630186. [CrossRef]
199. Cai, Z.; Yan, L.J.; Ratka, A. Telomere shortening and Alzheimer's disease. *Neuromol. Med.* **2013**, *15*, 25–48. [CrossRef] [PubMed]
200. Stock, C.J.W.; Renzoni, E.A. Telomeres in interstitial lung disease. *J. Clin. Med.* **2021**, *10*, 1384. [CrossRef]
201. Jenkins, E.C.; Tassone, F.; Ye, L.; Gu, H.; Xi, M.; Velinov, M.; Brown, W.T.; Hagerman, R.J.; Hagerman, P.J. Reduced telomere length in older men with premutation alleles of the fragile X mental retardation 1 gene. *Am. J. Med. Genet. A* **2008**, *146A*, 1543–1546. [CrossRef]
202. Jenkins, E.C.; Tassone, F.; Ye, L.; Hoogeveen, A.T.; Brown, W.T.; Hagerman, R.J.; Hagerman, P.J. Reduced telomere length in individuals with FMR1 premutations and full mutations. *Am. J. Med. Genet. A* **2012**, *158A*, 1060–1065. [CrossRef]
203. Albizua, I.; Chopra, P.; Allen, E.G.; He, W.; Amin, A.S.; Sherman, S.L. Study of telomere length in men who carry a fragile X premutation or full mutation allele. *Hum. Genet.* **2020**, *139*, 1531–1539. [CrossRef]
204. Albizua, I.; Rambo-Martin, B.L.; Allen, E.G.; He, W.; Amin, A.S.; Sherman, S.L. Women who carry a fragile X premutation are biologically older than noncarriers as measured by telomere length. *Am. J. Med. Genet. A* **2017**, *173*, 2985–2994. [CrossRef] [PubMed]
205. Miranda-Furtado, C.L.; Luchiari, H.R.; Chielli Pedroso, D.C.; Kogure, G.S.; Caetano, L.C.; Santana, B.A.; Santana, V.P.; Benetti-Pinto, C.L.; Reis, F.M.; Maciel, M.A.; et al. Skewed X-chromosome inactivation and shorter telomeres associate with idiopathic premature ovarian insufficiency. *Fertil. Steril.* **2018**, *110*, 476–485. [CrossRef] [PubMed]
206. Lee, D.C.; Im, J.A.; Kim, J.H.; Lee, H.R.; Shim, J.Y. Effect of long-term hormone therapy on telomere length in postmenopausal women. *Yonsei Med. J.* **2005**, *46*, 471–479. [CrossRef]
207. Levine, R.L.; Garland, D.; Oliver, C.N.; Amici, A.; Climent, I.; Lenz, A.G.; Ahn, B.W.; Shaltiel, S.; Stadtman, E.R. Determination of carbonyl content in oxidatively modified proteins. *Methods Enzym.* **1990**, *186*, 464–478. [CrossRef]
208. Martin, M. Cutadapt removes adapter sequences from high-throughput sequencing reads. *EMBnet J.* **2011**, *17*, 10–12. [CrossRef]
209. Love, M.I.; Huber, W.; Anders, S. Moderated estimation of fold change and dispersion for RNA-seq data with DESeq2. *Genome Biol.* **2014**, *15*, 550. [CrossRef]

International Journal of
Molecular Sciences

Review

Human α-Galactosidase A Mutants: Priceless Tools to Develop Novel Therapies for Fabry Disease

Andrea Modrego [1,2], Marilla Amaranto [3], Agustina Godino [3], Rosa Mendoza [1,4], José Luis Barra [3] and José Luis Corchero [1,4,5,*]

1 Institut de Biotecnologia i de Biomedicina, Universitat Autònoma de Barcelona, Bellaterra, 08193 Barcelona, Spain; modrego82@gmail.com (A.M.); rmendoza@ciber-bbn.es (R.M.)
2 Centro Nacional de Biotecnología, Consejo Superior de Investigaciones Científicas (CSIC), 28049 Madrid, Spain
3 Departamento de Química Biológica Ranwel Caputto, Centro de Investigaciones en Química Biológica de Córdoba, CONICET, Facultad de Ciencias Químicas, Universidad Nacional de Córdoba, Córdoba 5016, Argentina; marilla.amaranto@gmail.com (M.A.); agustinagodino@gmail.com (A.G.); jlbarra@fcq.unc.edu.ar (J.L.B.)
4 Centro de Investigación Biomédica en Red en Bioingeniería, Biomateriales y Nanomedicina (CIBER-BBN), c/Monforte de Lemos 3–5, 28029 Madrid, Spain
5 Departament de Genètica i de Microbiologia, Universitat Autònoma de Barcelona, Bellaterra, 08193 Barcelona, Spain
* Correspondence: jlcorchero@ciber-bbn.es

Abstract: Fabry disease (FD) is a lysosomal storage disease caused by mutations in the gene for the α-galactosidase A (GLA) enzyme. The absence of the enzyme or its activity results in the accumulation of glycosphingolipids, mainly globotriaosylceramide (Gb3), in different tissues, leading to a wide range of clinical manifestations. More than 1000 natural variants have been described in the GLA gene, most of them affecting proper protein folding and enzymatic activity. Currently, FD is treated by enzyme replacement therapy (ERT) or pharmacological chaperone therapy (PCT). However, as both approaches show specific drawbacks, new strategies (such as new forms of ERT, organ/cell transplant, substrate reduction therapy, or gene therapy) are under extensive study. In this review, we summarize GLA mutants described so far and discuss their putative application for the development of novel drugs for the treatment of FD. Unfavorable mutants with lower activities and stabilities than wild-type enzymes could serve as tools for the development of new pharmacological chaperones. On the other hand, GLA mutants showing improved enzymatic activity have been identified and produced in vitro. Such mutants could overcome several complications associated with current ERT, as lower-dose infusions of these mutants could achieve a therapeutic effect equivalent to that of the wild-type enzyme.

Keywords: alpha-galactosidase A; Fabry disease; pharmacological chaperones; rare diseases; enzyme replacement therapy

Citation: Modrego, A.; Amaranto, M.; Godino, A.; Mendoza, R.; Barra, J.L.; Corchero, J.L. Human α-Galactosidase A Mutants: Priceless Tools to Develop Novel Therapies for Fabry Disease. *Int. J. Mol. Sci.* **2021**, *22*, 6518. https://doi.org/10.3390/ijms22126518

Academic Editor: Ivano Condò

Received: 17 May 2021
Accepted: 14 June 2021
Published: 17 June 2021

1. Fabry Disease

Fabry disease (FD) is an X-linked inherited lysosomal storage disease (LSD). The cause of the disease is the mutation of the gene that encodes the enzyme α-galactosidase A (GLA), which leads to a total or partial loss of its function. GLA is a lysosomal enzyme that belongs to the glycoside hydrolase family, and it is responsible for hydrolyzing alpha-galactosyl terminal groups of glycolipids and glycoproteins. Its main substrate is globotriaosylceramide (Gb3), a glycosphingolipid consisting of a ceramide attached to a sugar chain of one glucose and two galactoses. When Gb3 is not properly metabolized, it mainly accumulates in the lysosomes, endothelial cells being one of the most affected cell types. As a result, different signs and symptoms associated with the disease begin to manifest [1]. Other minor substrates of GLA include a deacylated form of Gb3, called globotriaosylsphingosine (lyso-Gb3), digalactosylceramide, and blood group B glycosphingolipids.

FD is considered an attenuated disease as patients survive to adulthood. However, those who lack GLA activity completely have a shortened life expectancy. FD patients present a wide and diverse spectrum of clinical manifestations, varying from classic FD in males to asymptomatic disease in females, with several variants between these two extremes [2]. Currently, more than 1000 different mutations have been described [3], many of which are "private" as they only occur in one family, resulting in a high heterogeneity of the disease. Moreover, there is a high phenotypic variability even between individuals with the same pathogenic variant, which makes it very difficult to establish genotype-phenotype correlations. Having only one X chromosome, males are hemizygous for the pathogenic mutation, so they generally develop more severe signs and symptoms. As a result, the most severe phenotype or "classic phenotype" occurs predominantly in males and is characterized by zero or minimal residual activity of the enzyme. However, unlike other X-linked disorders, females may experience significant symptoms of the disease depending on their residual GLA activity. Clinical manifestations are present in two thirds of women, who generally suffer from a milder disease. This heterogeneity is explained by the process of lyonization, which consists of the random inactivation of one of the X chromosomes in the tissues and organs of the female [4].

Patients with the classic form of FD show no residual enzyme activity [5]. However, the residual activity threshold resulting in FD has not been clearly established. In this sense, it has been estimated that a residual activity of 30–35% of the mean normal α-galactosidase A activity is the cut-off to diagnose Fabry disease [6]. Moreover, such a threshold may vary between organs and even between patients. The nonspecific, multi-organic nature and the relatively high population frequency of FD are risks for the wrong attribution of pathogenicity to certain GLA mutants. This can lead to unnecessary, expensive, and invasive therapies such as ERT. Therefore, before concluding that a mutant variant is pathogenic, one must demonstrate evidence of altered sphingolipid homeostasis. Clinical symptoms are mild at first, which can make diagnosis difficult or delayed, especially if there is no previous family history [7]. Symptoms are usually progressive and affect multiple organs. However, due to the heterogeneity of the disease, there are patients who experience phenotypes in just a single organ (for example, cardiac variant). This occurs mainly in those patients with significant residual activity [1]. Among the characteristic symptoms of FD, there are neurological (pain), skin (angiokeratoma), kidney (proteinuria and kidney failure), cardiovascular (cardiomyopathy and arrhythmia), cochlear-vestibular, ocular, and cerebrovascular problems (transient ischemic attacks and strokes). In addition, anhidrosis or hypohidrosis can occur, causing intolerance to heat and exercise [1].

FD presents a complicated diagnosis, particularly in childhood, as the symptoms are often nonspecific, and the disease is not widely known [8]. Furthermore, some of the characteristic symptoms such as kidney and heart dysfunction appear only in more advanced stages of the disease. In males, hemizygous patients, enzyme activity can be used for diagnosis as it is drastically reduced or does not exist. For this, GLA activity in leukocytes is measured, and if it is significantly lower than normal values, FD can be diagnosed. However, in female patients and some variants of males, it is common that false negatives occur as activity can vary considerably in different cell and tissue types. Therefore, to diagnose a female, the most reliable method is the sequencing of the GLA gene [9]. This test can detect a disease-causing mutation in more than 97% of patients [2], but it has a high cost, making it difficult to use widely. Many attempts have been made over the years to find an ideal disease-specific marker that could serve as a rapid screening tool, as well as an indicator of response to the treatment. Gb3 was initially tested as a possible marker of disease; however, some healthy women showed elevated Gb3 levels in urine, making it difficult to differentiate FD patients from healthy ones. On the other hand, it was found that the levels of lyso-Gb3 (product of Gb3 degradation, as well as minor substrate of GLA) correlate with the clinical condition and the type of mutation, thus making it a reliable predictor of the disease [2]. Furthermore, various studies have demonstrated that the levels of lyso-Gb3 decrease in patients undergoing ERT, particularly in patients with the classic phenotype [10]. However, even lyso-Gb3 has proven not to be always reliable as a FD biomarker. For example, a recent study by Bichet et al. showed that lyso-Gb3 is not a prognostic biomarker of migalastat treatment response in patients with FD [11], a finding similar to another one published for ERT [12]. These findings would compromise the usefulness of lyso-Gb3 as a biomarker in treatment follow-up. In this context, other possible markers (microRNAs, new Gb3 isoforms, and abnormal protein excretion) have been studied but are not yet widely used due to lack of data [13–15].

2. The Human Gla Enzyme

2.1. Structure of GLA Enzyme

The GLA gene encoding the GLA enzyme is located in the long arm of the X chromosome at position Xq22, spans approximately 13 kb, and contains seven exons. The cDNA contains 1290 bases and, upon translation, results in a protein of 429 amino acids that include a 31 amino acid signal peptide [16].

In lysosomes, the glycoprotein is presented as a homodimer, in which each monomer contains two domains. The N-terminal domain spans from residue 32 to 330 and is a classic $(\beta/\alpha)8$ barrel with the active site in its center. The C-terminal domain extends from residue 331 to 429 and consists of an antiparallel β domain. Each monomer contains three N-linked carbohydrate sites, five disulfide bonds, two unpaired cysteine residues, and three cis proline residues (P210, P380, and P389) [17].

There are four potential N-glycosylation sites (N139, N192, N215, and N408), but only the first three (located in the first domain) are occupied by oligosaccharides. Of these, the third site N215 is significantly important for the solubility of the enzyme, as well as for its transport from the endoplasmic reticulum to the lysosomes [18]. In the structural analysis of the oligosaccharides linked to the glycosylation sites, it is observed that the oligosaccharides of the secreted enzyme are markedly heterogeneous, with structures containing high levels of mannose, as well as complex or hybrid structures. Furthermore, a significant number of the oligosaccharides have phosphate monoester groups [19]. The absorption of the enzyme in the lysosomes is mediated by mannose-6-phosphate receptors (MPRs), which explains the presence of high levels of mannose-6-phosphate.

Each GLA monomer contains 12 cysteine residues. Ten of them form disulfide bonds (C52-C94, C56-C63, C142-C172, C202-C223, and C378-C382) within the monomeric subunits, while the remaining two (C90 and C174) contain free sulfhydryl groups. Disulfide bonds play important roles for protein structure and stability. Furthermore, cysteine C142 is part of the active site of the enzyme, which is why the disulfide bond C142-C172 is relevant for the enzymatic activity. On the other hand, free cysteines play important roles in both protein structure and function, including dimerization, enzyme catalysis, redox regulation, and thermal stability [20]. The two free cysteines in GLA are found in different structural environments within the folded protein, which modulates the function of each one within it [21]. C90 is completely buried within the structure, while C174 is partially exposed on the surface of the protein. The active site is formed by the side chains of residues W47, D92, D93, Y134, C142, K168, D170, E203, L206, Y207, R227, D231, D266, and M267, with C172 making a disulfide bond with C142.

2.2. GLA Synthesis and Trafficking

The GLA enzyme precursor is synthesized in the rough endoplasmic reticulum and modified in the Golgi apparatus, where newly synthesized GLA acquires mannose-6-phosphate (M6P) moieties on its N-linked oligosaccharidic chains. These M6P residues serve as recognition signals for mannose-6-phosphate receptors (M6PRs) that transport GLA to endolysosomes in clathrin-coated vesicles through a cation-dependent (CD) and a cation-independent (CI) pathway. While the CD-M6PR appears to act intracellularly (i.e., in the trans-Golgi network, TGN, to endosome sorting), the CI-M6PR can mediate GLA transport from the TGN, as well as from the cell surface, helping to recapture secreted GLA. Once in the endosomes, the GLA-M6PR complex dissociates in the acidic environment, and the receptors recycle to their compartments of origin (plasma membrane and/or TGN) [22].

2.3. Catalytic Mechanism of GLA

The GLA enzyme is an α-retention glycoside hydrolase, because both the substrate and the product have anomeric carbons with α configurations. The mechanism of these enzymes occurs by means of a double displacement reaction, where two consecutive nucleophilic attacks on the anomeric carbon lead to the general retention of the anomeric configuration. Two carboxylates are required, one acting as a nucleophile and the other as an acid/base [23]. The first nucleophilic attack on the substrate comes from D170, which cleaves the glycosidic bond and results in the formation of a covalent enzyme-intermediate bond. In the second step of the reaction, a molecule of water (deprotonated by D231) attacks carbon 1 of the covalent intermediate, releasing the second half of the catalytic product and regenerating the enzyme to its initial state. The enzyme is more effective at low pH, according to its highly acidic composition and lysosomal location [24].

3. Current Treatments For FD

There have been numerous advances in the knowledge of the disease pathogenesis, as well as in its treatment. However, current available therapies are not curative, as they only delay the progression of the disease, in addition to presenting several limitations. Currently, two therapeutic approaches are approved for FD treatment: enzyme replacement therapy (ERT) and pharmacological chaperone therapy (PCT).

3.1. Enzyme Replacement Therapy

ERT consists of systemic infusions of recombinant GLA. In this way, it is possible to replace the enzyme that is absent or deficient and eliminate, or reduce, the accumulation of Gb3 [25]. Treatment should be started as soon as the diagnosis is made, regardless of whether there are clinical manifestations or not, as Gb3 deposits already begin in intrauterine life [2]. Two formulations of recombinant human GLA have been developed: alpha agalsidase (Replagal, by Shire) and beta agalsidase (Fabrazyme, by Sanofi Genzyme). The first one is produced by overexpression in human fibroblasts and the second one in CHO cells. They are administered by intravenous infusions every two weeks (agalsidase alpha at a dose of 0.2 mg/kg and agalsidase beta at a dose of 1 mg/kg). Both treatments result in the reduction of Gb3 plasma and urinary levels, as well as decreased storage of glycosphingolipids in lysosomes. The ERT response in patients can be influenced by several factors, an important one being the formation of anti-drug antibodies (ADAs). A high percentage of patients with Fabry disease lack the GLA enzyme completely; therefore, when the recombinant enzyme is administered, the immune system recognizes it as nonself, triggering an immune response. The apparition of ADAs may negatively influence the long-term efficacy of the treatment [26].

Apart from ADA formation, ERT for FD presents other potential limitations. It has a limited tissue penetration and lacks efficacy if started at advanced stages of the disease. GLA does not pass the blood–brain barrier and it may induce infusion adverse reactions. Finally, ERT is a lifelong therapy requiring intravenous administration every 2 weeks, where it is also associated with a high cost of the treatment.

3.2. Pharmacological Chaperones

On the other hand, PCT is based on the use of small molecules (pharmacological chaperones, PCs) capable of stabilizing the mutant enzyme to ensure its correct folding. This prevents the degradation of the enzyme by unproper folding, allowing trafficking to the lysosomes where the PC and mutant enzyme dissociate, so that the enzyme can exert its function. The term "pharmacological chaperones" appears to label those molecules that can act as drugs, facilitating the proper folding of certain mutant proteins with incorrect folding.

PCs should not be mistaken for molecular chaperones, as they are low-molecular-weight chemical molecules instead of proteins [27].

Lysosomal storage diseases have become an important target in the study and development of PCs, as the ER quality control often recognizes mutant forms of lysosomal enzymes that present minor modifications in their stability or conformation, but still retain catalytic activity. It results in a premature enzyme degradation, which prevents trafficking through the secretory pathway and causes the loss of function.

For PCT to be effective, patients must have susceptible (amenable) mutations, that is, those with certain GLA residual activity [28]. Currently, there are at least 367 amenable and 711 nonamenable mutations known, based on an in vitro assay [29]. The first studies on PCT were carried out for FD. In 1995, a study carried out by Okumya et al. [30] showed that the administration of galactose stabilized GLA mutants, resulting in a higher secretion and activity. Later, in 1999, a galactose analog (an iminosugar called 1-deoxygalactonojirimycin, DGJ) was found to be more efficient than galactose in stabilizing GLA mutants [31]. DGJ binds reversibly and selectively to the active site of GLA with high affinity. Iminosugars such as DGJ are charged molecules with a low capacity to cross membranes due to a negative octanol-water partition coefficient. Modifications to enhance lipophilicity with alkylation have been proposed but resulted in lowered DGJ efficacy [32]. DGJ is a potent GLA inhibitor, stabilizing the enzyme at both neutral and acidic pH. This means that it can stabilize the enzyme in the ER, but it can also cause a partial inhibition of the enzyme in the lysosome [33]. Due to this, for DGJ to be effective, its administration has to be intermittent as the half-life of most GLA mutants is longer than the half-life of DGJ in vivo. Nowadays, PCT with DGJ is approved for FD treatment in patients with amenable mutations under the name "Migalastat," a drug developed by Amicus Therapeutics.

For amenable mutations resulting in the synthesis of a defective enzyme in which activity and stability are altered, PCT by orally administered Migalastat might be a better choice for addressing certain unmet medical needs associated with ERT (like antibody formation). On the other hand, inhibitors are not the ideal drugs, as they may not be able to fully revert the phenotype of the disease. As seen, DGJ represents a good starting point but has several limitations that could be improved. Therefore, the study of novel PCs could overcome this limitation, increasing therapy effectiveness. Due to this, it has been proposed to modify first-generation PCs to enhance their therapeutic effects. To test these (and others) new potential PCs, the identification of amenable mutations and, ideally, their production as recombinant enzymes is needed.

4. GLA Mutations

Currently, more than 1000 different mutations have been described in the GLA gene. According to the Human Gene Mutation Database [3], such mutations can be sorted in different categories: missense/nonsense (~69%), deletions (~19%), insertions (~6%), splicing defects (~5%), and others (~2%). They are related to several abnormalities, such as irregular intracellular trafficking, altered protein folding, reduced activity of the active site, and lower affinity for substrates.

4.1. Most Frequent GLA Mutant Variants

Isolation and sequencing of the entire genomic sequence of the GLA gene allowed the detection and characterization of the mutations causing FD. Furthermore, diffusion of screening studies in the high-risk population and in newborns has contributed to the identification of numerous genetic variants. Among the most frequent variants described in the GLA gene, there are E66Q, A143T, D313Y, R118C, N215S, M296I, and R112H [4]. However, a percentage of mutations are classified by the scientific community as "genetic variants of unknown significance" (GVUS), as it cannot be stated with certainty whether they are benign or pathogenic [34].

The classic phenotype in males can be caused by a diversity of GLA variants, including genetic rearrangements of different sizes, splicing defects, and missense/nonsense mutations [35]. On the other hand, people with atypical late-onset variants (kidney, cardiac, or cerebrovascular disease) present mainly missense and splicing mutations that lead to residual enzymatic activity. However, several pathogenic variants (e.g., R301Q) have been identified to appear both in individuals with the classic phenotype and cardiac variant, indicating that there may be other important factors in the manifestation of the disease [36]. Due to this, the correlation between genotype and phenotype is complex as the same mutation can give rise to different clinical manifestations.

4.2. Most Frequently Mutated Residues

Mutations in the GLA gene can be found throughout the whole sequence. According to the Human Gene Mutation Database, 69% of the residues of GLA have been found to be mutated in patients with Fabry disease. There are certain amino acids in which mutations are most frequently located. Tryptophan is the amino acid with the highest mutation frequency in the GLA-described mutations. Of the 17 residues of tryptophan within the protein, 15 of them are found mutated in patients with Fabry disease (94% mutation rate). They are usually nonsense mutations, which can negatively affect the internal folding of the protein as tryptophan contributes to the formation of the hydrophobic nucleus.

The second most frequently mutated residue is cysteine. Of the 13 cysteines present in the protein, only the cysteine of the signal peptide has not been found in any patient with Fabry disease, which gives a mutation rate of 92%. Numerous mutations in cysteines involved in disulfide bond formation have been identified, while mutations in free cysteines are less common. Mutations in cysteine residues that form disulfide bonds result in incorrect folding of the protein and its complete or partial lack of function, indicating its importance in protein folding and thermodynamic stability. In most of the cases described, mutations affecting disulfide bonds lead to the severe form of Fabry disease [37]. Mutations in free cysteines (C90X, C174G, and C174R) have been described but appear much less frequently and are associated with a late, attenuated phenotype of the disease [38–40]. Consequently, free cysteine residues play a minor structural role within the protein versus residues that form disulfide bonds. Studies of recombinant GLA have revealed the tolerance of free cysteine residues to amino acid change [37]. C90 is completely buried inside the protein surrounded by relatively large hydrophobic and aromatic residues (L45, Y88, L166, and F295) and covered by K168. Therefore, this position only tolerates certain conservative mutations with small side chains such as C90S, C90A, C90T, and C90V. In the case of C174, the amino acid change causes loss of activity, indicating a greater importance in the protein. This residue is found on the surface of the enzyme with the sulfhydryl side chain partially buried. Moreover, it is close to the disulfide bond formed by C142 and C172, which is part of the active site. This explains the greater impact of mutations in this position compared to the other free cysteine. Other frequently mutated residues are glycine and arginine. As glycine is the smallest amino acid, changes to more voluminous amino acids could result in an alteration in the stability of the protein. Glycine has a mutation rate of 84% in described mutations, as, of the 33 residues of glycine, 28 have a described mutation in the Human Gene Mutation Database. Arginine can establish both hydrophobic and hydrophilic interactions. There are 21 residues of arginine within the sequence of the protein and 13 are found on the list of point mutations (62% mutation rate). Lysine is chemically similar to arginine but has a mutation rate significantly lower (35%), indicating the importance of arginine in the functioning of the enzyme. The explanation could be that arginine is buried or partially buried inside the protein more frequently than lysine, which is almost always exposed on the surface. Other residues commonly mutated are those that are part of the active site of the protein or are close to it. All residues that participate in the catalytic activity of the protein have been described with mutations in patients with FD; therefore, the active center is highly sensitive to mutations [41]. In the case of glycosylation sites, the main site affected by mutations that give rise to disease phenotypes is N215. However, different clinical manifestations have been described for the same mutation (N215S), including moderate phenotype [42], attenuated [43], and atypical variants [44]. This site, as previously mentioned, is important for the enzyme's solubility, as well as for its transport from the endoplasmic reticulum to the lysosomes, which explains the appearance of the disease. Mutations at this site would alter the correct folding and transport of GLA, resulting in a nonfunctional protein [45]. The presence of an adjacent hydrophobic sequence supports the importance of this site in the solubility of the enzyme, as the hydrophobic zone must be covered by the oligosaccharide to prevent the aggregation and degradation of the enzyme.

4.3. Nonsense Mutations

Finally, nonsense mutations, where the mutation produces a stop codon, are also common. Premature termination results in shortened proteins at the C-terminus, and sometimes, the protein cannot be produced, due to early degradation of the mRNA. Generally, this type of mutation is associated with a severe phenotype. However, removal of the first residues has been found to give rise to a protein with superior enzyme activity [46,47].

4.4. Location-Dependent Biological Effect of Mutations

Mutations in different locations of the protein can cause different effects. Mutations in enzyme active site residues result in a significant or total loss of enzymatic activity, while mutations in hidden residues inside the protein hamper the formation of the hydrophobic nucleus and affect correct folding. Both types of alterations can lead to the clinical manifestation of the disease. Figure 1 shows the location of described mutations throughout the GLA monomer. Mutations in the proximity of the active site are common. However, most of the residues affected by point mutations in FD are distributed throughout the protein sequence and, mainly, in the hydrophobic nucleus, where more than half of the described mutations are found [48]. These mutations, which do not directly affect the active site of the protein, often lead to proteins that retain some residual activity. Residues affecting the correct folding of the protein have been found throughout the full-length protein, while mutations directly affecting the activity of the protein occur preferentially within, or in close proximity to, the active site [49].

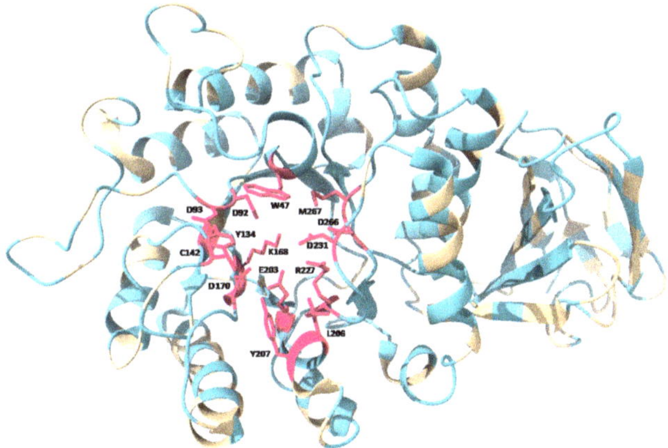

Figure 1. Tertiary structure of GLA monomer, showing (in blue) the location of point mutations described according to Human Gene Mutation Database [3], as well as the residues involved in the active site of the enzyme (in pink). Mutations for all residues in the active site have been found and described. Image obtained using UCSF ChimeraX software, based on PDB: 1R46 structure.

4.5. GLA Mutants with Low Activity and/or Stability

The majority of the above-mentioned GLA mutations are disease-causing. Taking into account the mechanism of pathogenesis, mutations can be divided into two groups: mutations that directly affect the active site ("catalytic") and mutations that affect the correct folding of the protein ("noncatalytic"). Between these two, "noncatalytic" mutations are more prevalent, a fact that leads to the conclusion that FD is mainly a protein folding disease [1]. Several studies analyzing the activity and stability of these mutants confirmed that mutations interfering with the correct folding lead to proteins with residual activity and that are prone to degradation. This results in a loss of function, as they are not able to escape the ER and travel to the lysosome [28]. Further characterization of the mutants will allow the development of improved treatments for the disease, particularly in the case of mutations concerning the structural conformation of the protein.

Mutants as Tools to Develop Pharmacological Chaperones

Pathogenic mutants affecting correct GLA folding are excellent tools to study and develop the previously mentioned PCT. The current PC used to treat patients with amenable mutations of GLA is DGJ, which represents a good starting point but has several limitations that could be improved. In this context, mutants with improper folding produced in vitro could have an important application used as a tool for the development of new PCs.

The residual activity and stability of GLA mutants have been measured in the presence or absence of pharmacological chaperones, mainly DGJ, by several authors, demonstrating the efficacy of this type of therapy in the treatment of FD. In Figure 2, some examples of known mutants with decreased activity are summarized, showing their enzymatic activity (in respect of that of the WT enzyme) both with and without DGJ.

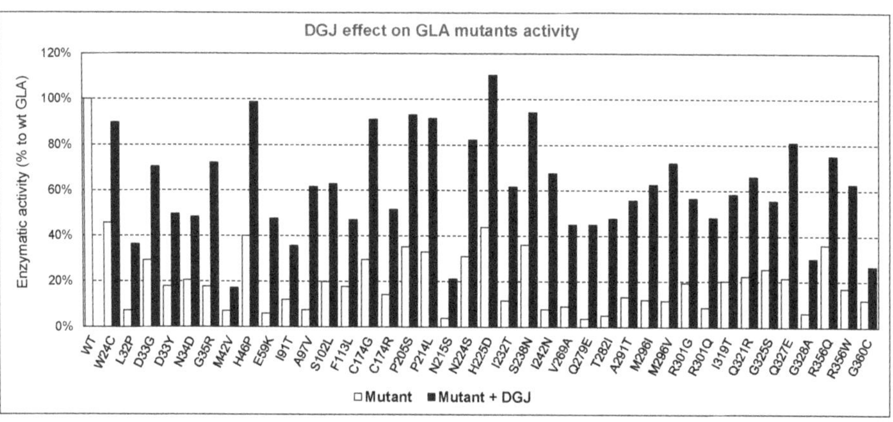

Figure 2. Enzymatic activity of GLA mutants compared with that of wild-type enzyme, and effect of DGJ addition. In white, residual enzymatic activity without DGJ; in black, enzymatic activity after adding DGJ. The mutants shown are examples described by several authors.

As can be seen, in all the GLA mutants tested, DGJ had a positive effect on the enzymatic activity displayed by such mutants. In some cases, enzymatic activity reached values very close (or even higher) compared to the wild-type enzyme. Moreover, in vitro prediction of the efficacy of PCs is possible, as the responses obtained by recombinant mutants of GLA have shown a high degree of consistency with the responses of Fabry patient's cells, as indicated in several studies [49,50]. Thus, it can be concluded that these DGJ in vitro tests are a reliable tool to measure the residual activity ex vivo and responsiveness to PCs.

Derivates of DGJ with improved therapeutic effect could also be examined. For example, Yu et al. synthetized a DGJ-thiourea derivate with high efficacy to enhance the residual activity of FD-associated GLA mutants, reducing the accumulation of the substrate Gb3 in FD cells. This chaperone, although inhibiting GLA at both neutral and acidic pH, showed a superior chaperoning effect than DGJ [51]. To avoid the inhibition of the enzyme in the lysosome, pH-sensitive derivates have been proposed. The strategy suggested by

Mena-Barragán et al. [52] is based on the incorporation of an orthoester segment into the iminosugar conjugate to switch the hydrophobic-hydrophilic balance of the molecule on going from the neutral ER to the acidic lysosome. This has a dramatic effect on the enzyme binding affinity, leading to irreversible dissociation of the PC-enzyme complex at the lysosome [52].

Although pH-responsive chaperones could correct the problem of lysosomal inhibition, the ideal PC would be an allosteric ligand in order to achieve stabilization of missense mutations without obstructing the active site of the enzyme.

In addition to modifying first-generation PCs, another option to search for new potential molecules with these characteristics is molecular docking. Citro et al. identified an allosteric hot-spot for ligand binding by in silico docking, where 2,6-dithiopurine (DTP), a drug-like compound also identified in the study, binds preferentially. DTP showed a stabilization of lysosomal GLA in vitro also in mutants that do not show responsiveness to DGJ [53].

Another interesting approach in the search for new molecules for PCT is "drug repositioning," which consists of using existing drugs for new therapeutic approaches. This strategy is particularly interesting for neglected diseases as it can speed up the development of the drug and lower the costs [54]. In this way, the screening of FDA-approved drugs could result in the discovery of novel molecules with the ability of assist lysosomal enzymes, such as GLA, to achieve the proper folding. As a matter of fact, there are several cases in which small, approved drugs have been successfully repositioned as PCs for rare diseases. An example is Diltiazem, an antihypertensive, that produces the restoration of mutant glucocerebrosidase activity in cells from patients with Gaucher disease [55]. In the case of FD, it was reported that Rosiglitazone (RSG), an antidiabetic approved by the FDA, when used in monotherapy or in conjunction with DGJ, resulted in a significant enhancement of mutant GLA activity [56].

Another small molecule identified through high-throughput screening is Ambroxol, an expectorant approved by the FDA. Ambroxol was found to be an inhibitor and stabilizer of lysosomal acid glucosylceramidase (Gaucher disease). However, its affinity is much lower in the acidic pH of the lysosome, allowing its dissociation [57]. Ambroxol has also showed efficacy for lysosomal GLA, but only in combination with DGJ [56]. This indicates the possibility of finding drugs that could act in different lysosomal storage diseases. The explanation for the wide-ranging effect of Ambroxol could be explained by the structural similarity of lysosomal enzymes, as well as the possible triggering of other stabilizing mechanisms apart from binding the enzyme.

There are other small molecules that can stabilize missense mutants, such as proteostasis regulators (PRs), which enhance the capacity of the proteostasis network [58]. PRs can activate the protein quality control of the cell so that the availability of molecular chaperones is increased. Additionally, they can directly enhance the function and activity of chaperones. This results in reduced protein misfolding due to a high protein folding capacity [59]. PRs were first tested as therapeutic drugs to treat lysosomal storage diseases. Seemann et al. investigated PRs for the treatment of FD, which resulted in the identification of small molecules (proteasome inhibitors such as MG132, BTZ, and CLC and an inhibitor of the ERAD) able to increase mutant GLA activity in patient-derived fibroblasts [60]. PRs can also work together with PCs in a synergistic mode of action to revert the disease of the phenotype, potentiating their therapeutic effect [61]. In the case of FD, the previously commented PRs identified also acted synergistically with DGJ, demonstrating the potential of combination treatment in a therapeutic application [60].

As already mentioned, not all GLA mutations can be corrected by PCT. Only patients with specific responsive, amenable mutations (those affecting correct protein folding and with residual activity) can benefit from this treatment [28]. Most GLA mutations leading to the manifestation of FD are private mutations, which makes it difficult to estimate the rate of responsive mutations. Due to this, further characterization of each described mutation is needed to determine which mutations are responsive and which are not. As different laboratories have developed their own assays to investigate lysosomal glycosidases, assay parameters (and conclusions) are often controversial. Therefore, a universal assay to check the responsiveness to PCT is needed. In any case, the current definition for a responsive mutation is with increases of 20% in the relative enzyme activity and 5% in the absolute enzyme activity compared to the wild-type enzyme [62].

There is a database, FabryCEP, that provides the comparison of results obtained by different experimental approaches for hundreds of GLA mutants in response to DGJ [63]. Additionally, if there are no experimental data for a specific mutation, it incorporates a predictive tool that provides a probability of a given mutation to be responsive to the drug based on a structural, functional, and evolutionary analysis.

Altogether, there are multiple ways in which PCT can be improved for the treatment of FD. In any case, the identification, production, characterization, and testing of unfavorable GLA mutants to test the efficacy of potential drugs with therapeutic effect are required.

4.6. GLA Mutants with Improved Activity and/or Stability

As previously described, most of the GLA mutations have been found in patients suffering from FD. However, mutations leading to a GLA with improved properties such as activity, stability, and/or secretion than the wild-type enzyme have also been described. Such mutations are not detected, as people expressing these "improved mutants" are expected to be phenotypically healthy and do not develop FD. However, different works involving the production of recombinant GLA mutants have found and described mutations rendering improved versions of the GLA enzyme.

For example, Qiu et al. produced in vitro recombinant mutants of GLA with the free cysteines (C90 and C174) mutated to understand their role in the structure and function of the protein [37]. Mutations in cysteines involved in disulfide bond formation are common in FD patients, but mutations in free cysteines appear less frequently. C90 is more tolerant to changes than C174, due to its structural environment. Conservative mutations at C90 do not affect protein function and showed even higher enzymatic activities than the wild-type enzyme did. These mutations were C90T, C90V, C90A, and C90S, and they showed 187%, 181%, 153%, and 119% of enzymatic activity compared to wild-type GLA, respectively (see Figure 3 panel A). The explanation for the increased catalytic activity could be that subtle modifications induced by conservative modifications lead to an enhancement of the conformational flexibility.

Lukas et al. analyzed a significant number of mutations to establish a correlation with clinical manifestations and identify novel mutations [62]. In this study, two-point mutations (N139S and R252T) showed higher enzymatic activities (see Figure 3 panel B) than the wild-type enzyme (148% and 117%, respectively) did. However, N139S was related to clinical manifestations of the disease in a study carried out by Havndrup et al. [9]. The mutation was found in a woman with asymmetric septal hypertrophy, a significant left ventricular outflow tract gradient, and chronic obstructive pulmonary disease cardiac symptoms. N139S is part of the 138GNK140 sequence, a signature sequence in the family 27 of glycoside hydrolases, which could explain the pathogenesis of the mutation. The discrepancy between the higher activity and the presence of possible FD symptoms demands for further characterization of the N139S mutation.

Furthermore, carboxyl-terminal deletions have been reported to increase GLA enzymatic activity. Miyamura et al. produced in vitro mutants with various deletions of the C-terminus in COS-1 cells to analyze their effect [47]. They reported an increase in enzymatic activity when the deletion occurs in the last ten residues. In contrast, if the deletion comprises more than those ten residues, it results in a drastic loss of enzymatic activity. The GLA activities observed were 420%, 620%, 560%, 230%, 460%, 480%, 270%, and 280% of those of WT GLA for Δ2, Δ4, Δ5, Δ6, Δ7, Δ8, Δ9, and Δ10, respectively (see Figure 3 panel C).

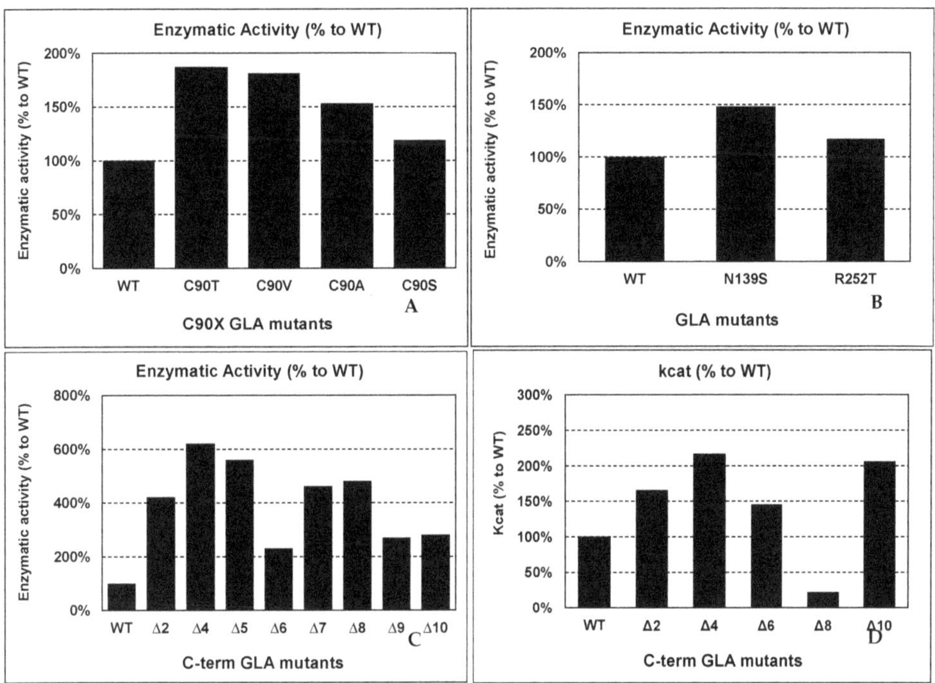

Figure 3. Enzymatic activity of several "improved" GLA mutants, compared to wild-type GLA enzyme. The mutants shown are examples described by several authors. Abbreviations: WT: wild-type GLA; ΔX: C-terminal deletions corresponding to "X" missing residues.

Further characterization of the carboxyl-terminal truncations was carried out by Meghdari et al. [46]. They produced (in the methylotrophic yeast *Pichia Pastoris*) and purified the mutant enzymes with deletions of 2, 4, 6, 8, and 10 C-terminal amino acids (Δ2, Δ4, Δ6, Δ8, Δ10, respectively) to perform quantitative enzyme assays. The results showed increases in the kcat and Vmax with deletions of 2, 4, 6, and 10 amino acids (see Figure 3 panel D). However, the deletion of 8 C-term amino acids decreased the kcat. Discrepancies in this Δ8 mutant with the previous report from Miyamura could be explained due to experimental differences between them. Meghdari et al. used *P. Pastoris* as the expression system, while, in the previous experiment, COS-1 cells were used. Therefore, there could be differences in GLA post-translational modifications that specifically affect the protein activity/stability of this mutant in each expression system, as well as the presence of other proteins in the cytoplasm of *P. pastoris* or COS-1 cells that could interact with the protein, affecting its catalytic activity.

The final residues of each monomer in GLA do not have an associated electron density, and it is likely that it is a disordered region. Furthermore, there is a low sequence homology between GLA from different species. The C-terminus is located far from the active site, so mutations in this position do not directly affect the active site. Altogether, it can be hypothesized that the impact of C-terminus deletions on the activity of the enzyme could be due to alterations of the enzyme's three-dimensional structure. This could influence enzyme dimerization, substrate binding, or potential interactions with other molecules in the cell.

Putative Applications of Improved Mutants in Innovative FD Therapies

As mentioned before, the current therapy to treat patients suffering from FD, and particularly those with mutations that are not responsive to PCT, is enzyme replacement therapy (ERT). The administration of recombinant GLA decreases Gb3 levels and alleviates clinical symptoms. However, the appearance of anti-drug antibodies (ADAs) against recombinant GLA can negatively influence the efficacy of the treatment by changing the distribution, cellular uptake, cellular localization, or catalytic activity of the administered enzyme [64]. Thus, further studies exploring the development of ADAs in FD patients using ERT and the potential impact of such antibodies on the efficacy of the therapy are necessary.

Vedder et al. reported that administering higher doses of the enzyme resulted in a stronger decrease in Gb3 with less impact of antibody formation [65]. Therefore, the use of mutated enzymes with increased activity would be expected to overcome the inhibitory effect of antibodies on treatment effectiveness. By using an enzyme that is more active on a per milligram basis, a therapeutic effect equivalent to the wild-type enzyme could be achievable through infusion of a lower dose. However, this approach is not being explored despite its apparent benefits. One possible explanation to this could be related to legal issues regarding the use of mutant enzymes instead of the wild-type enzyme, in terms of safety, side effects, and efficacy.

Nevertheless, mutants have already been approved for the treatment of other diseases, such as diabetes. Insulin lispro ([Lys(B28, Pro(B29)]-human insulin) is an insulin analogue in which the natural amino acid sequence of the B-chain at positions 28 and 29 is inverted, reducing potential dimerization of the molecule [66]. This allows larger amounts of active monomeric insulin to be available for postprandial injections. Insulin lispro, based on a mutation in the C-terminus, was first approved for use in the United States in 1996 and has proved to be efficient and safe.

As described above, removal of several residues from the C-terminal sequence of wild-type GLA results in a significant increase in enzymatic activity. Moreover, both agalsidase beta and agalsidase alfa display some degree of C-terminal heterogeneity with truncated species lacking either one or two C-terminal residues [67]. Agalsidase beta contains mainly full-length protein with 7.6% of Δ1 and 22.8% of Δ2, while agalsidase alfa contains only 5.7% full length, with 73.2% of Δ1 and 21.1% of Δ2. These differences most likely occur due to proteolytic processing of the mature full-length protein, according to the authors. The significantly increased activity of some of the mutants with C-terminal deletions suggests the starting point for an improved treatment for FD.

Furthermore, the existence of mutants with higher enzymatic activity raises the possibility of studying potential combinations of mutations that could lead to even higher activities. Therefore, further analysis and characterization of the mutations mentioned, as well as the investigation of novel mutations with similar properties, is required to determine the improvement that would mean the use of such mutants in ERT.

4.7. Other Modifications in Recombinant GLAs in Innovative FD Therapies

Other strategies to improve ERT have been studied, mainly regarding post-translational modifications, for example, the use of a recombinant GLA with enhanced sialylation obtained through in vitro glycosylation [68]. Sialic acid capping is essential for proper tissue targeting and increased half-life in serum as it masks the terminal galactose, preventing it from being recognized by the asialoglycoprotein receptor in the liver, which would end in clearance from the blood. Another strategy studied is covalent bonding between two GLA subunits. Ruderfer et al. studied this approach using PEG-based cross-linkers of various lengths and, currently, the developed enzyme (Pegunigalsidase alfa, by Protalix) is being tested in phase III of clinical trials [69]. These strategies could also be implemented with the use of improved mutants, which could result in a drug with increased stability and half-life in serum, as well as with the possibility to overcome the inhibitory effect of ADAs by applying lower doses of the enzyme.

5. Conclusions

Although important actions have been taken to improve the treatment of FD, a definitive cure is not yet available. An incomplete understanding of disease pathogenesis still limits the ability to effectively treat FD patients. ERT and PCT are currently approved for FD; however, these treatments are not curative and show several limitations. Further study of the mechanisms involved in the disease will allow the development of new therapies, as well as the improvement of the existing ones. The current availability of animal models together with the study and characterization of GLA mutants can be used as tools for the purpose.

In Figure 4, we summarize the putative use of GLA mutants as a tool for the study and development of novel drugs for the treatment of FD. For a small subset of patients with specific amenable mutations, treatment with PCs might be a suitable approach. Improvements in this therapy can be assessed by making use of unfavorable mutants with incorrect folding. Inhibitors such as DGJ (Migalastat) are not the ideal drugs, as they cannot fully revert the phenotype of the disease. Therefore, the screening of new molecules and in vitro analysis with GLA mutants could lead to the discovery of novel PCs for the treatment of FD. Concerning this matter, a useful approach is drug repositioning as it allows the lowering of the costs and shortening of the duration of the approval process. For the remaining patients, ERT is still the main treatment of the disease. It is not genotype-dependent, as PCT, but presents other limitations that influence its effectiveness. One of the main challenges with ERT (apart from targeted, specific delivery of GLA to difficult sites of pathology such as the kidney and heart in FD, infusion-associated reactions, and that it is a lifelong therapy requiring intravenous administration every 2 weeks associated with a high cost) is the associated formation of anti-drug antibodies (ADAs) against recombinant GLA. The use of mutated enzymes with increased activity could overcome the inhibitory effect of antibodies by achieving the same (or even higher) therapeutic effect with lower GLA doses. Such lower doses of infused enzyme would probably result in lower or reduced ADA formation. In this sense, the relationship between the amounts of infused GLA and ADAs titer has already been published—Vedder et al. reported that antibodies were more frequent in patients treated with agalsidase beta at 1.0 mg/kg than in patients treated with agalsidase alfa at 0.2 mg/kg ($p = 0.005$) [65]. Thus, GLA mutations involving C-terminal deletions seem to be a very promising approach to obtain "improved" versions of the therapeutic enzyme, allowing a similar therapeutic outcome with lower amounts of infused enzyme.

Enzymatic activity is not the only factor to take into account in order to obtain improved therapeutic enzymes. Other characteristics such as proper biodistribution and pharmacokinetics or plasma stability, among others, are key to achieve better therapeutic effects. Such parameters should be deeply characterized for every mutant, apart from their specific enzymatic activity. For example, Qui et al. showed that mutants C90S, C174S, and C90S/C174S are enzymatically active, structurally intact, and thermodynamically stable, as measured by circular dichroism and thermal denaturation [37]. In another study, C-terminal deletions of the GLA enzyme showed not only better enzymatic performance but also an equal thermal stability at 30 °C, 40 °C, and 50 °C for wild-type and C-terminal GLA mutants [46]. These studies show that researchers in the field are fully aware of the importance of other parameters than only enzymatic activity to check new putative candidates for ERT in FD.

Different approaches to improve ERT are being investigated. Therefore, the combination of these strategies with the exploitation of favorable mutations could lead to a recombinant GLA enzyme with much higher activity and stability. One concern about the use of GLA mutants as drugs for FD treatment is their approval by regulatory authorities. However, it has great potential to confer clinical benefits and could mean an improvement in the current methods. Therefore, it may be worth investing more time and efforts, despite the inconveniences, with the aim of achieving better results. Further investigation should be addressed to prove its safety and effectiveness. Our understanding of the pathogenesis of the disease is constantly changing and we should be ready to implement possible innovations.

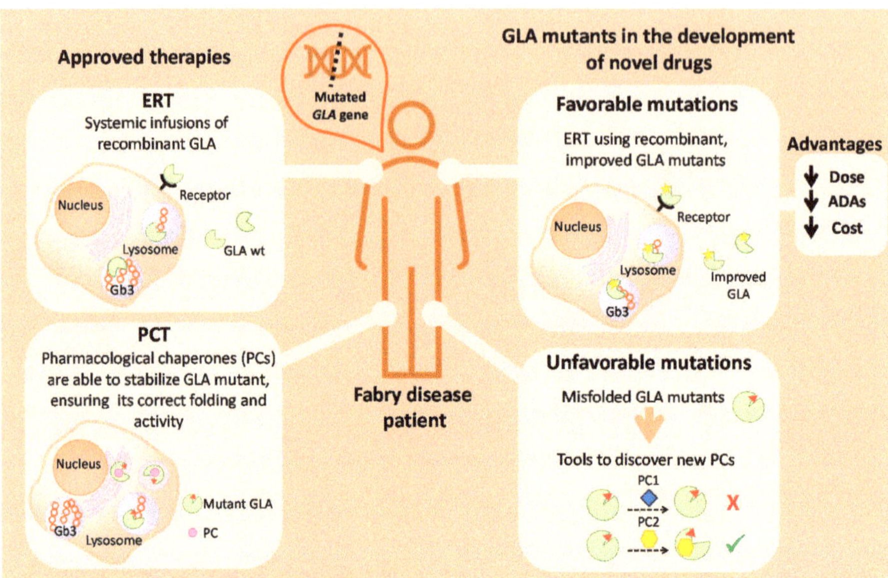

Figure 4. Schematic overview of current therapies for FD treatment, and the putative use of GLA enzyme mutants as tools to develop new therapeutic approaches for FD.

In the long-term, other strategies can be implemented, such as gene therapy. However, although a promising alternative, it still needs additional development. Therefore, for the moment, investing in improvements of current therapies is a more achievable option to develop new tools to treat FD patients.

Author Contributions: Conceptualization, A.M., J.L.B. and J.L.C.; Methodology, A.M., M.A., A.G. and R.M.; Software, A.M.; Formal Analysis, A.M., M.A. and A.G.; Writing—Original Draft Preparation, A.M.; Writing—Review and Editing, J.L.B. and J.L.C.; Supervision, J.L.B. and J.L.C.; Project Administration, R.M.; Funding Acquisition, J.L.B. and J.L.C. All authors have read and agreed to the published version of the manuscript.

Funding: We acknowledge financial support to our research in the field of new therapeutic approaches for lysosomal storage diseases received from Instituto de Salud Carlos III, through "Acciones CIBER". The Networking Research Center on Bioengineering, Biomaterials, and Nanomedicine (CIBER-BBN) is an initiative funded by the VI National R&D&I Plan 2008-2011, Iniciativa Ingenio 2010, Consolider Program, CIBER Actions and financed by the Instituto de Salud Carlos III with assistance from the European Regional Development Fund. The authors acknowledge the financial support from the European Commission through the Smart-4-Fabry project (ID 720942-2, H2020 program), and from CIBER-BBN through the projects "Improved drug delivery systems for enzyme replacement therapies in Fabry disease. Effect of targeting, pegylation and stabilization of the drug system" (PEGLyso), "Increase of the stability, activity and crossing of biological barriers of recombinant proteins through their vehiculization by exosomes and their potential biomedical and industrial use" (EXPLORE and EXPLORE2), and "Advanced Extracellular Vesicles for Enzyme Replacement Therapy" (ADVERT). We are indebted to AGAUR (2017 SGR-229, to Nanobiotechnology unit of IBB, UAB).

Institutional Review Board Statement: Not applicable.

Acknowledgments: Molecular graphics were performed with UCSF ChimeraX software, developed by the Resource for Biocomputing, Visualization, and Informatics at the University of California,

San Francisco, with support from the National Institutes of Health R01-GM129325 and the Office of Cyber Infrastructure and Computational Biology, and the National Institute of Allergy and Infectious Diseases. We also acknowledge the ICTS "NANBIOSIS", more specifically, the support from the Protein Production Platform of CIBER-BBN/IBB, at the UAB SepBioES scientific-technical service (http://www.nanbiosis.es/unit/u1-proteinproduction-platform-ppp/, accessed on 14 June 2021).

Conflicts of Interest: The authors declare no conflict of interest.

References

1. Germain, D.P. Fabry disease. *Orphanet J. Rare Dis.* **2010**, *5*, 30. [CrossRef] [PubMed]
2. Bernardes, T.P.; Foresto, R.D.; Kirsztajn, G.M. Fabry disease: Genetics, pathology, and treatment. *Rev. Assoc. Med. Bras.* **2020**, *66*, s10–s16. [CrossRef] [PubMed]
3. Stenson, P.D.; Ball, E.V.; Mort, M.; Phillips, A.D.; Shiel, J.A.; Thomas, N.S.T.; Abeysinghe, S.; Krawczak, M.; Cooper, D.N. Human Gene Mutation Database (HGMD®): 2003 update: HGMD 2003 UPDATE. *Hum. Mutat.* **2003**, *21*, 577–581. [CrossRef]
4. Simonetta, I.; Tuttolomondo, A.; Di Chiara, T.; Miceli, S.; Vogiatzis, D.; Corpora, F.; Pinto, A. Genetics and Gene Therapy of Anderson-Fabry Disease. *Curr. Gene Ther.* **2018**, *18*, 96–106. [CrossRef]
5. Sakuraba, H.; Oshima, A.; Fukuhara, Y.; Shimmoto, M.; Nagao, Y.; Bishop, D.F.; Desnick, R.J.; Suzuki, Y. Identification of point mutations in the alpha-galactosidase A gene in classical and atypical hemizygotes with Fabry disease. *Am. J. Hum. Genet.* **1990**, *47*, 784–789. [PubMed]
6. Andrade, J.; Waters, P.J.; Singh, R.S.; Levin, A.; Toh, B.-C.; Vallance, H.D.; Sirrs, S. Screening for Fabry disease in patients with chronic kidney disease: Limitations of plasma alpha-galactosidase assay as a screening test. *Clin. J. Am. Soc. Nephrol.* **2008**, *3*, 139–145. [CrossRef] [PubMed]
7. Beck, M. The Mainz Severity Score Index (MSSI): Development and validation of a system for scoring the signs and symptoms of Fabry disease. *Acta Paediatr.* **2006**, *95*, 43–46. [CrossRef] [PubMed]
8. Miller, J.J.; Kanack, A.J.; Dahms, N.M. Progress in the understanding and treatment of Fabry disease. *Biochim. Biophys. Acta Gen. Subj.* **2020**, *1864*, 129437. [CrossRef]
9. Havndrup, O.; Christiansen, M.; Stoevring, B.; Jensen, M.; Hoffman-Bang, J.; Andersen, P.S.; Hasholt, L.; Nørremølle, A.; Feldt-Rasmussen, U.; Køber, L.; et al. Fabry disease mimicking hypertrophic cardiomyopathy: Genetic screening needed for establishing the diagnosis in women. *Eur. J. Heart Fail.* **2010**, *12*, 535–540. [CrossRef]
10. Aerts, J.M.; Groener, J.E.; Kuiper, S.; Donker-Koopman, W.E.; Strijland, A.; Ottenhoff, R.; van Roomen, C.; Mirzaian, M.; Wijburg, F.A.; Linthorst, G.E.; et al. Elevated globotriaosylsphingosine is a hallmark of Fabry disease. *Proc. Natl. Acad. Sci. USA* **2008**, *105*, 2812–2817. [CrossRef]
11. Bichet, D.G.; Aerts, J.M.; Auray-Blais, C.; Maruyama, H.; Mehta, A.B.; Skuban, N.; Krusinska, E.; Schiffmann, R. Assessment of plasma lyso-Gb(3) for clinical monitoring of treatment response in migalastat-treated patients with Fabry disease. *Genet. Med.* **2021**, *23*, 192–201. [CrossRef]
12. Arends, M.; Biegstraaten, M.; Hughes, D.A.; Mehta, A.; Elliott, P.M.; Oder, D.; Watkinson, O.T.; Vaz, F.M.; van Kuilenburg, A.B.P.; Wanner, C.; et al. Retrospective study of long-term outcomes of enzyme replacement therapy in Fabry disease: Analysis of prognostic factors. *PLoS ONE* **2017**, *12*, e0182379. [CrossRef]
13. Cammarata, G.; Scalia, S.; Colomba, P.; Zizzo, C.; Pisani, A.; Riccio, E.; Montalbano, M.; Alessandro, R.; Giordano, A.; Duro, G. A pilot study of circulating microRNAs as potential biomarkers of Fabry disease. *Oncotarget* **2018**, *9*, 27333–27345. [CrossRef] [PubMed]
14. Boutin, M.; Auray-Blais, C. Metabolomic Discovery of Novel Urinary Galabiosylceramide Analogs as Fabry Disease Biomarkers. *J. Am. Soc. Mass Spectrom.* **2015**, *26*, 499–510. [CrossRef] [PubMed]
15. Matafora, V.; Cuccurullo, M.; Beneduci, A.; Petrazzuolo, O.; Simeone, A.; Anastasio, P.; Mignani, R.; Feriozzi, S.; Pisani, A.; Comotti, C.; et al. Early markers of Fabry disease revealed by proteomics. *Mol. Biosyst.* **2015**, *11*, 1543–1551. [CrossRef] [PubMed]
16. Bishop, D.F.; Kornreich, R.; Desnick, R.J. Structural organization of the human alpha-galactosidase A gene: Further evidence for the absence of a 3′ untranslated region. *Proc. Natl. Acad. Sci. USA* **1988**, *85*, 3903–3907. [CrossRef]
17. Garman, S.C.; Garboczi, D.N. The molecular defect leading to Fabry disease: Structure of human alpha-galactosidase. *J. Mol. Biol.* **2004**, *337*, 319–335. [CrossRef]
18. Ioannou, Y.A.; Zeidner, K.M.; Grace, M.E.; Desnick, R.J. Human alpha-galactosidase A: Glycosylation site 3 is essential for enzyme solubility. *Biochem. J.* **1998**, *332*, 789–797. [CrossRef]
19. Matsuura, F.; Ohta, M.; Ioannou, Y.A.; Desnick, R.J. Human -galactosidase A: Characterization of the N-linked oligosaccharides on the intracellular and secreted glycoforms overexpressed by Chinese hamster ovary cells. *Glycobiology* **1998**, *8*, 329–339. [CrossRef]
20. Poole, L.B. The basics of thiols and cysteines in redox biology and chemistry. *Free Radic. Biol. Med.* **2015**, *80*, 148–157. [CrossRef]
21. Nakaniwa, T.; Fukada, H.; Inoue, T.; Gouda, M.; Nakai, R.; Kirii, Y.; Adachi, M.; Tamada, T.; Segawa, S.; Kuroki, R.; et al. Seven Cysteine-Deficient Mutants Depict the Interplay between Thermal and Chemical Stabilities of Individual Cysteine Residues in Mitogen-Activated Protein Kinase c-Jun N-Terminal Kinase 1. *Biochemistry* **2012**, *51*, 8410–8421. [CrossRef]
22. Saftig, P. Physiology of the lysosome. In *Fabry Disease: Perspectives from 5 Years of FOS*; Mehta, A., Beck, M., Sunder-Plassmann, G., Eds.; Oxford PharmaGenesis: Oxford, UK, 2006.
23. Koshland, D.E. Stereochemistry and the Mechanism of Enzymatic Reactions. *Biol. Rev.* **1953**, *28*, 416–436. [CrossRef]

24. Guce, A.I.; Clark, N.E.; Salgado, E.N.; Ivanen, D.R.; Kulminskaya, A.A.; Brumer, H.; Garman, S.C. Catalytic Mechanism of Human α-Galactosidase. *J. Biol. Chem.* **2010**, *285*, 3625–3632. [CrossRef] [PubMed]
25. Ohashi, T. Current status and future prospect of enzyme replacement therapy for Fabry disease. *Rinsho Shinkeigaku* **2019**, *59*, 335–338. [CrossRef] [PubMed]
26. Stappers, F.; Scharnetzki, D.; Schmitz, B.; Manikowski, D.; Brand, S.; Grobe, K.; Lenders, M.; Brand, E. Neutralising anti-drug antibodies in Fabry disease can inhibit endothelial enzyme uptake and activity. *J. Inherit. Metab. Dis.* **2020**, *43*, 334–347. [CrossRef]
27. Liguori, L.; Monticelli, M.; Allocca, M.; Hay Mele, B.; Lukas, J.; Cubellis, M.V.; Andreotti, G. Pharmacological Chaperones: A Therapeutic Approach for Diseases Caused by Destabilizing Missense Mutations. *Int. J. Mol. Sci.* **2020**, *21*, 489. [CrossRef]
28. Ishii, S.; Chang, H.-H.; Kawasaki, K.; Yasuda, K.; Wu, H.-L.; Garman, S.C.; Fan, J.-Q. Mutant α-galactosidase A enzymes identified in Fabry disease patients with residual enzyme activity: Biochemical characterization and restoration of normal intracellular processing by 1-deoxygalactonojirimycin. *Biochem. J.* **2007**, *406*, 285–295. [CrossRef]
29. Lenders, M.; Stappers, F.; Brand, E. In Vitro and In Vivo Amenability to Migalastat in Fabry Disease. *Mol. Ther. Methods Clin. Dev.* **2020**, *19*, 24–34. [CrossRef]
30. Okumiya, T.; Ishii, S.; Takenaka, T.; Kase, R.; Kamei, S.; Sakuraba, H.; Suzuki, Y. Galactose Stabilizes Various Missense Mutants of α-Galactosidase in Fabry Disease. *Biochem. Biophys. Res. Commun.* **1995**, *214*, 1219–1224. [CrossRef]
31. Fan, J.-Q.; Ishii, S.; Asano, N.; Suzuki, Y. Accelerated transport and maturation of lysosomal α–galactosidase A in Fabry lymphoblasts by an enzyme inhibitor. *Nat. Med.* **1999**, *5*, 112–115. [CrossRef]
32. Asano, N.; Ishii, S.; Ikeda, K.; Kizu, H.; Yasuda, K.; Kato, A.; Martin, O.R.; Fan, J.-Q. *In vitro* inhibition and intracellular enhancement of lysosomal α-galactosidase A activity in Fabry lymphoblasts by 1-deoxygalactonojirimycin and its derivatives: Enhancement of α-Gal A in Fabry lymphoblasts. *Eur. J. Biochem.* **2000**, *267*, 4179–4186. [CrossRef]
33. Guce, A.I.; Clark, N.E.; Rogich, J.J.; Garman, S.C. The Molecular Basis of Pharmacological Chaperoning in Human α-Galactosidase. *Chem. Biol.* **2011**, *18*, 1521–1526. [CrossRef] [PubMed]
34. van der Tol, L.; Smid, B.E.; Poorthuis, B.J.H.M.; Biegstraaten, M.; Deprez, R.H.L.; Linthorst, G.E.; Hollak, C.E.M. A systematic review on screening for Fabry disease: Prevalence of individuals with genetic variants of unknown significance. *J. Med. Genet.* **2014**, *51*, 1–9. [CrossRef] [PubMed]
35. Mehta, A.; Hughes, D.A. Fabry Disease. In *GeneReviews®*; Adam, M.P., Ardinger, H.H., Pagon, R.A., Wallace, S.E., Bean, L.J.H., Stephens, K., Amemiya, A., Eds.; University of Washington: Seattle, WA, USA, 1993.
36. Germain, D.P.; Poenaru, L. Fabry Disease: Identification of Novel Alpha-Galactosidase A Mutations and Molecular Carrier Detection by Use of Fluorescent Chemical Cleavage of Mismatches. *Biochem. Biophys. Res. Commun.* **1999**, *257*, 708–713. [CrossRef]
37. Qiu, H.; Honey, D.M.; Kingsbury, J.S.; Park, A.; Boudanova, E.; Wei, R.R.; Pan, C.Q.; Edmunds, T. Impact of cysteine variants on the structure, activity, and stability of recombinant human α-galactosidase A: Cysteine Variants of Human α-Galactosidase A. *Protein Sci.* **2015**, *24*, 1401–1411. [CrossRef]
38. Lee, B.H.; Heo, S.H.; Kim, G.-H.; Park, J.-Y.; Kim, W.-S.; Kang, D.-H.; Choe, K.H.; Kim, W.-H.; Yang, S.H.; Yoo, H.-W. Mutations of the GLA gene in Korean patients with Fabry disease and frequency of the E66Q allele as a functional variant in Korean newborns. *J. Hum. Genet.* **2010**, *55*, 512–517. [CrossRef] [PubMed]
39. Serebrinsky, G.; Calvo, M.; Fernandez, S.; Saito, S.; Ohno, K.; Wallace, E.; Warnock, D.; Sakuraba, H.; Politei, J. Late onset variants in Fabry disease: Results in high risk population screenings in Argentina. *Mol. Genet. Metab. Rep.* **2015**, *4*, 19–24. [CrossRef] [PubMed]
40. Meng, Y.; Zhang, W.-M.; Shi, H.-P.; Wei, M.; Huang, S.-Z. Clinical manifestations and mutation study in 16 Chinese patients with Fabry disease. *Zhonghua Yi Xue Za Zhi* **2010**, *90*, 551–554.
41. Garman, S.C.; Garboczi, D.N. Structural basis of Fabry disease. *Mol. Genet. Metab.* **2002**, *77*, 3–11. [CrossRef]
42. Guffon, N.; Froissart, R.; Chevalier-Porst, F.; Maire, I. Mutation analysis in 11 French patients with Fabry disease. *Hum. Mutat.* **1998**, *11*, S288–S290. [CrossRef]
43. Ebrahim, H.Y.; Baker, R.J.; Mehta, A.B.; Hughes, D.A. Functional analysis of variant lysosomal acid glycosidases of Anderson-Fabry and Pompe disease in a human embryonic kidney epithelial cell line (HEK 293 T). *J. Inherit. Metab. Dis.* **2012**, *35*, 325–334. [CrossRef]
44. Nakao, S.; Takenaka, T.; Maeda, M.; Kodama, C.; Tanaka, A.; Tahara, M.; Yoshida, A.; Kuriyama, M.; Hayashibe, H.; Sakuraba, H.; et al. An Atypical Variant of Fabry's Disease in Men with Left Ventricular Hypertrophy. *N. Engl. J. Med.* **1995**, *333*, 288–293. [CrossRef]
45. Germain, D.P.; Brand, E.; Burlina, A.; Cecchi, F.; Garman, S.C.; Kempf, J.; Laney, D.A.; Linhart, A.; Maródi, L.; Nicholls, K.; et al. Phenotypic characteristics of the p.Asn215Ser (p.N215S) *GLA* mutation in male and female patients with Fabry disease: A multicenter Fabry Registry study. *Mol. Genet. Genom. Med.* **2018**, *6*, 492–503. [CrossRef] [PubMed]
46. Meghdari, M.; Gao, N.; Abdullahi, A.; Stokes, E.; Calhoun, D.H. Carboxyl-Terminal Truncations Alter the Activity of the Human α-Galactosidase A. *PLoS ONE* **2015**, *10*, e0118341. [CrossRef] [PubMed]
47. Miyamura, N.; Araki, E.; Matsuda, K.; Yoshimura, R.; Furukawa, N.; Tsuruzoe, K.; Shirotani, T.; Kishikawa, H.; Yamaguchi, K.; Shichiri, M. A carboxy-terminal truncation of human alpha-galactosidase A in a heterozygous female with Fabry disease and modification of the enzymatic activity by the carboxy-terminal domain. Increased, reduced, or absent enzyme activity depending on number of amino. *J. Clin. Investig.* **1996**, *98*, 1809–1817. [CrossRef] [PubMed]

48. Garman, S.C. Structure-function relationships in α-galactosidase A: Structure-function relationships in α-galactosidase A. *Acta Paediatr.* **2007**, *96*, 6–16. [CrossRef]
49. Wu, X.; Katz, E.; Valle, M.C.D.; Mascioli, K.; Flanagan, J.J.; Castelli, J.P.; Schiffmann, R.; Boudes, P.; Lockhart, D.J.; Valenzano, K.J.; et al. A pharmacogenetic approach to identify mutant forms of α-galactosidase a that respond to a pharmacological chaperone for Fabry disease. *Hum. Mutat.* **2011**, *32*, 965–977. [CrossRef] [PubMed]
50. Citro, V.; Cammisa, M.; Liguori, L.; Cimmaruta, C.; Lukas, J.; Cubellis, M.; Andreotti, G. The Large Phenotypic Spectrum of Fabry Disease Requires Graduated Diagnosis and Personalized Therapy: A Meta-Analysis Can Help to Differentiate Missense Mutations. *Int. J. Mol. Sci.* **2016**, *17*, 2010. [CrossRef]
51. Yu, Y.; Mena-Barragán, T.; Higaki, K.; Johnson, J.L.; Drury, J.E.; Lieberman, R.L.; Nakasone, N.; Ninomiya, H.; Tsukimura, T.; Sakuraba, H.; et al. Molecular Basis of 1-Deoxygalactonojirimycin Arylthiourea Binding to Human α-Galactosidase A: Pharmacological Chaperoning Efficacy on Fabry Disease Mutants. *ACS Chem. Biol.* **2014**, *9*, 1460–1469. [CrossRef]
52. Mena-Barragán, T.; Narita, A.; Matias, D.; Tiscornia, G.; Nanba, E.; Ohno, K.; Suzuki, Y.; Higaki, K.; Garcia Fernández, J.M.; Ortiz Mellet, C. pH-Responsive Pharmacological Chaperones for Rescuing Mutant Glycosidases. *Angew. Chem. Int. Ed.* **2015**, *54*, 11696–11700. [CrossRef]
53. Citro, V.; Peña-García, J.; den-Haan, H.; Pérez-Sánchez, H.; Del Prete, R.; Liguori, L.; Cimmaruta, C.; Lukas, J.; Cubellis, M.V.; Andreotti, G. Identification of an Allosteric Binding Site on Human Lysosomal Alpha-Galactosidase Opens the Way to New Pharmacological Chaperones for Fabry Disease. *PLoS ONE* **2016**, *11*, e0165463. [CrossRef] [PubMed]
54. Hay Mele, B.; Citro, V.; Andreotti, G.; Cubellis, M.V. Drug repositioning can accelerate discovery of pharmacological chaperones. *Orphanet J. Rare Dis.* **2015**, *10*, 55. [CrossRef]
55. Rigat, B.; Mahuran, D. Diltiazem, a L-type Ca^{2+} channel blocker, also acts as a pharmacological chaperone in Gaucher patient cells. *Mol. Genet. Metab.* **2009**, *96*, 225–232. [CrossRef] [PubMed]
56. Lukas, J.; Pockrandt, A.-M.; Seemann, S.; Sharif, M.; Runge, F.; Pohlers, S.; Zheng, C.; Gläser, A.; Beller, M.; Rolfs, A.; et al. Enzyme Enhancers for the Treatment of Fabry and Pompe Disease. *Mol. Ther.* **2015**, *23*, 456–464. [CrossRef] [PubMed]
57. Maegawa, G.H.B.; Tropak, M.B.; Buttner, J.D.; Rigat, B.A.; Fuller, M.; Pandit, D.; Tang, L.; Kornhaber, G.J.; Hamuro, Y.; Clarke, J.T.R.; et al. Identification and Characterization of Ambroxol as an Enzyme Enhancement Agent for Gaucher Disease. *J. Biol. Chem.* **2009**, *284*, 23502–23516. [CrossRef] [PubMed]
58. Calamini, B.; Silva, M.C.; Madoux, F.; Hutt, D.M.; Khanna, S.; Chalfant, M.A.; Saldanha, S.A.; Hodder, P.; Tait, B.D.; Garza, D.; et al. Small-molecule proteostasis regulators for protein conformational diseases. *Nat. Chem. Biol.* **2012**, *8*, 185–196. [CrossRef]
59. Muntau, A.C.; Leandro, J.; Staudigl, M.; Mayer, F.; Gersting, S.W. Innovative strategies to treat protein misfolding in inborn errors of metabolism: Pharmacological chaperones and proteostasis regulators. *J. Inherit. Metab. Dis.* **2014**, *37*, 505–523. [CrossRef]
60. Seemann, S.; Ernst, M.; Cimmaruta, C.; Struckmann, S.; Cozma, C.; Koczan, D.; Knospe, A.-M.; Haake, L.R.; Citro, V.; Bräuer, A.U.; et al. Proteostasis regulators modulate proteasomal activity and gene expression to attenuate multiple phenotypes in Fabry disease. *Biochem. J.* **2020**, *477*, 359–380. [CrossRef]
61. Mu, T.-W.; Ong, D.S.T.; Wang, Y.-J.; Balch, W.E.; Yates, J.R.; Segatori, L.; Kelly, J.W. Chemical and Biological Approaches Synergize to Ameliorate Protein-Folding Diseases. *Cell* **2008**, *134*, 769–781. [CrossRef]
62. Lukas, J.; Knospe, A.-M.; Seemann, S.; Citro, V.; Cubellis, M.V.; Rolfs, A. In Vitro Enzyme Measurement to Test Pharmacological Chaperone Responsiveness in Fabry and Pompe Disease. *J. Vis. Exp.* **2017**, *130*, 56550. [CrossRef]
63. Cammisa, M.; Correra, A.; Andreotti, G.; Cubellis, M. Fabry_CEP: A tool to identify Fabry mutations responsive to pharmacological chaperones. *Orphanet J. Rare Dis.* **2013**, *8*, 111. [CrossRef] [PubMed]
64. Deegan, P.B. Fabry disease, enzyme replacement therapy and the significance of antibody responses. *J. Inherit. Metab. Dis.* **2012**, *35*, 227–243. [CrossRef] [PubMed]
65. Vedder, A.C.; Breunig, F.; Donker-Koopman, W.E.; Mills, K.; Young, E.; Winchester, B.; Ten Berge, I.J.M.; Groener, J.E.M.; Aerts, J.M.F.G.; Wanner, C.; et al. Treatment of Fabry disease with different dosing regimens of agalsidase: Effects on antibody formation and GL-3. *Mol. Genet. Metab.* **2008**, *94*, 319–325. [CrossRef] [PubMed]
66. Howey, D.C.; Bowsher, R.R.; Brunelle, R.L.; Woodworth, J.R. [Lys(B28), Pro(B29)]-Human Insulin: A Rapidly Absorbed Analogue of Human Insulin. *Diabetes* **1994**, *43*, 396–402. [CrossRef]
67. Lee, K.; Jin, X.; Zhang, K.; Copertino, L.; Andrews, L.; Baker-Malcolm, J.; Geagan, L.; Qiu, H.; Seiger, K.; Barngrover, D.; et al. A biochemical and pharmacological comparison of enzyme replacement therapies for the glycolipid storage disorder Fabry disease. *Glycobiology* **2003**, *13*, 305–313. [CrossRef]
68. Sohn, Y.; Lee, J.M.; Park, H.-R.; Jung, S.-C.; Park, T.H.; Oh, D.-B. Enhanced sialylation and in vivo efficacy of recombinant human α-galactosidase through in vitro glycosylation. *BMB Rep.* **2013**, *46*, 157–162. [CrossRef]
69. Ruderfer, I.; Shulman, A.; Kizhner, T.; Azulay, Y.; Nataf, Y.; Tekoah, Y.; Shaaltiel, Y. Development and Analytical Characterization of Pegunigalsidase Alfa, a Chemically Cross-Linked Plant Recombinant Human α-Galactosidase-A for Treatment of Fabry Disease. *Bioconjug. Chem.* **2018**, *29*, 1630–1639. [CrossRef]

International Journal of
Molecular Sciences

MDPI

Review

Gene Therapy for Mucopolysaccharidosis Type II—A Review of the Current Possibilities

Paweł Zapolnik [1,*] and Antoni Pyrkosz [2]

1 Students' Scientific Association of Clinical Genetics, Department of Clinical Genetics, Medical College, University of Rzeszów, 35-959 Rzeszów, Poland
2 Department of Clinical Genetics, Medical College, University of Rzeszów, 35-959 Rzeszów, Poland; antoni.pyrkosz@gmail.com
* Correspondence: pawel.zapolnik@onet.pl

Abstract: Mucopolysaccharidosis type II (MPS II) is a lysosomal storage disorder based on a mutation in the *IDS* gene that encodes iduronate 2-sulphatase. As a result, there is an accumulation of glycosaminoglycans—heparan sulphate and dermatan sulphate—in almost all body tissues, which leads to their dysfunction. Currently, the primary treatment is enzyme replacement therapy, which improves the course of the disease by reducing somatic symptoms, including hepatomegaly and splenomegaly. The enzyme, however, does not cross the blood–brain barrier, and no improvement in the function of the central nervous system has been observed in patients with the severe form of the disease. An alternative method of treatment that solves typical problems of enzyme replacement therapy is gene therapy, i.e., delivery of the correct gene to target cells through an appropriate vector. Much progress has been made in applying gene therapy for MPS II, from cellular models to human clinical trials. In this article, we briefly present the history and basics of gene therapy and discuss the current state of knowledge about the methods of this therapy in mucopolysaccharidosis type II.

Keywords: mucopolysaccharidosis II; Hunter syndrome; adeno-associated viruses; genetic therapy; gene editing; review

Citation: Zapolnik, P.; Pyrkosz, A. Gene Therapy for Mucopolysaccharidosis Type II—A Review of the Current Possibilities. *Int. J. Mol. Sci.* **2021**, *22*, 5490. https://doi.org/10.3390/ijms22115490

Academic Editor: Ivano Condò

Received: 29 April 2021
Accepted: 21 May 2021
Published: 23 May 2021

1. Introduction

1.1. The Basics

Mucopolysaccharidosis type II (MPS II, OMIM #309900), also called Hunter syndrome, is a rare monogenic disease belonging to the group of lysosomal storage disorders (LSDs). The estimated incidence of MPS II is 0.3–0.7/100,000 births [1]. The condition occurs more often in the countries of East Asia than in Europe [2]. It is inherited in an X-linked recessive way as the only form among mucopolysaccharidoses. Almost exclusively males are affected, but there have also been reports of affected females, mainly due to non-random inactivation of the X chromosome [3–9]. Charles A. Hunter first described this entity in 1917 in two brothers [10].

MPS II is caused by a mutation in the *IDS* gene, located at Xq28 (OMIM *300823), which encodes iduronate 2-sulphatase. This enzyme is responsible for catalysing the hydrolysis of sulphate groups from dermatan sulphate (DS) and heparan sulphate (HS) molecules. Enzyme deficiency or decreased activity results in the accumulation of glycosaminoglycans (GAGs) in various tissues and organs, leading to dysfunction [11,12]. Dermatan sulphate and heparan sulphate accumulated in cells disrupt cellular processes such as endocytosis, ion balance, and cell movement. In addition, heparan sulphate promotes the accumulation of GM2 and GM3 gangliosides in the brain, due to which microglial cells are stimulated, and an inflammatory reaction occurs in the central nervous system (CNS) [13]. On the other hand, studies on model organisms (*Danio rerio*, *Mus musculus*) indicate that a mutation in the *IDS* gene may also affect the developmental process differently by influencing the signalling pathway of the fibroblast growth factor (FGF) [14]. Another factor contributing to the development of inflammation is the presence of incompletely degraded GAGs, which may structurally resemble lipopolysaccharide, an endotoxin of Gram-negative bacteria that activates the Toll-like receptor 4 (TLR4). This process leads to the secretion of pro-inflammatory cytokines and the activation of the STAT1/STAT3 protein pathway, increasing the concentration of tumour necrosis factor α (TNF-α) and inflammation in the affected tissues [15].

The animal models, mainly mouse models, of mucopolysaccharidosis type II significantly contributed to understanding the pathophysiology and the application of therapy in this disease, including gene therapy [13]. The first preclinical attempts of gene therapy were carried out in the 1990s [16]. It was also when the first clinical trials began, which resulted in the further development of Hunter syndrome treatment.

1.2. Clinical Features

Despite its heterogeneity, the disease is usually classified into two main forms: attenuated (without central nervous system involvement) and severe (with central nervous system involvement). Both conditions show signs of many organs' dysfunction, such as hepatosplenomegaly, coarse facial features, skeletal system abnormalities (dysostosis multiplex), joint stiffness, short stature, carpal tunnel syndrome, heart valve disease, hypertension and other cardiac abnormalities, communicating hydrocephalus and hearing loss. In addition, patients with MPS II also suffer from frequent respiratory infections and decreased exercise tolerance. The severe form of the disease, which accounts for about 60% of cases, is characterized by the child's normal development until about 2–4 years of age when cognitive functions deteriorate and the process of acquiring new skills is inhibited. Attention difficulties are also common in school-age patients with the severe form of Hunter syndrome. An additional manifestation of CNS damage is frequent epileptic seizures. Patients with deletions, recombination, frameshift mutations, or splicing abnormalities are more likely to develop a severe phenotype. In the case of missense mutations, some may lead to a severe phenotype but most likely predisposed to the attenuated form. Due to the heterogeneity, in many cases, it is difficult to determine the exact genotype/phenotype correlation and predict the nature of development in a particular patient [13,17].

1.3. Diagnosis and Management

For the final diagnosis of the disease, biochemical and molecular tests are performed. First, the level of glycosaminoglycans in a 24-h urine collection is determined and then the activity of the iduronate 2-sulphatase enzyme, e.g., in peripheral blood lymphocytes, skin fibroblasts or chorionic cells (as prenatal diagnosis) is assessed [11,13]. Sanger sequencing is usually performed to detect mutations in the *IDS*, but the development of Next Generation Sequencing (NGS) has paved the way to use gene panels to detect the mutation quickly and exclude other lysosomal storage disorders [11,18,19].

Patients with mucopolysaccharidosis type II, as a disease with a heterogeneous clinical picture, require multidisciplinary care. Historically, before the pathophysiology of the disease was known, management was limited to symptomatic treatment and palliative therapy. In studies conducted on fibroblasts collected from patients with mucopolysaccharidosis type I (MPS I) and mucopolysaccharidosis type II, normal levels of GAGs were observed. Researchers discovered the phenomenon, so-called cross-correction, i.e., enzyme secretion by some cells and its uptake by others via the mannose-6-phosphate receptor [20,21]. The discovery of this phenomenon contributed to the acceleration of studies on therapeutic agents. Currently, the mainstay of treatment is enzyme replacement therapy (ERT) using human recombinant iduronate 2-sulphatase administered intravenously [1]. Thanks to ERT, it is possible to reduce the concentration of GAGs in the urine, reduce the size of the liver and spleen, and improve physical tolerance and stabilize the bone and cardiac abnormalities.

Unfortunately, the enzyme itself cannot cross the blood–brain barrier (BBB). Attempts are made to administer the enzyme intrathecally or intraventricularly. These studies were successfully performed in the MPS II mouse model. Higuchi et al. [22] showed a reduced concentration of GAGs in the brain tissue of mice and other organs, such as the heart, lungs, kidneys, testes, and liver. In addition, they demonstrated an improvement in short-term memory and learning skills, and in brain autopsy, a reduction in cell vacuolization and the expression of the lysosomal marker protein LAMP2. Similar research results were obtained by Sohn et al. [23]. They additionally showed a correlation between the concentration of HS in the cerebrospinal fluid (CSF) and the GAGs levels in the brain. At a later stage, clinical trials were carried out in MPS II patients with the administration of the enzyme via the intrathecal or intraventricular route, along with the standard intravenous drug administration [24]. In a study conducted by Muenzer's group [25], the enzyme administered intrathecally in increasing doses made it possible to reduce the concentration of GAGs in CSF by 80–90%. Another clinical trial was conducted by the group of Seo et al. [26] in Japan (JMACCT CTR JMA-IIA00350) on a group of six patients with severe MPS II. They were administered idursulfase beta into the brain's lateral ventricle via a CSF reservoir system implanted under the scalp. The drug was administered in escalating doses from 1 to 30 mg every four weeks from 0 to 24 weeks and then 30 mg to 100 weeks of the study. At the same time, during the study, patients continued to receive the enzyme intravenously. The authors assessed the developmental age based on the Kyoto Scale of Psychological Development 2001 (KSPD). Patients receiving intraventricular idursulfase beta had stabilized CNS function decline compared to patients receiving the enzyme intravenously only. In addition, HS level in CSF at week 100 was significantly reduced compared to the baseline. Anti-idursulfase antibodies were not detected in cerebrospinal fluid during the trial. Apart from a few side effects (pyrexia, vomiting, and upper respiratory tract infection), the therapy was generally well tolerated. Thanks to this study, intraventricularly administered idursulfase beta was approved in Japan as Hunterase® [26]. The results of these studies are promising, but the procedure is more invasive than intravenous injection and still requires repeated administration of the enzyme.

Another way to bypass the difficulties associated with the blood–brain barrier is to create a system that allows the drug to pass through the barrier. For this purpose, two solutions are tried: receptor-mediated transcytosis and the use of carriers [24,27]. The first option uses a natural process by which proteins enter the central nervous system. The enzyme is conjugated with an antibody against a specific receptor and can defeat the BBB by transcytosis via epithelial cells. Clinical trials have been conducted using iduronate 2-sulphatase combined with an anti-human transferrin receptor antibody (J-Brain Cargo®, JR-141, JCR Pharmaceuticals) [28] and an anti-insulin receptor antibody (AGT-182, ArmaGen Technologies). The first outcomes of studies with JR-141 showed a positive result in reducing the concentration of HS and DS in the cerebrospinal fluid, urine, and plasma. The positive results of clinical trials conducted in Japan with pabinafusp alfa (JR-141) [29] enabled the approval of Izcargo®, a system based on J-Brain Cargo® containing 10 mL of pabinafusp alfa administered intravenously. It is the first agent on the market that allowed the enzyme to cross the blood–brain barrier [30]. The results of using AGT-182 have not been published yet [24]. A clinical trial is currently underway using an enzyme conjugated to an antibody binding site to the transferrin receptor (DNL-310, Denali Therapeutics). There are 16 patients with Hunter syndrome between the ages of 2 and 18 in the study, and recruitment is still open (ClinicalTrials.gov Identifier: NCT04251026) [31]. Another possibility to overcome the BBB is using a carrier that will deliver the enzyme to the cell through the surface heparan sulphate receptor. Such a system using a recombinant enzyme and a guanidinylated neomycin molecule, which has a high affinity for heparan sulphate proteoglycans, has so far been used in a mouse model of mucopolysaccharidosis type I (MPS I). Still, no trials have been carried out on MPS II models [32]. An additional limitation of ERT is the enzyme immunogenicity, against which antibodies in the IgG class are produced, even in about 50% of treated patients [11,33].

Another method of causal treatment is hematopoietic stem cell transplantation (HSCT). Peripheral blood monocytes can cross the blood–brain barrier and settle in the central nervous system as microglial cells [34]. This property, combined with a single transplant procedure, is a strong argument for the use of HSCT. However, complications associated with immunosuppressive therapy and graft versus host disease (GvHD) necessitate careful consideration of this method.

An alternative to the above therapeutic methods is the constantly developing gene therapy. Mucopolysaccharidosis type II is a monogenic disease with relatively well-known pathomechanisms, making it a good candidate as a target for gene therapy. In addition, it allows avoiding the characteristic difficulties of ERT and HSCT. A comparison of the three main treatments in MPS II is shown in Table 1. A more detailed discussion of the current possibilities of gene therapy in MPS II is provided in the next part of the review.

Table 1. Comparison of enzyme replacement therapy, gene therapy, and hematopoietic stem cell transplantation.

Features	Gene Therapy	Enzyme Replacement Therapy (ERT)	Hematopoietic Stem Cell Transplantation (HSCT)
The essence of the method	Delivery of the correct gene to cells by vector	Administration of the correct enzyme into the organism	Transplantation of donor cells with the correct gene
Number of therapeutic interventions	Single	Multiple	Single
The main advantages of the method	Stable gene expression, single application with a long-standing effect, improvement in the central nervous system (CNS) and other organs	Relatively safe route of administration (intravenous), the only Food and Drug Administration (FDA)-approved method so far	The ability to deliver the enzyme to the brain via monocytes, a single application
The main disadvantages of the method	The risk of developing an immune response against the vector and elements of the transgene, the risk of mutagenesis in the case of viral vectors	The need for multiple administration of the enzyme, the inability of the enzyme to cross the blood–brain barrier, the risk of developing neutralizing antibodies against the enzyme	Conditioning before HSCT, immunosuppression, the risk of graft vs. host disease (GvHD), limited gene expression in the central nervous system

2. Gene Therapy

2.1. The Basics

The term "gene therapy" includes the use of nucleic acids to treat human disease. Gene therapy is generally divided into in vivo therapy, with the direct administration of the vector to deliver the correct gene to the cells, and ex vivo therapy, in which cells are collected from the patient's body, applied with the vector and re-administered to the patient. Vectors are classified as viral (e.g., retroviruses, lentiviruses, adeno-associated viruses) and non-viral (e.g., Sleeping Beauty transposon system, nanoemulsions) [35,36]. Gene therapy was first used in the early 1990s in severe combined immunodeficiency due to adenosine deaminase deficiency [37]. The first available agent on the market was Gendicine®, an adenovirus with a human gene encoding p53 protein, introduced in 2004 in China, used in squamous cell carcinoma of the head and neck [38,39]. The first gene therapeutics approved in Europe was alipogene tiparvovec (Glybera®), introduced in 2012. It is an adeno-associated virus (AAV) vector expressing the human lipoprotein lipase gene used in familial lipoprotein lipase deficiency [40,41]. Currently, numerous clinical trials are conducted in many diseases, and several other drugs are available on the market, including voretigene neparvovec (Luxturna®) in congenital retinal dystrophy, onasemnogen abeparvovec (Zolgensma®) in spinal muscular atrophy or Strimvelis® in severe combined immunodeficiency caused by adenosine deaminase deficiency [42].

2.2. First Attempts Gene Therapy with Retroviruses

The first attempts to create gene therapy for Hunter syndrome were made in the 1990s using retroviral vectors. Initially, good results were obtained in preclinical trials using lymphocytes and hematopoietic stem cells CD34+ [16,43]. Braun et al. [16] used a Moloney murine leukaemia virus-derived retroviral vector containing an *IDS* complementary DNA (cDNA) under the transcriptional control of a long terminal repeat sequence or cytomegalovirus early promoter. The vector was applied to peripheral blood lymphocytes collected from affected persons. The authors assessed the normal activity of iduronate 2-sulphatase and observed the phenomenon of cross-correction between cells. A similar study using a modified virus was carried out by Hong et al. [43] using human hematopoietic cells as target cells. The *IDS* was highly expressed in cells. As a result, a clinical trial was opened but was discontinued due to insufficient gene expression and side effects [44,45]. The limitation of retroviruses is their integration into the host genome, the risk of mutagenesis and immunogenicity. However, research using other viral vectors on model organisms was continued and showed promising results [46–48]. Further clinical trials are conducted using various forms of gene therapy, including genome editing systems such as zinc finger nucleases (ZFNs) or Clustered Regularly Interspaced Short Palindromic Repeat/Cas9 (CRISPR/Cas9) [48].

2.3. Gene Therapy with Lentiviruses

One of the types of gene therapy is the ex vivo method of collecting cells from the patient to be treated. Usually, for this purpose, laboratories use hematopoietic stem cells or so-called induced pluripotent stem cells (iPSCs) obtained from already differentiated somatic cells, e.g., fibroblasts. iPSCs can later differentiate into many cell types, e.g., neurons, myocytes, hepatocytes, etc. [48,49]. After the cells are harvested, the appropriate genetic modification is made to produce and secrete the correct enzyme. For this purpose, lentiviruses (derived from retroviruses) are usually used, which provide stable gene expression and deliver the missing gene to non-dividing cells, e.g., neurons [50,51]. After the vector has been inserted, the cells are administered to the patient (autologous transplantation) [48]. This form of therapy was successfully applied in the MPS II mouse model by Wakabayashi et al. [52]. The administration of hematopoietic stem cells with the lentiviral vector containing the *IDS* to mice led to an increase in the activity of iduronate 2-sulphatase and a decrease in the accumulation of glycosaminoglycans in the central nervous system and other organs. In addition, the authors demonstrated a reduction in the accumulation of secondary substances related to autophagy, the p62 protein, and protein and ubiquitin conglomerates in the cells. Neither was there any deterioration in neuronal function [52]. As discussed above, lentiviral vectors have the advantages that make them frequently used in gene therapy. However, there is one major limitation. They integrate into the cell's genome in a random, non-targeted manner, which is associated with the potential risk of mutagenesis and neoplastic transformation. There are currently two clinical trials with this method in mucopolysaccharidosis type I and mucopolysaccharidosis type IIIA (MPS IIIA), but no clinical trials have been undertaken in Hunter syndrome [53,54].

2.4. Gene Therapy with Adeno-Associated Viruses

Another way to deliver the correct gene to cells is in vivo gene therapy with adeno-associated viruses (AAVs). They are small viruses (25 nm) belonging to the *Parvoviridae* family. They contain a 4.7 kbp genome as a single-stranded DNA molecule [55]. These viruses cannot replicate on their own and need assistance from another virus, such as an adenovirus or a herpes virus [56]. Human infection with AAVs occurs naturally. The first contact with the wild type of this virus appears around 1–3 years of age. Thus far, it has not been shown that they cause pathology in humans [57,58]. Unlike retroviruses and lentiviruses, they do not integrate with nuclear DNA but function in the form of extra-chromosomal material—episomes. This property reduces the potential risk of mutagenesis. They provide stable expression of the transgene in both dividing and non-dividing cells [11,48,59–61]. The limitation of the AAVs is immunogenicity and the risk of generating neutralizing antibodies. Still, of all currently available viral vectors, these appear to be the most advantageous and are most widely used in preclinical studies and clinical trials.

One of the first studies on using of AAVs in lysosomal storage disorders, including mucopolysaccharidoses, was carried out at the end of the 20th century by research groups led by Sands [62] and Davidson [63,64]. They conducted studies on mouse models, obtaining improvements in somatic symptoms and central nervous system function [65].

Cardone et al. conducted the first study using AAVs in an animal model (mouse) of mucopolysaccharidosis type II [66]. The authors used AAV type 2/8 with an inserted cDNA for the human *IDS* gene, under the transcriptional control of the liver-specific promoter of the *TBG* gene. Adult mice were injected intravenously with 1.0×10^{11} virus particles. The mice were evaluated for seven months. The authors showed that iduronate 2-sulphatase activity was restored in plasma and tissues (including the central nervous system), and the levels of GAGs in urine were normalized. Skeletal abnormalities also improved. A similar study on a mouse model was carried out by Jung et al. [47], who used the same vector—AAV type 2/8. In this study, the authors introduced *IDS* cDNA into the pAAV2-EF-eGFP-WPRE-polyA plasmid as a substitution for the eGFP element. As a result, they created pAAV2-EF-hIDS-WPRE-polyA plasmid, which, in addition to the coding sequence for iduronate 2-sulphatase, contains the human elongation factor 1-a promoter (EF), woodchuck hepatitis virus posttranscriptional element (WPRE) and a signal sequence of bovine growth hormone poly(A) chain. As in the study mentioned above, the viruses were administered intravenously at a dose of 1.0×10^{11} virus particles. The activity of iduronate 2-sulphatase and the concentration of GAGs were assessed 6 and 24 weeks after applying AAVs. Significantly increased activity of the enzyme and decreased concentration of GAGs in the liver, brain, kidneys and spleen have been demonstrated. Additionally, the authors observed a reduction in cell vacuolization in the histopathological examination of the brain [47].

Another type of adeno-associated virus, AAV type 9, was used in subsequent preclinical studies [67,68]. Motas et al. [69] focused on assessing the improvement in the central nervous system functions. They used AAV9 with the coding sequence of the *IDS*, under the control of the ubiquitous CAG promoter. In this case, a mouse model was also used, but the viruses were administered directly into the cerebrospinal fluid by injection into the cisterna magna at a dose of 5.0×10^{10} vector particles. Four months after the virus administration, the authors assessed the effects of the therapy. Iduronate 2-sulphatase activity in the central nervous system was determined to be 40% of that in wild-type mice. In addition, they showed that most vectors are located in neurons and not in microglia or astrocytes. The concentration of GAGs and the gene expression profile in the brain was normalized. In addition, the vectors permeated the blood and led to the improvement of somatic symptoms and the accumulation of GAGs in peripheral organs such as the liver, lungs and heart. The authors observed an improvement in motor and cognitive functions and increased overall survival compared to untreated individuals [69].

Promising results from preclinical studies have opened the way to first clinical trials using AAVs to improve central nervous system function in patients with Hunter syndrome. Two multicentre (United States, Brazil) phase I/II clinical trials are currently underway on RGX-121 (Regenxbio Inc.), containing a recombinant AAV9 vector with a cassette expressing the human iduronate 2-sulphatase gene (AAV9.CB7.hIDS) [70–73].

The first study (ClinicalTrials.gov Identifier: NCT03566043) included patients with the severe form of MPS II, with central nervous system involvement, from 4 months to 5 years of age. In the course of the study, a three-time single administration of RGX-121 to the cisterna magna or the lateral ventricle is planned in increasing doses, 1.3×10^{10} genome copies (GC)/g brain mass, 6.5×10^{10} GC/g brain mass, and 2.0×10^{11} GC/g brain mass, respectively [70]. The patients are assessed in this study for safety, tolerance and efficacy for 104 weeks after administration. Markers of the disease in serum, urine and cerebrospinal fluid are tested, and the development of the CNS (Bayley Scales of Infant and Toddler Development, BSID; Vineland Adaptive Behavior Scales, VABS) are assessed. Imaging diagnostics are also performed. The six patients who take RGX-121 in this study tolerate it well, and no severe drug-related adverse events were observed (as of September 2020) [73].

The second study (ClinicalTrials.gov Identifier: NCT04571970) targeted patients with severe Hunter syndrome aged 5 to 17. In this study, the vector administration route is also intracisternal or intracerebroventricular injection, but only once at a dose of 6.5×10^{10} GC/g brain mass. Similarly to the study mentioned earlier, the complete safety and efficacy evaluation will take 104 weeks from the agent's application [71]. Both of these trials are still actively recruiting participants.

Adeno-associated viruses are currently the most optimal choice for delivering the correct gene to affected cells among the viral vectors available. Their ability to provide stable gene expression, deliver it to both dividing and non-dividing cells, and a low risk of mutagenesis make their use in clinical trials likely to increase. The apparent limitation of all viral vectors administered into the body is the risk of developing an immune response against them. However, the above-mentioned benefits of using AAVs outweigh their immunogenicity, which is not major anyway.

2.5. Non-Viral Vectors

One way to overcome the limitations of viral vectors in gene therapy is using other vectors that are not based on viral particles. Non-viral vectors include nanoemulsions, the transposon system with the Sleeping Beauty transposase or the electro gene transfer method [35,74]. They are not genotoxic but have limited efficiency in delivering the gene to cells and maintaining its expression.

In research on MPS II using non-viral vectors, preclinical trials have been carried out using only gene electrotransfer, i.e., a method of delivering exogenous molecules into cells through electrical pulses [75,76]. Friso et al. administered 50 µg of plasmid DNA containing cDNA for the *IDS* in 50 µL solution into quadriceps of mice. Hyaluronidase was injected into the muscle before administration of the plasmid itself to increase the effectiveness. The muscle was electrically stimulated with a current of 75 mA. The procedure proved to be very effective in protein production in the muscle, unfortunately without the proper activity of iduronate 2-sulphatase in the blood. Additionally, the authors observed a strong immune response against the recombinant protein [76]. Thus far, no other research group have performed studies with non-viral vectors in MPS II. It seems that their use may be limited due to the low level of expression and difficulties in delivering them to target tissues, especially the central nervous system. For this mucopolysaccharidosis, the use of viral vectors has so far been more optimal.

2.6. Genome Editing Methods

Another promising way to obtain the correct gene into cells is genome editing methods that have been intensively developed in the last decade [77]. Genome editing platforms applicable to therapeutic trials include zinc finger nucleases, transcription activator-like effector nucleases (TALENs), and the CRISPR/Cas9 system, and based on it: CRISPR/Cas9-based editors and CRISPR/Cas9-prime editing. Thus far, only ZFNs and CRISPR/Cas9 have been used in preclinical studies on mucopolysaccharidoses. The essence of these methods is to make breaks in double-strand DNA in a specific site in the genome and then repair the damage by two mechanisms: nonhomologous end joining (NHEJ) or homologous recombination (HR). The first mechanism is characterized by frequent insertions or deletions and is mainly used for reading frame correction. The second mechanism requires the template with the intended sequence and homologous sequences to the target site. Thanks to this, it is possible to correct even a single base without errors (single nucleotide variant, SNV) [48]. Zinc finger nucleases contain a domain that binds specifically to a particular DNA sequence and a nuclease domain derived from the restriction enzyme FokI [78]. In the CRISPR/Cas9 system, Cas9 nuclease binds to a target site through a short

RNA sequence called guide RNA (gRNA). Additionally, in order to proper assembly, a short sequence called protospacer adjacent motif (PAM) is needed [79]. The most frequently used Cas9 nucleases in this technique come from the bacteria *Staphylococcus aureus* and *Streptococcus pyogenes* [80].

Laoharawee et al. conducted a preclinical study in a mouse MPS II model. They used zinc finger nucleases (ZFNs) to insert the human *IDS* gene in place of the first intron of the albumin gene *locus* in mouse hepatocytes. They applied AAV type 2/8 administered intravenously in increasing doses, 2.5×10^{11} vector genomes (vg), 5.0×10^{11} vg, and 1.5×10^{12} vg, respectively. The authors demonstrated normal activity of iduronate 2-sulphatase in blood and other tissues, as well as normalization of GAGs concentration in urine and organs. Additionally, they found that the applied therapy prevented the deterioration of neurocognitive functions in the examined mice [81].

Thanks to the positive results of research on model organisms, the first multicentre phase I/II clinical trial was launched in patients with MPS II (ClinicalTrials.gov Identifier: NCT03041324) using in vivo gene therapy based on genome editing [82]. This study uses SB-913 (Sangamo Therapeutics), which contains a zinc finger nuclease system, the correct *IDS* gene, and the AAV type 2/6 vector. The target site for *IDS* is the albumin *locus* in hepatocytes. The drug is administered intravenously in increasing doses. The study includes nine patients aged five years and older, divided into three cohorts depending on the dose. Currently, the recruitment for the trial is closed [83]. Sixteen weeks after the administration, in the middle dose group, a decrease in the concentration of GAGs in urine was observed, but no activity of the enzyme in the blood was found [84].

Genome editing methods are gaining significant interest due to their precision in reaching the target site in the genome. It seems that they will gain an advantage over other forms of gene therapy as they do not randomly integrate the correct gene but directly repair the defective sequence in its *locus* and provide long-stand gene expression. Genome editing systems, like their vectors, AAVs, are exogenous particles, and immune response may be developed against them. However, the benefits of using these precision genetic modification tools outweigh the potential limitations. A summary of the methods of gene therapy in mucopolysaccharidosis type II is presented in Table 2.

Table 2. Comparison of current methods of gene therapy in mucopolysaccharidosis type II.

Features	Adeno-Associated Viruses	Lentiviruses	Retroviruses	Genome Editing	Non-Viral Vectors
The main advantages	Possibility of delivering the gene to non-dividing cells, present in the cell in the form of an episome, low immunogenicity, stable gene expression	Possibility of delivering the gene to non-dividing cells, stable gene expression	Stable gene expression	Ability to precise correction of gene abnormalities	No risk of mutagenesis, lower costs of the method
The main limitations	Immunogenicity, minor risk of mutagenesis	Random integration with the genome, risk of mutagenesis, immunogenicity	Random integration with the genome, risk of mutagenesis, immunogenicity	High costs of the method, immunogenicity	Difficulty in delivering the gene to the target cell, low gene expression, immunogenicity
Use in clinical trials in patients with mucopolysaccharidosis type II	Yes, currently, active trials [70,71]	No	Yes, trials closed [45]	Yes, currently, active trials [83]	No

3. Challenges and Perspective

Gene therapy for mucopolysaccharidosis type II has made tremendous progress since the first studies on cell lines in the 1990s. However, there are still a few limitations that researchers dealing with this topic must face. First, many animal models differ from the in vivo conditions in the human body. The beneficial effects of the applied therapies in the mouse model, especially on the central nervous system, may not necessarily be reproduced in human studies due to the therapeutic threshold, which differs between species.

Another limitation is the risk of generating an immune response that mainly affects viruses but also elements of genome editing systems. AAVs seem to be the most optimal viral vector so far. However, neutralizing antibodies have been reported, and they lead the way to reduce the effectiveness of the entire therapy [85]. Additionally, cells that have already received the transgene can be eliminated by lymphocytes CD8+. Therefore, it is worth considering the use of immunosuppression in order to increase the effectiveness of subsequent gene therapy attempts [48,55]. A solution to this problem could be non-viral vectors, but so far, they have not been widely used in model organisms in MPS II, so further studies are needed to assess their efficacy and safety.

The use of retroviral and lentiviral vectors is likely to be abandoned in the near future due to their random integration into the genome and the risk of mutagenesis. AAVs seem to be the most optimal vectors. Combined with the precise genome editing method, they are currently the best form of gene therapy and the only one so far used in human clinical trials. It is possible that in the next few years, new clinical studies will be opened, and this process would be significantly accelerated by the positive results of phase I/II trials with RGX-121 and SB-913.

It is also worth mentioning additional therapy that may play a role in improving organ functions and quality of life in patients with Hunter syndrome soon. The process of autophagy, which is disturbed in lysosomal storage disorders, may become a potential therapeutic target. Maeda et al. [86] analysed the morphology of central nervous system cells in the MPS II mouse model. They showed an increased number of autophagy vesicles in an electron microscope and an increased concentration of the autophagy impairment markers—p62 protein and subunit c of mitochondrial ATP synthetase (SCMAS). The authors administered chloroquine orally at a dose of 10 mg/day for 25 weeks to improve the function of neurons. After this period, the vacuolization in neurons was significantly reduced compared to untreated individuals. Chloroquine can inhibit autophagy. Its use in MPS II patients may slow down neurodegenerative processes in the CNS. Researchers also tried to use verapamil and rapamycin (drugs that promote autophagy), but the study had to be discontinued due to serious adverse effects in mice [86].

4. Conclusions

Mucopolysaccharidosis type II, as a monogenic disease with relatively well-known pathophysiology, is an optimal example of the possibility of applying gene therapy. Over the past two decades, this technology has been advanced enough to open up clinical trials using the most promising methods—genome editing and adeno-associated viral vectors. Thus far, it has not been possible to come up with an ideal therapeutic agent. Additionally, gene therapy has some limitations: the immunogenicity of vectors and transgenic elements and the relatively high cost of developing this technology. However, comparing gene therapy to the current standard of treatment for Hunter syndrome—enzyme replacement therapy is essential. A single administration of a vector that will deliver the *IDS* gene to target cells and maintain a high expression level far exceeds the need for repeated intravenous administration of the enzyme. ERT applied intravenously is a heavy burden on the healthcare system and partially improves somatic symptoms without improving central nervous system function.

In conclusion, the further development of molecular engineering techniques and the ongoing clinical trials in humans will likely contribute to the broader use of gene therapy in mucopolysaccharidosis type II. Additionally, this will lead to faster achievement of therapeutic goals and improvement of patients' quality of life.

Author Contributions: P.Z. and A.P. conceived the idea; P.Z. wrote the manuscript and prepared the tables; A.P. supervised the work. Both authors have read and agreed to the published version of the manuscript.

Funding: This research received no external funding.

Institutional Review Board Statement: Not applicable.

Informed Consent Statement: Not applicable.

Data Availability Statement: No new data were created or analyzed in this study. Data sharing is not applicable to this article.

Conflicts of Interest: The authors declare no conflict of interest.

Abbreviations

AAV	Adeno-associated virus
AAVs	Adeno-associated viruses
BBB	Blood–brain barrier
CDNA	complementary DNA
CNS	Central nervous system
CRISPR	Clustered Regularly Interspaced Short Palindromic Repeat
CSF	Cerebrospinal fluid
DS	Dermatan sulphate
ERT	Enzyme replacement therapy
FDA	Food and Drug Administration
FGF	Fibroblast growth factor
GAGs	Glycosaminoglycans
GC	Genome copies
GvHD	Graft versus host disease
gRNA	guide RNA
HS	Heparan sulphate
HSCT	Hematopoietic stem cell transplantation
HR	Homologous recombination
iPSCs	induced pluripotent stem cells
KSPD	Kyoto Scale of Psychological Development 2001
LSDs	Lysosomal storage disorders
MPS I	Mucopolysaccharidosis type I
MPS II	Mucopolysaccharidosis type II
MPS IIIA	Mucopolysaccharidosis type IIIA
NGS	Next Generation Sequencing
NHEJ	Nonhomologous end joining
PAM	Protospacer adjacent motif
SCMAS	Subunit c of mitochondrial ATP synthetase
SNV	Single nucleotide variant
TALENs	Transcription activator-like effector nucleases
TLR4	Toll-like receptor 4
TNF-α	Tumour necrosis factor α
Vg	vector genomes
ZFNs	Zinc finger nucleases

References

1. Peters, H.; Ellaway, C.; Nicholls, K.; Reardon, K.; Szer, J. Treatable lysosomal storage diseases in the advent of disease-specific therapy. *Int. Med. J.* **2020**, *50* (Suppl. 4), 5–27. [CrossRef]
2. Khan, S.A.; Peracha, H.; Ballhausen, D.; Wiesbauer, A.; Rohrbach, M.; Gautschi, M.; Mason, R.W.; Giugliani, R.; Suzuki, Y.; Orii, K.E.; et al. Epidemiology of mucopolysaccharidoses. *Mol. Genet. Metab.* **2017**, *121*, 227–240. [CrossRef] [PubMed]

3. Manara, R.; Rampazzo, A.; Cananzi, M.; Salviati, L.; Mardari, R.; Drigo, P.; Tomanin, R.; Gasparotto, N.; Priante, E.; Scarpa, M. Hunter syndrome in an 11-year old girl on enzyme replacement therapy with idursulfase: Brain magnetic resonance imaging features and evolution. *J. Inherit. Metab. Dis.* **2010**, *33*, 67–72. [CrossRef] [PubMed]
4. Tuschl, K.; Gal, A.; Paschke, E.; Kircher, S.; Bodamer, O.A. Mucopolysaccharidosis type II in females: Case report and review of literature. *Pediatr. Neurol.* **2005**, *32*, 270–272. [CrossRef]
5. Sohn, Y.B.; Kim, S.J.; Park, S.W.; Park, H.-D.; Ki, C.-S.; Kim, C.H.; Huh, S.W.; Yeau, S.; Paik, K.-H.; Jin, D.-K. A mother and daughter with the p.R443X mutation of mucopolysaccharidosis type II: Genotype and phenotype analysis. *Am. J. Med. Genet. Part A* **2010**, *152A*, 3129–3132. [CrossRef] [PubMed]
6. Kloska, A.; Jakóbkiewicz-Banecka, J.; Tylki-Szymańska, A.; Czartoryska, B.; Węgrzyn, G. Female Hunter syndrome caused by a single mutation and familial XCI skewing: Implications for other X-linked disorders. *Clin. Genet.* **2011**, *80*, 459–465. [CrossRef] [PubMed]
7. Jurecka, A.; Krumina, Z.; Żuber, Z.; Różdżyńska-Świątkowska, A.; Kłoska, A.; Czartoryska, B.; Tylki-Szymańska, A. Mucopolysaccharidosis type II in females and response to enzyme replacement therapy. *Am. J. Med. Genet. Part A* **2012**, *158A*, 450–454. [CrossRef]
8. Piña-Aguilar, R.E.; Zaragoza-Arévalo, G.R.; Rau, I.; Gal, A.; Alcántara-Ortigoza, M.A.; López-Martínez, M.S.; Santillán-Hernández, Y. Mucopolysaccharidosis type II in a female carrying a heterozygous stop mutation of the iduronate-2-sulfatase gene and showing a skewed X chromosome inactivation. *Eur. J. Med. Genet.* **2013**, *56*, 159–162. [CrossRef]
9. Lonardo, F.; Di Natale, P.; Lualdi, S.; Acquaviva, F.; Cuoco, C.; Scarano, F.; Maioli, M.; Pavone, L.M.; Di Gregorio, G.; Filocamo, M.; et al. Mucopolysaccharidosis type II in a female patient with a reciprocal X;9 translocation and skewed X chromosome inactivation. *Am. J. Med. Genet. Part A* **2014**, *164*, 2627–2632. [CrossRef]
10. Hunter, C. A Rare Disease in Two Brothers. *Proc. R. Soc. Med.* **1917**, *10*, 104–116. [CrossRef]
11. D'Avanzo, F.; Rigon, L.; Zanetti, A.; Tomanin, R. Mucopolysaccharidosis Type II: One Hundred Years of Research, Diagnosis, and Treatment. *Int. J. Mol. Sci.* **2020**, *21*, 1258. [CrossRef]
12. McBride, K.L.; Berry, S.A.; Braverman, N. ACMG Therapeutics Committee. Treatment of mucopolysaccharidosis type II (Hunter syndrome): A Delphi derived practice resource of the American College of Medical Genetics and Genomics (ACMG). *Genet. Med.* **2020**, *22*, 1735–1742. [CrossRef] [PubMed]
13. Stapleton, M.; Kubaski, F.; Mason, R.W.; Yabe, H.; Suzuki, Y.; Orii, K.E.; Orii, T.; Tomatsu, S. Presentation and treatments for Mucopolysaccharidosis Type II (MPS II.; Hunter Syndrome). *Expert Opin. Orphan Drugs* **2017**, *5*, 295–307. [CrossRef] [PubMed]
14. Bellesso, S.; Salvalaio, M.; Lualdi, S.; Tognon, E.; Costa, R.; Braghetta, P.; Giraudo, C.; Stramare, R.; Rigon, L.; Filocamo, M.; et al. FGF signaling deregulation is associated with early developmental skeletal defects in animal models for mucopolysaccharidosis type II (MPSII). *Hum. Mol. Genet.* **2018**, *27*, 2262–2275. [CrossRef] [PubMed]
15. Fecarotta, S.; Gasperini, S.; Parenti, G. New treatments for the mucopolysaccharidoses: From pathophysiology to therapy. *Ital. J. Pediatr.* **2018**, *44*, 124. [CrossRef]
16. Braun, S.E.; Pan, D.; Aronovich, E.L.; Jonsson, J.J.; McIvor, R.S.; Whitley, C.B. Preclinical studies of lymphocyte gene therapy for mild Hunter syndrome (mucopolysaccharidosis type II). *Hum. Gene Ther.* **1996**, *7*, 283–290. [CrossRef] [PubMed]
17. Seo, J.H.; Okuyama, T.; Shapiro, E.; Fukuhara, Y.; Kosuga, M. Natural history of cognitive development in neuronopathic mucopolysaccharidosis type II (Hunter syndrome): Contribution of genotype to cognitive developmental course. *Mol. Genet. Metab. Rep.* **2020**, *24*, 100630. [CrossRef]
18. Zanetti, A.; D'Avanzo, F.; Bertoldi, L.; Zampieri, G.; Feltrin, E.; De Pascale, F.; Rampazzo, A.; Forzan, M.; Valle, G.; Tomanin, R. Setup and validation of a targeted NGS approach for the diagnosis of lysosomal storage disorders. *J. Mol. Diagn.* **2020**, *22*, 488–502. [CrossRef] [PubMed]
19. Brusius-Facchin, A.C.; Siebert, M.; Leão, D.; Malaga, D.R.; Pasqualim, G.; Trapp, F.; Matte, U.; Giugliani, R.; Leistner-Segal, S. Phenotype-oriented NGS panels for mucopolysaccharidoses: Validation and potential use in the diagnostic flowchart. *Genet. Mol. Biol.* **2019**, *42* (Suppl. 1), 207–214. [CrossRef]
20. Fratantoni, J.C.; Hall, C.W.; Neufeld, E.F. Hurler and Hunter syndromes: Mutual correction of the defect in cultured fibroblasts. *Science* **1968**, *162*, 570–572. [CrossRef]
21. Hasilik, A.; Neufeld, E.F. Biosynthesis of lysosomal enzymes in fibroblasts. Phosphorylation of mannose residues. *J. Biol. Chem.* **1980**, *255*, 4946–4950. [CrossRef]
22. Higuchi, T.; Shimizu, H.; Fukuda, T.; Kawagoe, S.; Matsumoto, J.; Shimada, Y.; Kobayashi, H.; Ida, H.; Ohashi, T.; Morimoto, H.; et al. Enzyme replacement therapy (ERT) procedure for mucopolysaccharidosis type II (MPS II) by intraventricular administration (IVA) in murine MPS II. *Mol. Genet. Metab.* **2012**, *107*, 122–128. [CrossRef] [PubMed]
23. Sohn, Y.B.; Ko, A.R.; Seong, M.R.; Lee, S.; Kim, M.R.; Cho, S.Y.; Kim, J.S.; Sakaguchi, M.; Nakazawa, T.; Kosuga, M.; et al. The efficacy of intracerebroventricular idursulfase-beta enzyme replacement therapy in mucopolysaccharidosis II murine model: Heparan sulfate in cerebrospinal fluid as a clinical biomarker of neuropathology. *J. Inherit. Metab. Dis.* **2018**, *41*, 1235–1246. [CrossRef]
24. Nan, H.; Park, C.; Maeng, S. Mucopolysaccharidoses I and II: Brief Review of Therapeutic Options and Supportive/Palliative Therapies. *BioMed Res. Int.* **2020**, *2020*, 2408402. [CrossRef]

25. Muenzer, J.; Hendriksz, C.J.; Fan, Z.; Vijayaraghavan, S.; Perry, V.; Santra, S.; Solanki, G.A.; Mascelli, M.A.; Pan, L.; Wang, N.; et al. A phase I/II study of intrathecal idursulfase-IT in children with severe mucopolysaccharidosis II. *Genet. Med.* **2016**, *18*, 73–81. [CrossRef]
26. Seo, J.H.; Kosuga, M.; Hamazaki, T.; Shintaku, H.; Okuyama, T. Impact of intracerebroventricular enzyme replacement therapy in patients with neuronopathic mucopolysaccharidosis type II. *Mol. Ther. Methods Clin. Dev.* **2021**, *21*, 67–75. [CrossRef] [PubMed]
27. Sato, Y.; Okuyama, T. Novel Enzyme Replacement Therapies for Neuropathic Mucopolysaccharidoses. *Int. J. Mol. Sci.* **2020**, *21*, 400. [CrossRef]
28. Kida, S.; Kinoshita, M.; Tanaka, S.; Okumura, M.; Koshimura, Y.; Morimoto, H. Non-clinical evaluation of a blood-brain barrier-penetrating enzyme for the treatment of mucopolysaccharidosis type I. *Mol. Genet. Metab.* **2019**, *126*, S83–S84. [CrossRef]
29. Okuyama, T.; Eto, Y.; Sakai, N.; Nakamura, K.; Yamamoto, T.; Yamaoka, M.; Ikeda, T.; So, S.; Tanizawa, K.; Sonoda, H.; et al. A Phase 2/3 Trial of Pabinafusp Alfa, IDS Fused with Anti-Human Transferrin Receptor Antibody, Targeting Neurodegeneration in MPS-II. *Mol. Ther.* **2021**, *29*, 671–679. [CrossRef]
30. JCR Pharma: Crossing Oceans and the Blood-Brain Barrier. Available online: https://invivo.pharmaintelligence.informa.com/IV124753/JCR-Pharma-Crossing-Oceans-And-The-Blood-Brain-Barrier (accessed on 19 May 2021).
31. ClinicalTrials.gov. A Study of DNL310 in Pediatric Subjects with Hunter Syndrome. Available online: https://clinicaltrials.gov/ct2/show/NCT04251026?term=NCT04251026&draw=2&rank=1 (accessed on 13 May 2021).
32. Tong, W.; Dwyer, C.A.; Thacker, B.E.; Glass, C.A.; Brown, J.R.; Hamill, K.; Moremen, K.W.; Sarrazin, S.; Gordts, P.; Dozier, L.E.; et al. Guanidinylated Neomycin Conjugation Enhances Intranasal Enzyme Replacement in the Brain. *Mol. Ther.* **2017**, *25*, 2743–2752. [CrossRef]
33. Lampe, C.; Bosserho, A.-K.; Burton, B.K.; Giugliani, R.; de Souza, C.F.; Bittar, C.; Muschol, N.; Olson, R.; Mendelsohn, N.J. Long-term experience with enzyme replacement therapy (ERT) in MPS II patients with a severe phenotype: An international case series. *J. Inherit. Metab. Dis.* **2014**, *37*, 823–829. [CrossRef] [PubMed]
34. Araya, K.; Sakai, N.; Mohri, I.; Kagitani-Shimono, K.; Okinaga, T.; Hashii, Y.; Ohta, H.; Nakamichi, I.; Aozasa, K.; Taniike, M.; et al. Localized donor cells in brain of a Hunter disease patient after cord blood stem cell transplantation. *Mol. Genet. Metab.* **2009**, *98*, 255–263. [CrossRef] [PubMed]
35. Kaufmann, K.B.; Büning, H.; Galy, A.; Schambach, A.; Grez, M. Gene therapy on the move. *EMBO Mol. Med.* **2013**, *5*, 1642–1661. [CrossRef]
36. Schuh, R.S.; de Carvalho, T.G.; Giugliani, R.; Matte, U.; Baldo, G.; Teixeira, H.F. Gene editing of MPS I human fibroblasts by co-delivery of a CRISPR/Cas9 plasmid and a donor oligonucleotide using nanoemulsions as nonviral carriers. *Eur. J. Pharm. Biopharm.* **2018**, *122*, 158–166. [CrossRef]
37. Blaese, R.M.; Culver, K.W.; Miller, A.D.; Carter, C.S.; Fleisher, T.; Clerici, M.; Shearer, G.; Chang, L.; Chiang, Y.; Tolstoshev, P.; et al. T lymphocyte-directed gene therapy for ADA- SCID: Initial trial results after 4 years. *Science* **1995**, *270*, 475–480. [CrossRef]
38. Wilson, J.M. Gendicine: The first commercial gene therapy product. *Hum. Gene Ther.* **2005**, *16*, 1014–1015. [CrossRef] [PubMed]
39. Zhang, W.W.; Li, L.; Li, D.; Liu, J.; Li, X.; Li, W.; Xu, X.; Zhang, M.J.; Chandler, L.A.; Lin, H.; et al. The First Approved Gene Therapy Product for Cancer Ad-p53 (Gendicine): 12 Years in the Clinic. *Hum. Gene Ther.* **2018**, *29*, 160–179. [CrossRef] [PubMed]
40. Miller, N. Glybera and the future of gene therapy in the European Union. *Nat. Rev. Drug Discov.* **2012**, *11*, 419. [CrossRef]
41. Ylä-Herttuala, S. Endgame: Glybera finally recommended for approval as the first gene therapy drug in the European union. *Mol. Ther.* **2012**, *20*, 1831–1832. [CrossRef]
42. Mendell, J.R.; Al-Zaidy, S.A.; Rodino-Klapac, L.R.; Goodspeed, K.; Gray, S.J.; Kay, C.N.; Boye, S.L.; Boye, S.E.; George, L.A.; Salabarria, S.; et al. Current Clinical Applications of In Vivo Gene Therapy with AAVs. *Mol. Ther.* **2021**, *29*, 464–488. [CrossRef]
43. Hong, Y.; Yu, S.S.; Kim, J.M.; Lee, K.; Na, Y.S.; Whitley, C.B.; Sugimoto, Y.; Kim, S. Construction of a high efficiency retroviral vector for gene therapy of Hunter's syndrome. *J. Gene Med.* **2003**, *5*, 18–29. [CrossRef]
44. Sawamoto, K.; Stapleton, M.; Alméciga-Díaz, C.J.; Espejo-Mojica, A.J.; Losada, J.C.; Suarez, D.A.; Tomatsu, S. Therapeutic Options for Mucopolysaccharidoses: Current and Emerging Treatments. *Drugs* **2019**, *79*, 1103–1134. [CrossRef] [PubMed]
45. ClinicalTrials.gov. Phase I/II Study of Retroviral-Mediated Transfer of Iduronate-2-Sulfatase Gene into Lymphocytes of Patients with Mucopolysaccharidosis II (Mild Hunter Syndrome). Available online: https://clinicaltrials.gov/ct2/show/NCT00004454 (accessed on 22 April 2021).
46. Wada, M.; Shimada, Y.; Iizuka, S.; Ishii, N.; Hiraki, H.; Tachibana, T.; Maeda, K.; Saito, M.; Arakawa, S.; Ishimoto, T.; et al. Gene Therapy Treats Bone Complications of Mucopolysaccharidosis Type II Mouse Models through Bone Remodeling Reactivation. *Mol. Ther. Methods Clin. Dev.* **2020**, *19*, 261–274. [CrossRef] [PubMed]
47. Jung, S.C.; Park, E.S.; Choi, E.N.; Kim, C.H.; Kim, S.J.; Jin, D.K. Characterization of a novel mucopolysaccharidosis type II mouse model and recombinant AAV2/8 vector-mediated gene therapy. *Mol. Cells* **2010**, *30*, 13–18. [CrossRef] [PubMed]
48. Poletto, E.; Baldo, G.; Gomez-Ospina, N. Genome Editing for Mucopolysaccharidoses. *Int. J. Mol. Sci.* **2020**, *21*, 500. [CrossRef] [PubMed]
49. Yamanaka, S. Induced pluripotent stem cells: Past, present, and future. *Cell Stem Cell* **2012**, *10*, 678–684. [CrossRef]
50. Sakuma, T.; Barry, M.A.; Ikeda, Y. Lentiviral vectors: Basic to translational. *Biochem. J.* **2012**, *443*, 603–618. [CrossRef] [PubMed]
51. Milone, M.C.; O'Doherty, U. Clinical use of lentiviral vectors. *Leukemia* **2018**, *32*, 1529–1541. [CrossRef]

52. Wakabayashi, T.; Shimada, Y.; Akiyama, K.; Higuchi, T.; Fukuda, T.; Kobayashi, H.; Eto, Y.; Ida, H.; Ohashi, T. Hematopoietic Stem Cell Gene Therapy Corrects Neuropathic Phenotype in Murine Model of Mucopolysaccharidosis Type II. *Hum. Gene Ther.* **2015**, *26*, 357–366. [CrossRef]
53. ClinicalTrials.gov. Gene Therapy with Modified Autologous Hematopoietic Stem Cells for the Treatment of Patients with Mucopolysaccharidosis Type I, Hurler Variant (TigetT10_MPSIH). Available online: https://clinicaltrials.gov/ct2/show/NCT03488394 (accessed on 22 April 2021).
54. ClinicalTrials.gov. Gene Therapy with Modified Autologous Hematopoietic Stem Cells for Patients with Mucopolysaccharidosis Type IIIA. Available online: https://clinicaltrials.gov/ct2/show/NCT04201405 (accessed on 22 April 2021).
55. Colella, P.; Ronzitti, G.; Mingozzi, F. Emerging Issues in AAV-Mediated In Vivo Gene Therapy. *Mol. Ther. Methods Clin. Dev.* **2018**, *8*, 87–104. [CrossRef]
56. Balakrishnan, B.; Jayandharan, G.R. Basic biology of adeno-associated virus (AAV) vectors used in gene therapy. *Curr. Gene Ther.* **2014**, *14*, 86–100. [CrossRef]
57. Calcedo, R.; Morizono, H.; Wang, L.; McCarter, R.; He, J.; Jones, D.; Batshaw, M.L.; Wilson, J.M. Adeno-associated virus antibody profiles in newborns, children, and adolescents. *Clin. Vaccine Immunol.* **2011**, *18*, 1586–1588. [CrossRef]
58. Smith, R.H. Adeno-associated virus integration: Virus versus vector. *Gene Ther.* **2008**, *15*, 817–822. [CrossRef]
59. McCarty, D.M.; Young, S.M.; Samulski, R.J. Integration of Adeno-Associated Virus (AAV) and Recombinant AAV Vectors. *Annu. Rev. Genet.* **2004**, *38*, 819–845. [CrossRef] [PubMed]
60. Donsante, A.; Vogler, C.; Muzyczka, N.; Crawford, J.M.; Barker, J.; Flotte, T.; Campbell-Thompson, M.; Daly, T.; Sands, M.S. Observed incidence of tumorigenesis in long-term rodent studies of rAAV vectors. *Gene Ther.* **2001**, *8*, 1343–1346. [CrossRef]
61. Rosas, L.E.; Grieves, J.L.; Zaraspe, K.; La Perle, K.M.D.; Fu, H.; McCarty, D.M. Patterns of scAAV vector insertion associated with oncogenic events in a mouse model for genotoxicity. *Mol. Ther.* **2012**, *20*, 2098–2110. [CrossRef] [PubMed]
62. Daly, T.M.; Okuyama, T.; Vogler, C.; Haskins, M.E.; Muzyczka, N.; Sands, M.S. Neonatal intramuscular injection with recombinant adeno-associated virus results in prolonged beta-glucuronidase expression in situ and correction of liver pathology in mucopolysaccharidosis type VII mice. *Hum. Gene Ther.* **1999**, *10*, 85–94. [CrossRef] [PubMed]
63. Davidson, B.L.; Stein, C.S.; Heth, J.A.; Martins, I.; Kotin, R.M.; Derksen, T.A.; Zabner, J.; Ghodsi, A.; Chiorini, J.A. Recombinant adeno-associated virus type 2, 4, and 5 vectors: Transduction of variant cell types and regions in the mammalian central nervous system. *Proc. Natl. Acad. Sci. USA* **2000**, *97*, 3428–3432. [CrossRef] [PubMed]
64. Stein, C.S.; Ghodsi, A.; Derksen, T.; Davidson, B.L. Systemic and central nervous system correction of lysosomal storage in mucopolysaccharidosis type VII mice. *J. Virol.* **1999**, *73*, 3424–3429. [CrossRef] [PubMed]
65. Sands, M.S.; Davidson, B.L. Gene therapy for lysosomal storage diseases. *Mol. Ther.* **2006**, *13*, 839–849. [CrossRef] [PubMed]
66. Cardone, M.; Polito, V.A.; Pepe, S.; Mann, L.; D'Azzo, A.; Auricchio, A.; Ballabio, A.; Cosma, M.P. Correction of Hunter syndrome in the MPSII mouse model by AAV2/8-mediated gene delivery. *Hum. Mol. Genet.* **2006**, *15*, 1225–1236. [CrossRef]
67. Hinderer, C.; Katz, N.; Louboutin, J.P.; Bell, P.; Yu, H.; Nayal, M.; Kozarsky, K.; O'Brien, W.T.; Goode, T.; Wilson, J.M. Delivery of an Adeno-associated virus vector into cerebrospinal fluid attenuates central nervous system disease in mucopolysaccharidosis type II mice. *Hum. Gene Ther.* **2016**, *27*, 906–915. [CrossRef] [PubMed]
68. Laoharawee, K.; Podetz-Pedersen, K.M.; Nguyen, T.T.; Evenstar, L.B.; Kitto, K.F.; Nan, Z.; Fairbanks, C.A.; Low, W.C.; Kozarsky, K.F.; McIvor, R.S. Prevention of neurocognitive deficiency in mucopolysaccharidosis type II mice by central nervous system-directed, AAV9-mediated iduronate sulfatase gene transfer. *Hum. Gene Ther.* **2017**, *28*, 626–638. [CrossRef] [PubMed]
69. Motas, S.; Haurigot, V.; Garcia, M.; Marcó, S.; Ribera, A.; Roca, C.; Sánchez, X.; Sánchez, V.; Molas, M.; Bertolin, J.; et al. CNS-directed gene therapy for the treatment of neurologic and somatic mucopolysaccharidosis type II (Hunter syndrome). *JCI Insight* **2016**, *1*, 1–18. [CrossRef]
70. ClinicalTrials.gov. RGX-121 Gene Therapy in Patients with MPS II (Hunter Syndrome). Available online: https://clinicaltrials.gov/ct2/show/NCT03566043 (accessed on 24 April 2021).
71. ClinicalTrials.gov. RGX-121 Gene Therapy in Children 5 Years of Age and Over with MPS II (Hunter Syndrome). Available online: https://clinicaltrials.gov/ct2/show/NCT04571970 (accessed on 24 April 2021).
72. Cowley, K.; Falabella, P.; Pakola, S.; Nevoret, M.-L. RGX-121 gene therapy for severe mucopolysaccharidosis type II (MPS II): A clinical program to address central nervous system manifestations. *Mol. Genet. Metab.* **2021**, *132*, S29. [CrossRef]
73. Nevoret, M.-L.; Escolar, M.; Ficicioglu, C.; Giugliani, R.; Harmatz, P.; Cho, Y.; Phillips, D.; Falabella, P. RGX-121 gene therapy for severe mucopolysaccharidosis type II (MPS II): Interim results of an ongoing first in human trial. *Mol. Genet. Metab.* **2021**, *132*, S76. [CrossRef]
74. Tomanin, R.; Friso, A.; Alba, S.; Piller Puicher, E.; Mennuni, C.; La Monica, N.; Hortelano, G.; Zacchello, F.; Scarpa, M. Non-viral transfer approaches for the gene therapy of mucopolysaccharidosis type II (Hunter syndrome). *Acta Paediatrica* **2002**, *91*, 100–104. [CrossRef] [PubMed]
75. Mir, L.M.; Moller, P.H.; André, F.; Gehl, J. Electric pulse-mediated gene delivery to various animal tissues. *Adv. Genet.* **2005**, *54*, 83–114. [CrossRef]
76. Friso, A.; Tomanin, R.; Zanetti, A.; Mennuni, C.; Calvaruso, F.; La Monica, N.; Marin, O.; Zacchello, F.; Scarpa, M. Gene therapy of Hunter syndrome: Evaluation of the efficiency of muscle electro gene transfer for the production and release of recombinant iduronate-2-sulfatase (IDS). *Biochim. Biophys. Acta Mol. Basis Dis.* **2008**, *1782*, 574–580. [CrossRef] [PubMed]
77. Bak, R.O.; Gomez-Ospina, N.; Porteus, M.H. Gene Editing on Center Stage. *Trends Genet.* **2018**, *34*, 600–611. [CrossRef]

78. Urnov, F.D.; Rebar, E.J.; Holmes, M.C.; Zhang, H.S.; Gregory, P.D. Genome editing with engineered zinc finger nucleases. *Nat. Rev. Genet.* **2010**, *11*, 636–646. [CrossRef] [PubMed]
79. Doudna, J.A.; Charpentier, E. The new frontier of genome engineering with CRISPR-Cas9. *Science* **2014**, *346*, 1258096. [CrossRef] [PubMed]
80. Ran, F.A.; Cong, L.; Yan, W.X.; Scott, D.A.; Gootenberg, J.S.; Kriz, A.J.; Zetsche, B.; Shalem, O.; Wu, X.; Makarova, K.S.; et al. In vivo genome editing using Staphylococcus aureus Cas9. *Nature* **2015**, *520*, 186–191. [CrossRef] [PubMed]
81. Laoharawee, K.; DeKelver, R.C.; Podetz-Pedersen, K.M.; Rohde, M.; Sproul, S.; Nguyen, H.O.; Nguyen, T.; St Martin, S.J.; Ou, L.; Tom, S.; et al. Dose-Dependent Prevention of Metabolic and Neurologic Disease in Murine MPS II by ZFN-Mediated In vivo Genome Editing. *Mol. Ther.* **2018**, *26*, 1127–1136. [CrossRef]
82. Nature America, Inc. First in vivo human genome editing trial. *Nat. Biotechnol.* **2018**, *36*, 5. [CrossRef]
83. ClinicalTrials.gov. Ascending Dose Study of Genome Editing by the Zinc Finger Nuclease (ZFN) Therapeutic SB-913 in Subjects with MPS II. Available online: https://clinicaltrials.gov/ct2/show/NCT03041324 (accessed on 26 April 2021).
84. Sheridan, C. Sangamo's landmark genome editing trial gets mixed reception. *Nat. Biotechnol.* **2018**, *36*, 907–908. [CrossRef]
85. Boutin, S.; Monteilhet, V.; Veron, P.; Leborgne, C.; Benveniste, O.; Montus, M.F.; Masurier, C. Prevalence of serum IgG and neutralizing factors against adeno-associated virus (AAV) types 1, 2, 5, 6, 8, and 9 in the healthy population: Implications for gene therapy using AAV vectors. *Hum. Gene Ther.* **2010**, *21*, 704–712. [CrossRef]
86. Maeda, M.; Seto, T.; Kadono, C.; Morimoto, H.; Kida, S.; Suga, M.; Nakamura, M.; Kataoka, Y.; Hamazaki, T.; Shintaku, H. Autophagy in the Central Nervous System and Effects of Chloroquine in Mucopolysaccharidosis Type II Mice. *Int. J. Mol. Sci.* **2019**, *20*, 5829. [CrossRef] [PubMed]

International Journal of *Molecular Sciences*

MDPI

Review

Gene Therapy in Hemophilia: Recent Advances

E. Carlos Rodríguez-Merchán [1], Juan Andres De Pablo-Moreno [2] and Antonio Liras [2,*]

[1] Osteoarticular Surgery Research, Hospital La Paz Institute for Health Research–IdiPAZ (La Paz University Hospital—Autonomous University of Madrid), 28046 Madrid, Spain; ecrmerchan@hotmail.com

[2] Department of Genetic, Physiology and Microbiology, Biology School, Complutense University of Madrid, 28040 Madrid, Spain; jdepablo@ucm.es

* Correspondence: aliras@ucm.es

Abstract: Hemophilia is a monogenic mutational disease affecting coagulation factor VIII or factor IX genes. The palliative treatment of choice is based on the use of safe and effective recombinant clotting factors. Advanced therapies will be curative, ensuring stable and durable concentrations of the defective circulating factor. Results have so far been encouraging in terms of levels and times of expression using mainly adeno-associated vectors. However, these therapies are associated with immunogenicity and hepatotoxicity. Optimizing the vector serotypes and the transgene (variants) will boost clotting efficacy, thus increasing the viability of these protocols. It is essential that both physicians and patients be informed about the potential benefits and risks of the new therapies, and a register of gene therapy patients be kept with information of the efficacy and long-term adverse events associated with the treatments administered. In the context of hemophilia, gene therapy may result in (particularly indirect) cost savings and in a more equitable allocation of treatments. In the case of hemophilia A, further research is needed into how to effectively package the large factor VIII gene into the vector; and in the case of hemophilia B, the priority should be to optimize both the vector serotype, reducing its immunogenicity and hepatotoxicity, and the transgene, boosting its clotting efficacy so as to minimize the amount of vector administered and decrease the incidence of adverse events without compromising the efficacy of the protein expressed.

Keywords: hemophilia; advanced therapies; gene therapy; FVIII transgene; FIX transgene; adeno-associated virus; lentiviral vectors

Citation: Rodríguez-Merchán, E.C.; De Pablo-Moreno, J.A.; Liras, A. Gene Therapy in Hemophilia: Recent Advances. *Int. J. Mol. Sci.* **2021**, *22*, 7647. https://doi.org/10.3390/ijms22147647

Academic Editor: Ivano Condò

Received: 30 June 2021
Accepted: 15 July 2021
Published: 17 July 2021

1. Introduction

Hemophilia A (HA) and hemophilia B (HB) are X-linked hereditary hemorrhagic disorders arising from mutations in the genes encoding coagulation factors VIII (FVIII) in HA and IX (FIX) in HB [1]. HA, with a prevalence of one in every 5000 live births, is more common than HB, which affects one in every 30,000 live births.

The severity of HA and HB is contingent on the functional levels of the corresponding circulating factors, with levels of FVIII or FIX below 1% considered diagnostic of severe hemophilia, levels between 1 and 5% diagnostic of moderate hemophilia and those between 5% and 40% of mild hemophilia.

Patients with severe hemophilia tend to experience bleeds into their joints, muscles or soft tissues following trauma or even without any apparent cause. They may also suffer life-threatening hemorrhagic episodes such as intercranial bleeding. Persons with mild or moderate factor deficiencies may experience spontaneous bleeding, with excessive bleeds occurring only after trauma or an invasive medical procedure.

Residual factor activity is usually well correlated with patients' clinical characteristics. However, individuals with the same coagulation factor levels may exhibit varying bleeding phenotypes. Although HA and HB are considered clinically indistinguishable, several recent studies have questioned this idea, suggesting that patients with HB could be less prone to severe bleeds than those with HA with the same plasma levels of residual factor (identical disease phenotypes) [2].

Treatment of hemophilia [3] has evolved very rapidly in the last few decades since the advent of plasma and cryoprecipitates; human plasma-derived clotting factor concentrates, and more recently recombinant factors, characterized by a high degree of purity and a longer half-life. These innovations have allowed the development of extremely safe against emerging pathogens and highly efficient treatments based on the replacement of the deficient clotting factor. Treatment protocols consist of intravenous administration of the deficient clotting factor either on-demand upon occurrence of bleeding episodes or prophylactically two or three times a week. Although they are generally highly effective, these replacement treatments may occasionally fail due to the presence of the so-called inhibitors present in up to 30% of hemophiliacs. Inhibitors are neutralizing antibodies that target exogenously administered coagulation factors.

Next-generation recombinant products offer a longer half-life, which allows for less frequent dosing and therefore higher adherence to treatment and a better quality of life. Recombinant FVIII/FIX concentrates are obtained by fusion to polyethylene glycol, IgG1-Fc or albumin [4].

New treatment strategies have recently been developed as alternatives to replacement protocols. One of these is emicizumab (Hemlibra®), a subcutaneously administered bispecific monoclonal antibody recently approved for treatment of HA with or without inhibitors against FVIII. This antibody binds to activated factor X (FXa) and to activated FIX (FIXa) simulating the function of FVIII. Given emicizumab's unique structure, its effect is not likely to be neutralized by inhibitors against FVIII. HAVEN 1, a non-interventional phase 3 trial involving patients with severe HA and inhibitors, demonstrated that patients on emicizumab had 87% less bleeding episodes than those without prophylaxis, and 79% less bleeding episodes than those on previous prophylaxis [5].

On the other hand, recent studies have shown that the natural anticoagulation pathway (antithrombin and tissue factor pathway inhibitors, protein S and protein C) is capable of restoring hemostasis in patients with a bleeding disorder [6]. Fitusiran is an RNA interference (RNAi) therapy that targets antithrombin (AT) in the liver and interferes with the translation of its messenger RNA, preventing AT synthesis and promoting hemostasis. Preclinical and clinical studies have found that AT suppression leads to a dose-dependent reduction in AT levels which results in a decreased severity of the bleeding phenotypes in patients with hemophilia [7,8]. Several phase 1 and phase 2 studies have tested the effectiveness of Fitusiran in patients with hemophilia with or without inhibitors; phase 3 trials are now underway with the same goal in mind. A phase 1 clinical trial (NCT02035605) demonstrated a 61–89% dose-dependent decrease in AT, which was correlated with increased thrombin generation in patients with HA and patients with HB without inhibitors.

Research is at present also focusing on the tissue factor pathway inhibitor (TFPI), the main inhibitor of the onset of the coagulation cascade, which has been shown to regulate the severity of a wide range of hemorrhagic and coagulation disorders. Multiple animal studies have shown that TFPI inhibition can reduce bleeding in the context of hemophilia [9]. In this respect, several strategies have been described with a view to inhibiting TFPI by means of aptamers, fucoidan, monoclonal antibodies and peptide agents [10]. Concizumab, a humanized monoclonal antibody targeted against TFPI, is the most highly developed of these strategies. Two phase 2/3 clinical trials are currently underway in patients with HA and HB with or without inhibitors. Specifically, TFPI inhibits activated FX (FXa) while activated protein C (APCa) degrades activated factor V (FVa) and activated FVIII (FVIIIa). APC is a serine protease that can bind to the endothelial protein C receptor, which approximates it to thrombin for its activation. This means that it acts by decreasing the amplification of FXa generation and intrinsic thrombin generation by FVa and FVIIIa proteolysis. APC is regulated by serine protease inhibitors (serpins), the protein C inhibitor (PCI), protein S and α1-antitrypsin. Available non-clinical data support the use of a serpin (serpinPC), a monoclonal antibody against APC, and protein S-targeted siRNA. SerpinPC is currently in the early stages of clinical investigation.

2. Advanced Therapies in Hemophilia

Advanced therapies comprise a set of novel and innovative strategies such as cell therapy, gene therapy and regenerative medicine or tissue engineering. They are aimed at conditions or disorders that currently lack a curative treatment of whose treatment requires optimization. According to international medicine agencies, the products used in the context of advanced therapies are drugs for human use that are based on genes, tissues or cells and that offer innovative solutions for the treatment of some diseases [11] (Figure 1).

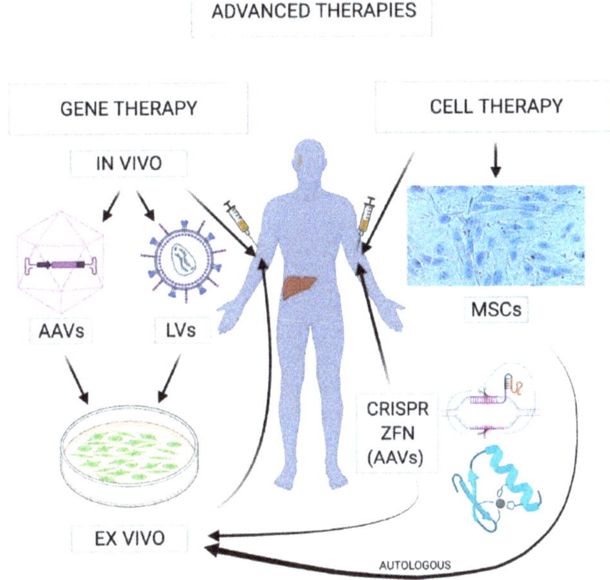

Figure 1. Advanced therapy strategies. In vivo gene therapy where the (typically organ-specific) adeno-associated viral (AAVs) or lentiviral (LVs) vectors carried by the "therapeutic gene" are administered systemically to the patient. In ex vivo gene therapy, autologous or allogenic cells are transfected in vitro and infused back into the patient following their expansion in the culture medium. Cell therapy uses stem cells, or already differentiated cells, transfected or otherwise, to correct a deficiency in the patient's physiological function. Another alternative is to correct the defective genes responsible for the disease. To do that, several gene-editing tools are available such as CRISPR/Cas9 or ZFNs (zinc finger nucleases). Adeno-associated viral vectors are used to insert guide RNAs and other components necessary to the correction. MSCs, mesenchymal stem cells. (Created with Biorender.com).

In hemophilia, protocols based on gene and cell therapy have been shown to hold great potential. Clearly viable cell and gene therapy approaches exist both to address monogenic and polygenic conditions, and to increase the effective duration of therapeutic proteins and boost their level of expression. This can be achieved thanks to the availability of a wide range of varieties and types of target cells and transfection vectors, and because it is now possible to regulate the gene expression and the characteristics of the transgene, and in the end, ensure its safety. Investments in investigating these new therapies are clearly justified as there are multiple chronic and severe conditions for which no treatment is currently available and others for which the existing treatment either results in severe side effects or is tedious and/or inconvenient, making adherence difficult.

Somatic cell therapy is defined as the administration of live autologous, allogenic or xenogenic non-germline cells that have been minimally manipulated or processed ex vivo through propagation, expansion, selection, pharmacological treatment or some other modification of their biological characteristics. The purpose of somatic therapy is to recover a nonexistent or lost function in a living organism. The cells used in cell therapy are generally stem cells equipped with a series of special characteristics, such as being undifferentiated, having the ability to self-renew and to differentiate to different cell lines or to different embryonic layers [12–15].

Use of induced pluripotent stem cells (iPSCs) in this context is also common as they present with similar characteristics to those of embryonic cells but without the bioethical problems associated with them. IPSCs have allowed significant progress in the realm of HA and HB [16,17].

Of all multipotent stem cells or adult cells, the ones presenting with the highest clinical application potential are the mesenchymal stem cells derived from adipose tissue, bone marrow, umbilical cord and placenta, due to their low immunogenicity, their immunomodulatory effect, and the fact that they do not induce an immunogenic response as they do not express the class II major histocompatibility complex (MHC-II) or T-lymphocyte costimulatory molecules such as CD40L, CD80 or CD86 [18,19].

As far as gene therapy protocols are concerned, these can be defined as the transfer to one or more cells of a DNA fragment that encodes a protein and may be able to cure a certain disease. This simple definition has opened up a whole new dimension in the treatment of a large variety of conditions, ranging from hereditary diseases to cancer and infectious, cardiovascular, hepatic and neurodegenerative conditions [20].

Even if gene therapy protocols are often more effective than those of cell therapy, they are associated with significant difficulties and a greater number of adverse events, mainly related to the transfer vector. An ideal transfer vector should be immunologically inert; highly tissue- and cell-specific; integrative; capable of sustainably producing the transgene and transducing genes to divided or undivided cells; easy to produce at large scale; and capable of maximum transgene loading. However, at the present time, immunogenicity and hepatotoxicity of the transfer vector as well as the problems associated with its integration (insertional mutagenesis) are the most important barriers that must be overcome [21].

Gene therapy protocols can be applied to somatic cells either in vivo, where correction of the gene, or its insertion into the cells, is performed in the body through a vector; or ex vivo, where correction of the gene, or its insertion, is carried out previously in a series of cells harvested from the patient then reimplanted following an expansion and selection process [20,22].

As regards transfection vectors, gene therapy can be classified into viral and non-viral. The higher efficacy of viral vectors as compared with non-viral vectors has made them the better candidates for clinical application [20,23,24]. However, it must be said that retrovirus (RNA virus)-based vectors are practically absent from clinical practice despite their high transduction efficacy and their ability to maintain transgene expression over time. Their drawbacks are related to their high capacity to integrate into the genome and their high insertional mutagenesis. Adenoviral (DNA virus) vectors are non-integrative and are able to efficiently package the "therapeutic" gene but present with a very high immunogenic capacity, which could trigger severe anaphylactic reactions. Adeno-associated virus (AAV) vectors for their part exhibit the greatest potential for clinical use. They are partially integrative vectors associated with a low risk of insertional mutagenesis. However, they are characterized by a high immunogenic response to their capsid and may, in some cases, be hepatotoxic. Lastly, lentiviral vectors (LVs), based on single-stranded RNA lentiviruses, are integrative and not very likely to result in insertional mutagenesis, nor do they induce a significant immune response or produce a hepatotoxic effect. Although, they exhibit the same low packaging ability as other retroviruses, next-generation LVs, which have been enhanced and endowed with protection from phagocytosis, have been shown to achieve optimal FVIII and FIX expression in the liver [25].

Another alternative to therapies consisting of transferring a therapeutic gene is correction of the defective genes causing the condition. Several gene-editing tools are currently in use, such as transcription activator-like effector nucleases (TALENs), zinc finger nucleases (ZFNs) and CRISPR/Cas9 [26,27]. These tools are based on the induction of DNA double-strand breaks and are equipped with a restriction endonuclease in charge of mediating the break and a variable part, which is designed as a complement to the target sequence guiding the endonuclease, where the break is performed. Subsequently, the cell repair system induces the potential correction into the DNA when repairing the break. As regards the CRISPR/Cas9 tool, the variable part is nucleotide-based (RNA) rather than protein-based.

As regards correction by gene editing, a phase 1 study has been initiated [28] to evaluate the long-term safety, tolerability and expression of FIX in patients with severe HB using an AAV vector. SB-FIX is a ZFN-mediated genome-editing therapy administered by means of AAV vectors. SB-FIX is intended to function by placing a corrective copy of the FIX transgene into the genome of the subject's own hepatocytes, under the control of the highly expressed endogenous albumin locus, thereby providing a permanent, liver-specific expression of FIX.

Gene therapy treatments based on siRNA (small interference RNA) are also being developed. These therapies are intended to modify the translation of messenger RNA, inhibiting or modifying production of the protein without affecting the encoding gene [29].

Recent work in gene therapy aimed at developing a treatment for congenital coagulopathies has instilled renewed hope in patients with HA and HB [4]. Use of different AAV vector genotypes has already yielded concrete, nd highly encouraging, results in clinical trials in terms of the expression levels (moderate phenotype) and expression times (several months) achieved [30]. However, the problems related to the vectors' immunogenicity and hepatotoxicity remain to be solved.

The first gene therapy clinical studies in coagulopathies, which used retroviral, adenoviral and non-viral (ex vivo) vectors were associated with transient expressions and low levels of the coagulation factor [31,32]. These results led to a change in strategy and to the use of "non integrative" recombinant viruses such as AAVs. This kind of vector derives from a native non-pathogenic and barely immunogenic AAV of the parvovirus family which, given its inability to replicate, requires an auxiliary virus to do so [33].

The DNA sequences transported by AAV vectors are stabilized in an episomal form, so that long-term expression is only possible through delivery into long-living postmitotic cells. The vector's DNA integrates at a very low rate and is typically lost from replicating cells. Recombinant vectors display tropism for a series of target tissues, such as the liver, which makes them the most commonly used gene therapy viral vectors in hemophilia (90%), followed by in vivo and ex vivo LVs (10%).

Liver-directed gene therapy may transform congenital hemophilia from an incurable disease characterized by a severe phenotype into a moderate or even mild type of hemophilia. However, it is difficult to anticipate what proportion of patients with hemophilia may benefit from the gene therapy, as multiple factors have been described which could restrict its application [34]. For example, very effective treatments such as extended half-life recombinant products, or the new non-replacement-based therapies, are palliative treatments which, when administered prophylactically, are capable of boosting patients' quality of life. A second hurdle to their implementation is related to the uncertainties around their potential adverse events, mainly derived from transfection vectors (unknown risks and long-term outcomes, persistence of the therapeutic effect, need to readminister the vector). To this should be added the lack of experience of healthcare providers in managing the adverse events derived from the new advanced therapies and their lack of experience in calculating the cost-effectiveness of gene therapy (eligibility criteria, predictability of response, unknown risks, long-term complications). There is also considerable uncertainty as to whether payers will be willing to defray the high upfront costs of the one-and-done cure promised by gene therapy, especially considering that reimbursements are normally made based on real world results. In short, the role of regulatory agencies, payers, patients and healthcare providers will be key in identifying the patient subpopulations most likely to benefit from gene therapy and, therefore, aligning patients' interests with those of pharmaceutical companies.

Pursuing this strategy will require a definition of variables (serotype/dose of the viral vector, manufacturing techniques) capable of controlling the integration capacity of the different vectors and/or the immune response against the AAV capsid, as well as an evaluation of the patients' quality of life defined as an improvement in those outcomes that are most relevant for the patients (chronic pain, yearly infusion rate, mental health). Educational programs should be introduced to provide all stakeholders with the information they require about gene therapy as a therapeutic option, including the management of adverse events and the strategies conducive to ensuring the cost-effectiveness of the treatment.

3. Adeno-Associated Virus, the Vectors of Choice. Optimization

AAVs are currently the vectors of choice in gene therapy, particularly in the context of congenital coagulopathies. This is due to their good safety profile and their capability for partial integration, which ensures long-term expression of the transgene.

Nonetheless, some serotypes may be hepatotoxic and more immunogenic [35]. For that reason, transient immunosuppressive techniques have been developed to blunt the cell's immune response against the virus. Paulk et al. [36] reported two AAV serotypes (NP40 and NP59) capable of improving in vivo transduction of human hepatocytes in murine models.

Moreover, pre-existing antibodies against AAVs significantly affect the use of AAV vectors even in the presence of relatively low levels of neutralizing antibodies from natural parvovirus infections. Such capsid-targeted antibodies have been shown to be capable of inhibiting transduction of the intravenously administered vector in animal models and have been associated with limited effectiveness in clinical trials in humans [37]. This is important to determine the primary eligibility of patients for AAV-based gene therapy clinical trials.

Immunosuppression using drugs such as corticosteroids is nowadays necessary to ensure the clinical benefits of AAV-mediated gene transfer. Nonetheless, immunosuppressive drugs are associated with metabolic side effects and with an increased overinfection risk. For this reason, further research is needed to find new, less immunogenic AAV serotypes.

One of the currently most widely accepted hypotheses is that $CD8^+$ lymphocytes may play an important role in the subject's immune response against the vector capsid, eliminating the hepatocytes infected by the virus and triggering an ensuing reduction in the levels of the transgene and of the circulating therapeutic protein. This has prompted the development of new strategies aimed at eradicating the parts of the viral vector responsible for inducing the immune and inflammatory response or providing $CD8^+$ lymphocytes with epitopes that enable them to bind to AAVs, thus attenuating the T-lymphocytes' response [38].

In sum, a great deal of research is still needed into the optimization of AAV vectors, along two different paths. One should be directed at the development of new, less immunogenic serotypes, and the other at the standardization of the clinical dose of vector to be administered. Selecting the first doses of investigative drugs to be administered to humans on the basis of the results of preclinical studies has been, and still is in the context of advanced therapies, an important challenge for translational medicine. Despite the substantial advances made in AAV optimization, significant limitations still exist to the use of preclinical data to guide dose selection in clinical trials.

Tang et al. recently introduced a series of new concepts and parameters aimed at optimizing selection of the AAV vector dose [39]. The novel gene efficiency factor (GEF) concept was developed to describe the efficiency of the in vivo AAV-mediated gene transfer system. It indicates the number of molecules of the therapeutic protein produced per day by each genome of the administered vector (vg). The concept encompasses both the transfection of the virus and the transcription/translation of the transgene required by the gene therapy product to exert its biological effect, and also allows prediction of the most effective dose of the AAV vector for each species. This makes it possible to translate the biological efficacy observed in animal studies to clinical trials in humans, which provides a rational basis to determine the initial dose of the new in vivo virus-mediated gene therapy products.

4. Gene Therapy Educational Programs for Patients

At present, patients with hemophilia benefit from palliative treatments providing unprecedented levels of safety and efficacy. These are based on recombinant DNA techniques that offer high levels of safety against emerging pathogens, as well as high efficacy and long half-lives, allowing administration of one single dose every 10 days in the case of FIX and two doses every 10 days in the case of FVIII. This treatment concept makes it difficult for both doctors and patients to switch to a different alternative such as gene therapy, particularly given the uncertainties around its medium- and long-term adverse events. Moreover, it must be remembered that patients with hemophilia tend to be well informed about their disease, the treatments and therapeutic options available and even the management of their disease, as most of them are on home self-treatment. For these reasons, and given the high prevalence of pre-existing AAV antibodies, recruiting patients for gene therapy trials is typically challenging. This usually results in small sample sizes and low statistical significance levels. This highlights the urgency of designing communication and information campaigns aimed at both the clinical staff practicing in the different hemophilia units, and at the patients themselves [40]. The information provided must be objective and realistic, as well as clear and simple, especially when discussing highly specialized concepts related to the fields of genetics and molecular biology. This will definitely contribute to promoting a better understanding of the possibilities, potential risks, benefits and limitations of these new therapeutic protocols.

The information should be conveyed in the form of semi-structured informative sessions where patients should also be asked open-ended questions about their perceptions about the new treatments. In a study by Overbeeke et al. [41], the majority of patients with hemophilia showed a positive attitude toward gene therapy and claimed to be keen to receive this treatment (40%; n = 8) o (35%; n = 7). The criteria influencing patients' decisions included the annual bleeding rate, the level of factor, uncertainty about long-term risks, the impact of treatment on their daily lives and the possibility that prophylaxis may be discontinued. Consistent use of specific and generally accepted terms and an optimized narrative (verbal, written and pictorial language) to convey the information could minimize confusion and facilitate the making of informed decisions on the potential offered by gene therapy [42].

Other educational programs geared at patients and healthcare providers include more specific training on matters such as the use of AAVs, whose application in patients with hemophilia is becoming increasingly common. To be effective, these programs require the participation of academically experienced specialists [43]. They are often based on an information handbook that includes information on AAV-mediated gene transfer, the eligibility criteria for participation in clinical trials, the possibility of anticipating the adverse events that could occur in the course of a clinical trial following gene therapy, the parameters that should be monitored during the clinical trial and the long-term effects of the therapy. Handbooks also include easy-to-read infographic materials that physicians can use as visual aids during patient consultations. They also review the basic principles of liver-targeted AAV-mediated gene transfer to facilitate discussions between physicians, or with patients and their families.

5. The World Federation of Hemophilia Gene Therapy Registry and the Global Multidisciplinary Consensus Framework on Hemophilia Gene Therapy

Since May 2021, the World Federation of Hemophilia (WFH) has been working on a Gene Therapy Registry [44,45] intended to put together a worldwide database of patients with hemophilia receiving gene therapy. The purpose of the Registry is to gain a comprehensive insight into the conditions under which gene therapy is administered; the expected, unexpected and unknown safety issues; and the duration of the efficacy of therapy over the long term. Against the background of the imminent approval and release of gene therapy drugs addressed to patients with hemophilia, the WFH saw the need to carry out this initiative at a global scale. Every patient will be monitored for 15 years, and any adverse events occurring from the start of administration of therapy will be recorded at 3, 6, 9, 12, 18 and 24 months, and yearly thereafter. Levels of circulating factor and any bleeding episodes will also be recorded from the start. The only inclusion criterion will be to have a diagnosis of HA or HB, with no age distinction, and to have been prescribed gene therapy.

Another important aspect will be reaching a consensus on the use of a multidisciplinary approach to the application of the new gene therapy protocols in hemophilia [46]. Treatment of hemophilia has always been multidisciplinary, with the patient at the center and practitioners from different departments of the hospital working as a team to ensure the delivery of an optimal standard of care [47]. In the case of gene therapy, this is more necessary, and the multidisciplinary team should include practitioners from departments such as molecular biology, clinical pharmacology and virology.

The new gene therapy protocols will require hospital pharmacy departments to revise their storage, handling and reconstitution processes. At the same time, the gastroenterology and hepatology departments will have to monitor the hepatic function of recipients of gene therapy, especially during the first few months of treatment. Physical therapists for their part will have to follow up the condition of the patients' joints by means of regular physical examinations and musculoskeletal ultrasounds.

In a nutshell, careful multidisciplinary planning will be required [48] before any gene therapy procedures can be applied. The procedure will include an assessment and informed consent phase, dosing, short- and long-term treatment and follow-up. Adverse events and fluctuations in hepatic function will have to be properly managed, and requirements in terms of immunosuppression and availability of resources will have to be met.

6. Cost-Effectiveness of Gene Therapy for Hemophilia

Treatment of hemophilia based on plasma-derived products can be very costly, especially if such products are of a recombinant nature [49]. In the United States, the annual cost exceeds USD 300,000 per adult patient. Indirect costs such as the loss of production capacity and the costs arising from the physical disability resulting from hemophilic arthropathy are also considerable. Factor concentrates account for 90% of all direct costs of treating hemophilia.

Certain strategies have been implemented to reduce the economic impact of hemophilia, such as disease management programs and programs to control the price of drugs. Production of longer half-life recombinant coagulation factors capable of reducing dosing frequency has also greatly contributed to reducing costs, with comparable clinical outcomes [50].

As a potential cure for hemophilia, the new advanced therapies are also associated with a high cost, which limits their applicability and raises doubts about their cost-effectiveness and their equitable administration to patients with hemophilia worldwide, ensuring what the WHO has called "the absence of avoidable and remediable differences between groups of people" [51].

Machin et al. [52] carried out a study to evaluate the economics of gene therapy in patients with severe HA as compared with prophylaxis with exogenous FVIII using a Markov state-transition model to estimate the costs and efficacy of treatment of severe HA in the United States. They performed several unidirectional and probabilistic sensitivity analyses to determine the soundness of their results. Over a 10-year period, the total per-patient cost of gene therapy was USD 1.0 million and 8.33 QALYs (quality-adjusted life-years, a parameter that measures the burden of disease, including quality of life and the number of years lived, and evaluates the economic efficiency of medical interventions). The cost of prophylaxis was USD 1.7 million and 6.62 QALYs. The better results obtained by gene therapy, which showed itself to be less costly and more effective than prophylaxis, could mean that gene therapy-based treatment of severe HA is more economically efficient than prophylaxis with FVIII.

In the case of hemophilia B, Bolous et al. [53] compared the potential economic efficiency of AAV-mediated Factor IX-Padua gene therapy in patients with severe HB in the United States with on-demand FIX replacement and primary prophylaxis with standard FIX or extended half-life FIX products. The authors built a Markov microsimulation model using published data as a basis for the transition probabilities between health states and utilities. Over an 18-year period, gene therapy also showed itself to be more economically efficient than on demand or prophylactic treatment in patients with severe HB.

Although further economic studies are required to address this dilemma, it would be misleading to conclude that gene therapy is inherently more costly [54]. A rigorous economic evaluation of the new therapies requires a careful comparison between the costs and benefits of gene therapy and the standard of care without adverse events throughout the patient's lifetime, including the relevant adjustments for price distortions. This would allow a better estimation of the cost-effectiveness ratio of every treatment.

7. Current Gene Therapy Clinical Trials in Hemophilia

An important feature of hemophilia treatment is that circulating factor concentrations need not reach normal concentrations for the patient to obtain a therapeutic effect. Indeed, a slight increase in plasma levels (to more than 1% of the reference value) is enough to reduce the morbidity and mortality risk. This reduces the expectations regarding the target factor correction as it is not necessary to restore circulating factor levels to 100%.

The analysis made in this article is based on some of the clinical trials in progress on HA or HB reported in the ClinicalTrials.gov repository [55] and the European Union Clinical Trials Register [56]. A total of over 40 active gene therapy clinical trials were found on the subject of hemophilia, with an approximate 50/50 split between HA and HB. Table 1 shows the recombinant virus vectors used in the different clinical trials for HA and HB, and Table 2 includes a list of the most significant clinical trials in progress on HA and HB.

One of these clinical trials, sponsored by the Oxford University Hospitals NHS Foundation Trust, will analyze the potential impact of gene therapy on the lives of persons with hemophilia and their family members [57]. It is a multiple-cohort research study to be conducted among diverse groups within the hemophilia community whose lives may have been impacted by gene therapy. The study has been designed to allow patients and their families to tell their own life stories through narrative accounts. The narratives will represent a true sharing of experiences and offer an insight into how these patients and families have coped with hemophilia following the application of the new therapies.

7.1. Hemophilia A

AAV-based gene therapies for the treatment of HA are at an earlier stage of development than those for HB. The main problem in the case of FVIII is related to the packaging in the vector as the gene of this protein is much larger (7 kb) and exceeds the packaging capacity of AAVs, which is around 5 kb.

BioMarin Pharmaceutical pioneered the first clinical trial devoted to HA, using hepatic gene transfer with an AAV serotype 5 (AAV5) vector expressing B-domain deleted FVIII (BDD-FVIII-DQ, BMN 270) [58]. Many of the patients participating in the trial (88.8%) experienced a slight increase in their alanine aminotransferase (ALT) levels, which was accompanied by a reduction in the activity of FVIII in one of the subjects. The company has started another two phase 3 trials with vector doses of 4×10^{13} and 6×10^{13} vg/kg [59,60].

Pasi et al. [61] obtained long-term efficacy and optimal clinical and histological results in 15 adult subjects with severe HA who had received different doses of a single infusion of AAV5-hFVIII-SQ. Three years following the infusion, FVIII expression in patients receiving the highest dose (6×10^{13} vg/kg) was 20% of the reference value, the median number of bleeding episodes treated annually was 0 and the use of exogenous FVIII decreased from a median of 138.5 to 0 infusions a year.

BioMarin Pharmaceutical also started a phase 1/2 clinical trial (results still unavailable) to evaluate the effect of administering high doses of the valoctocogene roxaparvovec gene to patients with HA with preexisting antibodies against the AAV5 capsid [62].

Table 1. Recombinant viral vectors used in the different clinical trials currently in progress on hemophilia A (HA) and hemophilia B (HB).

Recombinant Vector	Type of Hemophilia
Adeno-associated virus serotype 5 vector containing a variant of B-domain deleted human FVIII [a] known as BMN 270)	HA
Recombinant adeno-associated virus serotype 6 vector encoding B-domain deleted human FVIII (known as SB-525)	HA
Adeno-associated virus serotype 8 vector containing a functional copy of the codon-optimized FVIII's cDNA encoding B-domain deleted FVIII	HA
Recombinant adeno-associated virus vector containing a bioengineered capsid and a codon-optimized expression cassette to drive the expression of the SQ form of a B-domain deleted human FVIII (known as SPK-8011)	HA
Non-replicating adeno-associated virus serotype 2 vector expressing the Padua variant (R338L) of human FIX [b], under the control of the liver-specific apolipoproteín E/alpha1-antitrypsin (hAAT) enhancer	HB
Adeno-associated virus serotype rh10 vector containing the human FIX gene	HB
Recombinant adeno-associated virus serotype 3 vector containing a codon-optimized expression cassette encoding a variant of human FIX known as FLT180a	HB
Recombinant adeno-associated virus vector containing codon-optimized FIX-Padua (known as AMT-061)	HB
Lentiviral vector encoding human FVIII or FIX	HA and HB

[a] FVIII, factor VIII. [b] FIX, factor IX.

7.2. Hemophilia B

Generally, the results of the trials carried out so far on the use of AAVs in patients with HA and HB have yielded promising results. A clinical trial sponsored by the St. Jude Children's Research Hospital in patients with HB, which used a scAAV2/8-LP1-hFIXco vector, was the first to confirm the long-term benefit of gene therapy with this kind of AAV vector [63,64]. The authors evaluated the stability and long-term safety of the expression of the transgene in 10 patients with severe HB, all of whom were infused with a single dose of AAV8 vector. A vector dose-dependent increase in circulating FIX of 1% to 6% was observed over a mean 3-year period, which made it possible to reduce the frequency of prophylactic administration of FIX and, in some cases, even suspend it. All the patients, however, developed capsid-specific antibodies. The main adverse effect encountered was an increase in hepatic enzyme levels (ALT), which was treated with prednisolone.

Table 2. Main clinical trials underway in the field of hemophilia A (HA) and hemophilia B (HB).

Title	NCT Number	Intervention	Sponsor
Gene therapy study in patients with severe HA with antibodies against AAV5 [a]	NCT03520712	Adeno-associated virus serotype 5 vector containing a B-deleted variant of FVIII [b] (valoctocogene roxaparvovec)	BioMarin Pharmaceutical
Gene therapy study in patients with severe HA	NCT02576795	Valoctocogene roxaparvovec-BMN 270	BioMarin Pharmaceutical
Study evaluating the efficacy and safety of valoctocogene roxaparvovec in patients with HA	NCT03370913	Valoctocogene roxaparvovec	BioMarin Pharmaceutical
Study evaluating the efficacy and safety of volactocogene roxaparvovec combined with prophylactic administration of corticosteroids in HA	NCT04323098	Valoctocogene roxaparvovec	BioMarin Pharmaceutical
Single-arm study evaluating the efficacy and safety of a dose of 4×10^{13} vg/kg of valoctocogene roxaparvovec in patients with HA	NCT03391974	Valoctocogene roxaparvovec	BioMarin Pharmaceutical
Gene therapy for HA	NCT03001830	New adeno-associated virus serotype 8 capsid vector pseudotype encoding FVIII-V3 (AAV2/8-HLP-FVIII-V3)	University College, London/Medical Research Council
Safety and dose-escalation study of an adeno-associated virus vector used as gene therapy in HA	NCT03370172	Adeno-associated virus serotype 8 (AAV8) expressing FVIII Factor VIII (BDD-FVIII) (BAX 888)	Baxalta, now part of Shire
Study evaluating the efficacy and safety of PF-07055480 in adults with moderate or severe HA	NCT04370054	Recombinant AAV2/6 encoding B-domain deleted FVIII cDNA	UniQure Biopharma BV

Table 2. *Cont.*

Title	NCT Number	Intervention	Sponsor
Gene therapy study of recombinant AAV2/6 with the FVIII gene (SB-525) in patients with severe HA	NCT03061201	Recombinant adeno-associated virus serotype 6 (AAV6) encoding B-domain deleted human FVIII cDNA	Pfizer
Study of AAV5-hFIX [c] in patients with moderate or severe HB	NCT02396342	AAV5 containing the human FIX gene (AAV5-hFIX)	UniQure Biopharma BV
Dose confirmation trial of AAV5-hFIXco-Padua	NCT03489291	Recombinant adeno-associated virus serotype 5 (AAV5) vector containing the Padua variant of a codon-optimized complementary human FIX under the control of a liver-specific promoter (AAV5-hFIXco-Padua, AMT-061)	UniQure Biopharma BV
HOPE-B: Study of AMT-061 in patients with moderate or severe HB	NCT03569891	AAV5-hFIXco-Padua, AMT-061	UniQure Biopharma BV
Single ascending dose of adeno-associated virus serotype 8 of FIX in adults with HB	NCT01687608	Adeno-associated virus serotype 8 for FIX gene therapy (AskBio009)	Baxalta now part of Shire
Phase 1–2 study of SHP648, an adeno-associated virus vector for gene therapy in patients with HB	NCT04394286	Adeno-associated virus serotype 8 (AAV8) vector expressing FIX Padua (SHP648)	Baxalta now part of Shire
Dose-escalation study of a complementary adeno-associated virus gene therapy vector in patients with HB	NCT00979238	Self-complementary adeno-associated virus serotype 8 (AAV8) vector expressing a transgene of codon-optimized FIX (scAAV2/8-LP1-hFIXco)	St. Jude Children's Research Hospital/National Heart, Lung, and Blood Institute (NHLBI)/Hemophilia of Georgia, Inc./Children's Hospital of Philadelphia/University College, London

Table 2. *Cont.*

Title	NCT Number	Intervention	Sponsor
Gene therapy study of recombinant AAV2/6 with the FVIII gene (SB-525) in patients with severe HA	NCT03061201	Recombinant adeno-associated virus serotype 6 (AAV6) encoding B-domain deleted human FVIII cDNA	Pfizer
Study of AAV5-hFIX [c] in patients with moderate or severe HB	NCT02396342	AAV5 containing the human FIX gene (AAV5-hFIX)	UniQure Biopharma BV
Long-term safety and efficacy study of SPK-9001 in patients with HB	NCT03307980	Non-replicating adeno-associated virus serotype 2 (AAV2) vector expressing the Padua variant (R338L) of human FIX under the control of the liver-specific apolipoproteín E (Apo E) enhancer (PF-06838435/fidanacogene elaparvovec)	Pfizer
Study evaluating the efficacy and safety of gene therapy with PF-06838435 in adult males with moderate or severe HB	NCT03861273	Fidanacogene elaparvovec	Pfizer
Lentiviral FVIII Gene Therapy	NCT03217032	Lentiviral factor VIII gene. Modified autologous stem cells (YUVAGT-F801)	Shenzhen Geno-Immune Medical Institute
Lentiviral FIX Gene Therapy	NCT03961243	Lentiviral factor IX gene. Modified autologous stem cells (YUVA-GT-F901)	Shenzhen Geno-Immune Medical Institute

[a] AAV, adeno-associated viral vector. [b] FVIII, factor VIII. [c] FIX, factor IX.

University College London has started a new phase 1/2 clinical trial focused on HB, using the AAV-F9, FLT180a vector [65].

UniQure Biopharma B.V. recently published the results of a series of clinical trials on HB using an AAV5 vector that uses the same FIX expression cassette as the Jude Children's Research Hospital (AMT-060) but with the Padua variant of FIX [66,67]. The cassette used in this study was the same wild-type human FIX gene cassette tested previously [68], also with the AAV5 serotype. The study achieved similar steady-state FIX activity levels to a previous study by Nathwani et al. [63], who used an AAV8 vector. However, the highest vector dose used in the UniQure Biopharma BV study was higher than the AAV8 vector dose used by Nathwani et al. The authors concluded that pseudotyped scAAV5 vectors ought to be effective in the treatment of individuals with preexisting immunity to AAV5 [69,70].

In a recent clinical trial, Konkle et al. [71] analyzed safety, pharmacokinetic profile, FIX activity and immune response in patients undergoing a gene therapy protocol based on the AAV virus serotype 8 (AskBio009) [72]. Eight adult males were administered different doses of the gene therapy product BAX 335 IV (2.0×10^{11}; 1.0×10^{12}; or 3.0×10^{12} vg/kg), with three patients (37.5%) developing severe complications, all of them considered to be unrelated to BAX 335. No clinical thromboses, inhibitors or immunity reactions were observed against the Padua variant of FIX.

7.3. Padua Variants of Factor IX. An Exciting Alternative for Gene Therapy in HB

Although the results of AAV-based gene therapy using FIX as a transgene have been encouraging in the context of HB, increases in hepatotoxicity resulting from the loss of expression of the transgene are not uncommon, possibly due to adaptative or innate immune responses against the vector's capsid.

One alternative to reduce hepatotoxicity is to reduce the vector dose, but this usually results in a decreased expression efficacy. Another possibility is to modify the transgene itself to boost the efficiency of the coagulation factor once it has expressed itself. In this regard, Lombardi et al. [73] recently applied protein fusion engineering to design a FIX-HSA (human serum albumin) variant that lengthens the half-life of the FIX molecule. This combination could be used as a transgene in gene therapy protocols in the context of HB. Yet another alternative would be to use some variant of the protein endowed with greater clotting capacity, such as the Padua variant of FIX [74]. Juvenile thrombophilia is associated with the presence of the Padua variant of FIX, which is characterized by the substitution of leucine by arginine at position 338 (R338L) (FIX-R338L). Plasma concentrations of the FIX-R338L protein are normal but its clotting activity is approximately eight times higher than normal. In vitro, the specific activity of recombinant FX-R338L is 5–10 times higher than that of wild-type recombinant FIX (FIX-WT). Substitution of R338L leads to a gain-of-function mutation, which in turn results in a hyperfunctional FIX.

The clinical and preclinical experience with FIX-R338L gene therapy in the realm of HB has not provided evidence of a higher thrombogenic or immunogenic risk [68,75,76]. However, as the mechanism underlying R338L's clotting hyperactivity remains unknown, the potential adverse events associated with a random coagulation pattern and the possibility of developing thrombotic complications cannot be ascertained. Samelson-Jones et al. [77] showed that the high specific activity of FIX-R338L requires FVIIIa as a cofactor. The molecular regulation of the activation, inactivation and FVIIIa-dependence of FIX-R338L and FIX-WT are similar, but the FVIIIa-dependent hyperactivity of FIX-R338L is the result of a faster activation rate of FX. These findings helped dispel the fears associated with the unregulated clotting pattern of FIX-R338L and support its use in gene therapy protocols addressed at HB. Another study by Samelson-Jones et al. in dogs with HB [78] confirmed an increased specific in vivo activity of FIX-R338L as compared with FIX-WT following gene therapy with AAV6.

At present, Pfizer is conducting a clinical trial with the Padua variant of FIX [79,80]. The study is intended to evaluate the efficacy and safety of PF-06838435 (rAAV-SPARK100-HFIX-PADUA) in male adult subjects with moderate or severe HB, with circulating FIX concentrations <2%.

Nair et al. [81,82] tested a new modification of the Padua variant of FIX. In clinical trials with FIX-R338L, they found that some patients showed an increase in liver transaminase levels, which was correlated with the loss of expression of FIX, even in immunosuppressed subjects. These results underscore the urgency to look for new, more effective variants that may allow a reduction in the vector dose and, therefore, prevent potential adverse events. The study described how new variants could be generated.

Thus, dalcinonacog alpha, a new variant of R338L-Padua also known as CB 2679d-GT, containing three amino acid substitutions (R318Y, R338E, T343R), was shown to be more potent than R338L-Padua following AAV-based gene therapy in hemophilic mice. A significant (five to eight-fold) reduction in bleeding time and of the total amount of blood lost (approximately four-fold) was obtained compared to R338L-Padua, which allows a faster and more robust hemostatic correction. FIX expression remained stable for at least 20 weeks, and no significant increases in immunogenicity were observed. This novel gene therapy study demonstrated the superiority of CB 2679d-GT, highlighting its potential to obtain higher levels of FIX activity and superior hemostatic efficiency following AAV-based gene therapy in patients with HB.

8. Discussion

The current treatment of choice for hemophilia is fundamentally based on intravenous prophylactic replacement of the deficient factor by exogenous recombinant factors, characterized by their longer half-lives, their robust safety profile against emerging pathogens and their high efficacy. This has allowed patients to lead normal lives and resume their social and occupational activities.

The drawbacks of this treatment include its high administration frequency and inconvenient route of administration, as well as the development of inhibitors by up to 30% of patients against the infused coagulation factors.

Advanced (gene and cell) therapies could provide a "cure" for many hereditary conditions such as hemophilia. Gene and cell therapy protocols are appropriate for monogenic and polygenic diseases and should allow expression of the deficient factor over the long term and the maintenance of steady-state plasma concentrations. Protocols are currently highly variable on account of the wide range of target cell types and gene transfection vectors available and the possibility of modifying the transgene to boost its expression efficacy. The investment and dedication required by these procedures is fully justified as many of the conditions they are meant to treat are chronic and/or severe and either lack a curative treatment or are associated with tedious or inconvenient treatments or therapies, leading to significant side effects which hinder patient adherence.

Although gene therapy protocols are in general more efficient than cell therapy ones, they may be associated with a higher incidence of adverse events, derived mainly from the viral transfection vector used. The vector's immunogenicity and hepatotoxicity and the problems arising from its integration (insertional mutagenesis) are the main hurdles to overcome. The choice between a viral or non-viral vector, and between the different kinds of viral vectors available, must be based on a compromise between the phenotypic features of the disease and the therapeutic targets pursued. Highly effective vectors usually result in a faster and more efficient translation to clinical practice. AAV vectors are currently the ones presenting with the highest potential for clinical use given their partial integration capacity and their low insertional mutagenesis risk. It must, however, be remembered that they may lead to an intense immunogenic response and, in some cases, to hepatotoxicity.

One alternative to therapeutic gene-based gene therapy is correction of the defective genes causing the condition through a wide range of gene-editing tools (TALENs, ZFNs and CRISPR/Cas9).

The advances made in the last few years by gene therapy for the treatment of congenital coagulopathies have caused an unprecedented upheaval, specifically in the treatment of patients with HA and HB. Indeed, the use of AAV vectors has provided extraordinarily promising results in terms of the clotting factors' concentrations and expression times. However, further work will be necessary to address the side effects related to the immunogenicity and hepatotoxicity of these therapies, which usually make it necessary to prescribe concomitant administration of immunosuppressive corticosteroids.

As a result, switching the therapeutic strategy from the current optimized and safe standard-of-care treatment based on recombinant products to a gene therapy strategy is not easy, especially considering the uncertainties around the medium- and long-term effects of gene therapy protocols, the lack of experience with these new procedures and the potentially unfavorable cost-effectiveness ratio associated with them. For these reasons, further research must be conducted with a view to optimizing AAV vectors, and should come up with less immunogenic and hepatotoxic serotypes. Concomitant use of corticosteroid-based immunosuppressants should be avoided given their metabolic side effects and the fact that they may increase the risk of overinfection [83].

Providing patients with hemophilia with adequate information is now more necessary than ever. Information must be objective and realistic, but at the same time clear and easy to understand, especially when dealing with highly specialized concepts related to genetics and molecular biology. This will contribute to avoiding unrealistic patient expectations and promoting a deeper understanding of the potential risks, benefits and limitations of the new therapeutic protocols. It will also be necessary to keep a detailed record of patients with hemophilia receiving gene therapy across the world in order to track expected, unexpected and unknown safety issues and obtain a better insight into the maintenance of the protocols' efficacy in the long term. Appropriate multidisciplinary planning will also be necessary at the different stages of the procedure, including evaluation, informed consent, dosing and short- and long-term treatment and follow-up.

Although treatments based on advanced therapies are costly, several economic forecasting studies have demonstrated that they could become less costly than the currently used regimens. At the same time, every effort should be made to address the current inequalities in access to treatment by promoting social access and ensuring that patients receive the treatments they need regardless of geographical or economic factors.

As regards the efficacy of gene therapy protocols in the context of hemophilia, expectations can be relaxed to some extent as even a small increase in circulating factor concentrations to more than 1% of the reference value, which corresponds to a severe phenotype, would result in a significant reduction in morbidity and mortality.

In the case of HA, use of AAV vectors is at an earlier stage of development than in HB. Solid foundations should be laid to provide an effective solution to the issue of the FVIII gene that is too large to be packaged into a given vector, using a gene deleted from the B domain of FVIII. This is not an issue in HB as the FIX gene is smaller than the FVIII gene. The problem for both HA and HB remains the immunogenicity and hepatotoxicity of the AAV vector, which are related to the loss of expression of the transgene. Although one alternative is concomitant treatment with corticosteroid immunosuppressants, these drugs should be avoided given their poor side effect profile [83]. Reduction in the vector dose should also be avoided as it is often associated with a less efficient expression of the protein. A third alternative is to modify the transgene to boost the efficacy of the clotting factor once it has been expressed. This last possibility, together with the use of variants with a higher clotting ability, such as the Padua variant of FIX (factor IX-R338L), currently represents an exciting alternative for HB patients. The greater coagulating activity of the protein expressed from a transgene variant makes it possible to reduce the vector dose without affecting expression efficacy, thus minimizing the risk of adverse events.

More recently, a subvariant of FIX-R338L (CB 2679d-GT) that contains three amino acid substitutions (R318Y, R338E, T343R) has been shown to be more effective in reducing bleeding times and total blood loss in gene therapy protocols with AAV vectors. FIX expression persisted for at least 20 weeks and immunogenicity remained unchanged.

Gene therapy could cause a great upheaval in the treatment of patients with hemophilia and other congenital diseases. Although more work is needed to increase treatment efficacy and reduce adverse events such as immunogenicity and hepatotoxicity, these protocols could allow curation of the disease, thereby increasing the patients' quality of life, and a reduction in both direct and indirect costs throughout the lifetime of patients suffering from chronic lifelong diseases [54].

Protocols based on gene therapy have gone from being potentially applicable ideas to becoming tangible realities [84], and they are bound to herald a new era where analyzing the specific phenotype of a monogenic hereditary disease such as hemophilia may not be needed anymore [85]. Gene therapy is indeed a tangible reality and the first gene therapy-based products targeted at hemophilia will be available in the short term [86].

We are already well into the 21st century, an era in which precision and personalized medicine [87] are gaining huge momentum in the realm of pharmacology. The rise of gene therapy is a good example of that. Nevertheless, this has not happened without significant controversy between its advocates and its critics regarding its promises, limits, possibilities and opportunities. Although huge strides have been made in our understanding of the molecular mechanisms behind diseases and in the development of drugs that have had a hugely positive impact on the treatment of some types of cancer, many of the promises of gene therapy remain unfulfilled for other diseases as a result of the hurdles to the therapy's widespread use. The high cost of the new therapies could exacerbate existing inequalities and become a problem for the sustainability of healthcare systems, particularly in low- and medium-income countries. For that reason, it is of the essence for the international community to implement the advances made in an equitable way so as to reduce healthcare disparities [88].

The ultimate goal of the treatment of hemophilia should be a "functional cure", but also "healthcare equity". A "functional cure" means that patients should live as normal a life as possible, with minimum joint impairment, an absence of spontaneous bleeds, "normal" mobility, an ability to overcome minor trauma without requiring surgery or any other intervention and normal hemostasis. At the same time, the principle of healthcare equity should be upheld. By eliminating the need for (exogenous) plasma-derived concentrates or recombinant coagulation factors, gene therapy can make healthcare equality possible [89].

Author Contributions: E.C.R.-M. and A.L.: Bibliographic research, established the plan of the review and wrote the first draft of the manuscript. J.A.D.P.-M.: Bibliographic research and figure design. All authors have read and agreed to the published version of the manuscript.

Funding: This research received no external funding.

Conflicts of Interest: The authors declare no conflict of interest.

References

1. Castaman, G.; Matino, D. Hemophilia A and B: Molecular and Clinical Similarities and Differences. *Haematologica* **2019**, *104*, 1702–1709. [CrossRef]
2. Mannucci, P.M.; Franchini, M. Is Haemophilia B Less Severe than Haemophilia A? *Haemophilia* **2013**, *19*, 499–502. [CrossRef]
3. Trinchero, A.; Sholzberg, M.; Matino, D. The Evolution of Hemophilia Care: Clinical and Laboratory Advances, Opportunities, and Challenges. *Hämostaseologie* **2020**, *40*, 311–321. [CrossRef]
4. Tomeo, F.; Mariz, S.; Brunetta, A.L.; Stoyanova-Beninska, V.; Penttila, K.; Magrelli, A. Haemophilia, State of the Art and New Therapeutic Opportunities, a Regulatory Perspective. *Br. J. Clin. Pharmacol.* **2021**. [CrossRef]
5. Shima, M.; Hanabusa, H.; Taki, M.; Matsushita, T.; Sato, T.; Fukutake, K.; Fukazawa, N.; Yoneyama, K.; Yoshida, H.; Nogami, K. Factor VIII–Mimetic Function of Humanized Bispecific Antibody in Hemophilia A. *N. Engl. J. Med.* **2016**, *374*, 2044–2053. [CrossRef]
6. Shetty, S.; Vora, S.; Kulkarni, B.; Mota, L.; Vijapurkar, M.; Quadros, L.; Ghosh, K. Contribution of Natural Anticoagulant and Fibrinolytic Factors in Modulating the Clinical Severity of Haemophilia Patients. *Br. J. Haematol.* **2007**, *138*, 541–544. [CrossRef] [PubMed]
7. Pasi, K.J.; Rangarajan, S.; Georgiev, P.; Mant, T.; Creagh, M.D.; Lissitchkov, T.; Bevan, D.; Austin, S.; Hay, C.R.; Hegemann, I.; et al. Targeting of Antithrombin in Hemophilia A or B with RNAi Therapy. *N. Engl. J. Med.* **2017**, *377*, 819–828. [CrossRef]
8. Machin, N.; Ragni, M.V. An Investigational RNAi Therapeutic Targeting Antithrombin for the Treatment of Hemophilia A and B. *J. Blood Med.* **2018**, *9*, 135–140. [CrossRef] [PubMed]
9. Dockal, M.; Hartmann, R.; Fries, M.; Thomassen, M.C.L.G.D.; Heinzmann, A.; Ehrlich, H.; Rosing, J.; Osterkamp, F.; Polakowski, T.; Reineke, U.; et al. Small Peptides Blocking Inhibition of Factor Xa and Tissue Factor-Factor VIIa by Tissue Factor Pathway Inhibitor (TFPI). *J. Biol. Chem.* **2014**, *289*, 1732–1741. [CrossRef] [PubMed]
10. Weyand, A.C.; Grzegorski, S.J.; Rost, M.S.; Lavik, K.I.; Ferguson, A.C.; Menegatti, M.; Richter, C.E.; Asselta, R.; Duga, S.; Peyvandi, F.; et al. Analysis of Factor V in Zebrafish Demonstrates Minimal Levels Needed for Early Hemostasis. *Blood Adv.* **2019**, *3*, 1670–1680. [CrossRef] [PubMed]
11. Shukla, V.; Seoane-Vazquez, E.; Fawaz, S.; Brown, L.; Rodriguez-Monguio, R. The Landscape of Cellular and Gene Therapy Products: Authorization, Discontinuations, and Cost. *Hum. Gene Ther. Clin. Dev.* **2019**, *30*, 102–113. [CrossRef]
12. Khan, W.S.; Hardingham, T.E. Mesenchymal Stem Cells, Sources of Cells and Differentiation Potential. *J. Stem Cells* **2012**, *7*, 75–85. [PubMed]
13. Kolios, G.; Moodley, Y. Introduction to Stem Cells and Regenerative Medicine. *Respiration* **2013**, *85*, 3–10. [CrossRef] [PubMed]
14. Bacakova, L.; Zarubova, J.; Travnickova, M.; Musilkova, J.; Pajorova, J.; Slepicka, P.; Kasalkova, N.S.; Svorcik, V.; Kolska, Z.; Motarjemi, H.; et al. Stem Cells: Their Source, Potency and Use in Regenerative Therapies with Focus on Adipose-Derived Stem Cells—A Review. *Biotechnol. Adv.* **2018**, *36*, 1111–1126. [CrossRef] [PubMed]
15. Ma, Q.; Liao, J.; Cai, X. Different Sources of Stem Cells and Their Application in Cartilage Tissue Engineering. *Curr. Stem Cell Res. Ther.* **2018**, *13*, 568–575. [CrossRef] [PubMed]

16. Olgasi, C.; Talmon, M.; Merlin, S.; Cucci, A.; Richaud-Patin, Y.; Ranaldo, G.; Colangelo, D.; Di Scipio, F.; Berta, G.N.; Borsotti, C.; et al. Patient-Specific IPSC-Derived Endothelial Cells Provide Long-Term Phenotypic Correction of Hemophilia A. *Stem Cell Rep.* **2018**, *11*, 1391–1406. [CrossRef]
17. He, Q.; Wang, H.-H.; Cheng, T.; Yuan, W.-P.; Ma, Y.-P.; Jiang, Y.-P.; Ren, Z.-H. Genetic Correction and Hepatic Differentiation of Hemophilia B-Specific Human Induced Pluripotent Stem Cells. *Chin. Med. Sci. J.* **2017**, *32*, 135–144. [CrossRef]
18. Zhuang, W.-Z.; Lin, Y.-H.; Su, L.-J.; Wu, M.-S.; Jeng, H.-Y.; Chang, H.-C.; Huang, Y.-H.; Ling, T.-Y. Mesenchymal Stem/Stromal Cell-Based Therapy: Mechanism, Systemic Safety and Biodistribution for Precision Clinical Applications. *J. Biomed. Sci.* **2021**, *28*, 28. [CrossRef]
19. Wu, X.; Jiang, J.; Gu, Z.; Zhang, J.; Chen, Y.; Liu, X. Mesenchymal Stromal Cell Therapies: Immunomodulatory Properties and Clinical Progress. *Stem Cell Res. Ther.* **2020**, *11*, 345. [CrossRef]
20. High, K.A.; Roncarolo, M.G. Gene Therapy. *N. Engl. J. Med.* **2019**, *381*, 455–464. [CrossRef]
21. Bolt, M.W.; Brady, J.T.; Whiteley, L.O.; Khan, K.N. Development Challenges Associated with RAAV-Based Gene Therapies. *J. Toxicol. Sci.* **2021**, *46*, 57–68. [CrossRef]
22. Anguela, X.M.; High, K.A. Entering the Modern Era of Gene Therapy. *Annu. Rev. Med.* **2019**, *70*, 273–288. [CrossRef]
23. Wagner, H.J.; Weber, W.; Fussenegger, M. Synthetic Biology: Emerging Concepts to Design and Advance Adeno-Associated Viral Vectors for Gene Therapy. *Adv. Sci.* **2021**, *8*, 2004018. [CrossRef]
24. Toon, K.; Bentley, E.M.; Mattiuzzo, G. More than Just Gene Therapy Vectors: Lentiviral Vector Pseudotypes for Serological Investigation. *Viruses* **2021**, *13*, 217. [CrossRef]
25. Cantore, A.; Naldini, L. WFH State-of-the-art Paper 2020: In Vivo Lentiviral Vector Gene Therapy for Haemophilia. *Haemophilia* **2021**, *27*, 122–125. [CrossRef]
26. Barman, H.K.; Rasal, K.D.; Chakrapani, V.; Ninawe, A.S.; Vengayil, D.T.; Asrafuzzaman, S.; Sundaray, J.K.; Jayasankar, P. Gene Editing Tools: State-of-the-Art and the Road Ahead for the Model and Non-Model Fishes. *Transgenic Res.* **2017**, *26*, 577–589. [CrossRef] [PubMed]
27. Gupta, S.K.; Shukla, P. Gene Editing for Cell Engineering: Trends and Applications. *Crit. Rev. Biotechnol.* **2017**, *37*, 672–684. [CrossRef] [PubMed]
28. U.S. National Library of Medicine. Ascending Dose Study of Genome Editing by Zinc Finger Nuclease Therapeutic SB-FIX in Subjects with Severe Hemophilia B. ClinicalTrials.gov Identifier: NCT02695160. Available online: https://clinicaltrials.gov/ct2/show/NCT02695160?term=NCT02695160&draw=2&rank=1 (accessed on 28 June 2021).
29. Adachi, H.; Hengesbach, M.; Yu, Y.-T.; Morais, P. From Antisense RNA to RNA Modification: Therapeutic Potential of RNA-Based Technologies. *Biomedicines* **2021**, *9*, 550. [CrossRef] [PubMed]
30. Arruda, V.R.; Weber, J.; Samelson-Jones, B.J. Gene Therapy for Inherited Bleeding Disorders. *Semin. Thromb. Hemost.* **2021**, *47*, 161–173. [CrossRef]
31. Swystun, L.L.; Lillicrap, D. Gene Therapy for Coagulation Disorders. *Circ. Res.* **2016**, *118*, 1443–1452. [CrossRef]
32. Chapin, J.C.; Monahan, P.E. Gene Therapy for Hemophilia: Progress to Date. *BioDrugs* **2018**, *32*, 9–25. [CrossRef]
33. Atchison, R.W.; Casto, B.C.; Hammon, W. McD. Adenovirus-Associated Defective Virus Particles. *Science* **1965**, *149*, 754–755. [CrossRef]
34. Spadarella, G.; Di Minno, A.; Brunetti-Pierri, N.; Mahlangu, J.; Di Minno, G. The Evolving Landscape of Gene Therapy for Congenital Haemophilia: An Unprecedented, Problematic but Promising Opportunity for Worldwide Clinical Studies. *Blood Rev.* **2021**, *46*, 100737. [CrossRef]
35. Verdera, H.C.; Kuranda, K.; Mingozzi, F. AAV Vector Immunogenicity in Humans: A Long Journey to Successful Gene Transfer. *Mol. Ther.* **2020**, *28*, 723–746. [CrossRef]
36. Paulk, N.K.; Pekrun, K.; Zhu, E.; Nygaard, S.; Li, B.; Xu, J.; Chu, K.; Leborgne, C.; Dane, A.P.; Haft, A.; et al. Bioengineered AAV Capsids with Combined High Human Liver Transduction In Vivo and Unique Humoral Seroreactivity. *Mol. Ther.* **2018**, *26*, 289–303. [CrossRef]
37. Daniel, H.D.-J.; Kumar, S.; Kannangai, R.; Lakshmi, K.M.; Agbandje-Mckenna, M.; Coleman, K.; Srivastava, A.; Srivastava, A.; Abraham, A.M. Prevalence of Adeno-Associated Virus 3 Capsid Binding and Neutralizing Antibodies in Healthy and Hemophilia B Individuals from India. *Hum. Gene Ther.* **2021**, *32*, 451–457. [CrossRef]
38. Ertl, H.C.J. T Cell-Mediated Immune Responses to AAV and AAV Vectors. *Front. Immunol.* **2021**, *12*, 666666. [CrossRef] [PubMed]
39. Tang, F.; Wong, H.; Ng, C.M. Rational Clinical Dose Selection of Adeno-Associated Virus-Mediated Gene Therapy Based on Allometric Principles. *Clin. Pharmacol. Ther.* **2021**. [CrossRef]
40. Miesbach, W.; O'Mahony, B.; Key, N.S.; Makris, M. How to Discuss Gene Therapy for Haemophilia? A Patient and Physician Perspective. *Haemophilia* **2019**, *25*, 545–557. [CrossRef] [PubMed]
41. Overbeeke, E.; Michelsen, S.; Hauber, B.; Peerlinck, K.; Hermans, C.; Lambert, C.; Goldman, M.; Simoens, S.; Huys, I. Patient Perspectives Regarding Gene Therapy in Haemophilia: Interviews from the PAVING Study. *Haemophilia* **2021**, *27*, 129–136. [CrossRef] [PubMed]
42. Hart, D.P.; Branchford, B.R.; Hendry, S.; Ledniczky, R.; Sidonio, R.F.; Négrier, C.; Kim, M.; Rice, M.; Minshall, M.; Arcé, C.; et al. Optimizing Language for Effective Communication of Gene Therapy Concepts with Hemophilia Patients: A Qualitative Study. *Orphanet J. Rare Dis.* **2021**, *16*, 189. [CrossRef]

43. Sidonio, R.F.; Pipe, S.W.; Callaghan, M.U.; Valentino, L.A.; Monahan, P.E.; Croteau, S.E. Discussing Investigational AAV Gene Therapy with Hemophilia Patients: A Guide. *Blood Rev.* **2021**, *47*, 100759. [CrossRef]
44. Konkle, B.A.; Coffin, D.; Pierce, G.F.; Clark, C.; George, L.; Iorio, A.; Mahlangu, J.; Naccache, M.; O'Mahony, B.; Peyvandi, F.; et al. World Federation of Hemophilia Gene Therapy Registry. *Haemophilia* **2020**, *26*, 563–564. [CrossRef]
45. U.S. National Library of Medicine. The World Federation of Hemophilia Gene Therapy Registry (WFH GTR). ClinicalTrials.gov Identifier: NCT04883710. Available online: https://clinicaltrials.gov/ct2/show/record/NCT04883710?cond=Hemophilia&sort=nwst&draw=2&rank=2 (accessed on 28 June 2021).
46. Pierce, G.F.; Pasi, K.J.; Coffin, D.; Kaczmarek, R.; Lillicrap, D.; Mahlangu, J.; Rottellini, D.; Sannié, T.; Srivastava, A.; VandenDriessche, T.; et al. Towards a Global Multidisciplinary Consensus Framework on Haemophilia Gene Therapy: Report of the 2nd World Federation of Haemophilia Gene Therapy Round Table. *Haemophilia* **2020**, *26*, 443–449. [CrossRef]
47. Drayton Jackson, M.; Bartman, T.; McGinniss, J.; Widener, P.; Dunn, A.L. Optimizing Patient Flow in a Multidisciplinary Haemophilia Clinic Using Quality Improvement Methodology. *Haemophilia* **2019**, *25*, 626–632. [CrossRef] [PubMed]
48. Miesbach, W.; Pasi, K.J.; Pipe, S.W.; Hermans, C.; O'Mahony, B.; Guelcher, C.; Steiner, B.; Skinner, M.W. Evolution of Haemophilia Integrated Care in the Era of Gene Therapy: Treatment Centre's Readiness in United States and EU. *Haemophilia* **2021**. [CrossRef]
49. Rodriguez-Merchan, E.C. The Cost of Hemophilia Treatment: The Importance of Minimizing It without Detriment to Its Quality. *Expert Rev. Hematol.* **2020**, *13*, 269–274. [CrossRef] [PubMed]
50. Mannucci, P.M.; Cortesi, P.A.; Di Minno, M.N.D.; Sanò, M.; Mantovani, L.G.; Di Minno, G. Comparative Analysis of the Pivotal Studies of Extended Half-life Recombinant FVIII Products for Treatment of Haemophilia A. *Haemophilia* **2021**. [CrossRef] [PubMed]
51. Equity vs. Equality: What's the Difference? World Health Organization. Milken Institute School of Public Health. 2020. Available online: https://onlinepublichealth.gwu.edu/resources/equity-vs-equality/ (accessed on 28 June 2021).
52. Machin, N.; Ragni, M.V.; Smith, K.J. Gene Therapy in Hemophilia A: A Cost-Effectiveness Analysis. *Blood Adv.* **2018**, *2*, 1792–1798. [CrossRef] [PubMed]
53. Bolous, N.S.; Chen, Y.; Wang, H.; Davidoff, A.M.; Devidas, M.; Jacobs, T.W.; Meagher, M.M.; Nathwani, A.C.; Neufeld, E.J.; Piras, B.A.; et al. The Cost-Effectiveness of Gene Therapy for Severe Hemophilia B: Microsimulation Study from the United States Perspective. *Blood* **2021**. [CrossRef]
54. Garrison, L.P.; Jiao, B.; Dabbous, O. Gene Therapy May Not Be as Expensive as People Think: Challenges in Assessing the Value of Single and Short-Term Therapies. *J. Manag. Care Spec. Pharm.* **2021**, *27*, 674–681. [CrossRef] [PubMed]
55. ClinicalTrials.gov. Available online: https://clinicaltrials.gov/ct2/home (accessed on 28 June 2021).
56. EU Clinical Trials Register. Available online: https://www.clinicaltrialsregister.eu/ (accessed on 28 June 2021).
57. U.S. National Library of Medicine. An Exploration of the Impact of Gene Therapy on the Lives of People with Haemophilia and Their Families. ClinicalTrials.gov Identifier: NCT04723680. Available online: https://clinicaltrials.gov/ct2/show/NCT04723680?term=NCT04723680&draw=2&rank=1 (accessed on 28 June 2021).
58. U.S. National Library of Medicine. Gene Therapy Study in Severe Haemophilia a Patients (270-201). ClinicalTrials.gov Identifier: NCT02576795. Available online: https://clinicaltrials.gov/ct2/show/NCT02576795?term=NCT02576795&draw=2&rank=1 (accessed on 28 June 2021).
59. U.S. National Library of Medicine. Single-Arm Study to Evaluate the Efficacy and Safety of Valoctocogene Roxaparvovec in Hemophilia a Patients at a Dose of 4E13 vg/kg (BMN270-302). ClinicalTrials.gov Identifier: NCT03392974. Available online: https://clinicaltrials.gov/ct2/show/NCT03392974?term=NCT03392974&draw=2&rank=1 (accessed on 28 June 2021).
60. U.S. National Library of Medicine. Single-Arm Study to Evaluate the Efficacy and Safety of Valoctocogene Roxaparvovec in Hemophilia a Patients (BMN 270-301). ClinicalTrials.gov Identifier: NCT03370913. Available online: https://clinicaltrials.gov/ct2/show/NCT03370913?term=NCT03370913&draw=2&rank=1 (accessed on 28 June 2021).
61. Pasi, K.J.; Rangarajan, S.; Mitchell, N.; Lester, W.; Symington, E.; Madan, B.; Laffan, M.; Russell, C.B.; Li, M.; Pierce, G.F.; et al. Multiyear Follow-up of AAV5-HFVIII-SQ Gene Therapy for Hemophilia A. *N. Engl. J. Med.* **2020**, *382*, 29–40. [CrossRef]
62. U.S. National Library of Medicine. Gene Therapy Study in Severe Hemophilia a Patients with Antibodies against AAV5 (270-203). ClinicalTrials.gov Identifier: NCT03520712. Available online: https://clinicaltrials.gov/ct2/show/NCT03520712?term=NCT03520712&draw=2&rank=1 (accessed on 28 June 2021).
63. Nathwani, A.C.; Reiss, U.M.; Tuddenham, E.G.D.; Rosales, C.; Chowdary, P.; McIntosh, J.; Della Peruta, M.; Lheriteau, E.; Patel, N.; Raj, D.; et al. Long-Term Safety and Efficacy of Factor IX Gene Therapy in Hemophilia B. *N. Engl. J. Med.* **2014**, *371*, 1994–2004. [CrossRef]
64. U.S. National Library of Medicine. Dose-Escalation Study of a Self Complementary Adeno-Associated Viral Vector for Gene Transfer in Hemophilia B. ClinicalTrials.gov Identifier: NCT00979238. Available online: https://clinicaltrials.gov/ct2/show/NCT00979238?term=NCT00979238&draw=2&rank=1 (accessed on 28 June 2021).
65. U.S. National Library of Medicine. A Factor IX Gene Therapy Study (FIX-GT) (FIX-GT). ClinicalTrials.gov Identifier: NCT03369444. Available online: https://clinicaltrials.gov/ct2/show/NCT03369444?term=NCT03369444&draw=2&rank=1 (accessed on 28 June 2021).
66. U.S. National Library of Medicine. Trial of AAV5-hFIX in Severe or Moderately Severe Hemophilia B. ClinicalTrials.gov Identifier: NCT02396342. Available online: https://clinicaltrials.gov/ct2/show/NCT02396342?term=NCT02396342&draw=2&rank=1 (accessed on 28 June 2021).

67. U.S. National Library of Medicine. Dose Confirmation Trial of AAV5-hFIXco-Padua. ClinicalTrials.gov Identifier: NCT03489291. Available online: https://clinicaltrials.gov/ct2/show/NCT03489291?term=NCT03489291&draw=2&rank=1 (accessed on 28 June 2021).
68. Miesbach, W.; Meijer, K.; Coppens, M.; Kampmann, P.; Klamroth, R.; Schutgens, R.; Tangelder, M.; Castaman, G.; Schwäble, J.; Bonig, H.; et al. Gene Therapy with Adeno-Associated Virus Vector 5–Human Factor IX in Adults with Hemophilia B. *Blood* **2018**, *131*, 1022–1031. [CrossRef]
69. Von Drygalski, A.; Giermasz, A.; Castaman, G.; Key, N.S.; Lattimore, S.; Leebeek, F.W.G.; Miesbach, W.; Recht, M.; Long, A.; Gut, R.; et al. Etranacogene Dezaparvovec (AMT-061 Phase 2b): Normal/near Normal FIX Activity and Bleed Cessation in Hemophilia B. *Blood Adv.* **2019**, *3*, 3241–3247. [CrossRef]
70. U.S. National Library of Medicine. HOPE-B: Trial of AMT-061 in Severe or Moderately Severe Hemophilia B Patients. ClinicalTrials.gov Identifier: NCT03569891. Available online: https://clinicaltrials.gov/ct2/show/NCT03569891?term=NCT03569891&draw=2&rank=1 (accessed on 28 June 2021).
71. Konkle, B.A.; Walsh, C.E.; Escobar, M.A.; Josephson, N.C.; Young, G.; von Drygalski, A.; McPhee, S.W.J.; Samulski, R.J.; Bilic, I.; de la Rosa, M.; et al. BAX 335 Hemophilia B Gene Therapy Clinical Trial Results: Potential Impact of CpG Sequences on Gene Expression. *Blood* **2021**, *137*, 763–774. [CrossRef] [PubMed]
72. U.S. National Library of Medicine. HOPE-B: Open-Label Single Ascending Dose of Adeno-associated Virus Serotype 8 Factor IX Gene Therapy in Adults with Hemophilia B. ClinicalTrials.gov Identifier: NCT01687608. Available online: https://clinicaltrials.gov/ct2/show/NCT01687608?term=NCT01687608&draw=2&rank=1 (accessed on 28 June 2021).
73. Lombardi, S.; Aaen, K.H.; Nilsen, J.; Ferrarese, M.; Gjølberg, T.T.; Bernardi, F.; Pinotti, M.; Andersen, J.T.; Branchini, A. Fusion of Engineered Albumin with Factor IX Padua Extends Half-life and Improves Coagulant Activity. *Br. J. Haematol.* **2021**. [CrossRef] [PubMed]
74. Simioni, P.; Tormene, D.; Tognin, G.; Gavasso, S.; Bulato, C.; Iacobelli, N.P.; Finn, J.D.; Spiezia, L.; Radu, C.; Arruda, V.R. X-Linked Thrombophilia with a Mutant Factor IX (Factor IX Padua). *N. Engl. J. Med.* **2009**, *361*, 1671–1675. [CrossRef] [PubMed]
75. George, L.A.; Sullivan, S.K.; Giermasz, A.; Rasko, J.E.J.; Samelson-Jones, B.J.; Ducore, J.; Cuker, A.; Sullivan, L.M.; Majumdar, S.; Teitel, J.; et al. Hemophilia B Gene Therapy with a High-Specific-Activity Factor IX Variant. *N. Engl. J. Med.* **2017**, *377*, 2215–2227. [CrossRef]
76. Crudele, J.M.; Finn, J.D.; Siner, J.I.; Martin, N.B.; Niemeyer, G.P.; Zhou, S.; Mingozzi, F.; Lothrop, C.D.; Arruda, V.R. AAV Liver Expression of FIX-Padua Prevents and Eradicates FIX Inhibitor without Increasing Thrombogenicity in Hemophilia B Dogs and Mice. *Blood* **2015**, *125*, 1553–1561. [CrossRef]
77. Samelson-Jones, B.J.; Finn, J.D.; George, L.A.; Camire, R.M.; Arruda, V.R. Hyperactivity of Factor IX Padua (R338L) Depends on Factor VIIIa Cofactor Activity. *JCI Insight* **2019**, *4*, e128683. [CrossRef] [PubMed]
78. Samelson-Jones, B.J.; Finn, J.D.; Raffini, L.J.; Merricks, E.P.; Camire, R.M.; Nichols, T.C.; Arruda, V.R. Evolutionary Insights into Coagulation Factor IX Padua and Other High-Specific-Activity Variants. *Blood Adv.* **2021**, *5*, 1324–1332. [CrossRef] [PubMed]
79. Robinson, M.M.; George, L.A.; Carr, M.E.; Samelson-Jones, B.J.; Arruda, V.R.; Murphy, J.E.; Rybin, D.; Rupon, J.; High, K.A.; Tiefenbacher, S. Factor IX Assay Discrepancies in the Setting of Liver Gene Therapy Using a Hyperfunctional Variant Factor IX-Padua. *J. Thromb. Haemost.* **2021**, *19*, 1212–1218. [CrossRef] [PubMed]
80. U.S. National Library of Medicine. A Study to Evaluate the Efficacy and Safety of Factor IX Gene Therapy with PF-06838435 in Adult Males with Moderately Severe to Severe Hemophilia B (BENEGENE-2). ClinicalTrials.gov Identifier: NCT03861273. Available online: https://clinicaltrials.gov/ct2/show/NCT03861273?term=NCT03861273&draw=2&rank=1 (accessed on 28 June 2021).
81. Almeida-Porada, G. A New "FIX" for Hemophilia B Gene Therapy. *Blood* **2021**, *137*, 2860–2861. [CrossRef]
82. Nair, N.; De Wolf, D.; Nguyen, P.A.; Pham, Q.H.; Samara-Kuko, E.; Landau, J.; Blouse, G.E.; Chuah, M.K.; VandenDriessche, T. Gene Therapy for Hemophilia B Using CB 2679d-GT: A Novel Factor IX Variant with Higher Potency than Factor IX Padua. *Blood* **2021**, *137*, 2902–2906. [CrossRef]
83. Kapugi, M.; Cunningham, K. Corticosteroids. *Orthop. Nurs.* **2019**, *38*, 336–339. [CrossRef]
84. Smith, E.; Blomberg, P. Gene therapy—From idea to reality. *Lakartidningen* **2017**, *114*.
85. Lippi, G.; Favaloro, E.J. Gene therapy for hemophilias: The end of phenotypic testing or the start of a new era? *Blood Coagul. Fibrinolysis* **2020**, *31*, 237–242. [CrossRef]
86. Batty, P.; Lillicrap, D. Hemophilia Gene Therapy: Approaching the First Licensed Product. *Hemasphere* **2021**, *5*, e540. [CrossRef]
87. Iriart, J.A.B. Precision medicine/personalized medicine: A critical analysis of movements in the transformation of biomedicine in the early 21st century. *Cad. Saude Publica* **2019**, *35*, e00153118. [CrossRef]
88. Baynam, G.; Molster, C.; Bauskis, A.; Kowal, E.; Savarirayan, R.; Kelaher, M.; Easteal, S.; Massey, L.; Garvey, G.; Goldblatt, J.; et al. Indigenous Genetics and Rare Diseases: Harmony, Diversity and Equity. *Adv. Exp. Med. Biol.* **2017**, *1031*, 511–520. [CrossRef] [PubMed]
89. Skinner, M.W.; Nugent, D.; Wilton, P.; O'Mahony, B.; Dolan, G.; O'Hara, J.; Berntorp, E. Achieving the unimaginable: Health equity in haemophilia. *Haemophilia* **2020**, *26*, 17–24. [CrossRef] [PubMed]

International Journal of
Molecular Sciences

Article

High Mutational Heterogeneity, and New Mutations in the Human Coagulation Factor V Gene. Future Perspectives for Factor V Deficiency Using Recombinant and Advanced Therapies

Sara Bernal [1,2], Irene Pelaez [3], Laura Alias [1,2], Manel Baena [1], Juan A. De Pablo-Moreno [4], Luis J. Serrano [4], M. Dolores Camero [5], Eduardo F. Tizzano [6], Ruben Berrueco [7] and Antonio Liras [4,*]

1 Department of Genetics, Santa Creu i Sant Pau Hospital and IIB Sant Pau, 08041 Barcelona, Spain; sbernal@santpau.cat (S.B.); lalias@santpau.cat (L.A.); mbaena@santpau.cat (M.B.)
2 CIBERER. U-705, 18014 Barcelona, Spain
3 Department of Pediatric and Oncohematology, University Hospital Virgen de las Nieves, 18014 Granada, Spain; irene.pelaez.sspa@juntadeandalucia.es
4 Department of Genetic, Physiology and Microbiology, School of Biology, Complutense University, 28040 Madrid, Spain; jdepablo@ucm.es (J.A.D.P.-M.); luisserr@ucm.es (L.J.S.)
5 Association for the Investigation and Cure of Factor V Deficiency, 23002 Jaén, Spain; lolakamero5@gmail.com
6 Department of Clinical and Molecular Genetics, University Hospital Vall d'Hebron and Medicine Genetics Group, Vall d'Hebron Research Institute, 08035 Barcelona, Spain; etizzano@vhebron.net
7 Pediatric Hematology Department, Hospital Sant Joan de Déu, University of Barcelona and Research Institute Hospital Sant Joan de Déu, 08950 Barcelona, Spain; ruben.berrueco@sjd.es
* Correspondence: aliras@ucm.es

Citation: Bernal, S.; Pelaez, I.; Alias, L.; Baena, M.; De Pablo-Moreno, J.A.; Serrano, L.J.; Camero, M.D.; Tizzano, E.F.; Berrueco, R.; Liras, A. High Mutational Heterogeneity, and New Mutations in the Human Coagulation Factor V Gene. Future Perspectives for Factor V Deficiency Using Recombinant and Advanced Therapies. *Int. J. Mol. Sci.* **2021**, *22*, 9705. https://doi.org/10.3390/ijms22189705

Academic Editor: Ivano Condò

Received: 4 August 2021
Accepted: 4 September 2021
Published: 8 September 2021

Abstract: Factor V is an essential clotting factor that plays a key role in the blood coagulation cascade on account of its procoagulant and anticoagulant activity. Eighty percent of circulating factor V is produced in the liver and the remaining 20% originates in the α-granules of platelets. In humans, the factor V gene is about 80 kb in size; it is located on chromosome 1q24.2, and its cDNA is 6914 bp in length. Furthermore, nearly 190 mutations have been reported in the gene. Factor V deficiency is an autosomal recessive coagulation disorder associated with mutations in the factor V gene. This hereditary coagulation disorder is clinically characterized by a heterogeneous spectrum of hemorrhagic manifestations ranging from mucosal or soft-tissue bleeds to potentially fatal hemorrhages. Current treatment of this condition consists in the administration of fresh frozen plasma and platelet concentrates. This article describes the cases of two patients with severe factor V deficiency, and of their parents. A high level of mutational heterogeneity of factor V gene was identified, nonsense mutations, frameshift mutations, missense changes, synonymous sequence variants and intronic changes. These findings prompted the identification of a new mutation in the human factor V gene, designated as *Jaén-1*, which is capable of altering the procoagulant function of factor V. In addition, an update is provided on the prospects for the treatment of factor V deficiency on the basis of yet-to-be-developed recombinant products or advanced gene and cell therapies that could potentially correct this hereditary disorder.

Keywords: factor V deficiency; parahemophilia; Owren's disease; mutation analysis; advanced therapies

1. Introduction

Active factor V (FVa), sometimes referred to as labile factor or proaccelerin, is an essential clotting factor that plays a key role in the blood coagulation cascade on account of its procoagulant and anticoagulant activity [1–3]. Eighty percent of circulating FVa is produced in the liver and the remaining 20% originates in the α-granules of platelets [1,4,5]. This means that any disorder affecting the liver will significantly impact the concentration of FVa in plasma and thus disrupt the coagulation process. FVa shares approximately

40% functional and structural homology with coagulation factor VIII (FVIII), except in the B domain region where homology is around 15% [6–8]. Both proteins are structurally identical in their A1-A2-B-A3-C1-C2 domains and, in both, the large B-domain is cleaved off the full-length protein as a result of proteolytic activation [7,9].

Circulating FVa is a glycosylated 330kDa polypeptide which loses its B-domain on activation and, as a result, transforms itself into a dimeric molecule made up of a heavy 105 kDa chain that comprises domains A1 and A2 as well as a light 71–74 kDa chain comprising domains A3, C1 and C2. The heavy chain interacts with activated factor X (FXa) and with prothrombin, while the light chain interacts with membrane phospholipids. Activated protein C acts on three arginine residues of the heavy chain, inhibiting the procoagulant activity of FVa [10]. The presence of mutations at these cleavage sites gives rise to the so-called FVa Leiden, which creates a state of hypercoagulability [11].

Within the coagulation cascade, FVa is a non-enzymatic cofactor of the prothrombinase complex, which accelerates the conversion of prothrombin into thrombin [1]. This enzymatic complex is made up of activated FVa, calcium, phospholipids and FXa. FVa increases the concentration of FXa on the surface of the cell membrane and acts as an FXa co-receptor that allosterically modifies the active site of FXa, which enhances its ability to cleave prothrombin [12]. Figure 1 shows the functioning of the blood clotting mechanism according to Hoffman & Monroe's now generally accepted cell-based model [13,14] whereby coagulation factors assemble into complexes, whose stability is largely dependent on calcium and the endothelial membranes [15,16]. Following the purification of human plasma-derived factor V from plasma and platelets [17,18], recent efforts have resulted in a human recombinant FVa [19–21]. The possibility that a viable FVa deficient animal model may be developed in the medium- to short-term will greatly facilitate research into novel advanced therapies [22–24].

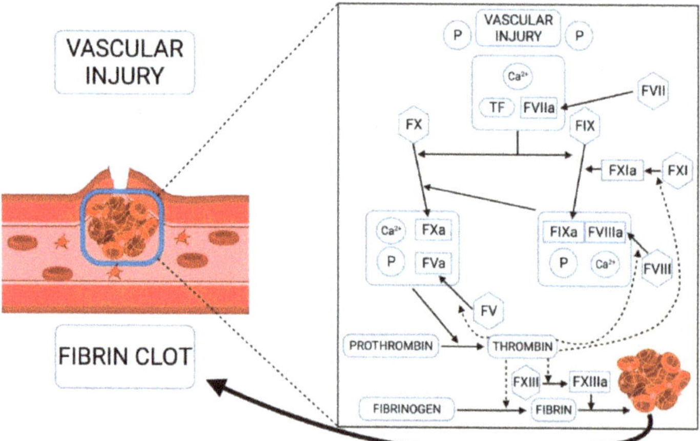

Figure 1. Blood clotting mechanism. Hoffman & Monroe's cell-based model of hemostasis according to which coagulation factors form multimolecular complexes, calcium and the endothelial membranes playing a key role in the stabilization of such complexes. Abbreviations: P, membrane phospholipids; TF, tissue factor; F, factor. Created in Biorender.com (accessed on 6 September 2021).

In humans, the factor V gene (*F5* gene) is about 80 kb in size and is located on chromosome 1q24.2. The fact that the *F5* and *F8* genes are homologous in terms of their structure and organization suggests that they could be derived from a common ancestor. The coding sequence of *F5* is divided into 25 exons and 24 introns and the length of its cDNA is 6914 bp [25].

The existence of multiple splicing alternatives was described in 2008 [26] and, ever since the first mutation in the *F5* gene was described in 1998 [27], a total of 190 mutations have been reported in the *F5* gene, the most numerous ones being missense and nonsense mutations, followed by small deletions, splicing mutations, small insertions, large insertions, complex rearrangements and a few large deletions. These mutations in the *F5* gene have been identified and subsequently registered in the 1000 Genomes Browser database [28]. The large number of mutations leading to a severe deficiency of FVa confirms the significant allelic heterogeneity of the disease.

Only two large-scale deletions have been reported: an interstitial one in chromosome 1q caused by a drastic reduction of FVa levels to 9%, which produces a nonsense mutation (p.Ser262Trp) in the other *F5* allele [29]; and a heterozygous 205 kb deletion spanning exons one to seven that is accompanied by a splicing mutation (F5 1975 + 5G > A) and results in FVa levels of 1% [30]. The striking scarcity of large-scale deletions in the mutational spectrum of the *F5* gene is probably due to the fact that the gene's rearrangements cannot be detected using conventional screening methods based on amplification and sequencing of individual exons.

FVa deficiency, also known as parahemophilia or Owren's disease, is an autosomal recessive coagulation disorder associated with mutations in the *F5* gene [31,32]. Homozygous and compound heterozygous FVa deficiency has been classified as an ultra-rare disease as it occurs in only one to nine per million live births (ORPHA:326) [33], although its incidence is 10 times higher in Western Asian countries where consanguinity is more common [34]. The disease is much less frequent in North America, but it is extremely prevalent in Caucasian countries, which are home to 67% of affected patients [35]. The greater ability to evaluate and detect patients in these countries as opposed to countries with limited health resources could constitute a bias in the calculation of the overall prevalence of the condition.

This hereditary coagulation disorder was first identified in Norway in 1943 by Owren [36], who posited the existence of a fifth essential component in the formation of fibrin, hence the term factor V. This is when the practice began of designating coagulation factors by Roman numerals. Clinically, the disease is characterized by the occurrence of mild to severe bleeding episodes [1]. Platelet-derived FVa is the best predictor of the severity of the condition in terms of circulating levels of FVa [37–39], and tissue factor pathway inhibition (TFPI) could act as an additional phenotypic modulator [40]. Consequently, no correlation can be established between FVa levels and the severity of clinical symptoms, although it is likely that the lower the FVa concentration the more severe the symptoms. Although the fastest and simplest procedure to evaluate potential symptoms or bleeding episodes is based on the measurement of plasma FVa levels, it is certainly not the best method to predict the evolution of the disease, which would require consideration of platelet FVa levels. However, plasma FVa measurement-based methods could provide a fairly accurate idea of the patient's potential clinical symptoms.

As with FVIII and factor IX in hemophilia, plasma levels of FVa determine the severity of the condition in terms of the risk of bleeding episodes and the seriousness of other symptoms. Here too, three levels of severity can be distinguished: mild (>10% of normally circulating FVa), moderate (<10% of normally circulating FVa) or severe (<1% of normally circulating FVa) [1,41,42]. Bleeding episodes usually begin before the age of six and are associated with a heterogeneous spectrum of hemorrhagic manifestations ranging from mucosal or soft tissue bleeds (such as epistaxis or hemarthrosis), to potentially fatal hemorrhages. Abundant nasal and menstrual bleeding are a hallmark in these patients; heavy bleeding is also common during minor and major surgery as well as in the course of dental procedures. Hemorrhagic arthropathy, hematomas and cranial or gastrointestinal hemorrhages are less common [1].

As no plasma-derived or recombinant factor V concentrates are available, current treatment of this condition consists either in administration of, preferably inactivated, fresh frozen plasma (FFP) or in the administration of platelet concentrates [43,44]. Octaplas®, a pharmaceutical product treated through a solvent/detergent process, is a good alternative to single donor-derived fresh frozen plasma that exhibits a high safety profile against emerging pathogens [45–47]. It is characterized by an optimal quantitative and qualitative composition, which, as a result of its high stability performance, remains unchanged from batch to batch for up to 4 years at a temperature of ≤−18 °C; its high fibrinogen concentration; and controlled levels of the different clotting and fibrinolytic factors and their inhibitors (FVa level ≥ 0.5 IU/mL). Apart from FFP, local and systemic antifibrinolytics such as tranexamic acid can be administered orally or intravenously in order to control bleeding (especially from mucosae) and prior to invasive surgical procedures. In patients with persistent menorrhagia, estrogen-progesterone replacement therapy may be used to reduce menstrual bleeding. Generally speaking, this treatment has not been associated with serious complications such as allergic reactions to plasma components, anaphylaxis or problems derived from perfusion of high levels of FFP. Inhibitors (anti- FVa antibodies) have also been reported, although their occurrence has been rare [48].

This article describes an exhaustive study of two patients with severe FVa deficiency, and of their parents. A high level of mutational heterogeneity of *F5* was identified and nonsense mutations, frameshift mutation, missense changes, synonymous sequence variants and intronic changes were found.

These findings prompted the identification of a new mutation in the human *F5* gene, designated as *Jaén-1*, which can alter the procoagulant function of FVa. The study also includes a comparative phylogenetic analysis between FVa and FVIII given their high degree of genetic homology and that they share biogenetic and coagulation cascade pathways. In addition, an update is provided of the prospects for the treatment of this condition on the basis of yet-to-be-developed recombinant products or advanced gene and cell therapies that could potentially correct this hereditary disorder.

2. Results

2.1. Clinical Characteristics of the Two Patients with Factor V Deficiency

2.1.1. Case A History

The first patient is a Caucasian 14-year-old girl from Jaén, in the south of Spain, who presented with severe FVa deficiency (FV:C <1%). The deficiency was diagnosed a few days after she was born, following administration of antibiotics to address a urinary infection which triggered mild gastric hemorrhages. Although her hemogram was normal with a normal platelet count, clotting tests showed abnormal values for both coagulation pathways. A determination of coagulation factor levels revealed a severe deficiency of FVa (<1%). The other clotting factors exhibited normal levels. Screening of the patient's parents showed that both were FVa deficient (mild phenotype). The patient did not experience significant hemorrhagic episodes until she was 8 months old, when she developed a hematoma in her right upper gingiva. As this occurred precisely when her teeth were erupting, she required administration of fresh frozen plasma (FFP), which resolved the hematoma within 24 h. At the age of one, two further bleeding episodes required a FFP transfusion, which immediately resolved the hematoma. Subsequently the patient was diagnosed with an atrial septal defect and was referred to the pediatric cardiology department. Corrective surgery was indicated at the age of two. Prophylactic FFP was administered preoperatively and one week after surgery. No intraoperative complications occurred, and correction of the cardiac defect was achieved. In the next few years, the patient presented only with mild episodes of epistaxis, which coexisted with upper airway infections. Local and oral administration of tranexamic acid resolved the hemorrhage in all but one of the cases, in which FFP had to be administered to address the epistaxis. At the age of six she presented with left hip pain of 12 hours' evolution. A diagnostic ultrasonogram revealed hemarthrosis in the left hip and no other previous trauma. Administration of FFP every 12 h during the first 24 h and every 24 h thereafter over 5 days resulted in progressive improvement. At age seven, she was admitted with vomiting, abdominal pain and mild yet progressive anemization. While in hospital, the patient experienced headache. Her family explained that on her birthday she had sustained a fall leading to mild cranial trauma without loss of consciousness. A cranial CT-scan revealed a predominantly dense extra-axial collection in the left frontal region extending to the cerebellar tentorium arising from a left subdural hematoma with a biconvex anterior portion, which suggested the possibility of an associated epidural component. Following examination by the neurosurgical department, it was decided to correct her coagulation disorder and adopt a watch-and-wait approach. Administration of FFP every 12 h for 5 days resulted in a slight resorption of the hematoma. The patient is now 14 years old and in the past 2 years has presented with various bleeding episodes, some of which required admission to hospital and administration of FFP and Octaplas® (14 mL/kg), which was initiated around 3 years ago.

2.1.2. Case B History

A 2-year-old Pakistani girl was admitted with cellulitis in the context of a severe FVa deficiency. There was a history of consanguinity in the family as her parents were cousins and had been previously diagnosed with mild FVa deficiency in Pakistan. She had two siblings with very mild FVa deficiency but no severe bleeding symptoms. At the age of 10 months, she had presented with a spontaneous nose-bleeding episode after which she was diagnosed with severe FVa deficiency. Up to her arrival in Spain, she had received FFP four times due to mucocutaneous bleeding episodes (two of them secondary to trauma). The child consulted to the hospital due to a suppurative lesion in her left leg and fever. There was a mosquito bite in that area, sustained three days before. A physical examination revealed an ulcerative, suppurative lesion with perilesional erythema and inflammatory symptoms. Blood cultures showed leukocytosis and neutrophilia; prothrombin time was 50.1 s, prothrombin activity was 10.7%, aPTT was 129.5 s and FVa expression levels were 0.7% of normal. The rest of the clotting factors were present at normal levels. Prior to wound debridement surgery, administration of FFP, antifibrinolytic and intravenous antibiotic therapy was determined. As blood cultures obtained at 48 h showed a methicillin-resistant staphylococcus aureus infection, tailored antibiotic therapy with intravenous clindamycin was initiated with excellent results. No bleeding problems were observed during admission, with no further FFP administration being necessary. The drainage was withdrawn after seven days. The patient was discharged on a four-week regimen of oral antibiotic therapy. At present, the patient is almost 3 years old and no further bleeding episodes have been reported.

2.2. Hematological and Coagulation Analysis

All hematological parameters in both the patients and their parents, including hemoglobin (Hb) concentration, mean corpuscular hemoglobin (MCH), mean corpuscular hemoglobin concentration (MCHC), total erythrocyte count (RBC), hematocrit (HCT) level, mean corpuscular volume (MCV), reticulocyte (RETIC), total white blood cell (WBC) count, WBC differential count, platelet (PLT) count, mean platelet volume (MPV), platelet distribution widths (PDW) and plateletcrit (PCT), were within the reference ranges.

A coagulation analysis (Table 1) showed altered values for both extrinsic and intrinsic coagulation pathways (prothrombin time, prothrombin activity, activated partial thromboplastin time (cephalin time)), the international normalized ratio, and FVa activity in both patients. Both presented severe FVa deficiency, with FVa activity under 1% of the reference value. In both patients' parents, alterations in the coagulation parameters and FVa activity were correlated with the type of mutation of the *F5* gene allele. The parents' heterozygous state induced a mild phenotype.

Table 1. Coagulation parameters and factor V activity.

	PT	PA	aPTT	F	INR	FVa
Patient A *	49.6	16.0	146.5	4.4	4.6	<1
Patient B **	50.1	10.7	129.5	3.7	5.8	<1
Father A	12.2	83.0	37.8	3.2	1.1	21.0
Mother A	11.0	96.0	34.4	3.6	1.0	62.9
Father B	11.5	86.8	30.3	3.3	1.1	45.9
Mother B	12.0	92.3	26.8	3.7	1.0	46.4

* Patient from Jaén, Spain. ** Patient from Pakistan. PT, prothrombin time (8–13 s); PA, prothrombin activity (75–120%); aPTT, activated partial thromboplastin time (cephalin time) (24–37 s); F, fibrinogen (2–6 g/L); INR, international normalized ratio (0.8–1.2); and FVa, factor V activity (60–130% or 0.6–1.3 IU).

2.3. Molecular Study of the F5 Gene in the Two Patients and Their Parents

All deleterious mutations and genetic variants were assigned a nucleotide number starting at the first translated base of the *F5* gene according to reference sequence *NM_000130.4*. The variant sequences were termed in accordance with the recommendations of the Human Genetic Variant Society (HGVS). Mutation screening of the *F5* gene was carried out in the two girls with a severe FVa deficiency. A total of 24 different sequence variants were identified in this study including 2 nonsense mutations, 1 frameshift mutation, 6 missense changes, 8 synonymous sequence variants and 7 intronic changes (Table 2). Two of them were identified in patient A, as nucleotide substitutions in the exon 13 of the *F5* gene involving an amino acid change (arginine or tryptophan) to a stop codon in the factor V protein: *NM_000130.4: c. 2218C>T* (*p.Arg740**) and *NM_000130.4: c.3279G>A* (*p.Trp1093**). The mutation found in patient B (*NM_000130.4: c.2862del*), also located in exon 13 of the *F5* gene, was a thymine nucleotide deletion. This deletion generated a frameshift mutation leading to an amino acid change from serine to alanine at position 955 in the protein encoded by the *F5* gene, resulting in the appearance of a premature stop codon (*p.Ser955Alafs*4*).

Seventeen out of the 24 variants have a minor allele frequency (MAF) $\geq$ 5% as described in the above mentioned 1000 Genomes Browser database (5 missense, 7 synonymous and 5 intronic variants) (Table 2). The remaining seven changes do not have a MAF value because they are rare sequence variants and have not been described previously (1 missense, 2 synonymous sequence variants and 2 intronic changes). Figures 2 and 3 show the genetic map of patient A's *F5* gene and the location of mutations Arg740* and Trp1093*. Figure 4 shows the genetic map of patient B's FVa gene and the location of mutation 2862del. Ser955Alafs*4.

In addition, a familial segregation pattern was detected in the mutations identified in patient A (Figure 5A). In this patient, both nonsense mutations (*p.Arg740** and *p.Trp1093**) were in a compound heterozygous state. One of them was inherited from her mother (*NM_000130.4: c. 2218C>T, p.Arg740**) and the other from her father (*NM_000130.4: c. 3279G>A, p.Trp1093**). Thus, each of her parents have one of these nonsense mutations in heterozygous state. In patient B, the frameshift mutation in the *F5* gene (*NM_000130.4: c.2862del, p.Ser955Alafs*4*) is in homozygous state (Figure 5B). Both her brother and her parents carried this mutation in heterozygous state.

Table 2. Genetic variants identified in the *F5* gene in patient A.

	F5 Gene †	Exon	Intron	FV Protein	SNP	MAF §	Polyphen	M-Taster	SIFT
1	c.237A>G	2		*p.Gln79Gln*	rs6028	23	**	**	**
2	c.952+76C>T		6	*	rs2239854	27	**	**	**
3	c.1238T>C	8		*p.Met413Thr*	rs6033	5	Benign	Polymorphism	Permissive
4	c.1242A>G	8		*p.Lys414Lys*	rs6035	7	**	**	**
5	c.1297-121T>C		8	*	rsl0800456	43	**	**	**
6	c.1380C>T	9		*p.Asn460Asn*	rs6015	5	**	**	**
7	c.1397-164C>T		9	*	rs7537742	50	**	**	**
8	c.1716G>A	11		*p.Glu572Glu*	rs6036	5	**	**	**
9	c.1926C>A	12		*p.Thr642Thr*	rs6037	5	**	**	**
10	**c.2218C>T**	**13**		***p.Arg740 ****	**	**	**	**	**
11	c.2289A>G	13		*p.Glu763Glu*	rs6024	5	**	**	**
12	c.2450A>C	13		*p.Asn817Thr*	rs6018	5	Benign	Polymorphism	Permissive
13	c.2758A>G	13		*p.Arg920Gly*	**	**	Benign	Polymorphism	Deleterious
14	**c.3279G>A**	**13**		***p.Trp1093 ****	**	**	**	**	**
15	c.3865T>C	13		*p.Phe1289Leu*	**	**	Benign	Polymorphism	Permissive
16	c.3939C>T	13		*p.Ser1313Ser*	**	**	**	**	**
17	c.3980A>G	13		*p.His1327Arg*	rs1800595	5	Benign	Polymorphism	Permissive
18	c.4095C>T	13		*p.Thr1365Thr*	rs9332607	26	**	**	**
19	c.4796+50A>C		13	*	**	**	**	**	**
20	c.5209-134T>C		15	*	**	**	**	**	**
21	c.5290A>G	16		*p.Met1764Val*	rs6030	29	Benign	Polymorphism	Permissive
22	c.6194-20C>A		22	*	rs6013	5	**	**	**
23	c.6529-65A>C		24	*	rs2227243	5	**	**	**
24	c.6665A>G	25		*p.Asp2222Gly*	rs6027	5	Benign	Polymorphism	Permissive

† The nomenclature follows the recommendations of the Human Genome Variant Society (HGVS). The cDNA reference sequence of the *F5* gene is *NM_000130.4*. § MAF: minor allele frequency obtained from dbSNP data base. The predictions were made in silico with the *Alamut visual v.2.6 software* that uses the prediction programs Polyphen, Mutation Taster and SIFT. * Not applicable. ** No data available.

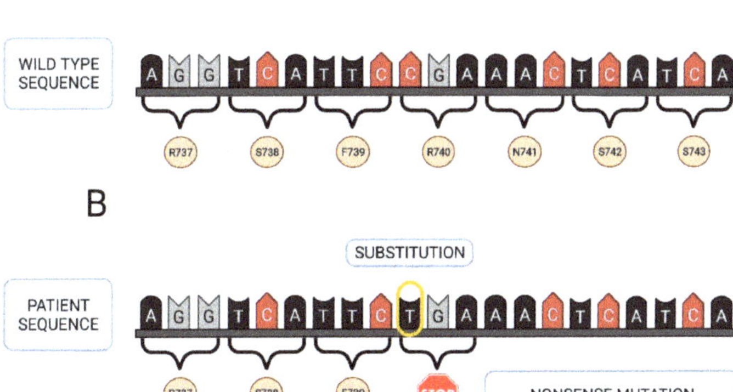

Figure 2. Genetic map of the *F5* gene of patient A. (**A**) Native sequence. (**B**) Sequence of the patient with the Arg740* mutation (premature stop codon, nonsense mutation). Created in Biorender.com (accessed on 6 September 2021).

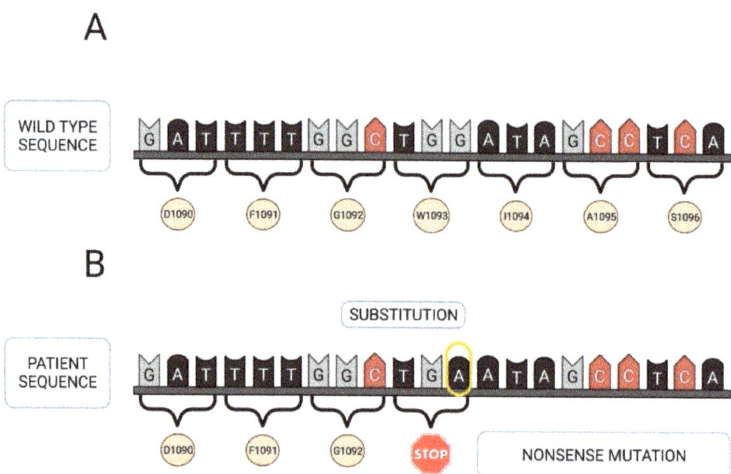

Figure 3. Genetic map of the *F5* gene of patient A. (**A**) Native sequence. (**B**) Sequence of the patient with the Trp1093* mutation (premature stop codon, nonsense mutation). Created in Biorender.com (accessed on 6 September 2021).

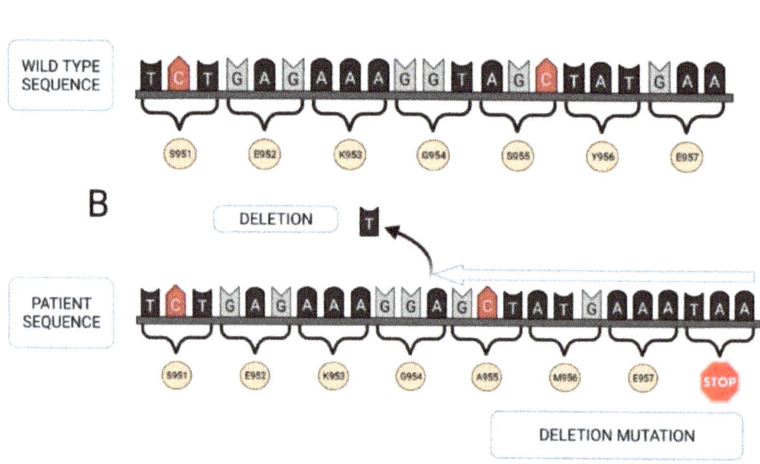

Figure 4. Genetic map of the *F5* gene of patient B. (**A**) Native sequence. (**B**) Sequence of the patient with the 2862del, Ser955Alafs*4 mutation (premature stop codon, deletion mutation). Created in Biorender.com (accessed on 6 September 2021).

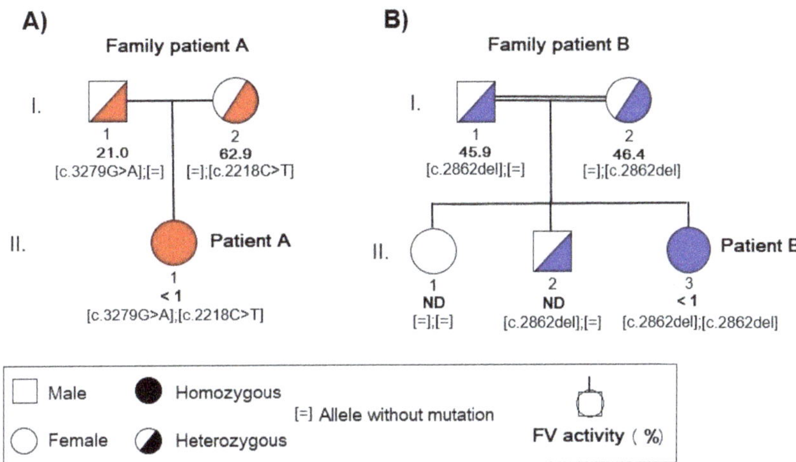

Figure 5. Family pedigrees of the FVa deficient Spanish and Pakistani families. FVa levels and segregation of the identified mutations are shown. (**A**) Patient A with FVa deficiency and two mutations in compound heterozygous state with familiar segregation (*c.3279G>A* mutation inherited from her father and *c.2218C>T* from her mother). (**B**) Patient B with FVa deficiency and mutation in homozygous state, the *c.2862del* mutation inherited from both, father and mother.

The type of mutation was correlated to FVa activity levels in both the patients and their parents. Mutation *c.3279G>A* in the father of patient A was related to FVa activity of 21% of normal; mutation *c. 2218C>T* in the mother of patient A was associated with FVa activity of 62.9% of normal; and mutation *c.2862del* in both parents of patient B exhibited a FVa activity of, approximately, 45% of normal.

Phylogenetic studies of different vertebrate species performed as part of this analysis showed that the three mutations described in the two patients are all located in well-conserved regions of exon 13 within the B domain of FVa (Figure 6). The zebrafish exhibits more differences than other species in their amino acid sequence probably due to the fact that, phylogenetically, the species is quite distant from humans.

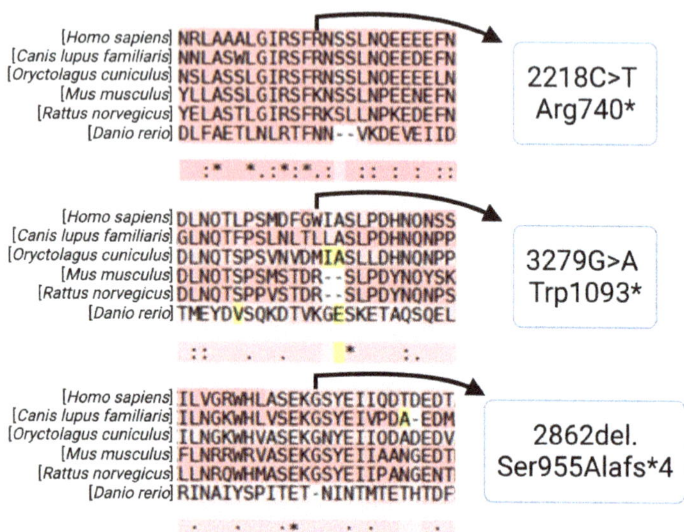

Figure 6. Multiple sequence alignment comparing the different regions of FVa corresponding to the 2218C>T Arg740, 3279G>A Trp1093* and 2862del, Ser955Alafs*4 mutations in *Homo sapiens*, *Canis lupus familiaris*, *Oryctolagus cuniculus*, *Mus musculus*, *Rattus norvegicus* and *Danio rerio*. The mutated amino acid is at the center, with an arrow pointing to it. The three mutations described in the two patients are in exon 13, corresponding to the protein's B domain, and in well-conserved regions across the different vertebrate species analyzed. Created in Biorender.com (accessed on 6 September 2021).

3. Discussion

Patients in this study suffered from severe FVa deficiency (<1% of normal factor levels), caused by mutations in both of the *F5* gene's alleles giving rise to the appearance of a premature stop codon. In the case of patient A, a compound heterozygous situation occurred, with one mutation in each allele, one inherited from the father and the other from the mother. These mutations induced a premature arrest in the synthesis of FVa. In the case of patient B, a homozygous situation occurred carrying the same mutation.

Of the 24 changes found in the *F5* gene mutation screening procedure carried out in the two girls affected with severe FVa deficiency, three were interpreted as pathogenic mutations (2 nonsense and 1 frameshift) because they met the pathogenicity criteria (the occurrence of a stop codon is in itself a pathogenic trait).

According to in silico studies with the Alamut visual v.2.6 software, all the missense mutations identified in this study should result in a neutral effect. MAF values $\geq$ 5% and the benign effect associated with the mutations in the *F5* gene were two important criteria that reinforced the idea that 19 of these 24 changes could be considered benign polymorphisms, only the remaining five changes being associated to an unknown clinical effect.

As regards familial segregation, patient A's mother was heterozygous for the already studied c.2218C>T, p.Arg740* mutation [49] with FVa plasma levels of 63%, indicative of a mild phenotype with no noteworthy clinical symptoms. The father, for his part, was heterozygous for the not-previously-reported c.3279G>A, p.Trp1093* mutation, and presented with FVa plasma levels of 21%, also pertaining to a mild phenotype without significant symptoms.

As regards the c.2218C>T, p.Arg740* mutation observed in patient A, it was first characterized by Lunghi et al. [49] in a patient with FV Leiden. This mutation results in a truncated non-functional protein lacking the full light chain (A3, C1 and C2 domains). Lunghi et al. observed the coexistence of the R506Q mutation, which gives rise to the

Leiden variant and to hypercoagulability, and the c.2218C>T, p.Arg740* mutation, which produces a deficiency of functional FVa. Although the latter mutation could in some cases compensate for the Leiden phenotype, the final result is usually plasma hypercoagulability with a significant resistance to APC. Our present study could be said to be the first to document the c.2218C>T, p.Arg740* mutation in a patient with a congenital FVa deficiency.

As far as the c.3279G>A, p.Trp1093* mutation is concerned (the second mutation exhibited by patient A), it is a new, not-yet-reported mutation, which results in a non-functional protein and in a congenital FVa deficiency. The mutation is currently being analyzed in our laboratory for possible correction by advanced therapies and gene editing using CRISPR/Cas9. Both the c.3279G>A, p.Trp1093* mutation and the CRISPR/Cas9 technology, developed to treat coagulopathies arising from this mutation, are protected by a Spanish patent (Ref. ES2785323B2) with international coverage [50,51].

Patient B's parents were heterozygous for the same mutation (the previously described c.2862del, p.Ser955Alafs*4) because there was a high degree of consanguinity between them [52]. They exhibited FVa plasma levels of 46% without any significant symptoms. Although both her brother and her parents carried this mutation in heterozygous state, her older sister was negative for this mutation in the *F5* gene.

Transmission of the FV deficiency in patient B is marked by very strong consanguinity as she is homozygous for the same mutation inherited from her parents who share a close family connection. In contrast, patient A presents with two different mutations inherited from her parents, each of whom comes from a close endogamous community. Homozygosis, particularly in the case of rare or ultra-rare diseases, generally occurs in countries in Northern Africa, Sub-Saharan Africa, the Arab world, Asia and especially India and Pakistan, where the culture favors consanguineous marriages for social, economic or political reasons. In India and Pakistan, consanguineous marriages account for 5–60% and for more than 73% of total, respectively [53–55]. According to some social science studies, there could be a direct relationship between a high rate of consanguinity and a higher incidence of hereditary and disabling conditions [53].

As shown by Naderi et al. [56], this circumstance is particularly striking in the case of FVa deficiency. These authors analyzed 23 patients with FVa deficiency in the Iranian Sistan and Baluchestan provinces, home to a population of 2.7 million and an incidence of FVa deficiency of one in every 50,000 live births, far above the global incidence of the condition, which is one to nine per million live births. Patients exhibited an overall 91.3% consanguinity rate; of them, 66.6% presented with second degree consanguinity and 33.3% were related by first-degree consanguinity. Although consanguinity may have a negative connotation from a healthcare perspective as compared with other hereditary conditions, the fact is that there are over one million people worldwide living in societies where consanguineous marriages are common. In a study of a wide sample of pregnant women, Raj et al. [57] discussed the benefits that consanguinity could entail considering the WHO's definition of health as a state of complete physical, mental and social wellbeing and not just the notion of absence of disease. In cases like that of patient B, consanguinity seems to have permitted a less pathological mutation in homozygosis than that observed in patient A.

For the reasons above, the screening of mutations in certain populations and regions is sometimes extremely challenging, as most mutations causing FVa deficiency have been identified in single families ("private mutations"). Only the Tyr1702Cys mutation has shown itself to be recurring, at least among the Italian population [58].

An interesting aspect of FVa deficiency, which separates it from other congenital coagulopathies like hemophilia, is the lack of a clear correlation between the disease's phenotype and its genotype; that is, between the clinical expression of the disease and the plasma levels of FVa and the mutation associated with the condition [59]. In fact, some patients present with a milder (or even nonexistent) hemorrhagic phenotype even if they carry the same mutation, or if they show similar FVa plasma levels. Although both patients in the present study had FVa levels below 1% (severe phenotype) and mutations resulting in a non-functional codon producing a truncated protein, the manifestation (phenotype) of the disease seemed more severe in the case of patient A, as the medical history seems to indicate with more frequent, and sometimes more severe hemorrhagic events. A potential explanation for the difference between both patients may have to do with the above-mentioned consanguinity factor, although it could also be attributed to the co-inheritance of FVa deficient alleles with risk factors for thrombophilia and/or with alleles of the modifier genes, which can influence the overall coagulation cascade [60].

This study has also demonstrated the high heterogeneity of the mutations affecting the *F5* gene. This finding warrants performing a full mutational screening of the gene to obtain a molecular diagnosis of the disease. This wide range of mutational changes observed in the *F5* gene is related to a large number of highly variable effects at a functional and molecular level. Some of the consequences of the different mutations include alterations in the stability of FVa [61]; alterations in the splicing [62,63] and in the biogenesis of FVa [64–67], mainly due to secretion disturbances; and mutations affecting the binding sites of the protein to certain activation or inactivation factors [68]. The mutations described in this study are related to the appearance of the stop codon, which results in a non-functional truncated protein. An unequal distribution is also observed among exons, with the largest number of mutations –both pathogenic and otherwise– corresponding to exon 13.

From a phylogenetic standpoint, the fact that two proteins share a common biogenetic pathway might indicate that the two molecules could present with similar characteristics. This could for example be the case of FVa and FVIII, both of which are encoded by very large genes and share a high percentage of sequence identity—approximately 40% in the A domain and 35–43% in the C domain [69,70]. The B domain is the one characterized by the lowest level of homology and, evolutionarily, the least preserved one. As a result, the proteins are very large (containing over 2000 amino acids each) and have a very similar domain structure, with the A1–A2 domains forming the heavy chain, the B domain exhibiting a large number of postranslational modifications, and the A3, C1 and C2 domains forming the light chain [71].

Similarities also exist in terms of functionality, which is in line with the phylogenetic evolution hypothesis as far as the conservation of these two proteins is concerned. In both cases, the mechanism of activation is mediated by thrombin, which binds to arginine residues residing in the B domain and leading to its elimination [71,72]. As regards the inactivation of FVa and FVIIIa, the process is identical in both cases and based on the cleavage and elimination of part of the A2 domain of the proteins by APC. Moreover, there is a functional interaction between both coagulation factors as FVa boosts the inactivation efficacy of FVIIIa [71–73]. For these reasons, it is not unconceivable that these two proteins may share one same biogenetic transport pathway, including a transport protein (LMAN1) and its cofactor (MCFD2) [74,75].

Genetic homology refers to the relationship between two genes that share a common ancestor; orthologous genes are homologous genes that diverged after a speciation event, and paralogous genes are homologous genes separated by a gene duplication event [76]. The degrees of conservation observed in the different regions of FVa analyzed in our study are in line with the findings of other authors [77,78], who have shown that the B domain tends to be the least conserved one, which is related to its importance in the function of FVa. Although the sequence alignment analyses indicate a lower homology in the case of the zebrafish, it must be noted that that is one of the teleost species where a greater number of orthologous genes are involved in hemostasis [79,80]. That is the reason why zebrafish has been postulated as a useful model for in vivo coagulation studies [81,82].

FVa deficiency is an ultra-rare disease, which has implications both in the social and the healthcare sphere. Indeed, research into novel treatments and diagnostic methods is a challenge for both the scientific community and for society, which are often mutually dependent. On the one hand every patient with a rare or an ultra-rare disease is considered unique and a rich source of valuable information, but on the other they are often neglected because research into these diseases seldom makes economic sense. This is the situation of patients with FVa deficiency, for whom no specific treatment is currently available [83]. This differs from the situation of patients with hemophilia for whom treatment with exogenous coagulation factors has been used for some time [84]. Current treatment of FVa deficiency is based on the administration of fresh frozen plasma, platelet concentrates and a pool of different coagulation factors.

Recombinant factors VIII and IX are nowadays the treatment of choice for hemophilia A and B, respectively [85]. Work is at present underway to develop a recombinant FVa, which will hopefully be available in the medium term for treating a wide range of hemorrhagic disorders, including FVa deficiency, although it is still in the early phases of preclinical development [19,20]. Recombinant DNA methods are safe and highly effective in alleviating the hemorrhagic symptoms and preventing the disabling articular effects experienced by patients.

Against this background, advanced gene and cell therapies are now able to offer a "curative," rather than merely substitutive treatment, as they are capable of correcting the mutational root cause of many conditions, particularly congenital ones, both monogenic and polygenic [86–89] Congenital coagulopathies, many of them rare diseases like hemophilia or ultra-rare diseases like FVa deficiency, may benefit from these developments.

The new advanced therapies are part of a new concept of customized or individualized pharmacology. At present, individualized pharmacology is gaining significant momentum in the treatment of many conditions as it increases the effectiveness of treatments, improves the patients' quality of life and contributes to a more rational use of medications, which leads to cost savings. With advanced therapies it is not only possible to individualize treatments when the products are already on the market but also during the design phase. This means that the therapeutic methodology best suited to a given patient can be selected according to the patient's clinical characteristics and to a very rigorous risk/benefit assessment. In fact, pharmacogenomics uses genomic information combined with other "omic" data to individualize the selection and use of medicines with a view to preventing adverse reactions and maximize their efficacy [90].

It would be ideal to develop a treatment offering maximum safety and efficacy. In the realm of advanced therapies, particularly gene therapy, increased efficacy is usually associated with a higher incidence of potential adverse events. Viral vectors [91], which are highly effective due to their total or partial ability to integrate into the genome, are prone to a high rate of adverse events. Conversely, non-viral vectors are extremely safe but not as effective as they rarely allow for therapeutic levels of the expressed protein to be achieved [92,93]. One could also classify cell therapy protocols as a function of the target cells used. Thus, pluripotent cells like embryonic stem cells or induced pluripotent stem cells [94] are the most effective, but at the same time the ones prone to a higher tumorigenicity risk, while mesenchymal stem cells for example, which could be somewhat less effective, have been shown to offer the highest levels of patient safety [95].

Very few cell-therapy approaches have been described in the context of congenital coagulopathies [96]. Our laboratory has obtained highly promising results with transplantable FVa-producing hepatocytes obtained from mesenchymal stem cells (results pending editorial acceptance).

The greatest strides have been taken in the field of gene therapy, where partially integrative adeno-associated viral vectors have been successfully used to treat hemophilia A and B [97]. However, no advanced-therapy approach has as yet been proposed to address other congenital coagulopathies. Studies are being conducted by our laboratory, using CRISPR/Cas9 technology, aimed at the correction of the mutations resulting in FVa deficiency (the results are at an advanced stage of development).

As for other types of therapeutic strategy, it is important to analyze their risk/benefit ratio. In the case of advanced (cell and gene) therapies, this criterion must be particularly demanding as their potential long-term adverse events are not known. Consequently, the inclusion of patients in a clinical trial is contingent on a series of factors. It is not the same to have an optimal substitutive treatment, such as that available for patients with hemophilia A and B, as it is to have no treatment whatsoever as in the case of FVa deficiency.

A compromise must be struck based on obtaining the greatest benefit at the lowest risk, considering not only whether a treatment is available or not, but taking into account the phenotype of the disease and the FVa expression levels and expression times sought. In the case of congenital coagulopathies, the success criteria are not as stringent as for other conditions. Indeed, converting patients from a severe to a moderate or mild phenotype should by itself be considered a successful result. This largely conditions the use of gene or cell therapies and viral or non-viral vectors.

4. Materials and Methods

4.1. Blood Collection

After the parents signed the informed consent documents, blood samples were drawn from both patients and their parents and placed in tubes containing 0.105 M sodium citrate (1:9 volume). Plasma was obtained following centrifugation at 2000× *g* for 20 min, and subsequently stored in aliquots at −80 °C until further analysis.

4.2. Hematological and Coagulation Tests

Hematimetric parameters included: hemoglobin (Hb) concentration, mean corpuscular hemoglobin (MCH), mean corpuscular hemoglobin concentration (MCHC), total erythrocyte count (RBC), hematocrit (HCT) level, mean corpuscular volume (MCV), reticulocyte (RETIC), total white blood cell (WBC) count, WBC differential count, platelet (PLT) count, mean platelet volume (MPV), platelet distribution widths (PDW) and plateletcrit (PCT). All of these parameters were obtained in an automated cell counter (ADVIA®120 Hematology System; Siemens Healthcare GmbH, Zurich, Switzerland). Blood samples for the analysis of platelet function and bleeding times were collected in evacuated tubes (Vacutainer, Becton Dickinson & Co, Franklin Lakes, NJ, USA), containing 3.8% citrate and processed in a Platelet Function Analyzer-100 (PFA-100®; Siemens Healthcare Diagnostics AG, Zurich, Switzerland) as described by Parri [98]. Collagen/adenosine-5-diphosphate (CAPD) cartridges (Dade® PFA Collagen/ADP Test Cartridge, Siemens Healthcare Diagnostics, AG, Zurich, Switzerland) were brought to room temperature and approximately 1 mL of the total citrated blood was added. Samples were aspirated under constant vacuum from a reservoir through a capillary and a microscopic opening in a collagen- and ADP-coated membrane. The PFA-100 test was used to measure the time needed for occlusion of the aperture by platelet plug formation.

All hemostasis-related parameters such as prothrombin time (PT), prothrombin activity (PA), activated partial thromboplastin time (aPTT) (cephalin time), fibrinogen (F) and international normalized ratio (INR), were determined in accordance with the standards defined by the Hematology Department of the Jaen University Hospital, Spain. HemosIL® kits were used (Instrumentation Laboratory; Bedford, MA, USA), as per the manufacturer's instructions.

HemosIL® RecombiPlasTin 2G was used to measure prothrombin time (PT) and fibrinogen levels as described by Tripodi [99]. A high-sensitivity thromboplastin reagent was used, based on recombinant human tissue factor (rhTF). Following reconstitution with RecombiPlasTin 2G diluents, the PT reagent included in the RecombiPlasTin 2G kit turned into a liposomal preparation containing rhTF relipidated in a synthetic phospholipid blend and combined with calcium chloride, buffer and a preservative. As rhTF does not contain any contaminating coagulation factor, RecombiPlasTin 2G is highly sensitive to extrinsic pathway coagulation factors, which makes it particularly suited to assays with such factors. The formulation of RecombiPlasTin 2G makes it insensitive to therapeutic heparin levels. In the PT test, addition of the reagent to the patient's plasma in the presence of calcium ions activates the extrinsic coagulation pathway. This eventually results in the conversion of fibrinogen into fibrin, with the formation of a solid gel. Fibrinogen must be quantitated (PT-based method) by relating the absorbance, or light scatter, during clot formation, to a calibrator. PT results are reported in seconds, activity percentage, or INR; fibrinogen levels, in g/L.

HemosIL® APTT-SP (liquid) was used to determine the aPTT value, as described by Van de Bresselaar [100,101]. The aPTT test utilizes a contact activator (a colloidal silica dispersion with synthetic phospholipids, buffer and a preservative) that stimulates factor XIIa production. This activator provides a contact surface for the interaction of high molecular weight kininogen, kallikrein and factor XIIa. Such contact-based activation occurs at 37 °C over a given period of time. Addition of 0.025 M calcium chloride with a preservative triggers a series of reactions that will result in clot formation. Phospholipids are required to generate the compounds that will activate factor X and prothrombin. Results are expressed in seconds.

HemosIL® FVa deficient plasma was used for the quantitative determination of FVa activity, as described by Tripodi [99]. FVa activity in plasma was determined using the modified prothrombin time test conducted in the patient's citrated diluted plasma in the presence of FVa immunodepleted human plasma ($\leq$1%). Correction of prolonged coagulation time for the deficient plasma was proportional to the concentration (activity percentage) of the specific factor (FVa) in the patient's plasma, which can be obtained by plotting a calibration curve. FVa reference levels ranged between 60 and 130% (0.60–1.30 IU).

4.3. DNA Extraction

Following informed consent, EDTA anticoagulated blood samples were collected. Genomic DNA was automatically extracted from peripheral leukocytes using the salting-out procedure (Qiagen, Düsseldorf, Germany).

4.4. Mutational Analysis

Primers corresponding to the complete coding sequence and intronic adjacent regions of the F5 gene were designed according to entry NC_000001.11 of the NCBI database to yield PCR products in a range between 300 and 600 bp approximately (Table 3). PCR amplification was performed in a final volume of 25 µL that contained 1.5 mM–2.5 mM $MgCl_2$ (depending on the fragment), 200 µM of each dNTP, 0.2 µM of each primer, 0.5 units of Taq DNA polymerase (Ecogen, Madrid, Spain) in the recommended buffer and 150–200 ng of genomic DNA. The thermocycling conditions used were as follows: 94 °C for 5 min, followed by 30 cycles at 94 °C for 30 s (denaturation step), 58–60 °C (depending on the fragment amplified) for 30 s and 72 °C for 2 min, followed by a 20 min final extension step at 72 °C. When PCR amplification products were completed, any inconsumable dNTPs and primers remaining in the PCR product mixture were removed by the action of the ExoSAP digestion enzymes (exonuclease I and shrimp alkaline phosphatase). After the clean-up procedure was completed, the bidirectional sequencing reaction (forward and reverse primers) was performed by the Sanger method using BigDye v.1.1. terminator chemistry. The sequencing reactions of the fragments amplified were cleaned by a Montage SEQ96 kit on a vacuum manifold, where the unincorporated dye terminators and salts were removed from sequencing products prior to their subjection to capillary electrophoresis on an ABI Prism 3500 genetic analyzer (Applied Biosystems, Massachusetts, USA). Sequence chromatograms were analyzed using SeqScape v.2.1.1. Mutations and genetic variants were described according to the Human Genome Variation Society (HGVS) nomenclature [102]. All nucleotide changes identified in the patients with FVa deficiency were analyzed by Alamut visual v.2.6. software (Sophia Genetics, Inc, Boston, USA), which integrates genetic and genomic information from different sources into one consistent and convenient environment to describe variants using HGVS nomenclature and help interpret their pathogenic status. Furthermore, all changes detected in the patients were studied in both their parents to determine family segregation.

Table 3. Primers design for human *F5* analysis.

Exon *	Primer	Sequence 5′ → 3′	Size (bp)
1	F5_1F	CACCTGCAGTAA AACAGTCAC	532
	F5_1R	AGCCATGACATTGCAAAGGG	
2	F5_2F	ACAGTTTGGGTTTCTACTGTG	461
	F5_2R	GCATGTGAATGCCAAATTACCC	
3	F5_3F	AAGTGAGTCAGCCTCAGGAC	451
	F5_3R	AATGCAGGTCTAGAGGACTC	
4	F5_4F	TACATGAGCATAGAAATGGGC	528
	F5_4R	TCAAACAATGATCTGGTCTCC	
5	F5_5F	CCCCAAAGCAAGAAGGTATC	488
	F5_5R	CCTTCTTGATAGGGAGTTGC	
6	F5_6F	AGGGCACAAACTACAACTGG	494
	F5_6R	TGAGGAAAGTTTGTCTGCGG	
7	F5_7F	TCTTGCCTTTTCTGGATGCC	463
	F5_7R	CCAATACATGTGTCCCCTTG	
8	F5_8F	ATGCAGGAGACAAATCAGAAG	572
	F5_8R	TTGAGAAACTGTCTCAGATCC	
9	F5_9F	AATGCTCCTGCCAAGTGATG	442
	F5_9R	AACTCCTGAAGTGAGAAGGG	
10	F5_10F	GCAATATTAATTGGTTCCAGCG	413
	F5_10R	TCTCTTGAAGGAAATGCCCC	
11	F5_11F	GGAATAGAGAATCCTTTCCC	433
	F5_11R	AAGTCTTTGGACTGGAAGTG	
12	F5_12F	AATCACTGCTTTGACACAACC	459
	F5_12R	TTGAAAGAAAAGCCTGCAGGG	
13	F5_13AF	TAGGTCACAGACAAGCAGTG	562
	F5_13AR	CTTTCTGAGGTTCTGCAAGG	
	F5_13BF	TGGCTGCAGCATTAGGAATC	549
	F5_13BR	CCAGTGTCTTGGCTAGGAAGG	
	F5_13CF	TAAGCATAAGGGACCCAAGG	568
	F5_13CR	TCTTAGAGGGTGAAAGGTCC	
	F5_13DF	AACAAGCCTGGAAAGCAGAG	501
	F5_13DR	TTCACTGAGCTCTGGAGAAG	
	F5_13EF	GACACTGGTCAGGCAAGCTG	651
	F5_13ER	TGGAGAAATGGGCATCTGAC	
	F5_13FF	AGATGCCCATTTCTCCAGAC	577
	F5_13FR	AGATCTGTCTCACCAAGGTC	
	F5_13GF	CAACCCTTTCTCTAGACCTC	520
	F5_13GR	AGTCATCTTCACTGCTCTGG	
	F5_13HF	CTTATCCAGACCTTGGTCAG	469
	F5_13HR	ATAGGGGAACCAGACTGTTC	

Table 3. *Cont.*

Exon *	Primer	Sequence 5′ → 3′	Size (bp)
14	F5_14F	AGGTCATAGGAAGACTTACC	480
	F5_14R	TCACCTATAGCTCTCTTGCC	
15	F5_15F	ACTTGGGCCATATCTCACAG	502
	F5_15R	GAAATAACCCCGACTCTTCC	
16	F5_16F	GATCAATCAGAGGAAGGAGG	506
	F5_16R	GTCTCAGAAGCATCTCATGTC	
17	F5_17F	GGGAATGCAGAATCATGAGG	449
	F5_17R	TTTGGGTCTATGGGTTTGCC	
18	F5_18F	GAAAGCCTCTTGTGAAGCAG	365
	F5_18R	CAATGCAATCAGACCATGGG	
19	F5_19F	TTAAGTCAGGGCCACACAAG	357
	F5_19R	CCCAAATGGAGCTGCTTCAC	
20	F5_20F	TACAACACAGGTCCTCCAAG	339
	F5_20R	GCCTCACACTTAGTACTTGC	
21	F5_21F	AGGCAGTGTGTGACTTGTTG	355
	F5_21R	CCATATGACCCTTAGAAAGCC	
22	F5_22F	CTGGAACTGGAATTATCCCC	425
	F5_22R	CAAAGGTTTTCCTAGGAGCC	
23	F5_23F	GCCTGAGAACAGTATTTGGC	414
	F5_23R	ATACTCCTGCTTCCCAGATC	
24	F5_24F	GAGACTGTGAATCCTAAGGG	420
	F5_24R	AGAGGTGGTACATGTCACTG	
25	F5_25F	GTTTAAGGCTGCAGTGAGCC	473
	F5_25R	CTTACTTACTGGTAGCAAGGAG	

* The coding region of the 25 exons and the corresponding exon–intron boundaries were analyzed in the *F5* gene.

4.5. Sequence Alignment

A multiple alignment of protein sequences from multiple vertebrate animal species (*Canis lupus familiaris*, *Oryctolagus cuniculus*, *Mus musculus*, *Rattus norvegicus* and *Danio rerio*) was carried out using the T-Coffee alignment tool [103], which combines different sequence-alignment methods to find out whether the mutations described are structurally and functionally important and whether they are located in conserved regions.

Author Contributions: S.B.: Design and data collection of the experiments, data analysis and interpretation. L.A. and M.B.: Support and technical assistance technical with regard to the experiments. I.P. and R.B.: Clinical monitoring of patient A and patient B respectively from diagnosis to treatment and collection of the samples of the patients and their parents once the informed consent forms were signed. E.F.T.: Critical review of the work. J.A.D.P.-M.: Data analysis and interpretation, design of figures related to factor V sequences, the location of described mutations and phylogenetic studies. L.J.S.: Data analysis and interpretation, and literature search for reports on consanguinity. M.D.C.: Contributions related to the social aspects and relevance of the studies; literature search for reports on consanguinity. A.L.: Principal investigator of the project, conception and design of the experiments, data analysis, contribution of reagents/materials/analytical tools, drafting of the manuscript, preparation of figures and/or tables, reviewing of drafts of the paper, and search for funding for the project. All authors have read and agreed to the published version of the manuscript.

Funding: This study was supported by the Andalusian Association of Hemophilia (ASANHEMO FV 2016–20 grant) and Octapharma S.A. (OCPH-2019-20 grant).

Institutional Review Board Statement: The study was conducted according to the guidelines of the Declaration of Helsinki and approved by the Institutional Ethics Committee of University Hospital of Jaen, 5 June 2015.

Informed Consent Statement: Informed consent for FVa testing was obtained from all patients involved the study at the time blood samples were taken for clinical purposes. This study was approved by the Medical Ethics Committee of the University Hospital of Jaen, Spain, and Santa Creu i Sant Pau Hospital. Barcelona, Spain, and was conducted in accordance with the Declaration of Helsinki. Blood donors agreed to participate by written informed consent.

Conflicts of Interest: The authors declare no conflict of interest.

References

1. Tabibian, S.; Shiravand, Y.; Shams, M.; Safa, M.; Gholami, M.S.; Heydari, F.; Ahmadi, A.; Rashidpanah, J.; Dorgalaleh, A. A Comprehensive Overview of Coagulation Factor V and Congenital Factor V Deficiency. *Semin. Thromb. Hemost.* **2019**, *45*, 523–543. [CrossRef] [PubMed]
2. Santamaria, S.; Reglińska-Matveyev, N.; Gierula, M.; Camire, R.M.; Crawley, J.T.B.; Lane, D.A.; Ahnström, J. Factor V has an anticoagulant cofactor activity that targets the early phase of coagulation. *J. Biol. Chem.* **2017**, *292*, 9335–9344. [CrossRef]
3. Van Doorn, P.; Rosing, J.; Duckers, C.; Hackeng, T.; Simioni, P.; Castoldi, E. Factor V Has Anticoagulant Activity in Plasma in the Presence of TFPIα: Difference between FV1 and FV2. *Thromb. Haemost.* **2018**, *118*, 1194–1202. [CrossRef]
4. Segers, K.; Dahlbäck, B.; Nicolaes, G. Coagulation Factor V and Thrombophilia: Background and Mechanisms. *Thromb. Haemost.* **2007**, *98*, 530–542. [CrossRef] [PubMed]
5. Hirbawi, J.; Vaughn, J.L.; Bukys, M.A.; Vos, H.L.; Kalafatis, M. Contribution of Amino Acid Region 659-663 of Factor Va Heavy Chain to the Activity of Factor Xa within Prothrombinase. *Biochemistry* **2010**, *49*, 8520–8534. [CrossRef]
6. Camire, R.M. A New Look at Blood Coagulation Factor V. *Curr. Opin. Hematol.* **2011**, *18*, 338–342. [CrossRef] [PubMed]
7. Mazurkiewicz-Pisarek, A.; Płucienniczak, G.; Ciach, T.; Płucienniczak, A. The Factor VIII Protein and Its Function. *Acta Biochim. Pol.* **2016**, *63*, 11–16. [CrossRef] [PubMed]
8. Vadivel, K.; Kumar, Y.; Bunce, M.W.; Camire, R.M.; Bajaj, M.S.; Bajaj, S.P. Interaction of Factor V B-Domain Acidic Region with Its Basic Region and with TFPI/TFPI2: Structural Insights from Molecular Modeling Studies. *Int. Biol. Rev.* **2017**, *1*. Available online: http://journals.ke-i.org/index.php/ibr/article/view/1334/975 (accessed on 2 September 2021).
9. Bos, M.H.A.; Camire, R.M. Blood Coagulation Factors V and VIII: Molecular Mechanisms of Procofactor Activation. *J. Coagul. Disord.* **2010**, *2*, 19–27.
10. Van Doorn, P.; Rosing, J.; Wielders, S.J.; Hackeng, T.M.; Castoldi, E. The C-Terminus of Tissue Factor Pathway Inhibitor-α Inhibits Factor V Activation by Protecting the Arg ^{1545} Cleavage Site. *J. Thromb. Haemost.* **2017**, *15*, 140–149. [CrossRef]
11. Favaloro, E.J. Genetic Testing for Thrombophilia-Related Genes: Observations of Testing Patterns for Factor V Leiden (G1691A) and Prothrombin Gene "Mutation" (G20210A). *Semin. Thromb. Hemost.* **2019**, *45*, 730–742. [CrossRef] [PubMed]
12. Ruben, E.A.; Rau, M.J.; Fitzpatrick, J.; Di Cera, E. Cryo-EM structures of human coagulation factors V and Va. *Blood* **2021**, *137*, 3137–3144. [CrossRef]
13. Hoffman, M.; Monroe, D.A. Cell-Based Model of Hemostasis. *Thromb. Haemost.* **2001**, *85*, 958–965. [CrossRef] [PubMed]
14. Smith, S.A. The Cell-Based Model of Coagulation. *J. Vet. Emerg. Crit. Care* **2009**, *19*, 3–10. [CrossRef] [PubMed]
15. Mann, K.G. Prothrombinase: The Paradigm for Membrane Bound Enzyme Complexes; a Memoir. *J. Thromb. Thrombolysis* **2021**. [CrossRef] [PubMed]
16. Stoilova-McPhie, S. Factor VIII and Factor V Membrane Bound Complexes. In *Macromolecular Protein Complexes III: Structure and Function*; Harris, J.R., Marles-Wright, J., Eds.; Subcellular Biochemistry; Springer International Publishing: Cham, Switzerland, 2021; Volume 96, pp. 153–175. [CrossRef]
17. Kane, W.H.; Majerus, P.W. Purification and Characterization of Human Coagulation Factor V. *J. Biol. Chem.* **1981**, *256*, 1002–1007. [CrossRef]
18. Gould, W.R.; Silveira, J.R.; Tracy, P.B. Unique in Vivo Modifications of Coagulation Factor V Produce a Physically and Functionally Distinct Platelet-Derived Cofactor. *J. Biol. Chem.* **2004**, *279*, 2383–2393. [CrossRef]
19. Von Drygalski, A.; Bhat, V.; Gale, A.J.; Burnier, L.; Cramer, T.J.; Griffin, J.H.; Mosnier, L.O. An Engineered Factor Va Prevents Bleeding Induced by Anticoagulant Wt Activated Protein C. *PLoS ONE* **2014**, *9*, e104304. [CrossRef]
20. Gale, A.J.; Bhat, V.; Pellequer, J.L.; Griffin, J.H.; Mosnier, L.O.; Von Drygalski, A. Safety, Stability and Pharmacokinetic Properties of SuperFactor Va, a Novel Engineered Coagulation Factor V for Treatment of Severe Bleeding. *Pharm. Res.* **2016**, *33*, 1517–1526. [CrossRef]
21. Bhat, V.; von Drygalski, A.; Gale, A.J.; Griffin, J.H.; Mosnier, L.O. Improved Coagulation and Haemostasis in Haemophilia with Inhibitors by Combinations of SuperFactor Va and Factor VIIa. *Thromb. Haemost.* **2016**, *115*, 551–561. [CrossRef]

22. Cui, J.; O'Shea, K.S.; Purkayastha, A.; Saunders, T.L.; Ginsburg, D. Fatal haemorrhage and incomplete block to embryogenesis in mice lacking coagulation factor V. *Nature* **1996**, *384*, 66–68. [CrossRef] [PubMed]
23. Yang, T.; Cui, J.; Taylor, J.; Yang, A.; Gruber, S.; Ginsburg, D. Rescue of Fatal Neonatal Hemorrhage in Factor V Deficient Mice by Low Level Transgene Expression. *Thromb. Haemost.* **2000**, *83*, 70–77. [CrossRef] [PubMed]
24. Weyand, A.C.; Grzegorski, S.J.; Rost, M.S.; Lavik, K.I.; Ferguson, A.C.; Menegatti, M.; Richter, C.E.; Asselta, R.; Duga, S.; Peyvandi, F.; et al. Analysis of Factor V in Zebrafish Demonstrates Minimal Levels Needed for Early Hemostasis. *Blood Adv.* **2019**, *3*, 1670–1680. [CrossRef]
25. Cripe, L.D.; Moore., K.D.; Kane, W.H. Structure of the gene for human coagulation factor V. *Biochemistry* **1992**, *31*, 3777–3785. [CrossRef]
26. Dall'Osso, C.; Guella, I.; Duga, S.; Locatelli, N.; Paraboschi, E.M.; Spreafico, M.; Afrasiabi, A.; Pechlaner, C.; Peyvandi, F.; Tenchini, M.L.; et al. Molecular Characterization of Three Novel Splicing Mutations Causing Factor V Deficiency and Analysis of the F5 Gene Splicing Pattern. *Haematologica* **2008**, *93*, 1505–1513. [CrossRef]
27. Guasch, J.F.; Cannegieter, S.; Reitsma, P.H.; Van 't Veer-Korthof, E.T.; Bertina, R.M. Severe Coagulation Factor V Deficiency Caused by a 4 Bp Deletion in the Factor V Gene: Genetic Defect in Severe Factor V Deficiency. *Br. J. Haematol.* **1998**, *101*, 32–39. [CrossRef]
28. The Human Gene Mutation Database at the Institute of Medical Genetics in Cardiff (HGMD). Available online: http://www.hgmd.cf.ac.uk/ac/index.php (accessed on 2 September 2021).
29. Caudill, J.S.; Sood, R.; Zehnder, J.L.; Pruthi, R.K.; Steensma, D.P. Severe Coagulation Factor V Deficiency Associated with an Interstitial Deletion of Chromosome 1q. *J. Thromb. Haemost.* **2007**, *5*, 626–628. [CrossRef]
30. Guella, I.; Paraboschi, E.M.; van Schalkwyk, W.A.; Asselta, R.; Duga, S. Identification of the First Alu-Mediated Large Deletion Involving the F5 Gene in a Compound Heterozygous Patient with Severe Factor V Deficiency. *Thromb. Haemost.* **2011**, *106*, 296–303. [CrossRef]
31. Lippi, G.; Favaloro, E.J.; Montagnana, M.; Manzato, F.; Guidi, G.C.; Franchini, M. Inherited and Acquired Factor V Deficiency. *Blood Coagul. Fibrinolysis* **2011**, *22*, 160–166. [CrossRef] [PubMed]
32. Thalji, N.; Camire, R. Parahemophilia: New Insights into Factor V Deficiency. *Semin. Thromb. Hemost.* **2013**, *39*, 607–612. [CrossRef]
33. Congenital Factor V Deficiency. The Portal for Rare Diseases and Orphan Drugs Orphanet. Available online: https://www.orpha.net/consor/cgi-bin/index.php?lng=EN (accessed on 2 September 2021).
34. Mannucci, P.M.; Duga, S.; Peyvandi, F. Recessively Inherited Coagulation Disorders. *Blood* **2004**, *104*, 1243–1252. [CrossRef] [PubMed]
35. Jenny, R.J.; Pittman, D.D.; Toole, J.J.; Kriz, R.W.; Aldape, R.A.; Hewick, R.M.; Kaufman, R.J.; Mann, K.G. Complete cDNA and Derived Amino Acid Sequence of Human Factor V. *Proc. Natl. Acad. Sci. USA* **1987**, *84*, 4846–4850. [CrossRef] [PubMed]
36. Owren, P. Parahaemophilia. *Lancet* **1947**, *249*, 446–448. [CrossRef]
37. Miletich, J.P.; Majerus, D.W.; Majerus, P.W. Patients with Congenital Factor V Deficiency Have Decreased Factor Xa Binding Sites on Their Platelets. *J. Clin. Investig.* **1978**, *62*, 824–831. [CrossRef]
38. Duckers, C.; Simioni, P.; Spiezia, L.; Radu, C.; Dabrilli, P.; Gavasso, S.; Rosing, J.; Castoldi, E. Residual Platelet Factor V Ensures Thrombin Generation in Patients with Severe Congenital Factor V Deficiency and Mild Bleeding Symptoms. *Blood* **2010**, *115*, 879–886. [CrossRef] [PubMed]
39. Castoldi, E.; Duckers, C.; Radu, C.; Spiezia, L.; Rossetto, V.; Tagariello, G.; Rosing, J.; Simioni, P. Homozygous F5 Deep-Intronic Splicing Mutation Resulting in Severe Factor V Deficiency and Undetectable Thrombin Generation in Platelet-Rich Plasma: Deep-Intronic Mutation Causing Severe FV Deficiency. *J. Thromb. Haemost.* **2011**, *9*, 959–968. [CrossRef] [PubMed]
40. Duckers, C.; Simioni, P.; Spiezia, L.; Radu, C.; Gavasso, S.; Rosing, J.; Castoldi, E. Low Plasma Levels of Tissue Factor Pathway Inhibitor in Patients with Congenital Factor V Deficiency. *Blood* **2008**, *112*, 3615–3623. [CrossRef]
41. Jain, S.; Acharya, S.S. Management of Rare Coagulation Disorders in 2018. *Transfus. Apher. Sci.* **2018**, *57*, 705–712. [CrossRef]
42. Peyvandi, F.; Di Michele, D.; Bolton-Maggs, P.H.B.; Lee, C.A.; Tripodi, A.; Srivastava, A. Classification of Rare Bleeding Disorders (RBDs) Based on the Association between Coagulant Factor Activity and Clinical Bleeding Severity: Classification of Rare Bleeding Disorders. *J. Thromb. Haemost.* **2012**, *10*, 1938–1943. [CrossRef] [PubMed]
43. Gupta, G.K.; Hendrickson, J.E.; Bahel, P.; Siddon, A.J.; Rinder, H.M.; Tormey, C.A. Factor V activity in apheresis platelets: Implications for management of FV deficiency. *Transfusion* **2021**, *61*, 405–409. [CrossRef]
44. Drzymalski, D.M.; Elsayes, A.H.; Ward, K.R.; House, M.; Manica, V.S. Platelet transfusion as treatment for factor V deficiency in the parturient: A case report. *Transfusion* **2019**, *59*, 2234–2237. [CrossRef]
45. Heger, A.; Svae, T.E.; Neisser-Svae, A.; Jordan, S.; Behizad, M.; Römisch, J. Biochemical Quality of the Pharmaceutically Licensed Plasma OctaplasLG® after Implementation of a Novel Prion Protein (PrP^{Sc}) Removal Technology and Reduction of the Solvent/Detergent (S/D) Process Time. *Vox Sang.* **2009**, *97*, 219–225. [CrossRef]
46. Cushing, M.M.; Asmis, L.; Calabia, C.; Rand, J.H.; Haas, T. Efficacy of Solvent/Detergent Plasma after Storage at 2–8 °C for 5 Days in Comparison to Other Plasma Products to Improve Factor V Levels in Factor V Deficient Plasma. *Transfus. Apher. Sci.* **2016**, *55*, 114–119. [CrossRef]
47. Spinella, P.C.; Borasino, S.; Alten, J. Solvent/Detergent-Treated Plasma in the Management of Pediatric Patients Who Require Replacement of Multiple Coagulation Factors: An Open-Label, Multicenter, Post-Marketing Study. *Front. Pediatr.* **2020**, *8*, 572. [CrossRef]

48. Goulenok, T.; Vasco, C.; Faille, D.; Ajzenberg, N.; De Raucourt, E.; Dupont, A.; Frere, C.; James, C.; Rabut, E.; Rugeri, L.; et al. RAVI study group. Acquired factor V inhibitor: A nation-wide study of 38 patients. *Br. J. Haematol.* **2021**, *192*, 892–899. [CrossRef]
49. Lunghi, B.; Castoldi, E.; Mingozzi, F.; Bernardi, F.; Castaman, G. A Novel Factor V Null Mutation Detected in a Thrombophilic Patient with Pseudo-Homozygous APC Resistance and in an Asymptomatic Unrelated Subject. *Blood* **1998**, *92*, 1463–1464. [CrossRef]
50. Patent "In Vitro Method to Recover the Expression of the F5 Gene Encoding Coagulation Factor V". Invention Patent with Examination. Patent recognition. Office for the Transfer of Research Results. 2021. Available online: https://consultas2.oepm.es/pdf/ES/0000/000/02/78/53/ES-2785323_B2.pdf (accessed on 2 September 2021).
51. In Vitro Method to Recover the Expression of the F5 Gene Encoding Coagulation Factor V. Transfer catalog. Office for the Transfer of Research Results. 2021. Available online: https://www.ucm.es/otri/complutransfer-metodo-in-vitro-para-recuperar-la-expresion-del-gen-f5-que-codifica-el-factor-v-de-la-coagulacion-1 (accessed on 2 September 2021).
52. Ajzner, E.; Balogh, I.; Szabó, T.; Marosi, A.; Haramura, G.; Muszbek, L. Severe Coagulation Factor V Deficiency Caused by 2 Novel Frameshift Mutations: 2952delT in Exon 13 and 5493insG in Exon 16 of Factor 5 Gene. *Blood* **2002**, *99*, 702–705. [CrossRef] [PubMed]
53. Maheswari, K.; Wadhwa, L. Role of consanguinity in paediatric neurological disorders. *Int. J. Contemp. Pediatrics* **2016**, *3*, 939–942. [CrossRef]
54. Bhinder, M.A.; Sadia, H.; Mahmood, N.; Qasim, M.; Hussain, Z.; Rashid, M.M.; Zahoor, M.Y.; Bhatti, R.; Shehzad, W.; Waryah, A.M.; et al. Consanguinity: A Blessing or Menace at Population Level? *Ann. Hum. Genet.* **2019**, *83*, 214–219. [CrossRef] [PubMed]
55. Oniya, O.; Neves, K.; Ahmed, B.; Konje, J.C. A Review of the Reproductive Consequences of Consanguinity. *Eur. J. Obstet. Gynecol. Reprod. Biol.* **2019**, *232*, 87–96. [CrossRef] [PubMed]
56. Naderi, M.; Tabibian, S.; Alizadeh, S.; Hosseini, S.; Zaker, F.; Bamedi, T.; Shamsizadeh, M.; Dorgalaleh, A. Congenital Factor V Deficiency: Comparison of the Severity of Clinical Presentations among Patients with Rare Bleeding Disorders. *Acta Haematol.* **2015**, *133*, 148–154. [CrossRef] [PubMed]
57. Bhopal, R.S.; Petherick, E.S.; Wright, J.; Small, N. Potential Social, Economic and General Health Benefits of Consanguineous Marriage: Results from the Born in Bradford Cohort Study. *Eur. J. Public Health* **2014**, *24*, 862–869. [CrossRef] [PubMed]
58. Castoldi, E.; Lunghi, B.; Mingozzi, F.; Muleo, G.; Redaelli, R.; Mariani, G.; Bernardi, F. A Missense Mutation (Y1702C) in the Coagulation Factor V Gene Is a Frequent Cause of Factor V Deficiency in the Italian Population. *Haematologica* **2001**, *86*, 629–633. [PubMed]
59. Kong, R.X.; Xie, Y.S.; Xie, H.X.; Luo, S.S.; Wang, M.S. Analysis of Phenotype and Genotype of a Family with Hereditary Coagulation Factor V Deficiency Caused by A Compound Heterozygous Mutation. *Zhongguo Shi Yan Xue Ye Xue Za Zhi* **2020**, *28*, 2033–2038. [CrossRef]
60. Asselta, R.; Tenchini, M.L.; Duga, S. Inherited Defects of Coagulation Factor V: The Hemorrhagic Side: Factor V and Bleeding Disorders. *J. Thromb. Haemost.* **2006**, *4*, 26–34. [CrossRef]
61. Duga, S.; Montefusco, M.C.; Asselta, R.; Malcovati, M.; Peyvandi, F.; Santagostino, E.; Mannucci, P.M.; Tenchini, M.L. Arg2074Cys Missense Mutation in the C2 Domain of Factor V Causing Moderately Severe Factor V Deficiency: Molecular Characterization by Expression of the Recombinant Protein. *Blood* **2003**, *101*, 173–177. [CrossRef]
62. Nuzzo, F.; Bulato, C.; Nielsen, B.I.; Lee, K.; Wielders, S.J.; Simioni, P.; Key, N.S.; Castoldi, E. Characterization of an Apparently Synonymous *F5* Mutation Causing Aberrant Splicing and Factor V Deficiency. *Haemophilia* **2015**, *21*, 241–248. [CrossRef]
63. Cunha, M.L.R.; Bakhtiari, K.; Peter, J.; Marquart, J.A.; Meijers, J.C.M.; Middeldorp, S. A Novel Mutation in the F5 Gene (Factor V Amsterdam) Associated with Bleeding Independent of Factor V Procoagulant Function. *Blood* **2015**, *125*, 1822–1825. [CrossRef]
64. Yamazaki, T.; Nicolaes, G.A.; Sørensen, K.W.; Dahlbäck, B. Molecular basis of quantitative factor V deficiency associated with factor V R2 haplotype. *Blood* **2002**, *100*, 2515–2521. [CrossRef]
65. Montefusco, M.C.; Duga, S.; Asselta, R.; Malcovati, M.; Peyvandi, F.; Santagostino, E.; Mannucci, P.M.; Tenchini, M.L. Clinical and Molecular Characterization of 6 Patients Affected by Severe Deficiency of Coagulation Factor V: Broadening of the Mutational Spectrum of Factor V Gene and in Vitro Analysis of the Newly Identified Missense Mutations. *Blood* **2003**, *102*, 3210–3216. [CrossRef]
66. Chen, T.Y.; Lin, T.M.; Chen, H.Y.; Wu, C.L.; Tsao, C.J. Gly392Cys Missense Mutation in the A2 Domain of Factor V Causing Severe Factor V Deficiency: Molecular Characterization by Expression of the Recombinant Protein. *Thromb. Haemost.* **2005**, *93*, 614–615. [CrossRef]
67. Liu, H.C.; Shen, M.C.; Eng, H.L.; Wang, C.H.; Lin, T.M. Asp68His Mutation in the A1 Domain of Human Factor V Causes Impaired Secretion and Ineffective Translocation. *Haemophilia* **2014**, *20*, e318–e326. [CrossRef]
68. Friedmann, A.P.; Koutychenko, A.; Wu, C.; Fredenburgh, J.C.; Weitz, J.I.; Gross, P.L.; Xu, P.; Ni, F.; Kim, P.Y. Identification and Characterization of a Factor Va-Binding Site on Human Prothrombin Fragment 2. *Sci. Rep.* **2019**, *9*, 2436. [CrossRef]
69. Peyvandi, F.; Garagiola, I.; Young, G. The Past and Future of Haemophilia: Diagnosis, Treatments, and Its Complications. *Lancet* **2016**, *388*, 187–197. [CrossRef]
70. Zhang, B.; Zheng, C.; Zhu, M.; Tao, J.; Vasievich, M.P.; Baines, A.; Kim, J.; Schekman, R.; Kaufman, R.J.; Ginsburg, D. Mice Deficient in LMAN1 Exhibit FV and FVIII Deficiencies and Liver Accumulation of A1-Antitrypsin. *Blood* **2011**, *118*, 3384–3391. [CrossRef] [PubMed]

71. Jenkins, P.V.; Dill, J.L.; Zhou, Q.; Fay, P.J. Contribution of Factor VIIIa A2 and A3-C1-C2 Subunits to the Affinity for Factor IXa in Factor Xase. *Biochemistry* **2004**, *43*, 5094–5101. [CrossRef]
72. DeAngelis, J.P.; Wakabayashi, H.; Fay, P.J. Sequences Flanking Arg336 in Factor VIIIa Modulate Factor Xa-Catalyzed Cleavage Rates at This Site and Cofactor Function. *J. Biol. Chem.* **2012**, *287*, 15409–15417. [CrossRef] [PubMed]
73. Williamson, D.; Brown, K.; Luddington, R.; Baglin, C.; Baglin, T. Factor V Cambridge: A New Mutation (Arg306→Thr) Associated With Resistance to Activated Protein C. *Blood* **1998**, *91*, 1140–1144. [CrossRef] [PubMed]
74. Zheng, C.; Zhang, B. Combined Deficiency of Coagulation Factors V and VIII: An Update. *Semin. Thromb. Hemost.* **2013**, *39*, 613–620. [CrossRef]
75. Everett, L.A.; Khoriaty, R.N.; Zhang, B.; Ginsburg, D. Altered Phenotype in LMAN1-Deficient Mice with Low Levels of Residual LMAN1 Expression. *Blood Adv.* **2020**, *4*, 5635–5643. [CrossRef] [PubMed]
76. Altenhoff, A.M.; Glover, N.M.; Dessimoz, C. Inferring Orthology and Paralogy. In *Evolutionary Genomics*; Anisimova, M., Ed.; Methods in Molecular Biology; Springer: New York, NY, USA, 2019; Volume 1910, pp. 149–175. [CrossRef]
77. Yang, T.L.; Cui, J.; Rehumtulla, A.; Yang, A.; Moussalli, M.; Kaufman, R.J.; Ginsburg, D. The Structure and Function of Murine Factor V and Its Inactivation by Protein C. *Blood* **1998**, *91*, 4593–4599. [CrossRef]
78. Rallapalli, P.M.; Orengo, C.A.; Studer, R.A.; Perkins, S.J. Positive Selection during the Evolution of the Blood Coagulation Factors in the Context of Their Disease-Causing Mutations. *Mol. Biol. Evol.* **2014**, *31*, 3040–3056. [CrossRef] [PubMed]
79. Mariz, J.P.V.; Nery, M.F. Unraveling the Molecular Evolution of Blood Coagulation Genes in Fishes and Cetaceans. *Front. Mar. Sci.* **2020**, *7*, 592383. [CrossRef]
80. Kretz, C.A.; Weyand, A.C.; Shavit, J.A. Modeling Disorders of Blood Coagulation in the Zebrafish. *Curr. Pathobiol. Rep.* **2015**, *3*, 155–161. [CrossRef] [PubMed]
81. Fish, R.J.; Freire, C.; Di Sanza, C.; Neerman-Arbez, M. Venous Thrombosis and Thrombocyte Activity in Zebrafish Models of Quantitative and Qualitative Fibrinogen Disorders. *Int. J. Mol. Sci.* **2021**, *22*, 655. [CrossRef] [PubMed]
82. Iyer, N.; Jagadeeswaran, P. Microkinetic Coagulation Assays for Human and Zebrafish Plasma. *Blood. Coagul. Fibrinolysis* **2021**, *32*, 50–56. [CrossRef]
83. Kuta, P.; Melling, N.; Zimmermann, R.; Achenbach, S.; Eckstein, R.; Strobel, J. Clotting Factor Activity in Fresh Frozen Plasma after Thawing with a New Radio Wave Thawing Device. *Transfusion* **2019**, *59*, 1857–1861. [CrossRef]
84. Marchesini, E.; Morfini, M.; Valentino, L. Recent Advances in the Treatment of Hemophilia: A Review. *BTT* **2021**, *15*, 221–235. [CrossRef]
85. Sankar, A.D.; Weyand, A.C.; Pipe, S.W. The Evolution of Recombinant Factor Replacement for Hemophilia. *Transfus. Apher. Sci.* **2019**, *58*, 596–600. [CrossRef] [PubMed]
86. High, K.A.; Roncarolo, M.G. Gene Therapy. *N. Engl. J. Med.* **2019**, *381*, 455–464. [CrossRef]
87. Anguela, X.M.; High, K.A. Entering the Modern Era of Gene Therapy. *Annu. Rev. Med.* **2019**, *70*, 273–288. [CrossRef]
88. Kamiyama, Y.; Naritomi, Y.; Moriya, Y.; Yamamoto, S.; Kitahashi, T.; Maekawa, T.; Yahata, M.; Hanada, T.; Uchiyama, A.; Noumaru, A.; et al. Biodistribution Studies for Cell Therapy Products: Current Status and Issues. *Regen. Ther.* **2021**, *18*, 202–216. [CrossRef]
89. Buzhor, E.; Leshansky, L.; Blumenthal, J.; Barash, H.; Warshawsky, D.; Mazor, Y.; Shtrichman, R. Cell-Based Therapy Approaches: The Hope for Incurable Diseases. *Regen. Med.* **2014**, *9*, 649–672. [CrossRef] [PubMed]
90. Weinshilboum, R.M.; Wang, L. Pharmacogenomics: Precision Medicine and Drug Response. *Mayo Clin. Proc.* **2017**, *92*, 1711–1722. [CrossRef] [PubMed]
91. De Haan, P.; Van Diemen, F.R.; Toscano, M.G. Viral Gene Delivery Vectors: The next Generation Medicines for Immune-Related Diseases. *Hum. Vaccin. Immunother.* **2021**, *17*, 14–21. [CrossRef]
92. Athanasopoulos, T.; Munye, M.M.; Yáñez-Muñoz, R.J. Nonintegrating Gene Therapy Vectors. *Hematol. Oncol. Clin. North Am.* **2017**, *31*, 753–770. [CrossRef]
93. Zu, H.; Gao, D. Non-Viral Vectors in Gene Therapy: Recent Development, Challenges, and Prospects. *AAPS J.* **2021**, *23*, 78. [CrossRef]
94. Gähwiler, E.K.N.; Motta, S.E.; Martin, M.; Nugraha, B.; Hoerstrup, S.P.; Emmert, M.Y. Human iPSCs and Genome Editing Technologies for Precision Cardiovascular Tissue Engineering. *Front. Cell Dev. Biol.* **2021**, *9*, 639699. [CrossRef]
95. García-Bernal, D.; García-Arranz, M.; Yáñez, R.M.; Hervás-Salcedo, R.; Cortés, A.; Fernández-García, M.; Hernando-Rodríguez, M.; Quintana-Bustamante, O.; Bueren, J.A.; García-Olmo, D.; et al. The Current Status of Mesenchymal Stromal Cells: Controversies, Unresolved Issues and Some Promising Solutions to Improve Their Therapeutic Efficacy. *Front. Cell Dev. Biol.* **2021**, *9*, 650664. [CrossRef] [PubMed]
96. Fomin, M.E.; Togarrati, P.P.; Muench, M.O. Progress and Challenges in the Development of a Cell-Based Therapy for Hemophilia A. *J. Thromb. Haemost.* **2014**, *12*, 1954–1965. [CrossRef] [PubMed]
97. Rodríguez-Merchán, E.C.; De Pablo-Moreno, J.A.; Liras, A. Gene Therapy in Hemophilia: Recent Advances. *Int. J. Mol. Sci.* **2021**, *22*, 7647. [CrossRef] [PubMed]
98. Parri, M.S.; Gianetti, J.; Dushpanova, A.; Della Pina, F.; Saracini, C.; Marcucci, R.; Giusti, B.; Berti, S. Pantoprazole Significantly Interferes with Antiplatelet Effect of Clopidogrel: Results of a Pilot Randomized Trial. *Int. J. Cardiol.* **2013**, *167*, 2177–2181. [CrossRef] [PubMed]

99. Tripodi, A.; Arbini, A.; Chantarangkul, V.; Mannucci, P.M. Recombinant Tissue Factor as Substitute for Conventional Thromboplastin in the Prothrombin Time Test. *Thromb. Haemost.* **1992**, *67*, 42–45. [CrossRef] [PubMed]
100. Van den Besselaar, A.M.; Neuteboom, J.; Bertina, R.M. Effect of Synthetic Phospholipids on the Response of the Activated Partial Thromboplastin Time to Heparin. *Blood Coagul. Fibrinolysis* **1993**, *4*, 895–903. [CrossRef] [PubMed]
101. Ray, M.J.; Hawson, G.A.T. A Comparison of Two APTT Reagents Which Use Silica Activators. *Clin. Lab. Haematol.* **1989**, *11*, 221–232. [CrossRef]
102. Den Dunnen, J.T.; Dalgleish, R.; Maglott, D.R.; Hart, R.K.; Greenblatt, M.S.; McGowan-Jordan, J.; Roux, A.F.; Smith, T.; Antonarakis, S.E.; Taschner, P.E.M.; et al. HGVS Recommendations for the Description of Sequence Variants: 2016 Update. *Hum. Mutat.* **2016**, *37*, 564–569. [CrossRef]
103. Di Tommaso, P.; Moretti, S.; Xenarios, I.; Orobitg, M.; Montanyola, A.; Chang, J.M.; Taly, J.F.; Notredame, C. T-Coffee: A Web Server for the Multiple Sequence Alignment of Protein and RNA Sequences Using Structural Information and Homology Extension. *Nucleic Acids Res.* **2011**, *39*, W13–W17. [CrossRef]

International Journal of
Molecular Sciences

MDPI

Article

Treatment of Erythroid Precursor Cells from β-Thalassemia Patients with *Cinchona* Alkaloids: Induction of Fetal Hemoglobin Production

Cristina Zuccato [1], Lucia Carmela Cosenza [1], Matteo Zurlo [1], Ilaria Lampronti [1,2], Monica Borgatti [1,2], Chiara Scapoli [3], Roberto Gambari [1,2,4,*] and Alessia Finotti [1,2,4,*]

1. Section of Biochemistry and Molecular Biology, Department of Life Sciences and Biotechnology, University of Ferrara, 44121 Ferrara, Italy; cristina.zuccato@unife.it (C.Z.); luciacarmela.cosenza@unife.it (L.C.C.); matteo.zurlo@unife.it (M.Z.); ilaria.lampronti@unife.it (I.L.); monica.borgatti@unife.it (M.B.)
2. Research Laboratory "Elio Zago" on the Pharmacologic and Pharmacogenomic Therapy of Thalassemia (Thal-LAB), University of Ferrara, 44121 Ferrara, Italy
3. Section of Biology and Evolution, Department of Life Sciences and Biotechnology, University of Ferrara, 44121 Ferrara, Italy; chiara.scapoli@unife.it
4. Interuniversity Consortium for Biotechnology (C.I.B.), 34148 Trieste, Italy

* Correspondence: gam@unife.it (R.G.); alessia.finotti@unife.it (A.F.); Tel.: +39-0532-974443 (R.G.); +39-0532-974510 (A.F.); Fax: +39-0532-974500 (R.G. & A.F.)

Abstract: β-thalassemias are among the most common inherited hemoglobinopathies worldwide and are the result of autosomal mutations in the gene encoding β-globin, causing an absence or low-level production of adult hemoglobin (HbA). Induction of fetal hemoglobin (HbF) is considered to be of key importance for the development of therapeutic protocols for β-thalassemia and novel HbF inducers need to be proposed for pre-clinical development. The main purpose on this study was to analyze *Cinchona* alkaloids (cinchonidine, quinidine and cinchonine) as natural HbF-inducing agents in human erythroid cells. The analytical methods employed were Reverse Transcription quantitative real-time PCR (RT-qPCR) (for quantification of *γ-globin* mRNA) and High Performance Liquid Chromatography (HPLC) (for analysis of the hemoglobin pattern). After an initial analysis using the K562 cell line as an experimental model system, showing induction of hemoglobin and *γ-globin* mRNA, we verified whether the two more active compounds, cinchonidine and quinidine, were able to induce HbF in erythroid progenitor cells isolated from β-thalassemia patients. The data obtained demonstrate that cinchonidine and quinidine are potent inducers of *γ-globin* mRNA and HbF in erythroid progenitor cells isolated from nine β-thalassemia patients. In addition, both compounds were found to synergize with the HbF inducer sirolimus for maximal production of HbF. The data obtained strongly indicate that these compounds deserve consideration in the development of pre-clinical approaches for therapeutic protocols of β-thalassemia.

Keywords: β-thalassemia; fetal hemoglobin; γ-globin; HbF induction; K562 cells; *Cinchona* alkaloids; cinchonidine; quinidine; cinchonine; combined treatments

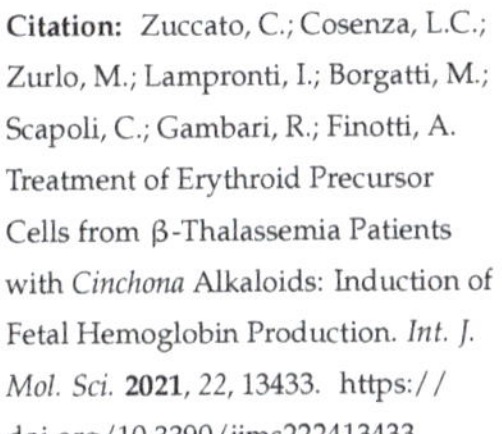

Citation: Zuccato, C.; Cosenza, L.C.; Zurlo, M.; Lampronti, I.; Borgatti, M.; Scapoli, C.; Gambari, R.; Finotti, A. Treatment of Erythroid Precursor Cells from β-Thalassemia Patients with *Cinchona* Alkaloids: Induction of Fetal Hemoglobin Production. *Int. J. Mol. Sci.* **2021**, *22*, 13433. https://doi.org/10.3390/ijms222413433

Academic Editor: Ivano Condò

Received: 4 October 2021
Accepted: 9 December 2021
Published: 14 December 2021

1. Introduction

β-thalassemias are among the most common inherited hemoglobinopathies worldwide, and are the result of more than 300 autosomal mutations of the gene encoding β-globin, causing an absence or low-level synthesis of this protein (and consequently of adult hemoglobin, HbA) in erythropoietic cells [1–5]. The phenotypes range widely from asymptomatic (β-thalassemia trait or carrier) to clinically relevant anemia, which is categorized as transfusion-dependent β-thalassemia (TDT, including thalassemia major) and non-transfusion-dependent β-thalassemia (NTDT, thalassemia intermedia) [1].

In the therapy of β-thalassemia, the induction of fetal hemoglobin (HbF) is considered to be of key importance for several concurrent reasons. In rare forms of β^0-thalassemia, particularly those with large deletions responsible for $\delta\beta^0$-thalassemia or hereditary persistence of fetal hemoglobin (HPFH), high γ-globin chain production results in high levels of HbF, which is associated with a relatively benign phenotype [6–9]. More-recent clinical studies have disclosed that the naturally higher production of HbF improves the clinical course in a variety of patients with β-thalassemia [10–14]. Furthermore, approaches using genome editing are available in a form that is finalized to the induction of HbF following the elimination of genomic sequences encoding for *γ-globin* gene transcriptional repressors or genomic sequences targeted by these regulatory factors [15–19]. In this context, clinical trials are ongoing, such as NCT03655678 (A Safety and Efficacy Study Evaluating CTX001 in Subjects with Transfusion-Dependent β-Thalassemia), based on the use of autologous CRISPR-Cas9 Modified $CD34^+$ Human Hematopoietic Stem and Progenitor Cells (hHSPCs) using CTX001 [20].

While some positive results have been described based on HbF induction by gene editing, the safety of this approach is still to be determined and, even in the case that its safety is demonstrated, the expected costs of this therapeutic approach would still be very high [21–23].

Accordingly, the validation of clinical relevance of already known HbF inducers and the characterization of novel HbF inducers are still projects of great interest [24–27].

In this context, a recent paper was published on the effects of *Cinchona* alkaloids (cinchonidine and quinidine) as natural fetal hemoglobin-inducing agents, using the human erythroleukemia K562 cell line as in vitro experimental model system [28]. In this study, *Cinchona* alkaloids showed dose-dependent induction of erythroid differentiation, increased production of HbF and high contents of *γ-globin* mRNA. Despite the fact that this study was based only on the use of K562 cells as a model system, it demonstrated that *Cinchona* alkaloids should be considered for the development of therapeutic protocols for β-thalassemia [28]. In addition, it should be underlined that molecules belonging to this family have been extensively used in therapy for other indications than β-thalassemia and, therefore, might be considered as "repurposed drugs" with a facilitated strategy to reach technology transfer [29–34]. For instance, quinine, quinidine, cinchonidine and cinchonine alkaloids had a powerful bioimpact as anti-malarial drugs [29–31]. Quinidine is also used to treat cardiac arrhythmias because it inhibits fibrillation [32–34].

Three are the main objectives of our study: (a) to confirm the data already published on the effects of *Cinchona* alkaloids (cinchonidine, quinidine and cinchonine) on K562 erythroleukemia cells; (b) to verify whether these molecules might potentiate the activity of other HbF inducers; (c) to verify whether the more promising *Cinchona* alkaloids tested induce HbF in erythroid progenitor cells isolated from β-thalassemia patients.

In terms of combined treatments, we decided to employ sirolimus (SIR, rapamycin) [35] in the co-treatment experiments for the following reasons: (a) sirolimus increases HbF in cultures from β-thalassemia patients with different basal HbF levels [36–39]; (b) sirolimus increases the overall Hb content per cell [37]; (c) sirolimus selectively induces *γ-globin* mRNA accumulation, with only minor effects on β-globin and *α-globin* mRNAs [36,37]; (d) sirolimus was found to induce HbF in vivo in mouse model systems [40–42]; and, more importantly, (e) sirolimus was able to increase HbF in sickle-cell disease (SCD) patients [43,44]. For these reasons, sirolimus obtained the Orphan Drug Designation by the European Medicinal Agency (EMA, Europe) and by the Food and Drug Administration (FDA, USA) for both β-thalassemia and SCD. Of relevance for this study, two ongoing clinical trials are based on sirolimus: NCT03877809 (A Personalized Medicine Approach for β-thalassemia Transfusion Dependent Patients: Testing sirolimus in a First Pilot Clinical Trial) and NCT04247750 (Treatment of β-thalassemia Patients with Rapamycin: From Pre-clinical Research to a Clinical Trial) [45]. Finally, it should be noted that sirolimus is a well-known drug, since it is employed for other therapeutic indications, such as kidney transplantation [46,47], cardiac [48] and liver [49] transplantation, lupus erythematosus

(SLE) [50], lymphangioleiomyomatosis (LAM) [51], tuberous sclerosis complex [52] and different types of cancers [53–55].

2. Results

2.1. Cinchonidine, Quinidine and Cinchonine Induce Differentiation of K562 Erythroleukemia Cells

Figure 1 shows that cinchonidine (CincD), quinidine (QuinD) and cinchonine (CincN) are all able to induce erythroid differentiation of human K562 cells in a concentration-dependent fashion. Determinations were performed after 5, 6 and 7 days of differentiation induction.

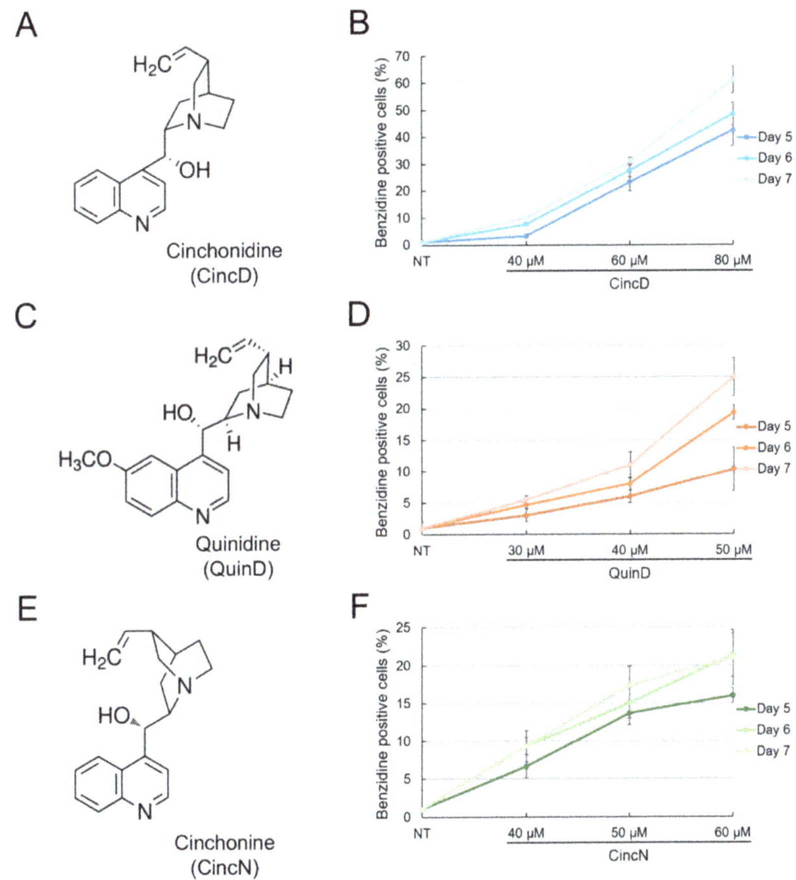

Figure 1. Effects of cinchonidine, quinidine and cinchonine on erythroid differentiation of K562 cells. (**A**,**C**,**E**) Structure of cinchonidine (CincD) (**A**), quinidine (QuinD) (**C**) and cinchonine (CincN) (**E**). (**B**,**D**,**F**) Effects on erythroid differentiation of K562 cells. K562 cells were cultured in the presence of the indicated concentrations of cinchonidine (**B**), quinidine (**D**) and cinchonine (**F**) and analysis of the proportion of benzidine-positive hemoglobin-containing cells was performed after 5, 6 and 7 days of cell culture. The data represent the average ± S.E.M. ($n = 3$).

After this induction period, cells were stained with benzidine in order to identify the hemoglobin-containing cells [26,36–38]. These data confirm the already-reported effects of these *Cinchona* alkaloids on the K562 cell system [28]. The induction of K562 differentiation was found to be associated, as expected, with the inhibition of cell proliferation, as reported in the K562 cell system for several other HbF inducers, such as hydroxyurea, mithramycin, rapamycin, cisplatin analogues and trimethylangelicin [25–27,37,56–59]. This effect of cinchonidine, quinidine and cinchonine on K562 cell proliferation is presented and comparatively analyzed in Supplementary Figures S1 and S2. The induction of K562 erythroid differentiation by the studied *Cinchona* alkaloids was similar (with respect to the extent of the induced proportion of benzidine-positive cells) to that of sirolimus (rapamycin), despite the fact that the effects of sirolimus were appreciable at 100–200 nM (Supplementary Figure S3). These differences among putative erythroid inducers were fully expected [60]. For example, butyrates are active at mM concentrations [61].

2.2. *Cinchonidine, Quinidine and Cinchonine Potentiate Sirolimus-Induced Differentiation of K562 Erythroleukemia Cells*

In order to verify possible combined effects of *Cinchona* alkaloids and sirolimus, we treated K562 cells simultaneously with different concentrations of these molecules. The obtained results of the treatment, reported in Figure 2, show that CincD, QuinD and CincN were able to further increase the erythroid differentiation activity of sirolimus, when this HbF inducer was used at 100 nM (Figure 2, panels A, C and E) and 200 nM (Figure 2, panels B, D and F).

In this experiment, K562 cells were cultured with increasing concentrations of cinchonidine, quinidine and cinchonine in the presence of 100 nM and 200 nM sirolimus, as indicated. The proportion of benzidine-positive K562 cells was determined after 5, 6 and 7 days of treatment. The data presented in Figure 2G are related to the use of suboptimal concentrations of CincD, QuinD and CincN (60, 40 and 50 μM, respectively) in the presence of 100 and 200 nM sirolimus. When the inducers (either sirolimus, or cinchonidine, quinidine and cinchonine) were used alone, the proportion of benzidine-positive hemoglobin-containing K562 cells was about 10–30%. On the contrary, when the inducers were used in combination, the proportion of benzidine-positive hemoglobin-containing K562 cells was always found to exceed 50% (the maximum level was reached using cinchonidine in combination with 200 nM sirolimus). The increased % of benzidine-positive cells compared with single administrations was always found to be statistically highly significant ($p < 0.001$). Untreated K562 displayed a proportion of benzidine-positive cells which never exceeded 2–5% (see, for instance, the microphotographs presented in the left part of Figure 3).

Figure 3 shows a representative microscopic analysis of the benzidine assays performed under some of the different experimental conditions described in Figure 2. The data give clear evidence that, in addition to the already reported increase in the proportion of benzidine-positive cells, and additional feature of combined treatments was evident, i.e., that all the benzidine-positive cells were brightly stained with benzidine-hydrogen peroxide solution (black arrowheads).

On the contrary, the presence of slightly stained benzidine-positive cells was clearly appreciable in singularly treated K562 cells (white arrowheads). While the representative data shown in Figure 3 were obtained using CincD plus SIR combination, the data obtained using QuinD/SIR and CincN/SIR combinations were found to be very similar (as shown in Supplementary Figure S4). Although this assay was not quantitative, the results obtained sustain the concept that the combined treatments lead to the highest levels of increase in the proportion of brightly stained benzidine-positive K562 cells. In order to verify this hypothesis using a more quantitative assay, RT-qPCR was performed on isolated RNA.

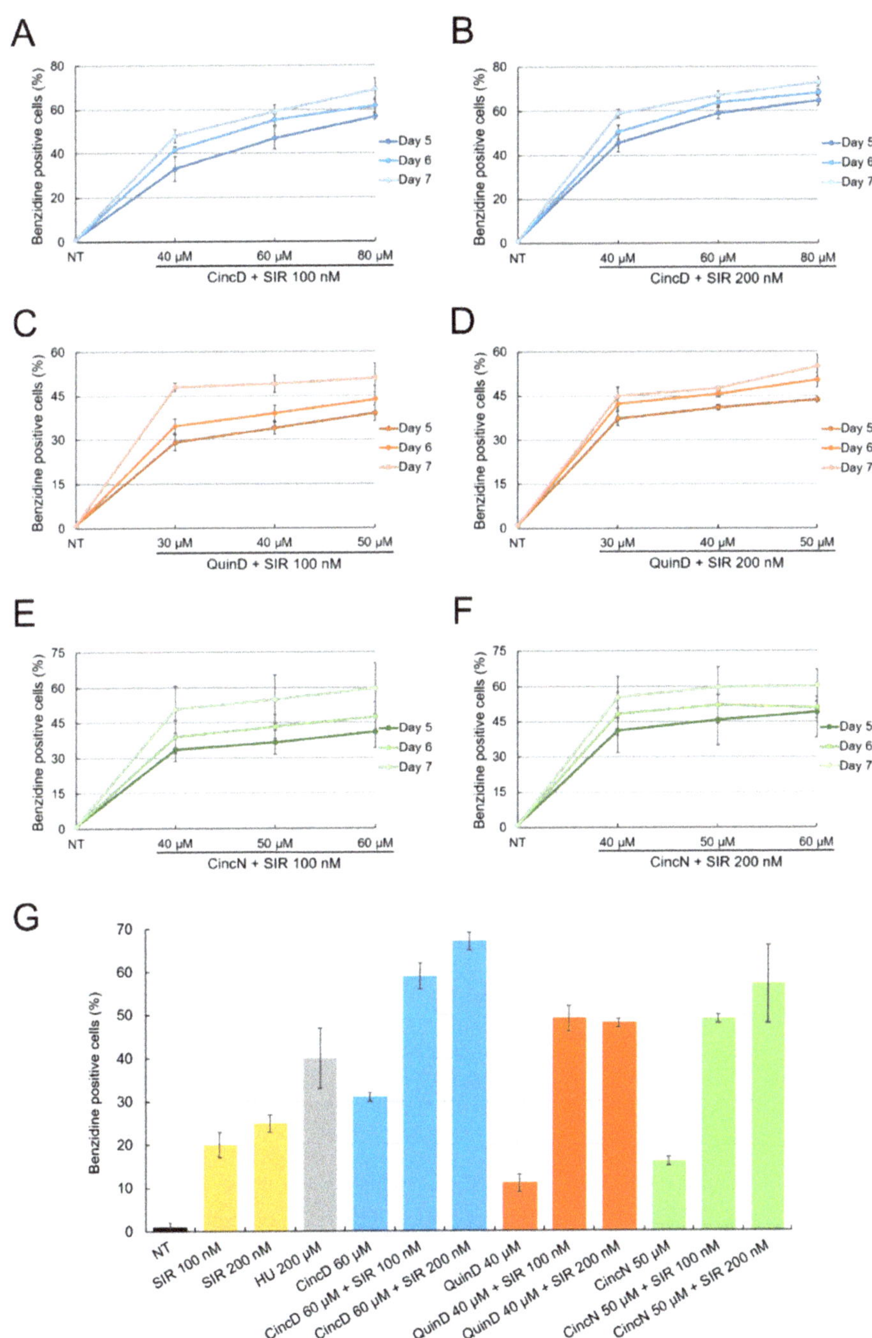

Figure 2. Effects of cinchonidine, quinidine and cinchonine on K562 erythroid differentiation induced by sirolimus. K562 cells were cultured with increasing concentrations of CincD (**A**,**B**), QuinD (**C**,**D**) and CincN (**E**,**F**) in the presence of 100 nM (**A**,**C**,**E**) and 200 nM (**B**,**D**,**F**) sirolimus (SIR). The proportion of benzidine-positive K562 cells was determined after 5, 6 and 7 days of treatment, as indicated. (**G**) Comparative analysis of the data obtained at day 7 of treatment. The data represent the average ± S.E.M. (n = 3).

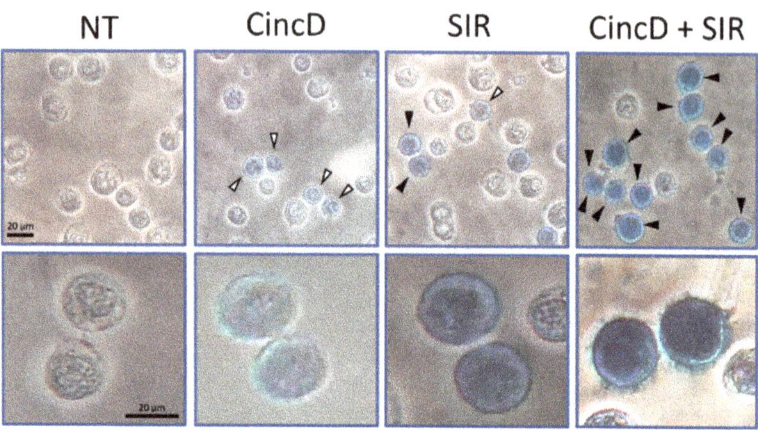

Figure 3. Benzidine-positive cells in untreated K562 cell cultures (NT) or in K562 cell cultures treated with CincD, SIR and CincD + SIR, as indicated. White arrowheads = slightly benzidine-stained cells; black arrowheads = brightly benzidine-stained cells. Magnitude: 20× (upper panels) and 40× (lower panels). Scale bar, 20 µm.

2.3. *The Effects of Cinchonidine, Quinidine and Cinchonine on K562 Erythroid Differentiation Are Associated with a Modulation of Expression of α-Globin and γ-Globin Genes*

Figure 4 shows that the erythroid differentiation induced by cinchonidine, quinidine and cinchonine was associated with an increase in the production of *α-globin* (Figure 4A,C,E) and *γ-globin* mRNAs (Figure 4B,D,F). The analysis of the expression of *α-globin* and *γ-globin* mRNAs was performed by RT-qPCR.

Interestingly, when used singularly, the increases in the production of both mRNAs induced by CincD (Figure 4A,B), QuinD (Figure 4C,D), and CincN (Figure 4E,F) were found to be similar to those found when 100 nM and 200 nM sirolimus was employed ($p > 0.05$). In addition, the increased levels of *γ-globin* mRNAs were similar to those originally described by Iftikhar et al. [28]. On the contrary, when cinchonidine, quinidine and cinchonine were used in combination with 100 nM and 200 nM sirolimus, a sharp increase in the content of *α-globin* mRNA ($p < 0.01$) and a less extensive but still significant ($p < 0.05$) increase in *γ-globin* mRNAs were found, fully in agreement with the effects of these compounds on sirolimus-induced K562 erythroid differentiation (Figures 2 and 3). In Figure 4, relevant examples of the p values obtained are shown, while the complete statistical analysis is presented in Supplementary Figures S5–S7.

The effects of CincD, QuinD, and CincN were dose-dependent, but clearly evident even when sub-optimal concentrations of compounds were employed.

Among the different combinations studied, those based on cinchonidine and quinidine were found to be the most effective in terms of inducing increases in the proportion of benzidine-positive cells and increased expression of globin genes (Supplementary Figures S5 and S6). Therefore, these two compounds were selected for further studies using erythroid precursor cells (ErPCs) from β-thalassemia patients as an experimental model system.

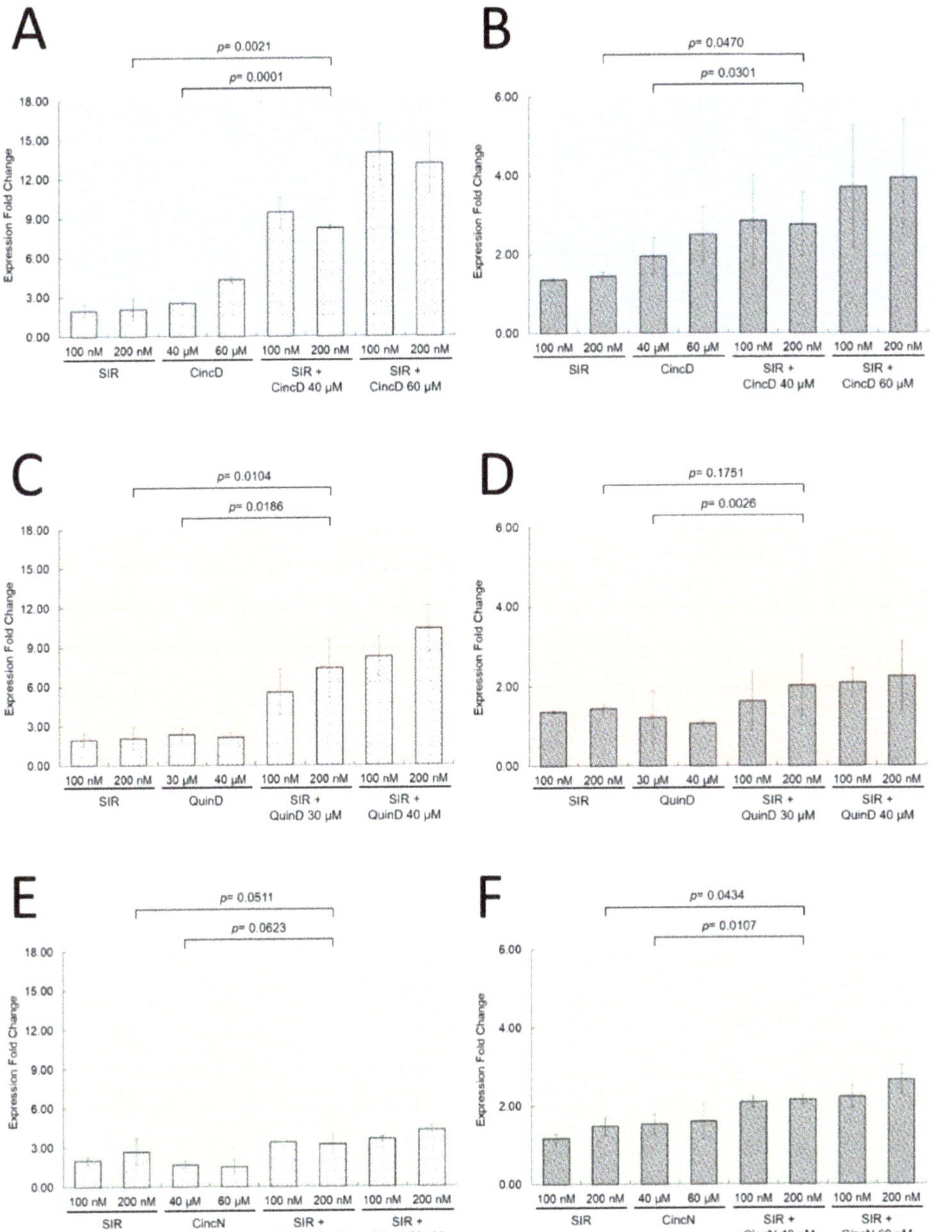

Figure 4. Effects of cinchonidine, quinidine and cinchonine on *α-globin* (left) and *γ-globin* (right) mRNAs. K562 cells were cultured with increasing concentrations of CincD (**A**,**B**), QuinD (**C**,**D**) and CincN (**E**,**F**) in the presence of 100 nM and 200 nM sirolimus (SIR) as indicated. After 5 days of treatment, RNA was isolated and RT-qPCR was performed using primers amplifying *α-globin* (**A**,**C**,**E**) and *γ-globin* (**B**,**D**,**F**) sequences. The increase in expression of *α-globin* and *γ-globin* mRNAs is presented as fold change with respect to control untreated cells. The data represent the average ± S.E.M. (n = 3).

2.4. Cinchonidine and Quinidine Induce HbF and γ-Globin mRNA in Erythroid Precursor Cells (ErPCs) from β-Thalassemia Patients

Patients were recruited at the Thalassemia Centre of Azienda Ospedaliera-Universitaria S. Anna (Ferrara, Italy). In total, 10 patients were enrolled. Informed written consent from all participants was obtained before recruiting them into the study. Different genotypes were present in the recruited cohort: five patients were $\beta^0 39/\beta^+$IVSI-110, two patients were β^+IVSI-110/β^+IVSI-110 and three patients were $\beta^0 39/\beta^0 39$. ErPCs were isolated from the β-thalassemia patients and cultured, as described elsewhere [59], with erythropoietin in the presence of sirolimus, cinchonidine and quinidine administered alone (this was performed in ErPC cultures from all the ten patients) or in combination (this was performed in ErPC cultures from five patients).

In Figure 5, the HPLC profiles of two representative ErPC populations are shown, one isolated from patient #10 (Figure 5A–C) and the other from patient #5 (Figure 5D–F), exhibiting a differential response to cinchonidine (Figure 5B,E) and quinidine (Figure 5C,F) treatment.

The ErPCs from patient #10 exhibited a 69.57% and 118.04% increase in HbF after treatment with cinchonidine and quinidine, respectively (see the raw data shown in Supplementary Table S1). A lower increase in HbF was obtained when the ErPCs from patient #1 were employed (in this case, the HbF increase was 5.40% and 9.53% for cinchonidine and quinidine, respectively).

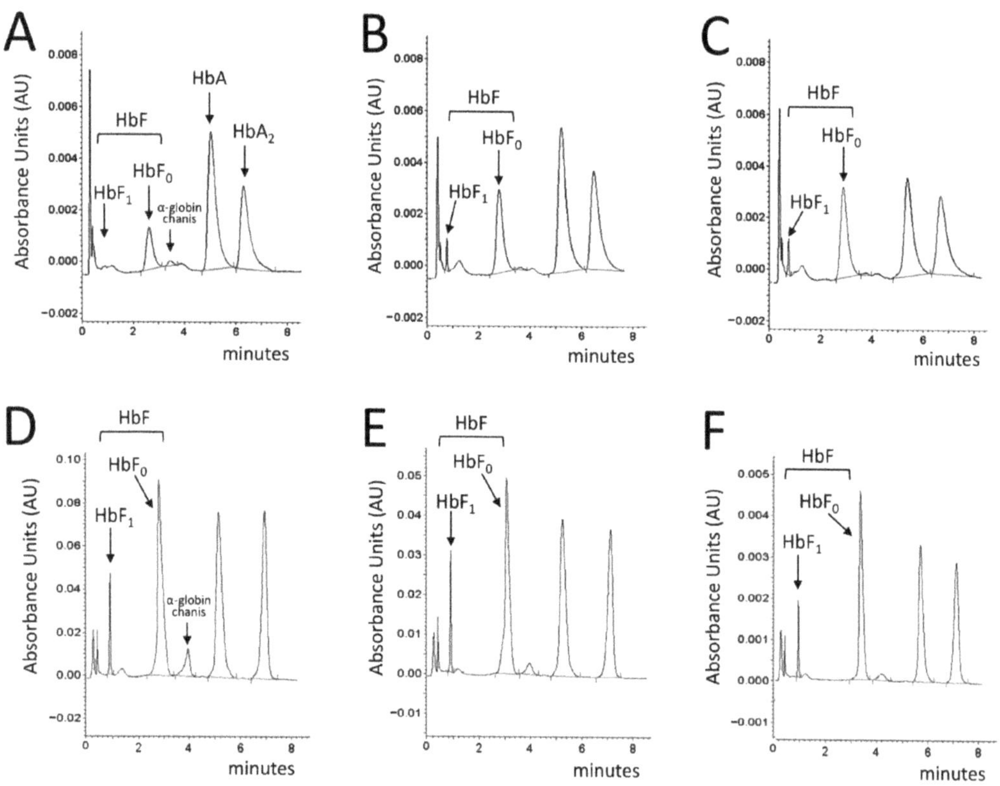

Figure 5. HPLC profile of cytoplasm isolated from ErPCs treated with cinchonidine and quinidine. ErPCs isolated from patient #10 (**A–C**) and patient #1 (**D–F**) were treated with 60 μM cinchonidine (**B**,**E**) and 30 μM quinidine (**C**,**F**) for 7 days and the lysate was analyzed by HPLC. HPLC profiles of untreated ErPCs are depicted in panels (**A**,**D**). The peaks corresponding to HbF, free α-globin chains, HbA and HbA_2 are shown.

Figure 6 shows the effects of cinchonidine and quinidine on the ErPCs from all the 10 recruited β-thalassemia patients. All the raw data are reported in Supplementary Table S1. As clearly evident, increases in the proportion of HbF (% of all the accumulated hemoglobins) were found in the treated ErPCs from most of the recruited β-thalassemia patients (Figure 6A).

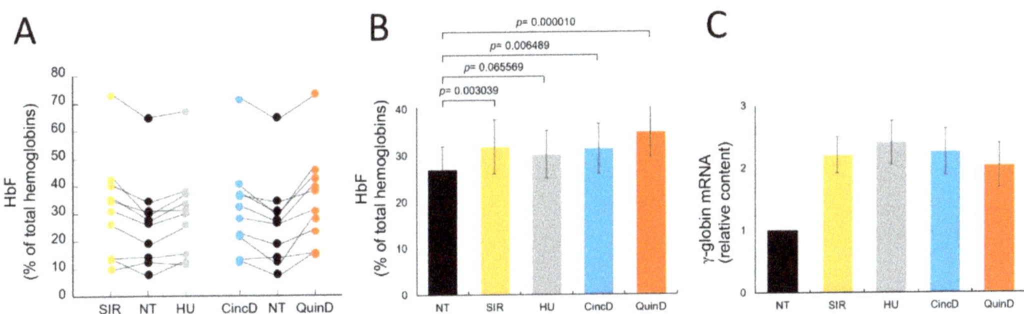

Figure 6. Effects of cinchonidine and quinidine on the HbF production and *γ-globin* gene expression in treated ErPCs. ErPcs isolated from 10 β-thalassemia patients were treated with 100 nM sirolimus (SIR), 100 μM hydroxyurea (HU), 60 μM cinchonidine (CincD) and 30 μM quinidine (QuinD) for 7 days. The lysate was analyzed by HPLC for HbF quantification (% with repect to total HbF produced) and the RNA was analyzed by RT-qPCR for determination of the relative *γ-globin* mRNA content. The raw data of HbF content are shown in the Supplementary Table S1. The % of HbF following the different treatments are reported in (**A**,**B**). The increase in *γ-globin* mRNA content is shown in (**C**). The results of panels (**B**,**C**) are reported as mean ± S.E.M. (n = 10).

Figure 6B presented this increase in the % of HbF with respect to untreated ErPCs and gives evidence for an HbF increase comparable to that of sirolimus and the established HbF inducer hydroxyurea.

In fact, the HbF increase was significant in the treated cells when the obtained values were compared to those found in untreated cells (Figure 6B).

As expected for HbF inducers, the differences were not statistically significant when the data of CincD- and/or QuinD-treated ErPCs were compared with those obtained using SIR or HU.

The HbF increase was, as expected, associated with increase in *γ-globin* mRNA (Figure 6C). Moreover, in this case, the increases in *γ-globin* mRNA (2.26 ± 0.37 and 2.04 ± 0.36 folds for cinchonidine and quinidine, respectively) were similar to the increases found in sirolimus-treated (2.21 ± 0.29) and hydroxyurea-treated (2.41 ± 0.35) ErPCs. The expected slight inhibitory effects of CincD and QuinD on ErPC cell proliferation were similar to those of the validated HbF inducers SIR and HU (Supplementary Figure S8).

2.5. Cinchonidine and Quinidine Potentiate Sirolimus-Mediated Induction of HbF and γ-Globin mRNA in ErPCs from β-Thalassemia Patients

Figure 7 shows the data obtained when ErPCs from five patients were treated, in addition to the treatments already mentioned in Section 2.4, with cinchonidine and quinidine in the presence of 100 nM sirolimus. The data reported are related to increase in the % of HbF and of *γ-globin* mRNA content. As clearly evident, the ErPC cultures exhibiting the highest levels of % of HbF and increased *γ-globin* mRNA are those treated with the two alkaloids and sirolimus (the raw data are reported in Supplementary Table S2). For instance, the % increase in HbF was 66.30 ± 25.09 and 82.30 ± 37.18 in ErPCs co-treated with sirolimus plus cinchonidine or sirolimus plus quinidine, respectively. These values were higher than those found when single drugs were added (20.45 ± 13.82, 31.61 ± 11.62 and 54.83 ± 19.52 for sirolimus, cinchonidine and quinidine, respectively). When the values relative to the treatments with CincD plus SIR and with QuinD plus SIR were compared to the treatment with the reference HbF inducers HU and SIR, the differences in the % of HbF incresae were found to be highly significant (Figure 7A, upper part of the panel). On the contrary, when the values of CincD or QuinD are compared to HU or SIR, the differences were not significant ($p > 0.2$).

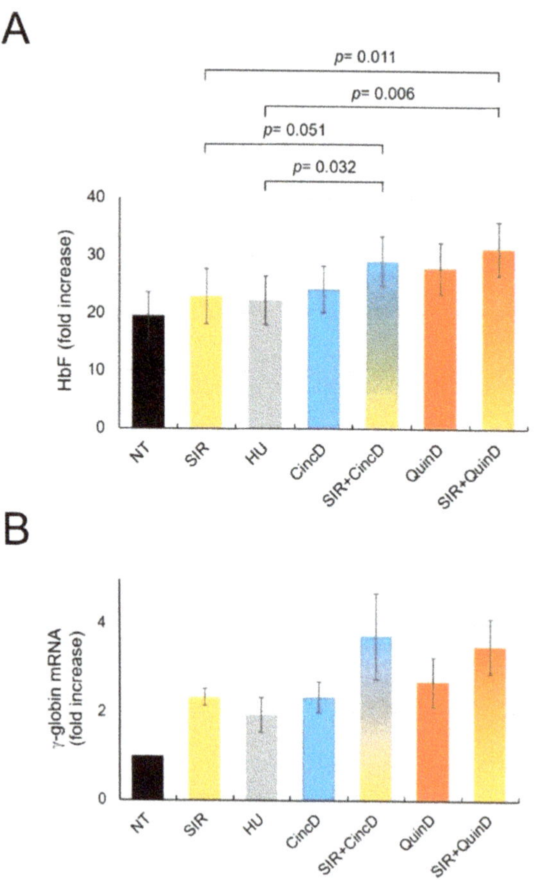

Figure 7. Effects of co-treatment of ErPCs with sirolimus and cinchonidine or quinidine. ErPCs isolated from five β-thalassemia patients were treated, as indicated, for 7 days with 100 nM sirolimus (SIR), 100 µM hydroxyurea (HU), 60 µM cinchonidine (CincD), 30 µM quinidine (QuinD) or with SIR + CincD or SIR + QuinD. The lysates were analyzed by HPLC for HbF quantification and the RNA by RT-qPCR for determination of the *γ-globin* mRNA content. The average increases in HbF are reported in (**A**), the increase in *γ-globin* mRNA content is shown in (**B**). The data represent the mean ± S.E.M. ($n = 5$).

2.6. Treatment of ErPCs from β-Thalassemia Patients with Cinchonidine and Quinidine Is Associated with a Sharp Decrease in the Free α-Globin Chains

Reduction in the excess α-globin should be considered as an important objective in the development of therapeutic interventions of β-thalassemia, since the excess α-globin decreases the lifespan of the red-blood cells, causes ineffective erythropoiesis and is a major determinant of the clinical severity of β-thalassemia [42].

The effects of cinchonidine and quinidine on the free α-globin chains produced by the ErPCs from the recruited β-thalassemia patients displaying α-globin > 2.5 are summarized in Figure 8. All the raw data are reported in Supplementary Table S3.

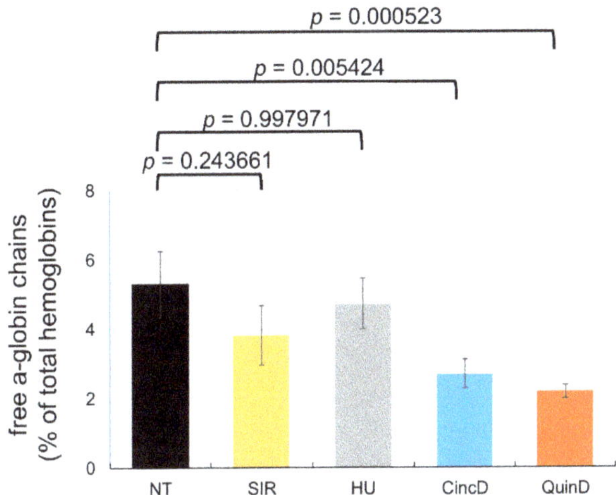

Figure 8. Effects of cinchonidine and quinidine on the free α-globin chains. ErPCs isolated from β-thalassemia patients were treated as indicated for 7 days with 100 nM sirolimus (SIR), 100 μM hydroxyurea (HU), 60 μM cinchonidine (CincD) and 30 μM quinidine (QuinD). The lysates were analyzed by HPLC (see Figure 5 for representative profiles) for quantification of free α-globin chains. The data represent the mean ± S.E.M. (n = 10).

As clearly evident, a decrease in the % of the free α-globin peak was found in the treated ErPCs from most of the recruited β-thalassemia patients (see Supplementary Table S3). Of great interest is a lower reduction in the free α-globin chains when ErPCs were treated with sirolimus or hydroxyurea. An average α-globin peak reduction of 27.98% was found with sirolimus, while the % average reduction with hydroxyurea was only 10.78%. Higher and more significant reductions (p = 0.024504 and 0.006129, respectively) were found with cinchonidine and quinidine (49.15% and 58.79%, respectively). These data suggest that, in addition to increased expression of γ-globin genes and HbF production, cinchonidine and quinidine might exert their beneficial effects on ErPCs through a decrease in the excess free α-globin chains.

3. Discussion

Induction of fetal hemoglobin (HbF) is considered a very promising strategy in the therapy of β-thalassemia and sickle-cell disease (SCD). The gene-therapy-mediated induction of γ-globin gene expression and HbF production in erythroid cells was described. In addition, approaches using genome editing are available in forms that are finalized to the induction of HbF following the elimination of genomic sequences encoding for transcriptional repressors or genomic sequences targeted by these regulatory factors [15–19].

While some positive results have been described based on HbF induction by gene therapy and gene editing, the safety of these approaches are still to be determined. In addition, it is expected that the costs of these therapeutic interventions will be very high [21–23].

Accordingly, the validation of the clinical relevance of already-known HbF inducers and the characterization of novel HbF inducers are still projects of great interest [24–27]. In this context, a paper was recently published by Iftikhar et al. on the effects of *Cinchona* alkaloids (cinchonidine and quinidine) as natural fetal hemoglobin-inducing agents in human erythroleukemia cells [28]. This study was very interesting, despite the fact that the key results were obtained using only K562 cells as an in vitro experimental model system.

The key conclusions of our study are the following: (a) cinchonidine and quinidine are inducers of an increase in the % of HbF in erythroid progenitor cells isolated from β-thalassemia patients; (b) cinchonidine and quinidine potentiate the activity of sirolimus (a HbF inducer employed in clinical trials). These data sustain the concept that cinchonidine and quinidine should be considered for further studies aimed at developing protocols for the treatment of β-thalassemia patients. Our data show that the HbF induction efficiency of cinchonidine and quinidine is similar to that of hydroxyurea (the reference HbF inducer) and sirolimus (Figure 6).

In addition to the induction of changes in HbF expression, cinchonidine and quinidine might also act through a reduction in the excess free α-globins present in the erythroid cells of β-thalassemia patients. This reduction should be considered a key objective in the use of molecules for therapeutic interventions in the management of β-thalassemia, since the excess α-globin is one of key factors causing short lifespans of the red-blood cells with associated ineffective erythropoiesis [42]. Interestingly, this therapeutic relevant target is reached very efficiently using cinchonidine and quinidine, as both are more efficient in reducing the excess free α-globin than hydroxyurea and sirolimus (Figure 8). Further studies will clarify whether the reduction in the excess α-globin is associated with the activation of autophagy, as proposed elsewhere [42].

In terms of combined treatments, we decided to employ sirolimus [35] as this HbF inducer is, at present, employed in two ongoing clinical trials: NCT03877809 (A Personalized Medicine Approach for β-thalassemia Transfusion Dependent Patients: Testing sirolimus in a First Pilot Clinical Trial) and NCT04247750 (Treatment of β-thalassemia Patients with Rapamycin: From Pre-clinical Research to a Clinical Trial) [45]. Further studies will verify whether the *Cinchona* alkaloids employed in this study potentiate the activity of other HbF inducers, including hydroxyurea, that are extensively employed in the treatment of β-thalassemia and sickle-cell disease [62,63]. One of the limits of our study is that the mechanism(s) of action was not experimentally evaluated. Further studies are required to understand this specific issue in our ErPC model system. However, published studies support the hypothesis that the mechanism(s) of action of *Cinchona* alkaloids and sirolimus are sharply different. In fact, *Cinchona* alkaloids are reported to inhibit cytochrome P450 enzyme 2D6 and the transport protein P-glycoprotein [64,65], while sirolimus is firmly established as an mTOR inhibitor [35,66].

The data obtained in our study strongly support the concept that cinchonidine and quinidine might be employed in combination with sirolimus in order to maximize its effects on in vivo-treated β-thalassemia patients.

In this respect, it is interesting to observe that cinchonidine and quinidine might be more active than hydroxyurea, which is one of the most important reference compounds when clinical treatment of β-thalassemia and sickle-cell disease is considered [62,63]. Further studies employing analyses of the effects on transcriptome and proteome, as well as confirming the presence of increased HbF within selected cell populations, are needed in order to verify whether the increase in the % of HbF reported in the present study is accompanied by a clinically relevant increase in the content of HbF in each treated erythroid cell.

Moreover, in order to propose a possible protocol for therapeutic purposes, a proof-of-principle showing in vivo effects on animal model systems is highly recommended, as well as a careful analysis of the relationship between the effects on HbF and the presence of DNA polymorphisms associated with the predisposition of patients to high HbF induction.

4. Materials and Methods

4.1. Patients Recruitment

Cultures of erythroid progenitors were derived from the peripheral blood of β-thalassemia patients. Patients were recruited at the Day Hospital Thalassemia and Hemoglobinophaties of Azienda Ospedaliera-Universitaria S. Anna (Ferrara, Italy). All the patients received a patient information sheet to read and time to clarify doubts with investigators before consenting. All the participants signed an informed consent form on the basis of approvals of the Ethical Committee in charge of human studies at the University Hospital. The recruited patients were all transfusion dependent and not under hydroxyurea therapy. Treatments were performed on cultured ErPCs derived from patients blood isolated just before transfusion.

4.2. Chemical Reagents for Cell Culture Treatments

The reagents used for K562 treatments (rapamycin (sirolimus, SIR), cat. R0395; hydroxyurea (HU), cat. H8627; cinchonine (CincN), cat. 27370; cinchonidine (CincD), cat. C80407; quinidine (QuinD), cat. 22600) were purchased from Sigma Aldrich (St. Louis, MO, USA). HU was solubilized in sterile deionized H_2O, whereas rapamycin, cinchonine, cinchonidine and quinidine were solubilized in ethanol and stored at −20 °C. Stock solution of rapamycin was prepared at 5 mM and 20 mM for each of the *Cinchona* alkaloids used. We used a concentration of sirolimus known to induce both K562 and erythroid precursor cells from β-thalassemia patients [37]. These stocks were further diluted to the indicated concentrations in culture medium prior to experimentation. All the treatments were performed by adding the compounds once at the beginning of the culturing period.

4.3. Human K562 Cell Cultures

The human leukemia K562 [26,36] cells were cultured in a humidified atmosphere of 5% CO_2/air in RPMI 1640 medium (RPMI 1640 medium; Lonza, Verviers, Belgium) supplemented with 10% (vol/vol) fetal bovine serum (FBS; Biowest, Nuaille, France) and 1% penicillin–streptomycin (Euroclone, Milano, Italy). Cell growth was studied by determining the cell number per ml with a Z2 Coulter Counter (Beckman Coulter, Fullerton, CA, USA).

4.4. In Vitro Culture of Erythroid Progenitors from β-Thalassemia Patients

The two-phase liquid culture procedure was employed as previously described [37,59]. Mononuclear cells were isolated from peripheral blood samples of β-thalassemia patients: 20–25 mL of peripheral blood were collected before transfusion from patients who gave informed consent. A mixture of blood and PBS 1× at a 1:1 ratio was stratified on top of Lympholyte®-H Cell Separation Media (Cedarlane, Burlington, NC, USA). After isolation, the mononuclear cell layer was washed three times by adding 1× PBS solution and seeded in α-minimal essential medium (α-MEM; Sigma Aldrich, St. Louis, MO, USA) supplemented with 10% FBS (FBS; Biowest, Nuaille, France), 1 μg/mL cyclosporine A (Sigma Aldrich, St. Louis, MO, USA), 10% conditioned medium from the 5637 bladder carcinoma cell line culture and stem cell factor (SCF, Life Technologies, Monza, MB, Italy) at the final concentration of 10 ng/ml. The cultures were incubated at 37 °C, under an atmosphere of 5% CO_2, with extra humidity. After 7 days in this phase-I culture, the non-adherent cells were harvested from the flask, washed in 1x PBS, and then cultured in phase-II medium, composed of α-MEM medium (Sigma Aldrich, St. Louis, MO, USA), 30% FBS (FBS; Biowest, Nuaille, France), 1% deionized bovine serum albumin (BSA, Sigma Aldrich, St. Louis, MO, USA), 10^{-5} M β-mercaptoethanol (Sigma Aldrich, St. Louis, MO, USA), 2 mM L-glutamine (Sigma Aldrich, St. Louis, MO, USA), 10^{-6} M dexamethasone (Dexamethasone 21-phosphate disodium salt; Sigma Aldrich, St. Louis, MO, USA) and 1 U/mL human recombinant erythropoietin (EPO) (Tebu-bio, Magenta, Milano, Italy) and stem cell factor (SCF) at the final concentration of 10 ng/mL. Erythroid precursor cells' differentiation was assessed by benzidine staining [59].

4.5. Reverse Transcription and Quantitative Real-Time PCR (RT-qPCR)

For gene expression analysis, 500 ng of total RNA were reverse transcribed using the TaqMan® Reverse Transcription Reagents kit and random hexamers (Applied Biosystems, Foster City, CA, USA). The RT-qPCR assay was carried out using gene-specific double fluorescently labeled probes. The reaction mixture had a final volume of 25 μL and was composed of Prime Time® Gene Expression Master Mix 1× (IDT, Tema Research, Castenaso, BO, Italy), the pairs of forward and reverse primers (α, β, γ, together or *GAPDH*, *RPL13A*, *ACTB* together) used at 500 nM concentration and the probes (α, β, γ, together or *GAPDH*, *RPL13A*, *ACTB* together) used at 250 nM concentration. The probes that contained 6-carboxyfluorescein (FAM) and hexachloro-6-carboxyfluorescein (HEX) as chromogenic molecules at 5′ were quenched by the Iowa Black® FQ molecule at 3′, while probes that contained indocarbocyanine (Cy5) were quenched by Iowa Black® RQ. After an initial step for the denaturation at 95 °C for 2 min, the reactions were performed for 50 cycles consisting of two phases, 95 °C for 10 s and 60 °C for 45 s.

4.6. HPLC Analysis of Hemoglobins

To evaluate the effective quantity of the various types of hemoglobin produced by the cultured erythroid cells after treatment, High-Performance Liquid Chromatography was performed. The ErPCs were centrifuged at 2000 rpm for 6 minutes and washed with PBS (Phosphate buffered saline). The pellet was then resuspended in a predefined volume of water for HPLC (Sigma-Aldrich, St. Louis, MO, USA). This was followed by 3 freeze/thaw cycles on dry ice in order to lyse the cells and obtain the protein extracts. Lysates were centrifuged for 5 min at 14,000 rpm and the supernatant was collected. Hemoglobin analysis was performed by loading the protein extracts into a PolyCAT-A cation exchange column, and they were then eluted in a sodium-chloride-BisTris-KCN aqueous mobile phase using the HPLC Beckman Coulter Instrument System Gold 126 Solvent Module-166 Detector, which allowed us to obtain a quantification of the hemoglobins present in the sample. The reading was performed at a wavelength of 415nm, and a commercial solution of purified human HbAF (Sigma-Aldrich) extracts was used as a standard. The values thus obtained were processed using "32 Karat software".

4.7. Statistical Analysis

All the data were normally distributed and presented as mean ± S.E.M. Statistical differences between groups were compared using a paired *t*-test or a one-way repeated measures ANOVA (ANalyses of VAriance between groups) followed by LSD post-hoc tests. Statistical differences were considered significant when $p < 0.05$ (*) and highly significant when $p < 0.01$ (**).

Supplementary Materials: The following are available online at https://www.mdpi.com/article/10.3390/ijms222413433/s1.

Author Contributions: Conceptualization, A.F. and R.G.; methodology, validation, C.Z., L.C.C., M.Z. and I.L.; formal analysis, R.G. and A.F.; investigation, A.F. and C.Z.; lab resources, R.G., I.L. and M.B.; data curation, C.Z. and A.F.; statistical analysis, C.S.; writing—original draft preparation, R.G., C.Z. and A.F.; writing—review and editing, R.G. and A.F.; supervision, R.G. and A.F.; project administration, A.F.; funding acquisition, R.G. and A.F. All authors have read and agreed to the published version of the manuscript.

Funding: This study was sponsored by the Wellcome Trust (Innovator Award 208872/Z/17/Z) and AIFA (AIFA-2016-02364887). The research leading to these results has also received funding by the UE THALAMOSS Project (Thalassemia Modular Stratification System for Personalized Therapy of Beta-Thalassemia; no. 306201-FP7- HEALTH-2012-INNOVATION-1) and FIR and FAR funds from the University of Ferrara. This research was also supported by ALT (Associazione per la lotta alla Talassemia "Rino Vullo"—Ferrara, AVLT (Associazione Veneta per la Lotta alla Talassemia "Elio Zago" - APS), and by the Interuniversity Consortium for Biotechnology (C.I.B.), Italy. C.Z. was supported by a fellowship from "Tutti per Chiara Onlus".

Institutional Review Board Statement: The study was conducted according to the guidelines of the Declaration of Helsinki and the use of human material was approved by the Ethics Committee of Ferrara's District, protocol name: THAL-THER, document number 533/2018/Sper/AOUFe, approved on 14 November 2018. All samples of peripheral blood were obtained after receiving written informed consent from donor patients or their legal representatives.

Informed Consent Statement: Informed consent for the testing of HbF inducers in erythroid progenitors derived from peripheral blood was obtained from all β-thalassemia patients involved in the study before the blood was drawn.

Data Availability Statement: Most of the raw data are included in Supplementary Materials. Additional information will be freely available upon request to the correspondence authors.

Acknowledgments: We thank Marco Prosdocimi (Rare Partners srl) for helpful suggestions. This study is dedicated to the memory of Chiara Gemmo and Elio Zago. The authors wish to thank all β-thalassemia patients who participated in this study.

Conflicts of Interest: The authors declare no conflict of interest.

Abbreviations

Hb, hemoglobin; HbA, adult hemoglobin; HbF, fetal hemoglobin; HPFH, hereditary persistence of fetal hemoglobin; SCD, sickle-cell disease; ErPCs, erythroid precursor cells, CincD, cinchonidine, QuinD, quinidine; CincN, cinchonine, SIR, Sirolimus; HU, hydroxyurea; RT-qPCR, Reverse Transcription quantitative real-time PCR; HPLC, High Performance Liquid Chromatography.

References

1. Weatherall, D.J. Phenotype-genotype relationships in monogenic disease: Lessons from the thalassaemias. *Nat. Rev. Genet.* **2001**, *2*, 245–255. [CrossRef]
2. Origa, R. β-Thalassemia. *Genet. Med.* **2017**, *19*, 609–619. [CrossRef] [PubMed]
3. Fucharoen, S.; Weatherall, D.J. Progress Toward the Control and Management of the Thalassemias. *Hematol. Oncol. Clin. N. Am.* **2016**, *30*, 359–371. [CrossRef]
4. Modell, B.; Darlison, M.; Birgens, H.; Cario, H.; Faustino, P.; Giordano, P.C.; Gulbis, B.; Hopmeier, P.; Lena-Russo, D.; Romao, L.; et al. Epidemiology of haemoglobin disorders in Europe: An overview. *Scand. J. Clin. Lab. Investig.* **2007**, *67*, 39–69. [CrossRef] [PubMed]
5. Galanello, R.; Origa, R. B-thalassemia. *Orphanet. J. Rare Dis.* **2010**, *5*, 11. [CrossRef] [PubMed]
6. Thein, S.L. The molecular basis of β-thalassemia. *Cold Spring Harb. Perspect. Med.* **2013**, *3*, a011700.
7. Sripichai, O.; Fucharoen, S. Fetal hemoglobin regulation in β-thalassemia: Heterogeneity, modifiers and therapeutic approaches. *Expert Rev. Hematol.* **2016**, *9*, 1129–1137. [CrossRef]
8. Forget, B.G. Molecular basis of hereditary persistence of fetal hemoglobin. *Ann. N. Y. Acad. Sci.* **1998**, *850*, 38–44. [CrossRef]
9. Musallam, K.M.; Sankaran, V.G.; Cappellini, M.D.; Duca, L.; Nathan, D.G.; Taher, A.T. Fetal hemoglobin levels and morbidity in untransfused patients with β-thalassemia intermedia. *Blood* **2012**, *119*, 364–367. [CrossRef]
10. Nuinoon, M.; Makarasara, W.; Mushiroda, T.; Setianingsih, I.; Wahidiyat, P.A.; Sripichai, O.; Kumasaka, N.; Takahashi, A.; Svasti, S.; Munkongdee, T.; et al. A genome-wide association identified the common genetic variants influence disease severity in β0-thalassemia/hemoglobin E. *Hum. Genet.* **2010**, *127*, 303–314. [CrossRef]
11. Danjou, F.; Anni, F.; Perseu, L.; Satta, S.; Dessì, C.; Lai, M.E.; Fortina, P.; Devoto, M.; Galanello, R. Genetic modifiers of b-thalassemia and clinical severity as assessed by age at first transfusion. *Haematologica* **2012**, *97*, 989–993. [CrossRef] [PubMed]
12. Badens, C.; Joly, P.; Agouti, I.; Thuret, I.; Gonnet, K.; Fattoum, S.; Francina, A.; Simeoni, M.C.; Loundou, A.; Pissard, S. Variants in genetic modifiers of β-thalassemia can help to predict the major or intermedia type of the disease. *Haematologica* **2011**, *96*, 1712–1714. [CrossRef] [PubMed]
13. Uda, M.; Galanello, R.; Sanna, S.; Lettre, G.; Sankaran, V.G.; Chen, W.; Usala, G.; Busonero, F.; Maschio, A.; Albai, G.; et al. Genome wide association study shows BCL11A associated with persistent fetal hemoglobin and amelioration of the phenotype of β-thalassemia. *Proc. Natl. Acad. Sci. USA* **2008**, *105*, 1620–1625. [CrossRef] [PubMed]
14. Breveglieri, G.; Bianchi, N.; Cosenza, L.C.; Gamberini, M.R.; Chiavilli, F.; Zuccato, C.; Montagner, G.; Borgatti, M.; Lampronti, I.; Finotti, A.; et al. An Agamma-globin G->A gene polymorphism associated with beta(0)39 thalassemia globin gene and high fetal hemoglobin production. *BMC Med. Genet.* **2017**, *18*, 93. [CrossRef]
15. Antoniani, C.; Meneghini, V.; Lattanzi, A.; Felix, T.; Romano, O.; Magrin, E.; Weber, L.; Pavani, G.; El Hoss, S.; Kurita, R.; et al. Induction of fetal hemoglobin synthesis by CRISPR/Cas9-mediated editing of the human β-globin locus. *Blood* **2018**, *131*, 1960–1973. [CrossRef]

16. Wu, Y.; Zeng, J.; Roscoe, B.P.; Liu, P.; Yao, Q.; Lazzarotto, C.R.; Clement, K.; Cole, M.A.; Luk, K.; Baricordi, C.; et al. Highly efficient therapeutic gene editing of human hematopoietic stem cells. *Nat. Med.* **2019**, *25*, 776–783. [CrossRef]
17. Métais, J.Y.; Doerfler, P.A.; Mayuranathan, T.; Bauer, D.E.; Fowler, S.C.; Hsieh, M.M.; Katta, V.; Keriwala, S.; Lazzarotto, C.R.; Luk, K.; et al. Genome editing of HBG1 and HBG2 to induce fetal hemoglobin. *Blood Adv.* **2019**, *3*, 3379–3392. [CrossRef]
18. Mingoia, M.; Caria, C.A.; Ye, L.; Asunis, I.; Marongiu, M.F.; Manunza, L.; Sollaino, M.C.; Wang, J.; Cabriolu, A.; Kurita, R.; et al. Induction of therapeutic levels of HbF in genome-edited primary β(0) 39-thalassaemia haematopoietic stem and progenitor cells. *Br. J. Haematol.* **2021**, *192*, 395–404. [CrossRef]
19. Frangou, H.; Altshuler, D.; Cappellini, M.D.; Chen, Y.S.; Domm, J.; Eustace, B.K.; Foell, J.; de la Fuente, J.; Grupp, S.; Handgretinger, R.; et al. CRISPR-Cas9 Gene Editing for Sickle Cell Disease and β-Thalassemia. *N. Engl. J. Med.* **2021**, *384*, 252–260. [CrossRef]
20. NCT03655678 (A Safety and Efficacy Study Evaluating CTX001 in Subjects with Transfusion-Dependent β-Thalassemia). Available online: https://clinicaltrials.gov/ct2/show/NCT03655678 (accessed on 28 November 2021).
21. Sherkow, J.S. CRISPR, Patents, and the Public Health. *Yale J. Biol. Med.* **2017**, *90*, 667–672.
22. Rigter, T.; Klein, D.; Weinreich, S.S.; Cornel, M.C. Moving somatic gene editing to the clinic: Routes to market access and reimbursement in Europe. *Eur. J. Hum. Genet.* **2021**, *29*, 1477–1484. [CrossRef]
23. Cornel, M.C.; Howard, H.C.; Lim, D.; Bonham, V.L.; Wartiovaara, K. Moving towards a cure in genetics: What is needed to bring somatic gene therapy to the clinic? *Eur. J. Hum. Genet.* **2019**, *27*, 484–487. [CrossRef]
24. Finotti, A.; Gambari, R. Recent trends for novel options in experimental biological therapy of β-thalassemia. *Expert Opin. Biol. Ther.* **2014**, *14*, 1443–1454. [CrossRef]
25. Mukherjee, M.; Rahaman, M.; Ray, S.K.; Shukla, P.C.; Dolai, T.K.; Chakravorty, N. Revisiting fetal hemoglobin inducers in beta-hemoglobinopathies: A review of natural products, conventional and combinatorial therapies. *Mol. Biol. Rep.* 2021. Online ahead of print.
26. Lampronti, I.; Bianchi, N.; Zuccato, C.; Dall'Acqua, F.; Vedaldi, D.; Viola, G.; Potenza, R.; Chiavilli, F.; Breveglieri, G.; Borgatti, M.; et al. Increase in gamma-globin mRNA content in human erythroid cells treated with angelicin analogs. *Int. J. Hematol.* **2009**, *90*, 318–327. [CrossRef]
27. Finotti, A.; Bianchi, N.; Fabbri, E.; Borgatti, M.; Breveglieri, G.; Gasparello, J.; Gambari, R. Erythroid induction of K562 cells treated with mithramycin is associated with inhibition of raptor gene transcription and mammalian target of rapamycin complex 1 (mTORC1) functions. *Pharmacol. Res.* **2015**, *91*, 57–68. [CrossRef] [PubMed]
28. Iftikhar, F.; Ali, H.; Musharraf, S.G. Cinchona alkaloids as natural fetal hemoglobin inducing agents in human erythroleukemia cells. *RSC Adv.* **2019**, *9*, 17551. [CrossRef]
29. Warhurst, D.C. Cinchona alkaloids and malaria. *Lancet* **1981**, *2*, 1346.
30. Uzor, P.F. Alkaloids from Plants with Antimalarial Activity: A Review of Recent Studies. *Evid. Based Complement. Alternat. Med.* **2020**, *2020*, 8749083. [CrossRef] [PubMed]
31. Maldonado, C.; Barnes, C.J.; Cornett, C.; Holmfred, E.; Hansen, S.H.; Persson, C.; Antonelli, A.; Rønsted, N. Phylogeny Predicts the Quantity of Antimalarial Alkaloids within the Iconic Yellow Cinchona Bark (Rubiaceae: Cinchona calisaya). *Front. Plant. Sci.* **2017**, *8*, 391. [CrossRef]
32. Deshmukh, A.; Larson, J.; Ghannam, M.; Saeed, M.; Cunnane, R.; Ghanbari, H.; Latchamsetty, R.; Crawford, T.; Jongnarangsin, K.; Pelosi, F.; et al. Efficacy and tolerability of quinidine as salvage therapy for monomorphic ventricular tachycardia in patients with structural heart disease. *J. Cardiovasc. Electrophysiol.* 2021. Online ahead of print.
33. Vitali Serdoz, L.; Rittger, H.; Furlanello, F.; Bastian, D. Quinidine-A legacy within the modern era of antiarrhythmic therapy. *Pharmacol. Res.* **2019**, *144*, 257–263. [CrossRef] [PubMed]
34. Li, D.L.; Cox, Z.L.; Richardson, T.D.; Kanagasundram, A.N.; Saavedra, P.J.; Shen, S.T.; Montgomery, J.A.; Murray, K.T.; Roden, D.M.; Stevenson, W.G. Quinidine in the Management of Recurrent Ventricular Arrhythmias: A Reappraisal. *JACC Clin. Electrophysiol.* **2021**, *7*, 1254–1263. [CrossRef]
35. Sehgal, S.N. Sirolimus: Its discovery, biological properties, and mechanism of action. *Transplant. Proc.* **2003**, *35*, 7S–14S. [CrossRef]
36. Mischiati, C.; Sereni, A.; Lampronti, I.; Bianchi, N.; Borgatti, M.; Prus, E.; Fibach, E.; Gambari, R. Rapamycin-mediated induction of gamma-globin mRNA accumulation in human erythroid cells. *Br. J. Haematol.* **2004**, *126*, 612–621. [CrossRef] [PubMed]
37. Fibach, E.; Bianchi, N.; Borgatti, M.; Zuccato, C.; Finotti, A.; Lampronti, I.; Prus, E.; Mischiati, C.; Gambari, R. Effects of rapamycin on accumulation of alpha-, β- and gamma-globin mRNAs in erythroid precursor cells from β-thalassaemia patients. *Eur. J. Haematol.* **2006**, *77*, 437–441. [CrossRef] [PubMed]
38. Zuccato, C.; Bianchi, N.; Borgatti, M.; Lampronti, I.; Massei, F.; Favre, C.; Gambari, R. Everolimus is a potent inducer of erythroid differentiation and gamma-globin gene expression in human erythroid cells. *Acta Haematol.* **2007**, *117*, 168–176. [CrossRef]
39. Pecoraro, A.; Troia, A.; Calzolari, R.; Scazzone, C.; Rigano, P.; Martorana, A.; Sacco, M.; Maggio, A.; Di Marzo, R. Efficacy of Rapamycin as Inducer of Hb F in Primary Erythroid Cultures from Sickle Cell Disease and β-Thalassemia Patients. *Hemoglobin* **2015**, *39*, 225–229. [CrossRef]
40. Khaibullina, A.; Almeida, L.E.; Wang, L.; Kamimura, S.; Wong, E.C.; Nouraie, M.; Maric, I.; Albani, S.; Finkel, J.; Quezado, Z.M. Rapamycin increases fetal hemoglobin and ameliorates the nociception phenotype in sickle cell mice. *Blood Cells Mol. Dis.* **2015**, *55*, 363–372. [CrossRef]
41. Wang, J.; Tran, J.; Wang, H.; Guo, C.; Harro, D.; Campbell, A.D.; Eitzman, D.T. mTOR Inhibition improves anaemia and reduces organ damage in a murine model of sickle cell disease. *Br. J. Haematol.* **2016**, *174*, 461–469. [CrossRef]

42. Lechauve, C.; Keith, J.; Khandros, E.; Fowler, S.; Mayberry, K.; Freiwan, A.; Thom, C.S.; Delbini, P.; Romero, E.B.; Zhang, J.; et al. The autophagy-activating kinase ULK1 mediates clearance of free α-globin in β-thalassemia. *Sci. Transl. Med.* **2019**, *11*, eaav4881. [CrossRef]
43. Gaudre, N.; Cougoul, P.; Bartolucci, P.; Dörr, G.; Bura-Riviere, A.; Kamar, N.; Del Bello, A. Improved Fetal Hemoglobin With mTOR Inhibitor-Based Immunosuppression in a Kidney Transplant Recipient with Sickle Cell Disease. *Am. J. Transplant.* **2017**, *17*, 2212–2214. [CrossRef]
44. Al-Khatti, A.A.; Alkhunaizi, A.M. Additive effect of sirolimus and hydroxycarbamide on fetal haemoglobin level in kidney transplant patients with sickle cell disease. *Br. J. Haematol.* **2019**, *185*, 959–961. [CrossRef]
45. Gamberini, M.R.; Prosdocimi, M.; Gambari, R. Sirolimus for Treatment of β-Thalassemia: From Pre-Clinical Studies to the Design of Clinical Trials. *Health Educ. Public Health* **2021**, *4*, 425–435.
46. Kahan, B.D. Sirolimus: A new agent for clinical renal transplantation. *Transplant. Proc.* **1997**, *29*, 48–50. [CrossRef]
47. Vasquez, E.M. Sirolimus: A new agent for prevention of renal allograft rejection. *Am. J. Health Syst. Pharm.* **2000**, *57*, 437–448. [CrossRef]
48. Schaffer, S.A.; Ross, H.J. Everolimus: Efficacy and safety in cardiac transplantation. *Expert Opin. Drug Saf.* **2010**, *9*, 843–854. [CrossRef]
49. Tang, C.Y.; Shen, A.; Wei, X.F.; Li, Q.D.; Liu, R.; Deng, H.J.; Wu, Y.Z.; Wu, Z.J. Everolimus in de novo liver transplant recipients: A systematic review. *Hepatobiliary Pancreat. Dis. Int.* **2015**, *14*, 461–469. [CrossRef]
50. Ji, L.; Xie, W.; Zhang, Z. Efficacy and safety of sirolimus in patients with systemic lupus erythematosus: A systematic review and meta-analysis. *Semin. Arthritis Rheum.* **2020**, *50*, 1073–1080. [CrossRef]
51. Wang, Q.; Luo, M.; Xiang, B.; Chen, S.; Ji, Y. The efficacy and safety of pharmacological treatments for lymphangioleiomyomatosis. *Respir. Res.* **2020**, *21*, 55. [CrossRef]
52. Sasongko, T.H.; Ismail, N.F.; Zabidi-Hussin, Z. Rapamycin and rapalogs for tuberous sclerosis complex. *Cochrane Database Syst. Rev.* **2016**, *7*, CD011272. [CrossRef] [PubMed]
53. Graillon, T.; Sanson, M.; Campello, C.; Idbaih, A.; Peyre, M.; Peyrière, H.; Basset, N.; Autran, D.; Roche, C.; Kalamarides, M.; et al. Everolimus and Octreotide for Patients with Recurrent Meningioma: Results from the Phase II CEVOREM Trial. *Clin. Cancer Res.* **2020**, *26*, 552–557. [CrossRef]
54. Gallo, M.; Malandrino, P.; Fanciulli, G.; Rota, F.; Faggiano, A.; Colao, A.; NIKE Group. Everolimus as first line therapy for pancreatic neuroendocrine tumours: Current knowledge and future perspectives. *J. Cancer Res. Clin. Oncol.* **2017**, *143*, 1209–1224. [CrossRef]
55. Motzer, R.J.; Escudier, B.; Oudard, S.; Hutson, T.E.; Porta, C.; Bracarda, S.; Grünwald, V.; Thompson, J.A.; Figlin, R.A.; Hollaender, N.; et al. Phase 3 trial of everolimus for metastatic renal cell carcinoma: Final results and analysis of prognostic factors. *Cancer* **2010**, *116*, 4256–4265. [CrossRef] [PubMed]
56. Algiraigri, A.H.; Wright, N.A.M.; Paolucci, E.O.; Kassam, A. Hydroxyurea for lifelong transfusion-dependent β-thalassemia: A meta-analysis. *Pediatr. Hematol. Oncol.* **2017**, *34*, 435–448. [CrossRef] [PubMed]
57. Bianchi, N.; Finotti, A.; Ferracin, M.; Lampronti, I.; Zuccato, C.; Breveglieri, G.; Brognara, E.; Fabbri, E.; Borgatti, M.; Negrini, M.; et al. Increase of microRNA-210, decrease of raptor gene expression and alteration of mammalian target of rapamycin regulated proteins following mithramycin treatment of human erythroid cells. *PLoS ONE* **2015**, *10*, e0121567. [CrossRef]
58. Bianchi, N.; Ongaro, F.; Chiarabelli, C.; Gualandi, L.; Mischiati, C.; Bergamini, P.; Gambari, R. Induction of erythroid differentiation of human K562 cells by cisplatin analogs. *Biochem. Pharmacol.* **2000**, *60*, 31–40. [CrossRef]
59. Cosenza, L.C.; Breda, L.; Breveglieri, G.; Zuccato, C.; Finotti, A.; Lampronti, I.; Borgatti, M.; Chiavilli, F.; Gamberini, M.R.; Satta, S.; et al. A validated cellular biobank for β-thalassemia. *J. Transl. Med.* **2016**, *14*, 255. [CrossRef] [PubMed]
60. Gambari, R.; Fibach, E. Medicinal chemistry of fetal hemoglobin inducers for treatment of beta-thalassemia. *Curr. Med. Chem.* **2007**, *14*, 199–212. [CrossRef]
61. Bianchi, N.; Chiarabelli, C.; Zuccato, C.; Lampronti, I.; Borgatti, M.; Amari, G.; Delcanale, M.; Chiavilli, F.; Prus, E.; Fibach, E.; et al. Erythroid differentiation ability of butyric acid analogues: Identification of basal chemical structures of new inducers of foetal haemoglobin. *Eur. J. Pharmacol.* **2015**, *752*, 84–91. [CrossRef] [PubMed]
62. Hashemi, Z.; Ebrahimzadeh, M.A. Hemoglobin F (HbF) inducers; History, Structure and Efficacies. *Mini Rev. Med. Chem.* 2021. Online ahead of print.
63. Ansari, S.H.; Lassi, Z.S.; Khowaja, S.M.; Adil, S.O.; Shamsi, T.S. Hydroxyurea (hydroxycarbamide) for transfusion-dependent beta-thalassaemia. *Cochrane Database Syst. Rev.* **2019**, *3*, CD012064.
64. McLaughlin, L.A.; Paine, M.J.; Kemp, C.A.; Maréchal, J.D.; Flanagan, J.U.; Ward, C.J.; Sutcliffe, M.J.; Roberts, G.C.; Wolf, C.R. Why is quinidine an inhibitor of cytochrome P450 2D6? The role of key active-site residues in quinidine binding. *J. Biol. Chem.* **2005**, *280*, 38617–38624. [CrossRef]
65. Fromm, M.F.; Kim, R.B.; Stein, C.M.; Wilkinson, G.R.; Roden, D.M. Inhibition of P-glycoprotein-mediated drug transport: A unifying mechanism to explain the interaction between digoxin and quinidine [see comments]. *Circulation* **1999**, *99*, 552–557. [CrossRef] [PubMed]
66. Ballou, L.M.; Lin, R.Z. Rapamycin and mTOR kinase inhibitors. *J. Chem. Biol.* **2008**, *1*, 27–36. [CrossRef] [PubMed]

International Journal of
Molecular Sciences

MDPI

Article

Gender-Dependent Phenotype in Polycystic Kidney Disease Is Determined by Differential Intracellular Ca^{2+} Signals

Khaoula Talbi, Inês Cabrita , Rainer Schreiber and Karl Kunzelmann *

Physiological Institute, University of Regensburg, University Street 31, D-93053 Regensburg, Germany; khaoula.talbi@vkl.uni-regensburg.de (K.T.); ines.cabrita@vkl.uni-regensburg.de (I.C.); rainer.schreiber@vkl.uni-regensburg.de (R.S.)

* Correspondence: karl.kunzelmann@ur.de; Tel.: +49-(0)941-943-4302; Fax: +49-(0)941-943-4315

Abstract: Autosomal dominant polycystic kidney disease (ADPKD) is caused by loss of function of PKD1 (polycystin 1) or PKD2 (polycystin 2). The Ca^{2+}-activated Cl^- channel TMEM16A has a central role in ADPKD. Expression and function of TMEM16A is upregulated in ADPKD which causes enhanced intracellular Ca^{2+} signaling, cell proliferation, and ion secretion. We analyzed kidneys from Pkd1 knockout mice and found a more pronounced phenotype in males compared to females, despite similar levels of expression for renal tubular TMEM16A. Cell proliferation, which is known to be enhanced with loss of $Pkd1^{-/-}$, was larger in male when compared to female $Pkd1^{-/-}$ cells. This was paralleled by higher basal intracellular Ca^{2+} concentrations in primary renal epithelial cells isolated from $Pkd1^{-/-}$ males. The results suggest enhanced intracellular Ca^{2+} levels contributing to augmented cell proliferation and cyst development in male kidneys. Enhanced resting Ca^{2+} also caused larger basal chloride currents in male primary cells, as detected in patch clamp recordings. Incubation of mouse primary cells, mCCDcl1 collecting duct cells or M1 collecting duct cells with dihydrotestosterone (DHT) enhanced basal Ca^{2+} levels and increased basal and ATP-stimulated TMEM16A chloride currents. Taken together, the more severe cystic phenotype in males is likely to be caused by enhanced cell proliferation, possibly due to enhanced basal and ATP-induced intracellular Ca^{2+} levels, leading to enhanced TMEM16A currents. Augmented Ca^{2+} signaling is possibly due to enhanced expression of Ca^{2+} transporting/regulating proteins.

Keywords: TMEM16A; ADPKD; polycystic kidneys; androgen; estrogen; CFTR

Citation: Talbi, K.; Cabrita, I.; Schreiber, R.; Kunzelmann, K. Gender-Dependent Phenotype in Polycystic Kidney Disease Is Determined by Differential Intracellular Ca^{2+} Signals. *Int. J. Mol. Sci.* **2021**, *22*, 6019. https://doi.org/10.3390/ijms22116019

Academic Editor: Ivano Condò

Received: 7 May 2021
Accepted: 31 May 2021
Published: 2 June 2021

1. Introduction

Male gender is a risk factor for progression of autosomal-dominant polycystic kidney disease (ADPKD) [1,2]. Affected men demonstrate faster loss of renal function and earlier onset of end stage renal disease, when compared to women [3]. Orchiectomy led to reduced renal size and cyst volume density, indicating attenuation of renal disease. In contrast, testosterone substitution was shown to antagonize the protective effect of gonadal ablation [4]. Moreover, in females, testosterone increased kidney size and cyst growth, clearly identifying androgens as a progression factor. On the other hand, estrogens have been proposed as protective hormones [5]. These results have been confirmed in additional experiments with rats [6,7]. Thus both, androgens and estrogens have an impact on cyst growth and disease progression in ADPKD.

Evidence has been provided for a crucial role of two chloride ion channels in the pathology of ADPKD, namely protein kinase A-regulated cystic fibrosis transmembrane conductance regulator (CFTR) and the Ca^{2+}-activated Cl^- channel transmembrane 16A (TMEM16A). Support for a pro-secretory role of CFTR in ADPKD came from a number of in vitro studies [8–11], ex vivo experiments in embryonic renal cysts in metanephric organ culture [12,13], and observations in vivo in animals and humans [12,14,15]. In contrast, another in vivo study could not confirm a protective effect of missing CFTR-function in cystic fibrosis for ADPKD [16].

Subsequent work identified a contribution of purinergic Ca^{2+} signaling to ADPKD [17–19]. In a series of studies in vitro, in metanephric renal organ cultures, and in mice in vivo, we demonstrated the essential contribution of TMEM16A to renal cyst formation in ADPKD [20–24]. We showed that knockout of Tmem16a or inhibition of TMEM16A in vivo by the FDA-approved drugs such as niclosamide, benzbromarone, and the TMEM16A-specific inhibitor Ani9 largely reduced cyst enlargement and abnormal cyst cell proliferation. Based on these results, we proposed a novel therapeutic concept for the treatment of ADPKD, based on the inhibition of TMEM16A.

In the present study we asked whether enhanced expression or function of TMEM16A, and/or hormonal regulation may account for the more severe phenotype in male ADPKD. Typically, loss of PKD1 leads to a more severe phenotype than loss of PKD2 and accounts for about 85% of all ADPKD. We therefore examined $Pkd1^{-/-}$ animals in the present study. While loss of PKD1 does not affect expression of PKD2, loss of either PKD1 or PKD2 leads to similar changes in intracellular Ca^{2+} signaling [23,25].

In a previous study, androgen-response elements were found in the TMEM16A promoter region, and were shown to be relevant for testosterone-dependent induction of TMEM16A [26]. Moreover, TMEM16A expression was found to be enhanced in male when compared to female sympathetic ganglia [27], but lower levels were detected in male than in female urethral smooth muscle [28]. Apart from the differences in TMEM16A-expression, CFTR might be expressed at lower levels in female ADPKD individuals, which could contribute to reduced renal cyst growth in females. In fact, a so-called cystic fibrosis (CF) gender gap describes the higher mortality in females with CF, due to lower expression of CFTR [29]. Lower CFTR-expression may be caused by estrogen-dependent regulation of CFTR [30,31].

In ADPKD, cysts occur in different renal tubular segments, yet it is assumed that most cysts are derived from the collecting duct [32]. In the present study we analyzed primary epithelial cells isolated from renal medulla and mouse mCCDcl1 collecting duct cells. We compared the properties of primary cells isolated from male and female mice, and examined whether gender differences can be reproduced in the mCCDcl1 cell line by treatment with male (dihydrotestosterone) and female (estrogen) hormones. We detected enhanced renal cyst growth and cell proliferation in male mice lacking expression of the polycystic kidney disease gene Pkd1 (polcycystin 1), when compared to female $Pkd1^{-/-}$ mice. The data suggest enhanced proliferation, increased basal Ca^{2+} levels, and larger secretory chloride currents in cells derived from male $Pkd1^{-/-}$ kidneys, which is likely to contribute to the enhanced progression in male ADPKD patients.

2. Results

2.1. *Male Mice Lacking Expression of Pkd1 Show a Larger Number of Renal Cysts, But Similar Levels of Expression of TMEM16A*

The KspCreERT2;Pkd1lox;lox system was used to obtain a tamoxifen inducible tubule-specific knockout of Pkd1, as described previously [24,33]. Knockdown of Pkd1 in male and female mice was validated by RT-PCR (Figure 1A,B). Ten weeks old $Pkd1^{-/-}$ mice demonstrated multiple renal cysts, which were more pronounced in males when compared to females (Figure 1C,D).

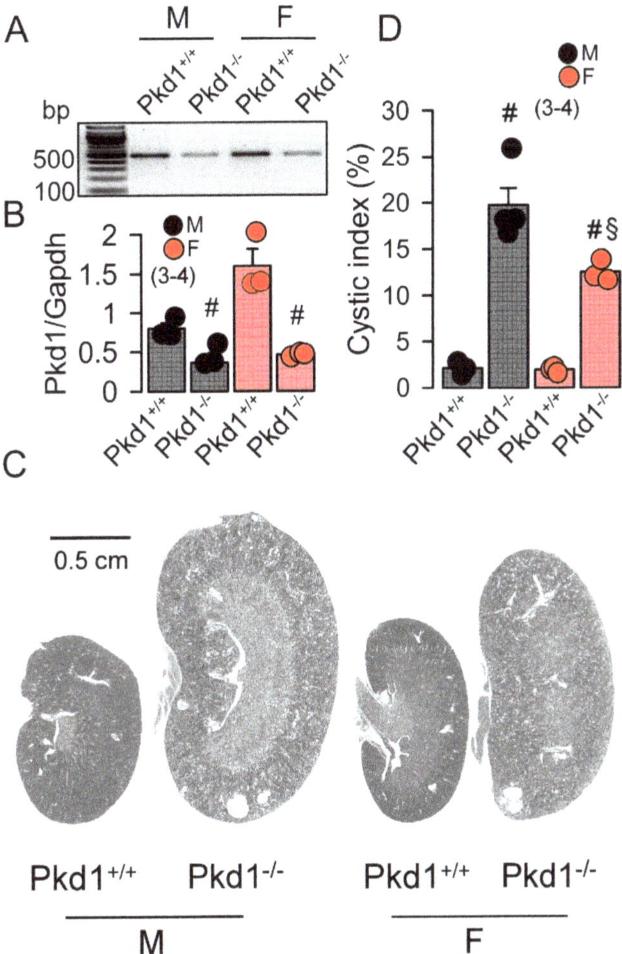

Figure 1. Renal cysts in male and female PKD1-knockout mice. (**A**,**B**) RT-PCR of Pkd1 in renal primary epithelial cells from male (M) and female (F) Pkd1$^{+/+}$ and Pkd1$^{-/-}$ mice, and semiquantitative analysis of expression. (**C**) HE staining of whole kidneys and analysis of renal cysts by stitching microscopy. Male Pkd1$^{-/-}$ kidneys had larger sizes when compared to kidneys from female Pkd1$^{-/-}$ mice. (**D**) Cystic index in male and female Pkd1$^{+/+}$ and Pkd1$^{-/-}$ kidneys. Mean ± SEM (number of animals in each series). # significant difference when compared to Pkd1$^{+/+}$ ($p < 0.05$; unpaired t-test). § significant difference when compared to male ($p < 0.05$; ANOVA with Tukey's post-hoc test).

In our previous study we reported a crucial role of the Ca^{2+}-activated Cl^- channel TMEM16A for the development of polycystic kidneys in Pkd1$^{-/-}$ mice and in additional in vitro models for ADPKD [20,23,24]. We therefore expected to find higher levels of TMEM16A-expression in kidneys from male Pkd1$^{-/-}$ mice. However, both Western blotting of TMEM16A from renal lysates, as well as immunohistochemistry suggested similar levels of TMEM16A-expression in kidneys from male and female Pkd1$^{-/-}$ mice (Figure 2). Similarly, mRNA-expression for the epithelial Cl^- channel cystic fibrosis transmembrane conductance regulator (CFTR) was not different between males and females (Figure S1).

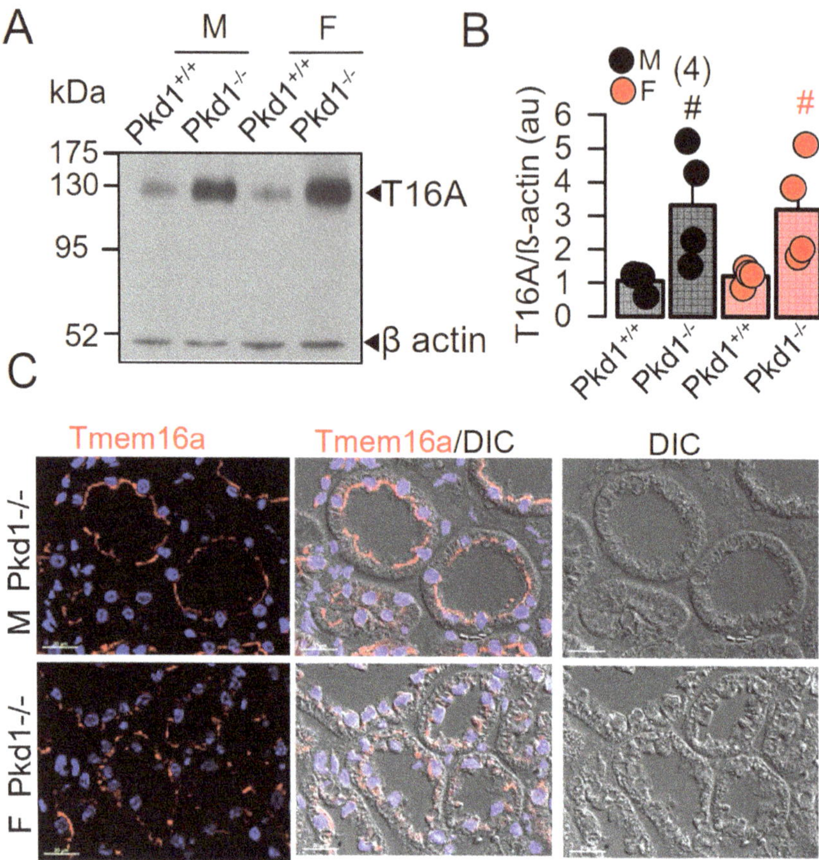

Figure 2. Expression of TMEM16A in male and female PKD1$^{-/-}$ mice. (**A**,**B**) Western blotting of TMEM16A-expression in primary renal epithelial cells from kidneys of male and female Pkd1$^{+/+}$ and Pkd1$^{-/-}$ mice. (**C**) Immunocytochemistry of Tmem16a in kidneys from male and female Pkd1$^{-/-}$ mice. Bar = 20 μm. Representative images from three mice each. Mean ± SEM (number of animals in each series). $^{\#}$ significant difference when compared to Pkd1$^{+/+}$ ($p < 0.05$; unpaired t-test).

2.2. Cell Proliferation and Basal Ca^{2+} Levels Are More Enhanced in Renal Epithelial Cells from Pkd1$^{-/-}$ Males Than Pkd1$^{-/-}$ Females

Cell proliferation is enhanced in kidneys from Pkd1$^{-/-}$ mice [24]. We compared the cell proliferation in males and females using Ki-67 staining. The results show that cell proliferation is enhanced in kidneys of both male and female Pkd1$^{-/-}$ mice, however, proliferation was more enhanced in males (Figure 3A,B). Cellular Ca^{2+} levels are intimately related with cell proliferation, and have been shown to be augmented in renal epithelial cells from mice lacking expression of Pkd1, which leads to enhanced expression of TMEM16A [23–25]. We therefore compared the basal Ca^{2+} levels in primary renal epithelial cells from male and female mice. Remarkably, knockout of Pkd1 caused higher basal Ca^{2+} levels in renal epithelial cells from males, but not in cells from female kidneys (Figure 3C,D). These results suggest a contribution of enhanced intracellular Ca^{2+} levels to augmented cell proliferation and cyst development in male kidneys.

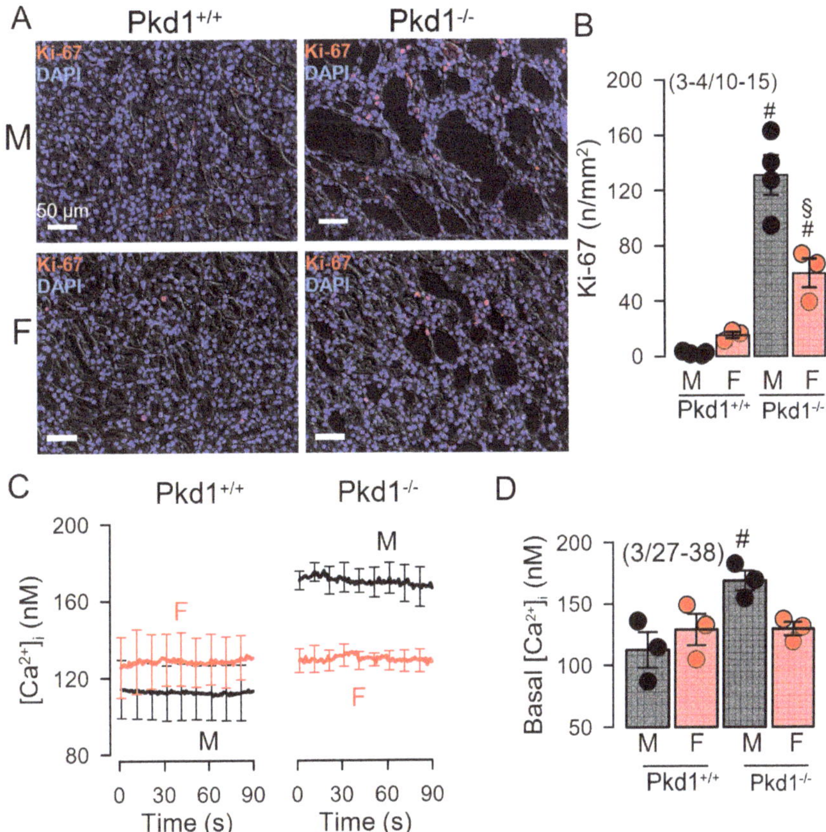

Figure 3. Cell proliferation and intracellular Ca^{2+} concentrations in renal epithelial cells from Pkd1$^{+/+}$ and Pkd1$^{-/-}$ mice. (**A**) Analysis of tubular epithelial cell proliferation in kidney sections from male and female Pkd1$^{+/+}$ and Pkd1$^{-/-}$ mice as indicates by Ki-67 staining. Bars = 50 µm. (**B**) Summary of Ki-67 positive cells/mm^2 tissue area in kidneys from male and female Pkd1$^{+/+}$ and Pkd1$^{-/-}$ mice. Bars = 50 µm. (**C**,**D**) Intracellular basal Ca^{2+} concentrations ($[Ca^{2+}]_i$) as assessed by Fura2 indicates higher basal $[Ca^{2+}]i_I$ in male Pkd1$^{-/-}$ mice. Mean ± SEM (number of animals/number of experiments in each series). # Significant difference when compared to Pkd1$^{+/+}$ ($p < 0.05$; unpaired t-test). § Significant difference when compared to male ($p < 0.05$; ANOVA and Tukey's post-hoc test).

We further compared the whole cell currents measured in primary renal epithelial cells isolated from male and female kidneys. Primary cells were isolated from male and female Pkd1$^{+/+}$ and Pkd1$^{-/-}$ mice. As reported previously, in cells isolated from male or female Pkd1$^{+/+}$ animals, we found little activation of TMEM16A currents by increase of intracellular Ca^{2+}, using the purinergic ligand ATP (Figure 4A,B, left panels). In contrast, cells isolated from Pkd1$^{-/-}$ animals showed large ATP-activated whole cell currents. We noticed that the basal current in male Pkd1$^{-/-}$ cells was significantly larger than that measured in female Pkd1$^{-/-}$ cells (Figure 4A,B, right panels). This may be explained by the fact that male Pkd1$^{-/-}$ cells showed a higher basal intracellular Ca^{2+} concentration (Figure 3C,D). We also examined the whole cell currents activated by IBMX and forskolin (IF), which both increase intracellular cAMP. No activation of whole cell currents was observed in Pkd1$^{+/+}$ cells, while a small but significant current was activated in male Pkd1$^{-/-}$ cells (Figure 4C,D). Patch clamp ion current data were supported by additional iodide quenching data, which also indicated a larger halide permeability in male when compared to female Pkd1$^{-/-}$ cells (Figure S2). Taken together, higher basal Ca^{2+} concentrations, enhanced proliferation and larger basal Cl^- currents may explain the more pronounced cystic phenotype in male Pkd1$^{-/-}$ kidneys.

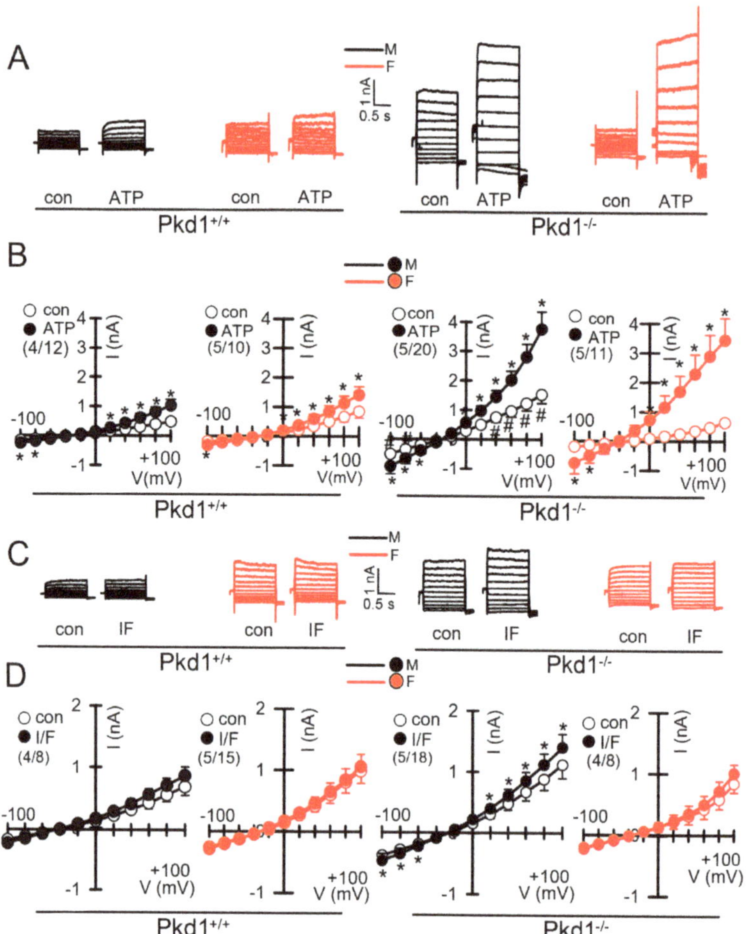

Figure 4. Whole cell patch clamp experiments in primary renal epithelial cells from male and female Pkd1$^{+/+}$ and Pkd1$^{-/-}$ mice. (**A**) Original whole cell overlay currents of basal and ATP (50 μM)—activated currents in male and female Pkd1$^{+/+}$ and Pkd1$^{-/-}$ mice. (**B**) Summary i/v curves of basal and ATP-activated currents indicating larger ATP-activated currents in Pkd1$^{-/-}$ mice and larger basal currents in male Pkd1$^{-/-}$ mice. (**C**) Original overlay currents of basal and IF (100 μM IBMX and 2 μM forskolin)—activated currents in male and female Pkd1$^{+/+}$ and Pkd1$^{-/-}$ mice. (**D**) Summary i/v curves of basal and IF-activated currents indicating small but significant IF-activated currents in male Pkd1$^{-/-}$ mice. Mean ± SEM (number of animals/number of experiments in each series). * Significant activation by ATP and IF, respectively ($p < 0.05$; paired t-test). # Significant difference compared to other basal currents ($p < 0.05$; ANOVA and Tukey's post-hoc test).

2.3. *Testosterone Augments ATP-Induced Whole Cell Currents in Female Pkd1$^{+/+}$ Cells*

Androgens have been implicated in the severity of the disease phenotype in ADPKD [34]. We therefore examined the impact of dihydrotestosterone (DHT) on the ion currents in primary renal epithelial cells, isolated from female Pkd$^{+/+}$ mice. Expression of androgen receptors in primary renal epithelial cells was determined by RT-PCR (Figure 5A,B). Moreover, different types of estrogen receptors were found to be expressed in these cell (Figure 5C). As expected, ATP-activated whole cell currents were moderate in control cells, but were significantly enhanced in cells treated with DHT. In contrast, in cells treated with the inhibitor of androgen receptors, cyproterone acetate (CA), no whole cell currents could be activated by ATP (Figure 5D,E).

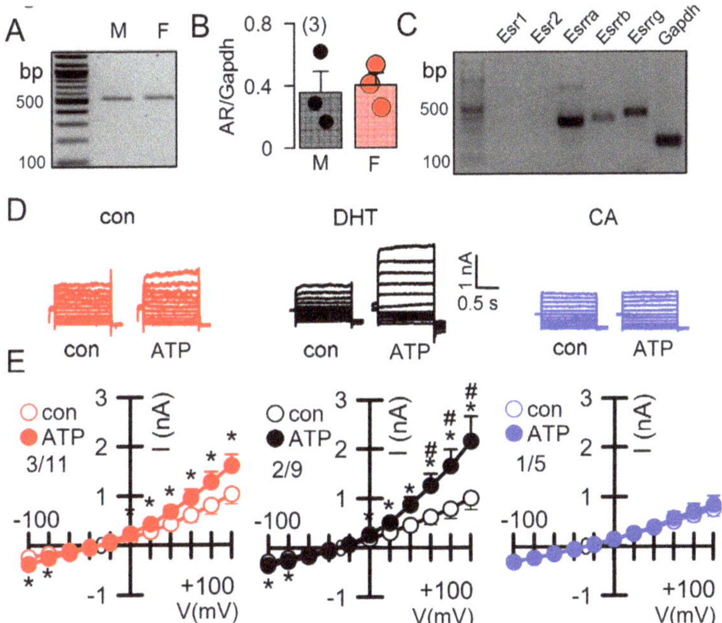

Figure 5. Effect of dihydrotestosterone on whole cell currents in primary renal epithelial cells from female Pkd1$^{+/+}$ mice. (**A**,**B**) RT-PCR analysis of expression of the androgen receptor (AR) in primary renal epithelial cells from male and female kidneys. (**C**) Expression of estrogen receptors in primary renal epithelial cells from female Pkd1$^{+/+}$ mice. (**D**,**E**) Whole cell current overlays and corresponding I/V-curves from primary renal epithelial cells of female Pkd1$^{+/+}$ mice. ATP (50 µM) -activated currents were augmented in cells-incubated with dihydrotestosterone (DHT; 10 µM, 24 h), but were absent in cyproterone acetate (CA, 10 µM, 24 h) incubated cells. Mean ± SEM (number of animals/number of experiments in each series). * Significant activation by ATP ($p < 0.05$; paired t-test). # Significant difference when compared to con (no DHT) ($p < 0.05$; ANOVA and Tukey's post-hoc test).

DHT-dependent regulation of Ca^{2+}-activated Cl^- currents was further examined in mCCDcl1 mouse renal cortical collecting duct cells [35]. These cells express TMEM16A, CFTR, and receptors for testosterone and estrogen (Figure 6A,B). However, the main estrogene receptors Esr1 and Esr2 are not expressed in renal epithelial cells, suggesting that estrogen-dependent regulation of protein expression is not dominant in kidney.

Treatment with DHT-enhanced expression of TMEM16A in mCCDcl1 cells (Figure 6C). In these cells, ATP concentrations as low as 0.1 µM stimulated a whole cell current that was not observed in the absence of DHT (Figure 6D–F). Upregulation of ATP-activated whole cell currents by DHT was also observed in mouse M1 collecting duct cells, which also showed a slight but detectable increase in cAMP-activated currents upon DHT treatment (Figure S3). No whole Cl^- currents could be activated in cells incubated with estrogen.

2.4. Testosterone Enhances Intracellular Ca^{2+} Signals

DHT increased the expression of TMEM16A and enhanced ATP-activated whole cell currents, which may be due to enhanced intracellular Ca^{2+} signaling. We measured the intracellular Ca^{2+} concentrations in control mCCDcl1 cells, and in cells treated with DHT. Cells were stimulated by different concentrations of ATP, which caused a concentration-dependent increase in intracellular Ca^{2+} concentration (Figure 6G). Notably, DHT did not further enhance transient Ca^{2+} increase induced by ATP, but shifted the basal Ca^{2+} concentration to higher values. This suggests a Ca^{2+} leakage out of the endoplasmic reticulum (ER) Ca^{2+} store, or enhanced Ca^{2+} influx through plasma membrane Ca^{2+} channels (Figure 6G).

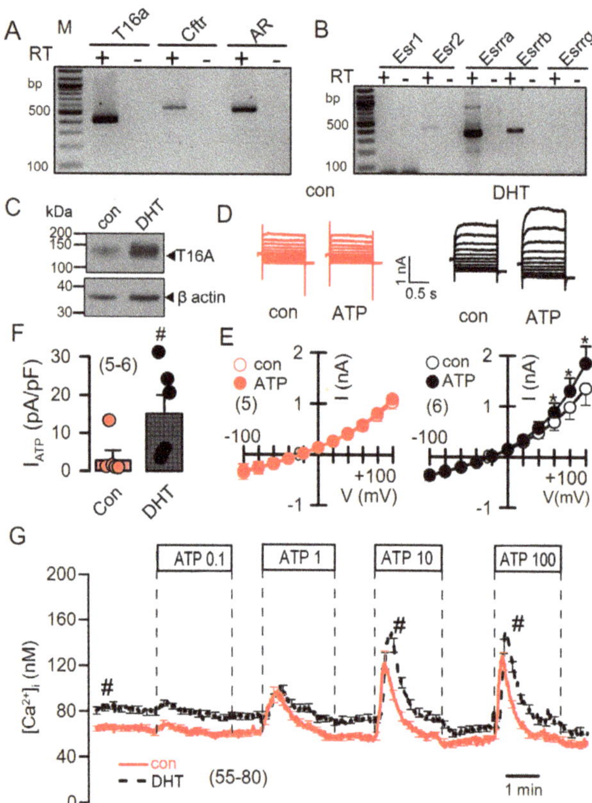

Figure 6. Effect of dihydrotestosterone on whole cell currents and intracellular Ca^{2+} concentrations in mouse mCCDcl1 cortical collecting duct cells. (**A**,**B**) RT-PCR analysis of expression of Tmem16a, Cftr, androgen receptors (**A**), and estrogen receptors (**B**) in Pkd1$^{+/+}$ mCCDcl1 mouse collecting duct cells. (**C**) Western blot from mCCDcl1 cells, indicating upregulation of Tmem16a-expression by DHT. (**D**,**E**) Whole cell current overlays (**D**), corresponding I/V-curves (**E**), and summary of ATP (0.1 µM) -activated whole cell currents (**F**) from mCCDcl1 cells, indicating larger basal and ATP-activated currents in DHT (10 µM; 24 h) treated cells. Current densities (pA/pF) were assessed at the clamp voltage of +100 mV. (**G**) Basal and ATP (0.1–100 µM)–induced increase in intracellular Ca^{2+} concentration in control (con) and DHT-incubated mCCDcl1 cells. Mean ± SEM (number of experiments). * Significant activation by ATP ($p < 0.05$; paired t-test). # Significant difference compared to con ($p < 0.05$; unpaired t-test).

We applied a Ca^{2+} store release protocol to empty the ER Ca^{2+} store by inhibiting the sarcoplasmic endoplasmic reticulum ATPase (SERCA), using cyclopiazonic acid (CPA). Remarkably, removal of extracellular Ca^{2+} lead to a more pronounced decrease in basal cytosolic Ca^{2+} in DHT, but not in EST-treated cells (Figure 7A, left and middle panel). Subsequent addition of CPA induced a more pronounced Ca^{2+} store release in DHT-treated cells, and a largely augmented Ca^{2+} influx upon re-addition of extracellular Ca^{2+} (Figure 7A, left and right panel). Thus, Ca^{2+} store release and store operated Ca^{2+} influx were augmented in DHT-treated cells. We analyzed the expression of a number of Ca^{2+} transporting proteins such as inositol trisphosphate receptors (IP3R1-3), the Ca^{2+} influx channels Orai1, TRPC1, TRPV4, and the ER Ca^{2+} sensor stromal interaction molecule 1 (Stim1). Notably, expression of a number of Ca^{2+} channels and Stim1 were augmented by both DHT and EST incubated cells. As shown in Figure 6C, DHT enhanced the expression of TMEM16A, while EST had no significant effect on TMEM16A-expression (data not shown). By comparing Ca^{2+} signals obtained from male and female primary renal epithelial cells, we also found higher basal Ca^{2+} levels in male cells, along with larger ATP-induced Ca^{2+} transients (Figure S4). Taken together, the more severe cystic phenotype found in males is likely to be caused by enhanced cell proliferation possibly due to enhanced basal intracellular Ca^{2+} levels, which is probably due to enhanced expression of Ca^{2+} transporting/regulating proteins.

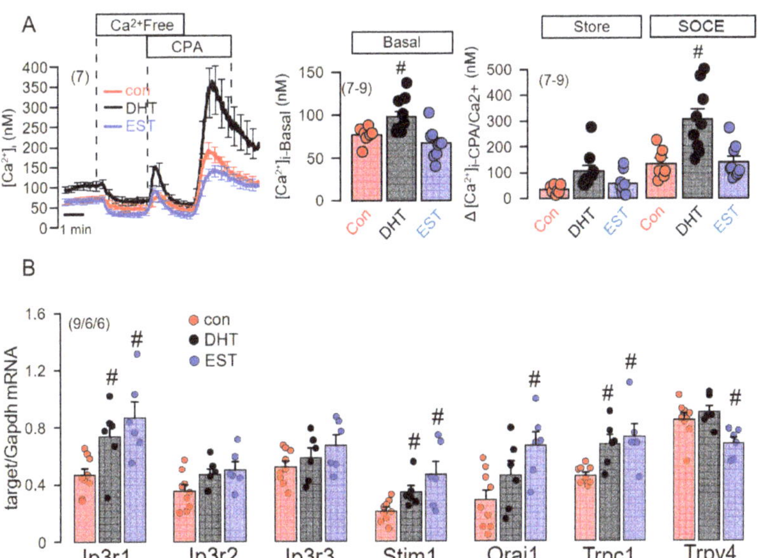

Figure 7. Effect of dihydrotestosterone and estrogen on intracellular Ca^{2+} concentrations and expression of Ca^{2+}-regulating proteins in mouse mCCDcl1 cortical collecting duct cells. (**A**) Summary time course of the effects of extracellular Ca^{2+}-free buffer and cyclopiazonic acid (10 μM) on intracellular Ca^{2+} concentrations in control mCCDcl1 cells and mCCDcl1 cells treated with dihydrotestosterone (DHT; 10 μM, 24 h) and estrogen (EST, 10 μM; 24 h). Summary of basal intracellular Ca^{2+} concentration and changes in intracellular Ca^{2+} induced by CPA (Ca^{2+} store release through leakage channels; Store) and re-addition of Ca^{2+} to the extracellular buffer (store operated Ca^{2+} entry; SOCE). (**B**) RT-PCR analysis of the expression of Ca^{2+}-regulating proteins and effects of DHT and EST. Mean ± SEM (number of experiments in each series). # Significant increase when compared with con ($p < 0.05$; ANOVA).

3. Discussion

In the present study we analyzed the primary epithelial cells isolated from renal medulla and mouse mCCDcl1 collecting duct cells. We compared the properties of primary cells isolated from male and female mice, and examined whether gender differences can be reproduced in the mCCDcl1 cell line by treatment with DHT and EST. We investigated whether enhanced expression or function of TMEM16A, and/or hormonal regulation may account for the more severe ADPKD phenotype caused by knockout of Pkd1. As shown previously, loss of PKD1 or PKD2 led to similar changes in intracellular Ca^{2+} signaling which could be reproduced in the present study [23,25]. The present data show a higher cystic index in kidneys from male Pkd1$^{-/-}$ mice. Although expression of the disease-associated Ca^{2+}-activated Cl^- channel TMEM16A was not different between male and female Pkd1$^{-/-}$ kidneys, cell proliferation, basal cytosolic Ca^{2+} levels, and basal Cl^- currents were larger in renal epithelial cells derived from males (Figures 1–4).

Proliferation of renal tubular epithelial cells is enhanced in both male and female kidneys from Pkd1$^{-/-}$ knockout mice, however, the effect is more pronounced in male kidneys. Proliferation is predominantly due to the upregulation of TMEM16A-expression as demonstrated in our recent study [24]. TMEM16A is well-known to enhance proliferation in different types of cells including cancer cells. This is due to the upregulation of Ca^{2+} (ATP^-) activated chloride currents and upregulated ATP-induced Ca^{2+} store release [24]. Enhanced Ca^{2+} signaling is also demonstrated by augmented store release triggered by cyclopiazonic acid with consecutive increase in store-operated Ca^{2+} entry. These changes are observed in both primary renal epithelial cells from male and female Pkd1$^{-/-}$ mice, however, upregulation of these Ca^{2+} signals is more pronounced in epithelial cells from males (Figure S5). Furthermore, a small cAMP-regulated CFTR current was detected in the cells from male Pkd1$^{-/-}$ kidneys, but was not found in cells from female kidneys. Expression levels for CFTR were not different between male and female kidneys as detected by semiquantitative RT-PCR or Western blotting.

Higher TMEM16A currents and small but detectable CFTR currents were also found in renal epithelial cells upon exposure to dihydrotestosterone, which also enhanced expression of TMEM16A. Importantly, Cha et al. reported three androgen-response elements in the TMEM16A promoter region, which are relevant for the DHT-dependent induction of TMEM16A [26]. Comparable to our previous report showing inhibition of cell proliferation and renal cyst growth [24], Cha et al. reported inhibition of prostate hyperplasia by siRNA-knockdown of TMEM16A [26]. In contrast, androgens potentiated renal cell proliferation and cyst enlargement through ERK1/2-dependent and ERK1/2-independent signaling in another study [34]. However, in our study we were unable to detect the different levels for TMEM16A expression in male and female kidneys (Figure 2). A likely explanation might be that knockout of Pkd1 already induced a strong upregulation of TMEM16A-expression [24], thereby overrunning androgen-dependent regulation of expression. Notably, in cultured mCCDcl1 collecting duct cells, where Pkd1 is not knocked out, androgen-dependent upregulation of TMEM16A is detected (Figure 6).

Enhanced basal (Figures 3, 6 and 7) and ATP-activated (Figures 6 and S4) Ca^{2+} levels were clearly detectable in male primary renal epithelial cells and androgen-treated mCCDcl1 cells. Androgens also caused increased intracellular Ca^{2+} levels in prostate and skeletal muscle cells [36,37]. A number of previous studies examined the effects of androgens on cytosolic Ca^{2+} signaling, and expression of proteins that regulate intracellular Ca^{2+} levels. Thus androgens upregulated the Ca^{2+} influx channel Orai1 in MCF-7 breast tumor cells [38], while in LNCaP cells, androgens were shown to increase cytosolic Ca^{2+} by enhancing Ca^{2+} influx through L-type channels [39], and via G-protein-coupled receptors [40,41]. Moreover, expression of the Ca^{2+} sensor in the endoplasmic reticulum, stromal interaction molecule 1 (STIM1), was shown to be regulated by androgens [42]. Conversely, estradiol was shown to inhibit the phosphorylation of STIM1 which attenuated SOCE [43].

We found lower expression of TRPV4 upon treatment with estradiol, which may suggest a role of TRPV4 in differential Ca^{2+} signaling in male vs. female renal epithelial cells. Notably, using RT-PCR we also found a reduced expression of TRPV4 in female Pkd1$^{-/-}$ mice, when compared to male Pkd1$^{-/-}$ mice (data not shown). Interestingly, enhanced expression of TRPV4 was also found in male hypertensive rats when compared to female animals [44]. However, given the number of Ca^{2+} transporting proteins that differ between cells derived from males and females, enhanced basal Ca^{2+} levels in males are likely due to a combination of differentially expressed Ca^{2+}-transporting proteins. We believe that the differences in basal Ca^{2+} concentrations could be sufficient to explain the different basal activity of TMEM16A in male and female cells. Activation of TMEM16A depends largely on the membrane voltage [45]. Activation of inward currents at hyperpolarized (physiological) voltages require higher $[Ca^{2+}]_i$ than outward currents and are in the range of 1 μM and higher. However, Ca^{2+} concentrations related to channel activity are typically obtained by Ca^{2+}-sensing dyes, which measure global cytosolic Ca^{2+} concentrations, such as Fura-2. These global $[Ca^{2+}]_i$ do not reflect the true Ca^{2+} levels in the TMEM16A-containing subapical compartment. We now know that TMEM16A is a membrane tether that binds the receptor for inositol trisphosphate (IP3) and thereby enhances local Ca^{2+} levels in close proximity to TMEM16A [46]. Using membrane-bound Ca^{2+} sensors, remarkably higher Ca^{2+} concentrations were found in TMEM16A-containing compartments underneath the cell membrane, causing more pronounced activation of TMEM16A [47]. Finally, differential expression of splice variants for TMEM16A could be an additional reason for the larger basal Cl^- currents detected in male renal epithelial cells. For example, Ferrera and collaborators reported differential Ca^{2+} sensitivity for the isoforms TMEM16A(a,b,c), and TMEM16A(a,c) [48]. Differential expression of TMEM16A isoforms should be therefore examined in subsequent studies.

The results of the present study may be summarized as follows: (i) Renal cysts found in ADPKD caused by knockout of Pkd1, are associated with enhanced basal intracellular Ca^{2+} levels and enhanced agonist (ATP)$^-$ and CPA-induced Ca^{2+} store release and SOCE. (ii) The more severe phenotype in males is related to larger store release and SOCE detected in cells derived from males when compared to females (Figure S5). (iii) Stimulation of the androgen receptor by DHT in cells derived from primary female cells and in the mCCDcl1 cell line reproduces the findings obtained in primary male tissues. (iv) Gender-dependent phenotype differences are not explained by different levels of expression of the ADPKD-relevant ion channel TMEM16A. (v) Differential Ca^{2+} homeostasis in kidneys of male and female ADPKD patients is suggested. (vi) Higher Ca^{2+} levels not only enhance TMEM16A-activity, but may also contribute to higher activity of CFTR [49]. (vii) Additional work will be required to identify the molecular mechanisms underlying the gender-dependent differences of Ca^{2+} signaling in ADPKD.

4. Materials and Methods

4.1. RT-PCR

For semi-quantitative RT-PCR total RNA from primary medullary kidney epithelial cells were isolated using NucleoSpin RNA II columns (Macherey-Nagel, Düren, Germany). Total RNA (0.5 μg/25 μL reaction) was reverse-transcribed using random primer (Promega, Mannheim, Germany) and M-MLV Reverse Transcriptase RNase H Minus (Promega, Mannheim, Germany). Each RT-PCR reaction contained sense (0.5 μM) and antisense primer (0.5 μM) (Table 1), 0.5 μL cDNA and GoTaq Polymerase (Promega, Mannheim, Germany). After 2 min at 95 °C cDNA was amplified (targets 30 cycles, reference Gapdh 25 cycles) for 30 s at 95 °C, 30 s at 56 °C, and 1 min at 72 °C. PCR products were visualized by loading on Midori Green Xtra (Nippon Genetics Europe) containing agarose gels and analyzed using Image J 1.52r (NIH, Bethesda, MD, USA).

Table 1. Primers used for PCR-analysis.

Gene Accession Number	Primer	Size (bp)
Tmem16a NM_001242349.2	s: 5′-GTGACAAGACCTGCAGCTAC as: 5′-GCTGCAGCTGTGGAGATTC	406
Cftr NM_021050.2	s: 5′-GAATCCCCAGCTTATCCACG as: 5′-CTTCACCATCATCTTCCCTAG	544
Pkd1 NM_013630.2	s: 5′-CTTCTACTTTGCCCATGAGG as: 5′-CTTCTACTTGCACCTCTGTC	473
Esr1 NM_007956.5	s: 5′-CTCAAGATGCCCATGGAGAG as: 5′-GTTTCCTTTCTCGTTACTGCTG	441
Esr2 NM_207707.1	s: 5′-GACCTACGCAAGACATGGAG as: 5′-CTTGGACTAGTAACAGGGCTG	436
Esrra NM_007953.2	s: 5′-CAGGGCAGTGGGAAGCTAG as: 5′-GCTACTGCCAGAGGTCCAG	362
Esrrb NM_011934.4	s: 5′-GTGGTATCATGGAGGACTCC as: 5′-GTCAATGGCTTTTTAGCAGGTG	388
Esrrg NM_011935.3	s: 5′-CAGCACCATCGTAGAGGATC as: 5′-CATGGCATAGATCTTCTCTGG	442
Gapdh NM_001289726	s: 5′-GTATTGGGCGCCTGGTCAC as: 5′-CTCCTGGAAGATGGTGATGG	200

4.2. Cell Culture

The mouse collecting duct cell line mCCDcl1 was kindly provided by Prof. Dr. Johannes Loffing (Institute of Anatomy, University of Zurich, Zurich, Switzerland) and were grown as described previously [35]. M1 mouse collecting duct cells were grown as outlined in an earlier publication [50].

4.3. Renal Medullary Primary Cells Isolation

Mice were sacrificed through CO_2 inhalation and cervical dislocation and kidneys were kept in ice-cold DMEM/F12 medium (Thermo Fisher Scientific, Darmstadt, Germany). The renal capsule was removed under sterile conditions. The medulla was separated from the cortex and chopped into smaller pieces of tissue using a sharp razor blade (Heinz Herenz, Hamburg, Germany). Tissues were incubated in Hanks balanced salt solution/DMEM/F12 (Life Technologies/Gibco®, Karlsruhe, Germany) containing 1 mg/ml collagenase type 2 (Worthington, Lakewood, NJ, USA) for 20 min at 37 °C. The digested tissue was passed through a 100 µm cell strainer (Merck KGaA, Darmstadt, Germany), transferred to a 50 mL falcon tube and washed with ice-cold PBS. After centrifugation at 1880 rpm for 4 min/4 °C, cells were resuspended and centrifuged 3X at 2260 rpm for 4 min at 4 °C. After washing with ice-cold PBS, tubular preparations were cultured at 37 °C/5% CO_2 in DMEM/F12 supplemented with 1% FBS, 1% Pen/Strep, 1% L-glutamine (200 mM), 1% ITS (100×), 50 nM hydrocortisone, 5 nM triiodothyronine, and 5 nM epidermal growth factor (Sigma Aldrich, Taufkirchen, Germany). After washing with ice-cold PBS, tubular preparations were maintained at 37 °C/5% CO_2 in DMEM/F12 supplemented with 1% FBS, 1% Pen/Strep, 1% L-glutamine (200 mM), 1% ITS (100×), 50 nM hydrocortisone, 5 nM triiodothyronine, and 5 nM epidermal growth factor (Sigma Taufkirchen, Germany). After 24 h, primary cells grew out from isolated tubules.

4.4. Animals and Treatments

Animal experiments were approved by the local institutional review board and all animal experiments complied with the with the United Kingdom Animals Act, 1986, and associated guidelines, EU Directive 2010/63/EU for animal experiments. Experiments were approved by the local Ethics Committee of the Government of Unterfranken/Wuerzburg (AZ: 55.2-2532-2-823). Mice with a floxed *PKD1* allele were generously provided by Prof. Dr. Dorien J.M. Peters (Department of Human Genetics, Leiden University Medical Center, Leiden, The Netherlands) [51]. Animals were hosted on a 12:12 h light:dark cycle under constant temperature (24 $\pm$ 1 °C) in standard cages. They were fed a standard diet with free access to tap water. Generation of mice with a tamoxifen inducible, kidney epithelium-specific *Pkd1*-deletion was done as previously described [24]. Mice carrying loxP-flanked conditional alleles of *Pkd1* were crossed with KSP-Cre mice in a C57BL/6 background (KspCreER^{T2}; *Pkd1*$^{lox;lox}$; abbreviated as *Pkd1*$^{-/-}$). Mice of age 8–10 weeks were used in the experiments.

4.5. Histologic Analysis, Cystic Index

Photographs from hematoxylin and eosin-stained kidney sections were taken at a magnification of $\times$23 and stitched to obtain a single photograph of the whole transverse kidney sections using a Axiovert 200 microscope (Zeiss, München, Germany). The whole kidney cortex was defined as the region of interest using ImageJ (version 1.48). Next, we used an algorithm (ImageJ software version 1.48, NIH, Bethesda, MD, USA) that separates normal tubule space from cystic area by defining diameters of noncystic tubules < 50 mm [32]. The whole cortex cyst area was divided by the whole cortex area and defined as the cystic index.

4.6. Immunohistochemistry

Five-micron thick transverse kidney sections were stained. Anti-TMEM16A (rabbit; 1:100; Abcam, Berlin, Germany) antibody was used as described previously [20]. As secondary antibodies, anti-rabbit IgG Alexa Fluor 546 (1:300; Thermo Fisher Scientific, Inc., Erlangen, Germany) was used. Immunofluorescence was detected using an Axiovert 200 microscope equipped with ApoTome and AxioVision (Zeiss, Germany).

4.7. Western Blotting

Proteins were isolated from perfused whole kidneys and from mCCDcl1 cells using a sample buffer containing 25 mM Tris–HCl, 150 mM NaCl, 1% Nonidet P-40, 5% glycerol, 1 mM EDTA, and 1% protease inhibitor mixture (Roche, Mannheim, Germany). Equal amounts of protein were separated using 8.5% sodium dodecyl sulfate (SDS) polyacrylamide gel. Proteins were transferred to a polyvinylidene difluoride membrane (GE Healthcare Europe GmbH, Munich, Germany) using a semi-dry transfer unit (Bio-Rad, Feldkirchen, Germany). Membranes were incubated with primary anti-TMEM16A (rabbit 1:500; Alomone, Jerusalem, Israel) mouse antibody overnight at 4 °C. Proteins were visualized using horseradish peroxidase-conjugated secondary antibody and ECL detection. Beta-Actin was used as a loading control.

4.8. Ki-67 Assay

Ki-67 staining was performed using a monoclonal anti-ki-67 antibody (rabbit; 1:100, Linaris, Dossenheim, Germany). Signals were amplified by the use of the Vectastain Elite ABC Kit (Vector Laboratories, Burlingame, CA, USA) according to the manufacturer's instructions. Signals were analyzed with a Axiovert 200 microscope (Zeiss, Germany).

4.9. Patch Clamp

Patch-clamp experiments were performed in the fast whole-cell configuration. Patch pipettes had an input resistance of 4–6 MΩ, when filled with a cytosolic-like pipette filling solution [52] containing (mM) KCl 30, K-gluconate 95, NaH_2PO_4 1.2, Na_2HPO_4 4.8, EGTA 1, Ca-gluconate 0.758, $MgCl_2$ 1.034, D-glucose 5, ATP 3. The pH was 7.2, and the Ca^{2+} activity was 0.1 μM. The extracellular bath perfusion was a Ringer solution containing (mmol/L) NaCl 145; KH_2PO_4 0.4; K_2HPO_4 1.6; glucose 5; $MgCl_2$ 1; Ca^{2+}-Gluconat 1.3. The access conductance was measured continuously and was 30–140 nS. Currents (voltage clamp) and voltages (current clamp) were recorded using a patch-clamp amplifier (EPC 9, List Medical Electronics, Darmstadt, Germany) and PULSE software (HEKA, Lambrecht, Germany) as well as Chart software (AD-Instruments, Spechbach, Germany). In intervals, membrane capacitance was measured using the EPC9 device. Data were stored continuously on a computer hard disc and were analyzed using PULSE software. In regular intervals, membrane voltages (V_c) were clamped in steps of 20 mV from −100 to +100 mV relative to resting potential. Current densities (pA/pF) were assessed at the clamp voltage of +100 mV.

4.10. Iodide Quenching Experiments

For YFP-quenching assays, primary renal cells were infected with lentiviral vectors to express halide-sensitive YFP_{I152L}, as previously described [53]. Primary renal cells were isolated and cultured and for each mouse 40 cells were measured. Quenching of the intracellular fluorescence generated by the iodide sensitive enhanced yellow fluorescent protein (EYFP-I152L) was used to measure the anion conductance. YFP-I152L fluorescence was excited at 500 nm using a polychromatic illumination system for microscopic fluorescence measurement (Visitron Systems, Puchheim, Germany) and the emitted light was measured at 535 ± 15 nm with a Coolsnap HQ CCD camera (Roper Scientific, Tucson, AZ, USA). Quenching of YFP-I152L fluorescence by I^- influx was induced by replacing 5 mM extracellular Cl^- with I^-. Cells were grown on coverslips and mounted in a thermostatically controlled imaging chamber maintained at 37 °C. Cells were continuously perfused at 8 mL/min with Ringer solution and exposed to I^- concentration of 5 mM by replacing same amount of NaCl with equimolar NaI. Background fluorescence was subtracted, while auto-fluorescence was negligible. Changes in fluorescence induced by I^- are expressed as initial rates of maximal fluorescence decrease ($\Delta F/\Delta t$). For quantitative analysis, cells with low or excessively high fluorescence were discarded.

4.11. Measurement of $[Ca^{2+}]_i$

Cells were loaded with 2 μM Fura-2/AM and 0.02% Pluronic F-127 (Invitrogen, Darmstadt, Germany) in Ringer solution (mmol/L: NaCl 145; KH_2PO_4 0.4; K_2HPO_4 1.6; Glucose 5; $MgCl_2$ 1; Ca^{2+}-Gluconat 1.3) for 1 h in the dark at room temperature. Cells were perfused with Ringer solution at 37 °C and fluorescence was detected using an inverted microscope (Axiovert S100, Zeiss, Germany) and a high-speed polychromator system (VisiChrome, Puchheim, Germany). Fura-2 was excited at 340/380 nm, and emission was recorded between 470 and 550 nm using a CoolSnap camera (CoolSnap HQ, Roper Scientific, Tucson, AZ, USA). $[Ca^{2+}]_i$ was calculated from the 340/380 nm fluorescence ratio after background subtraction. The formula used to calculate $[Ca^{2+}]_i$ was $[Ca^{2+}]_i = Kd \times (R - R_{min})/(R_{max} - R) \times (S_{f2}/S_{b2})$, where R is the observed fluorescence ratio. The values R_{max} and R_{min} (maximum and minimum ratios) and the constant S_{f2}/S_{b2} (fluorescence of free and Ca^{2+}-bound Fura-2 at 380 nm) were calculated using 1 μmol/L ionomycin (Calbiochem, Merck, Darmstadt, Germany), 5 μmol/L nigericin, 10 μmol/L monensin (Sigma Aldrich, Taufkirchen, German), and 5 mmol/L EGTA to equilibrate intracellular and extracellular Ca^{2+} in intact Fura-2-loaded cells. The dissociation constant for the Fura-2•Ca^{2+} complex was taken as 224 nmol/L.

4.12. Materials and Statistical Analysis

All compounds used were of highest available grade of purity. Data are reported as mean ± SEM. Student's *t* test for unpaired samples and ANOVA were used for statistical analysis. $p < 0.05$ was accepted as the significant difference. Data are expressed as mean ± SEM. Differences among groups were analyzed using one-way ANOVA, followed by a Bonferroni test for multiple comparisons. An unpaired *t* test was applied to compare the differences between the two groups.

Supplementary Materials: The following are available online at https://www.mdpi.com/article/10.3390/ijms22116019/s1. Supplementary Figure S1. CFTR mRNA expression in primary renal epithelial cells. Supplementary Figure S2. YFP-quenching to assess anion permeability. Supplementary Figure S3. Ion currents in M1 mouse collecting duct cells. Supplementary Figure S4. Basal and ATP-induced intracellular Ca^{2+} in primary renal epithelial cells. Supplementary Figure S5. CPA-induced store release in male and female primary renal epithelial cells.

Author Contributions: K.T., I.C. and R.S. performed the experiments, analyzed data, and wrote the manuscript. K.T., R.S., and K.K. wrote the manuscript. All authors have read and agreed to the published version of the manuscript.

Funding: Supported by the Deutsche Forschungsgemeinschaft (DFG, German Research Foundation), project number 387509280, SFB 1350 (project A3).

Institutional Review Board Statement: Animal experiments were approved by the local institutional review board and all animal experiments complied with the with the United Kingdom Animals Act, 1986, and associated guidelines, EU Directive 2010/63/EU for animal experiments. Experiments were approved by the local Ethics Committee of the Government of Unterfranken/Wuerzburg (AZ: 55.2-2532-2-823).

Acknowledgments: We are grateful to Dorien J.M. Peters (Department of Human Genetics, Leiden University Medical Center, Leiden, The Netherlands) for providing us with the ADPKD mouse model and Johannes Loffing (Institute of Anatomy, University of Zurich, Zurich, Switzerland) for providing the mouse renal cortical collecting duct cell line mCCDcl1. We thank Sascha Hofmann, Helena Lowak, and Patricia Seeberger for experimental and technical assistance.

Conflicts of Interest: The authors declare no conflict of interest.

References

1. Gabow, P.A.; Johnson, A.M.; Kaehny, W.D.; Kimberling, W.J.; Lezotte, D.C.; Duley, I.T.; Jones, R.H. Factors affecting the progression of renal disease in autosomal-dominant polycystic kidney disease. *Kidney Int.* **1992**, *41*, 1311–1319. [CrossRef]
2. Stewart, J.H. End-stage renal failure appears earlier in men than in women with polycystic kidney disease. *Am. J. Kidney Dis.* **1994**, *24*, 181–183. [CrossRef]
3. Johnson, A.M.; Gabow, P.A. Identification of patients with autosomal dominant polycystic kidney disease at highest risk for end-stage renal disease. *J. Am. Soc. Nephrol.* **1997**, *8*, 1560–1567. [CrossRef] [PubMed]
4. Cowley, B.D., Jr.; Rupp, J.C.; Muessel, M.J.; Gattone, V.H., 2nd. Gender and the effect of gonadal hormones on the progression of inherited polycystic kidney disease in rats. *Am. J. Kidney Dis.* **1997**, *29*, 265–272. [CrossRef]
5. Dubey, R.K.; Jackson, E.K. Estrogen-induced cardiorenal protection: Potential cellular, biochemical, and molecular mechanisms. *Am. J. Physiol. Ren. Physiol.* **2001**, *280*, F365–F388. [CrossRef] [PubMed]
6. Stringer, K.D.; Komers, R.; Osman, S.A.; Oyama, T.T.; Lindsley, J.N.; Anderson, S. Gender hormones and the progression of experimental polycystic kidney disease. *Kidney Int.* **2005**, *68*, 1729–1739. [CrossRef] [PubMed]
7. Gretz, N.; Ceccherini, I.; Kränzlin, B.; Klöting, I.; Devoto, M.; Rohmeiss, P.; Hocher, B.; Waldherr, R.; Romeo, G. Gender-dependent disease severity in autosomal polycystic kidney disease of rats. *Kidney Int.* **1995**, *48*, 496–500. [CrossRef]
8. Grantham, J.J.; Ye, M.; Gattone, V.H., 2nd; Sullivan, L.P. In vitro fluid secretion by epithelium from polycystic kidneys. *J Clin. Invest.* **1995**, *95*, 195–202. [CrossRef] [PubMed]
9. Hanaoka, K.; Devuyst, O.; Schwiebert, E.M.; Wilson, P.D.; Guggino, W.B. A role for CFTR in human autosomal dominant polycystic kidney disease. *Am. J. Physiol.* **1996**, *270*, C389–C399. [CrossRef]
10. Davidow, C.J.; Maser, R.L.; Rome, L.A.; Calvet, J.P.; Grantham, J.J. The cystic fibrosis transmembrane conductance regulator mediates transepithelial fluid secretion by human autosomal dominant polycystic kidney disease epithelium in vitro. *Kidney Int.* **1996**, *50*, 208–218. [CrossRef] [PubMed]
11. Buchholz, B.; Teschemacher, B.; Schley, G.; Schillers, H.; Eckardt, K.U. Formation of cysts by principal-like MDCK cells depends on the synergy of cAMP- and ATP-mediated fluid secretion. *J. Mol. Med.* **2011**, *89*, 251–261. [CrossRef]
12. Yang, B.; Sonawane, N.D.; Zhao, D.; Somlo, S.; Verkman, A.S. Small-Molecule CFTR Inhibitors Slow Cyst Growth in Polycystic Kidney Disease. *J. Am. Soc. Nephrol.* **2008**, *19*, 1300–1310. [CrossRef]
13. Magenheimer, B.S.; St John, P.L.; Isom, K.S.; Abrahamson, D.R.; De Lisle, R.C.; Wallace, D.P.; Maser, R.L.; Grantham, J.J.; Calvet, J.P. Early embryonic renal tubules of wild-type and polycystic kidney disease kidneys respond to cAMP stimulation with cystic fibrosis transmembrane conductance regulator/Na(+),K(+),2Cl(-) Co-transporter-dependent cystic dilation. *J. Am. Soc. Nephrol.* **2006**, *17*, 3424–3437. [CrossRef] [PubMed]
14. Xu, N.; Glockner, J.F.; Rossetti, S.; Babovich-Vuksanovic, D.; Harris, P.C.; Torres, V.E. Autosomal dominant polycystic kidney disease coexisting with cystic fibrosis. *J. Nephrol.* **2006**, *19*, 529–534. [PubMed]

15. O'Sullivan, D.A.; Torres, V.E.; Gabow, P.A.; Thibodeau, S.N.; King, B.F.; Bergstralh, E.J. Cystic fibrosis and the phenotypic expression of autosomal dominant polycystic kidney disease. *Am. J. Kidney Dis.* **1998**, *32*, 976–983. [CrossRef]
16. Persu, A.; Devuyst, O.; Lannoy, N.; Materne, R.; Brosnahan, G.; Gabow, P.A.; Pirson, Y.; Verellen-Dumoulin, C. CF gene and cystic fibrosis transmembrane conductance regulator expression in autosomal dominant polycystic kidney disease. *J. Am. Soc. Nephrol.* **2000**, *11*, 2285–2296. [CrossRef]
17. Wilson, P.D.; Hovater, J.S.; Casey, C.C.; Fortenberry, J.A.; Schwiebert, E.M. ATP release mechanisms in primary cultures of epithelia derived from the cysts of polycystic kidneys. *J. Am. Soc. Nephrol.* **1999**, *10*, 218–229. [CrossRef]
18. Schwiebert, E.M.; Wallace, D.P.; Braunstein, G.M.; King, S.R.; Peti-Peterdi, J.; Hanaoka, K.; Guggino, W.B.; Guay-Woodford, L.M.; Bell, P.D.; Sullivan, L.P.; et al. Autocrine extracellular purinergic signaling in epithelial cells derived from polycystic kidneys. *Am. J. Physiol. Ren. Physiol.* **2002**, *282*, F763–F775. [CrossRef]
19. Kraus, A.; Grampp, S.; Goppelt-Struebe, M.; Schreiber, R.; Kunzelmann, K.; Peters, D.J.; Leipziger, J.; Schley, G.; Schodel, J.; Eckardt, K.U.; et al. P2Y2R is a direct target of HIF-1alpha and mediates secretion-dependent cyst growth of renal cyst-forming epithelial cells. *Purinergic Signal.* **2016**, *12*, 687–695. [CrossRef]
20. Buchholz, B.; Faria, D.; Schley, G.; Schreiber, R.; Eckardt, K.U.; Kunzelmann, K. Anoctamin 1 induces calcium-activated chloride secretion and tissue proliferation in polycystic kidney disease. *Kidney Int.* **2014**, *85*, 1058–1067. [CrossRef]
21. Buchholz, B.; Schley, G.; Faria, D.; Kroening, S.; Willam, C.; Schreiber, R.; Klanke, B.; Burzlaff, N.; Kunzelmann, K.; Eckardt, K.U. Hypoxia-Inducible Factor-1a Causes Renal Cyst Expansion through Calcium-Activated Chloride Secretion. *J. Am. Soc. Nephrol.* **2014**, *25*, 465–474. [CrossRef] [PubMed]
22. Schreiber, R.; Buchholz, B.; Kraus, A.; Schley, G.; Scholz, J.; Ousingsawat, J.; Kunzelmann, K. Lipid peroxidation drives renal cyst growth in vitro through activation of TMEM16A. *J. Am. Soc. Nephrol.* **2019**, *30*, 228–242. [CrossRef] [PubMed]
23. Cabrita, I.; Buchholz, B.; Schreiber, R.; Kunzelmann, K. TMEM16A drives renal cyst growth by augmenting Ca(2+) signaling in M1 cells. *J. Mol. Med.* **2020**, *98*, 659–671. [CrossRef]
24. Cabrita, I.; Kraus, A.; Scholz, J.K.; Skoczynski, K.; Schreiber, R.; Kunzelmann, K.; Buchholz, B. Cyst growth in ADPKD is prevented by pharmacological and genetic inhibition of TMEM16A in vivo. *Nat. Commun.* **2020**, *11*, 4320. [CrossRef]
25. Cabrita, I.; Talbi, K.; Kunzelmann, K.; Schreiber, R. Loss of PKD1 and PKD2 share common effects on intracellular Ca2+ signaling. *Cell Calcium* **2021**, *97*, 102413. [CrossRef]
26. Cha, J.Y.; Wee, J.; Jung, J.; Jang, Y.; Lee, B.; Hong, G.S.; Chang, B.C.; Choi, Y.L.; Shin, Y.K.; Min, H.Y.; et al. Anoctamin 1 (TMEM16A) is essential for testosterone-induced prostate hyperplasia. *Proc. Natl. Acad. Sci. USA* **2015**, *112*, 9722–9727. [CrossRef]
27. Martinez-Pinna, J.; Soriano, S.; Tudurí, E.; Nadal, A.; de Castro, F. A Calcium-Dependent Chloride Current Increases Repetitive Firing in Mouse Sympathetic Neurons. *Front. Physiol.* **2018**, *9*, 508. [CrossRef]
28. Chen, D.; Meng, W.; Shu, L.; Liu, S.; Gu, Y.; Wang, X.; Feng, M. ANO1 in urethral SMCs contributes to sex differences in urethral spontaneous tone. *Am. J. Physiol Ren. Physiol* **2020**. [CrossRef] [PubMed]
29. Davis, P.B. The gender gap in cystic fibrosis survival. *J. Gend. Specif. Med. JGSM Off. J. Partnersh. Women's Health Columbia* **1999**, *2*, 47–51.
30. Sweezey, N.; Tchepichev, S.; Gagnon, S.; Fertuck, K.; O'Brodovich, H. Female gender hormones regulate mRNA levels and function of the rat lung epithelial Na channel. *Am. J. Physiol* **1998**, *274*, C379–C386. [CrossRef]
31. Saint-Criq, V.; Harvey, B.J. Estrogen and the cystic fibrosis gender gap. *Steroids* **2014**, *81*, 4–8. [CrossRef]
32. Kraus, A.; Peters, D.J.M.; Klanke, B.; Weidemann, A.; Willam, C.; Schley, G.; Kunzelmann, K.; Eckardt, K.U.; Buchholz, B. HIF-1alpha promotes cyst progression in a mouse model of autosomal dominant polycystic kidney disease. *Kidney Int.* **2018**, *94*, 887–899. [CrossRef]
33. Lantinga-van Leeuwen, I.S.; Leonhard, W.N.; van de Wal, A.; Breuning, M.H.; Verbeek, S.; de Heer, E.; Peters, D.J. Transgenic mice expressing tamoxifen-inducible Cre for somatic gene modification in renal epithelial cells. *Genesis* **2006**, *44*, 225–232. [CrossRef] [PubMed]
34. Nagao, S.; Kusaka, M.; Nishii, K.; Marunouchi, T.; Kurahashi, H.; Takahashi, H.; Grantham, J. Androgen receptor pathway in rats with autosomal dominant polycystic kidney disease. *J. Am. Soc. Nephrol.* **2005**, *16*, 2052–2062. [CrossRef]
35. Gaeggeler, H.P.; Gonzalez-Rodriguez, E.; Jaeger, N.F.; Loffing-Cueni, D.; Norregaard, R.; Loffing, J.; Horisberger, J.D.; Rossier, B.C. Mineralocorticoid versus glucocorticoid receptor occupancy mediating aldosterone-stimulated sodium transport in a novel renal cell line. *J. Am. Soc. Nephrol.* **2005**, *16*, 878–891. [CrossRef] [PubMed]
36. Gong, Y.; Blok, L.J.; Perry, J.E.; Lindzey, J.K.; Tindall, D.J. Calcium regulation of androgen receptor expression in the human prostate cancer cell line LNCaP. *Endocrinology* **1995**, *136*, 2172–2178. [CrossRef]
37. Estrada, M.; Liberona, J.L.; Miranda, M.; Jaimovich, E. Aldosterone- and testosterone-mediated intracellular calcium response in skeletal muscle cell cultures. *Am. J. Physiol. Endocrinol. Metab.* **2000**, *279*, E132–E139. [CrossRef] [PubMed]
38. Liu, G.; Honisch, S.; Liu, G.; Schmidt, S.; Alkahtani, S.; AlKahtane, A.A.; Stournaras, C.; Lang, F. Up-regulation of Orai1 expression and store operated Ca(2+) entry following activation of membrane androgen receptors in MCF-7 breast tumor cells. *BMC Cancer* **2015**, *15*, 995. [CrossRef]
39. Steinsapir, J.; Socci, R.; Reinach, P. Effects of androgen on intracellular calcium of LNCaP cells. *Biochem. Biophys. Res. Commun.* **1991**, *179*, 90–96. [CrossRef]
40. Sun, Y.H.; Gao, X.; Tang, Y.J.; Xu, C.L.; Wang, L.H. Androgens induce increases in intracellular calcium via a G protein-coupled receptor in LNCaP prostate cancer cells. *J. Androl.* **2006**, *27*, 671–678. [CrossRef]

41. Vicencio, J.M.; Ibarra, C.; Estrada, M.; Chiong, M.; Soto, D.; Parra, V.; Diaz-Araya, G.; Jaimovich, E.; Lavandero, S. Testosterone induces an intracellular calcium increase by a nongenomic mechanism in cultured rat cardiac myocytes. *Endocrinology* **2006**, *147*, 1386–1395. [CrossRef] [PubMed]
42. Berry, P.A.; Birnie, R.; Droop, A.P.; Maitland, N.J.; Collins, A.T. The calcium sensor STIM1 is regulated by androgens in prostate stromal cells. *Prostate* **2011**, *71*, 1646–1655. [CrossRef]
43. Sheridan, J.T.; Gilmore, R.C.; Watson, M.J.; Archer, C.B.; Tarran, R. 17β-Estradiol inhibits phosphorylation of stromal interaction molecule 1 (STIM1) protein: Implication for store-operated calcium entry and chronic lung diseases. *J. Biol. Chem.* **2013**, *288*, 33509–33518. [CrossRef] [PubMed]
44. Onishi, M.; Yamanaka, K.; Miyamoto, Y.; Waki, H.; Gouraud, S. Trpv4 involvement in the sex differences in blood pressure regulation in spontaneously hypertensive rats. *Physiol. Genom.* **2018**, *50*, 272–286. [CrossRef] [PubMed]
45. Xiao, Q.; Yu, K.; Perez-Cornejo, P.; Cui, Y.; Arreola, J.; Hartzell, H.C. Voltage- and calcium-dependent gating of TMEM16A/Ano1 chloride channels are physically coupled by the first intracellular loop. *Proc. Natl. Acad. Sci. USA* **2011**, *108*, 8891–8896. [CrossRef]
46. Jin, X.; Shah, S.; Liu, Y.; Zhang, H.; Lees, M.; Fu, Z.; Lippiat, J.D.; Beech, D.J.; Sivaprasadarao, A.; Baldwin, S.A.; et al. Activation of the Cl- Channel ANO1 by Localized Calcium Signals in Nociceptive Sensory Neurons Requires Coupling with the IP3 Receptor. *Sci. Signal.* **2013**, *6*, ra73. [CrossRef] [PubMed]
47. Cabrita, I.; Benedetto, R.; Fonseca, A.; Wanitchakool, P.; Sirianant, L.; Skryabin, B.V.; Schenk, L.K.; Pavenstadt, H.; Schreiber, R.; Kunzelmann, K. Differential effects of anoctamins on intracellular calcium signals. *FASEB J.* **2017**, *31*, 2123–2134. [CrossRef] [PubMed]
48. Ferrera, L.; Caputo, A.; Ubby, I.; Bussani, E.; Zegarra-Moran, O.; Ravazzolo, R.; Pagani, F.; Galietta, L.J. Regulation of TMEM16A chloride channel properties by alternative splicing. *J. Biol. Chem.* **2009**, *284*, 33360–33368. [CrossRef] [PubMed]
49. Lerias, J.; Pinto, M.; Benedetto, R.; Schreiber, R.; Amaral, M.; Aureli, M.; Kunzelmann, K. Compartmentalized crosstalk of CFTR and TMEM16A (ANO1) through EPAC1 and ADCY1. *Cell Signal.* **2018**, *44*, 10–19. [CrossRef] [PubMed]
50. AlDehni, F.; Spitzner, M.; Martins, J.R.; Barro Soria, R.; Schreiber, R.; Kunzelmann, K. Role of bestrophin for proliferation and in epithelial to mesenchymal transition. *J. Am. Soc. Nephrol.* **2009**, *20*, 1556–1564. [CrossRef] [PubMed]
51. Lantinga-van Leeuwen, I.S.; Leonhard, W.N.; van der Wal, A.; Breuning, M.H.; de Heer, E.; Peters, D.J. Kidney-specific inactivation of the Pkd1 gene induces rapid cyst formation in developing kidneys and a slow onset of disease in adult mice. *Hum. Mol. Genet.* **2007**, *16*, 3188–3196. [CrossRef] [PubMed]
52. Greger, R.; Kunzelmann, K. Simultaneous recording of the cell membrane potential and properties of the cell attached membrane of HT_{29} colon carcinoma and CF-PAC cells. *Pflügers Arch.* **1991**, *419*, 209–211. [CrossRef] [PubMed]
53. Schreiber, R.; Uliyakina, I.; Kongsuphol, P.; Warth, R.; Mirza, M.; Martins, J.R.; Kunzelmann, K. Expression and Function of Epithelial Anoctamins. *J. Biol. Chem.* **2010**, *285*, 7838–7845. [CrossRef] [PubMed]

International Journal of
Molecular Sciences

MDPI

Review

Molecular Pathophysiology of Autosomal Recessive Polycystic Kidney Disease

Adrian Cordido [1,2,†], Marta Vizoso-Gonzalez [1,2,†] and Miguel A. Garcia-Gonzalez [1,2,3,*]

1 Grupo de Xenética e Bioloxía do Desenvolvemento das Enfermidades Renais, Laboratorio de Nefroloxía (No. 11), Instituto de Investigación Sanitaria de Santiago (IDIS), Complexo Hospitalario de Santiago de Compostela (CHUS), 15706 Santiago de Compostela, Spain; adriancosman@hotmail.es (A.C.); martavizosoglez@gmail.com (M.V.-G.)
2 Grupo de Medicina Xenómica, Complexo Hospitalario de Santiago de Compostela (CHUS), 15706 Santiago de Compostela, Spain
3 Fundación Publica Galega de Medicina Xenómica-SERGAS, Complexo Hospitalario de Santiago de Compostela (CHUS), 15706 Santiago de Compostela, Spain
* Correspondence: miguel.garcia.gonzalez@sergas.es; Tel.: +34-981-555-197
† Equal contribution.

Abstract: Autosomal recessive polycystic kidney disease (ARPKD) is a rare disorder and one of the most severe forms of polycystic kidney disease, leading to end-stage renal disease (ESRD) in childhood. *PKHD1* is the gene that is responsible for the vast majority of ARPKD. However, some cases have been related to a new gene that was recently identified (*DZIP1L* gene), as well as several ciliary genes that can mimic a ARPKD-like phenotypic spectrum. In addition, a number of molecular pathways involved in the ARPKD pathogenesis and progression were elucidated using cellular and animal models. However, the function of the ARPKD proteins and the molecular mechanism of the disease currently remain incompletely understood. Here, we review the clinics, treatment, genetics, and molecular basis of ARPKD, highlighting the most recent findings in the field.

Keywords: ARPKD; cyst; rare monogenic disease; nephrology

Citation: Cordido, A.; Vizoso-Gonzalez, M.; Garcia-Gonzalez, M.A. Molecular Pathophysiology of Autosomal Recessive Polycystic Kidney Disease. *Int. J. Mol. Sci.* **2021**, *22*, 6523. https://doi.org/10.3390/ijms22126523

Academic Editor: Ivano Condò

Received: 31 May 2021
Accepted: 16 June 2021
Published: 17 June 2021

1. Introduction

Autosomal recessive polycystic kidney disease (ARPKD) is a severe inherited cystic disease characterized by the combination of bilateral renal cystic disease and congenital hepatic fibrosis. ARPKD manifests perinatally, or in childhood, as an important cause of pediatric morbidity and mortality [1]. ARPKD is a rare disease; an incidence of 1 in 8000 births was calculated in an isolated and inbred population from Finland [2]. However, in the Americas (North, Central, and South), the reported incidence is 1 in 26:485 births, and the annualized prevalence is 1.17 per 100,000 [3]. The widespread prevalence of ARPKD is estimated to be 1 in 20:000 births [4].

ARPKD is the recessive form of a group of heterogeneous monogenic disorders named polycystic kidney disease (PKD). The dominant form, autosomal dominant polycystic kidney disease (ADPKD), has a higher epidemiological prevalence and is typically diagnosed in adults [5].

2. Autosomal Recessive Polycystic Kidney Disease Clinical Presentation

ARPKD is phenotypically highly variable; it can present as a disease of perinatal, neonatal, infantile, juvenile, or young adult-onset disease [6], with no known gender or ethnic bias [7]. Typically, the most severe cases of ARPKD present in the late gestational or neonatal stage, with bilateral massively enlarged and echogenic kidneys with poor corticomedullary differentiation, retained reniform contour, and multiple tiny cysts [8–10]. In addition, they can present with oligo- or anhydramnios, resulting in the typical "Potter sequence" phenotype with pulmonary hypoplasia, characteristic facial features, and clubfoot contracted limbs [9,11,12]. In addition to the Potter sequence, the presence of other extrarenal manifestations is not common [13]. There are no documented hepatic phenotypes [14], although some associated, such as abdominal dystocia, have been reported [15].

Detection of severe oligohydramnios is associated with worse outcome due to the high risk of associated pulmonary hypoplasia. Up to 50% of ARPKD neonates die of respiratory insufficiency due to pulmonary hypoplasia and thoracic compression. However, after the perinatal period, survival is high, reaching 1-year and 10-year survival rates of 85% and 82%, respectively [7,8,12]. The patients who survive the perinatal period require extensive care by specialists in internal medicine [16]. Note that prenatal diagnosis and termination of pregnancy are factors to consider in the epidemiology of the disease [15,17]. Another problem derived from kidney enlargement and pulmonary hypoplasia, in addition to early uremia and pulmonary immaturity, includes the difficulty of enteral feeding that could complicate nutrition, requiring persistent nasogastric feeding [11,18].

In most cases, the severe problem of ARPKD during the first few months of life is arterial hypertension, which requires treatment with multiple classes of pharmacological interventions. Tight blood pressure control is crucial to prevent further kidney damage from hypertension [11,12]. However, renal failure is not a very common cause of neonatal demise [19]. Renal disease manifestations include urinary tract infections, macroscopic hematuria, as well as renal osteopathy early in childhood. In regards to sonography anomalies, increased echogenicity and renal cysts are usually reported prior to kidney enlargement [14]. Kidney enlargement is due to extensive dilatations of the distal nephron, typically in the collecting duct. As the disease progresses, the kidney structure gradually resembles the pattern seen in ADPKD with renal macrocysts of different size and appearance, often accompanied by interstitial fibrosis [11,20,21]. This situation leads to almost 50% of end-stage renal disease (ESRD) in the first decades of life, requiring renal replacement therapy [9,14,17].

Although the disease is called "autosomal recessive polycystic kidney disease," a cystic liver phenotype plays a significant role in the disease, which explains why the primary gene is entitled "polycystic kidney and hepatic disease 1 (*PKHD1*)", with polycystic liver disease (PLD) being the principal extrarenal manifestation [22]. These histological changes are the consequences of a developmental defect of the hepatic ductal plate termed, ductal plate malformation (DPM) [23,24]. DPM is also a common feature of other ciliopathies [4]. Liver disease appears with increasing age of the patient. The first manifestation appears as congenital hepatic fibrosis (CHF) with variable dilatations of both intra- and extrahepatic bile ducts (Caroli syndrome) [23]. Over the course of ARPKD, liver disease presents with two main manifestations: portal hypertension due to progressive hepatic fibrosis; and cholangitis [14]. Complications related to portal hypertension can include splenomegaly, thrombocytopenia, and esophageal varices, producing severe bleeding complications [14,25]. There is some link between adult patients with ARPKD (more than 40 years) and the risk of developing hepatic tumors, especially cholangiocarcinoma [24]. Interestingly, the hepatocellular function normally remains stable, with serum liver enzymes in the normal range, except for cholestasis parameters [4,12,23,26]. With the current clinical methods, this fact makes it difficult to monitor the severity and progression of liver disease [25].

Although cerebral aneurysms can occur in ~10% of ADPKD patients, only a few cases have been described in ARPKD [27]. The highest risk factor for an intracranial aneurysm is hypertension, a condition shared in patients with ADPKD and ARPKD [28]. The early age at which some of these aneurysms are diagnosed is remarkable since pediatric aneurysms are very rare [27]. Moreover, other extrarenal and extrahepatic manifestations can be present in ARPKD patients, including left ventricular hypertrophy, recurrent respiratory infections, neurological abnormalities, abnormal ocular fundus, abdominal pain, septic episodes, and deformities of the spine and limbs [14,29].

While a majority of ARPKD patients show similar disease progression, there are atypical phenotypes; among the elderly population, some ARPKD cases were reported as moderately affected or even exclusive or predominant phenotypes of either the liver or kidneys [4,21]. This is related to the fact that although ARPKD is a recessive disease, in which heterozygous carriers should not show any clinical manifestation of the disease, the data suggest heterozygous for *PKHD1* mutations have an increased risk of PLD and mild PKD [30,31].

3. Diagnosis

ARPKD is frequently diagnosed in the prenatal period due to its early and severe manifestations. In prenatal diagnosis, an ultrasound from second/third trimester can detect enlarged, echogenic kidneys, and medullary hyperechogenicity, due to the loss of corticomedullary differentiation and diffusively increased hepatic parenchymal echogenicity with fibrous tissue. The presence of oligohydramnios can make it challenging, so ultrasonography and MRI are required. The finding of microcysts (5–7 mm) was reported in 30% of ARPKD cases, but macrocysts (>10 mm) are rare and could indicate another different ciliopathy. These ultrasound findings are common in other pathologies, like Meckel syndrome, and mild forms of the disease may not be detected by prenatal ultrasounds. In these cases, the genetic test offers the possibility of providing an accurate diagnosis [32,33].

Identifying the *PKHD1* gene made it possible to perform the genetic diagnosis by direct DNA sequencing (Sanger method). However, the genetic test for the *PKHD1* mutation is complicated due to the large genomic size and the allelic heterogeneity of the disease-associated mutations [34]. However, according to the Genetics Work Group, single-gene testing should be avoided in cases of suspected ARPKD due to its broad overlapping phenotypic spectrum. As an alternative, methods such as next generation sequencing have become of interest as techniques that can simultaneously and efficiently analyze multiple candidate genes in a unique test, at relative low cost. In rare cases, mutations in two genes can even be observed in children with severe neonatal clinical phenotype [4,33,35].

The outcome of the genetic testing is essential for clinical management of comorbidities and complications associated with each disease, allowing informed genetic counselling and, in the future, precision medicine on a more specific basis [4].

4. Differential Diagnosis

The ARPKD phenotype is not only caused by mutations in *PKHD1*. This makes diagnosis and management, including care during the perinatal period, a difficult task. A number of other recessive and dominant genes need to be considered (Figure 1) [4,21,36].

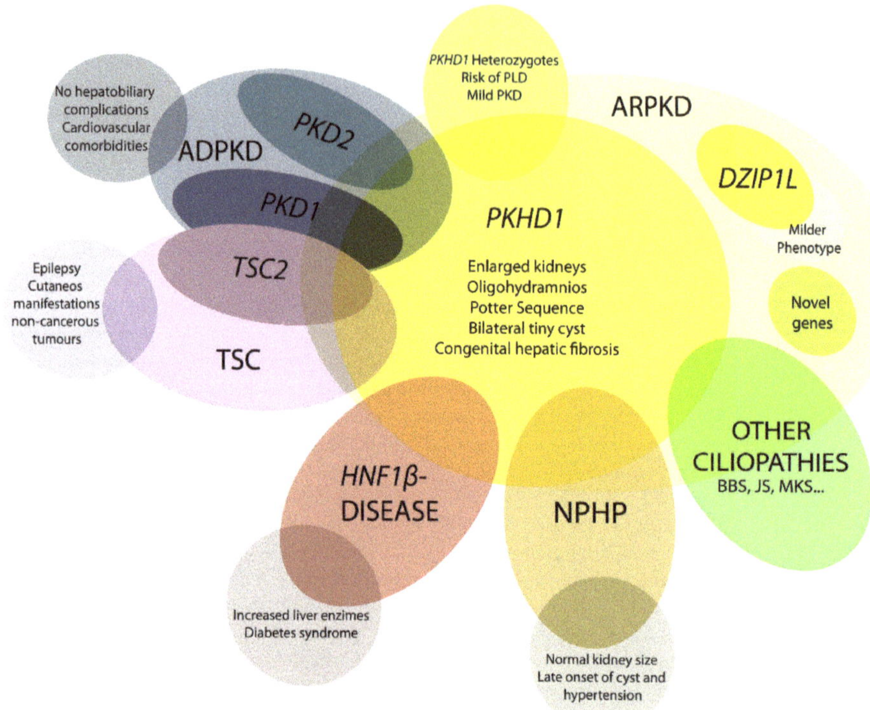

Figure 1. Schematic representation of ARPKD differential diagnosis. *PHKD1* is the main causative gene in ARPKD, where *DZIP1L* present milder phenotype. Mutations in other genes can overlap clinical manifestations of ARPKD, such as *PKD1* and *PKD2*, the main causative genes of autosomal dominant polycystic kidney disease (ADPKD); TSC2, that causes tuberous sclerosis (TSC); and others for instance *HNF1β*, nephronophthisis (NPHP) genes and other ciliopathies as Bardet–Biedl (BBS), Joubert (JS), and Meckel syndrome (MKS). These overlapping phenotypes manifest the physiologic complex and functional interactions that occur among ciliopathy genes.

4.1. *DZIP1L-Related Polycystic Kidney Disease*

A moderate manifestation of ARPKD has been described in patients with *DZIP1L* mutations [37]. The first reported cases associated with this gene are in the prenatal stage or in early childhood; however, no cases of perinatal death have been described [4,21,37].

4.2. *Early and Severe Autosomal Dominant Polycystic Kidney Disease (ADPKD)*

In most cases, ADPKD does not appear clinically before adulthood, but there is a small portion of cases (up to 5%) where the disease can manifest during childhood or before birth. There is no established reason that stochastic, epigenetic, and environmental aspects are believed to alter the phenotype. Families with one child showing early-onset ADPKD usually have high repetition in other siblings, leading us to believe a familial modifier, such as complex genetic interactions with second modifiers. A "dosage-sensitive network" could worsen cell integrity and could explain the more severe clinical course of these patients [4,21,38,39].

Clinically, there is a significant overlap between ARPKD and ADPKD, but there are also several characteristic manifestations associated with each disease. ADPKD patients rarely develop hepatobiliary complications, unlike patients with ARPKD. In addition, ADPKD patients are more likely to have cardiovascular comorbidities, particularly intracranial aneurysms [21].

4.3. Early and Severe PKD due to TSC2-PKD1

Mutations in *TSC1* and *TSC2* genes cause tuberous sclerosis (TSC). This multiorgan disorder is predominantly associated with epilepsy and cutaneous manifestations. Patients may develop non-cancerous tumors in the kidney, heart, and brain. Renal manifestations are the main cause of death in adult patients [40,41]. When a deletion in chromosome 16p affects the two adjacent genes (*PKD1* and *TSC2*), it usually causes an early onset of PKD. Moreover, their close physiologic interrelations can explain the clinical overlap between PKD and TSC; *TSC2* protein function has been shown to play a role in assisting polycystin 1 localization [21].

4.4. HNF1β-Related Disease

Mutations in the *HNF1β/TCF2* gene cause prenatal hyperechogenic kidneys and Potter's sequence, as well as enlarged polycystic kidneys that can be confused with ARPKD. One explanation linking these similar phenotypes is that *HNF1β* is a transcription factor that regulates *PKHD1* expression, in addition to other polycystic kidney genes. However, mutations in this gene can cause a wide range of manifestations: renal cysts and diabetes syndrome; defects in the genital tract; endocrine/exocrine gland disorders, hypomagnesemia, and an increase in liver enzymes [21,42,43].

4.5. Nephronophthisis (NPHP)

NPHP is classified as an autosomal recessive cystic kidney disease that is characterized by tubulointerstitial cysts accompanied with fibrosis. To date, about 20 genes are related with recessive NPHP [36]. Unlike ARPKD, NPHP kidneys remain small. It is common that cysts and hypertension only manifest in late disease and it is one of the main causes of ESRD in patients under 25. In some cases, the manifestations may mimic ARPKD with enlarged kidneys or Potter's sequence. NPHP proteins work in functional networks with other ciliopathy proteins; the identification and characterization of NPHP genes has contributed to the understanding of the molecular mechanisms of cystogenesis [4,21,44,45].

Medullary cystic kidney disease (recently named tubulointerstitial kidney disease TKD), caused by mutations in *MUC1* and *UMOD*, is considered the autosomal dominant NPHP, with a later onset than the recessive form [21].

4.6. Mutations in Other Ciliary Genes

Mutations may mimic ARPKD in genes that typically cause other (usually more complex) ciliopathies, such as Bardet–Biedl (BBS), Joubert (JS), and Meckel syndrome (MKS). BBS phenotype can be heterogeneous, but often presents enlarged and hyperechogenic kidneys with a loss of corticomedullary differentiation. The most severe ciliopathies are MKS and JS, characterized by early-onset developmental disorder and neurological problems, and many features of ciliopathy, such as liver fibrosis, polydactyly, and cystic kidneys [4,21].

Another example is Von Hippel–Lindau, an autosomal dominant disease caused by mutations in the VHL gene, a tumor suppressor gene. This causes hemangioblastoma in the central nervous system accompanied by renal tumors. Affected people also have a high probability of renal and pancreatic cysts. The similarity between the manifestations between TSC, VHL, and PKD suggests a functional connection: primary cilium and mTOR signaling pathway [4,46].

5. Genetics of ARPKD

As we mentioned earlier, ARPKD is caused by mutations in *PKHD1* and, the recently discovered, *DZIP1L* [37,47,48]. *PKHD1*, located on chromosome 6 (6p12.3-p12.2) (Figure 2A) [34,48–50], is one of the largest human genes with a genomic segment of ~500 kb. It is predicted to have a minimum of 86 exons assembled in a complicated pattern of alternative splice variants, transcribing a large full-length mRNA of approximately 8.5 kb–13 kb [49]. Multiple types of mutations characterized as pathogenic have been identified across the gene. Currently, approximately 750 *PKHD1* mutations have been identified, of which approximately half are missense changes. A missense mutation in exon 3 (c. 107C>T; p.Thr36Me) is the most common mutation described, accounting for more than 20% of all cases [34,51]. This mutation has been observed in the context of heterozygotes, with a second distinct mutant allele [29]. Most cases are familial, but de novo mutations are also reported and account for 2 to 5% of cases [34]. Interestingly, in the context of isolated autosomal dominant polycystic liver disease (ADPLD), Besse and colleagues have reported several individuals with *PKHD1* mutations in heterozygote carriers, 10 of 102 ADPLD patients of their cohort were explained by *PKHD1* mutations, one of them presented the p.Thr36Me missense variant [30]. According to the clinical observation, it is a genetic fact that 10% of ARPKD patients present innumerable asymptomatic liver cysts [31]. However, the data are not sufficient to explain why *PKHD1* in ARPKD leads to severe hepatic and renal phenotype, but not in ADPLD; in this regard, more studies are needed [52].

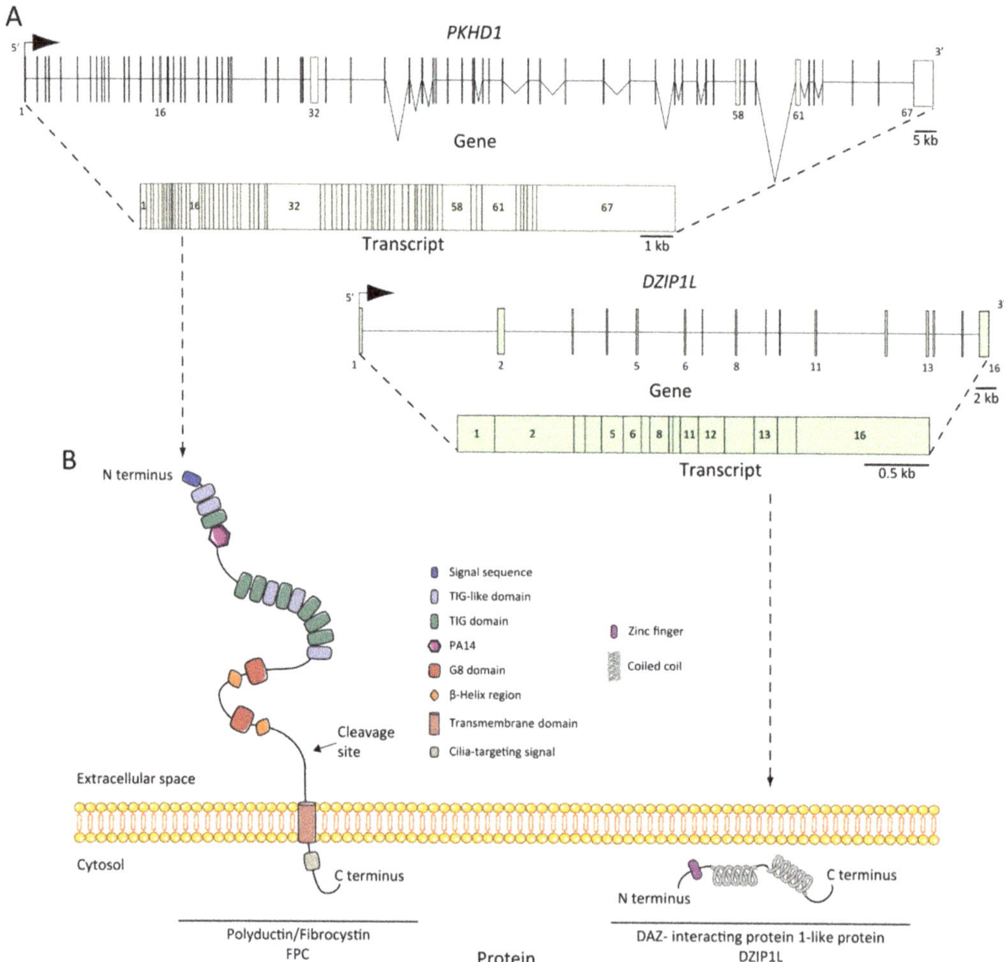

Figure 2. ARPKD genes, transcripts, and proteins: (**A**) *PKHD1* and *DZIP1L* genes and transcripts. The positions of the exons are illustrated and numbered, and the longest transcript are shown from both: 67 exons for *PKHD1* and 16 for *DZIP1L*; (**B**) structure of fibrocystin/polyductin (FPC) and DAZ-interacting protein 1-like protein (DZIP1L). Proteins are not to scale.

Recently, using genome-wide analysis of SNPs, whole-exome sequencing, and Sanger sequencing, Lu and colleagues [37] established *DZIP1L*, located on chromosome 3 (3q22.1-q23) (Figure 2A), and encoding a 767-amino acid protein as a new gene related to ARPKD pathogenesis. The authors identified mutations in seven patients (~0.3% of their cohort) with the ARPKD phenotype without evidence of another mutation in the PKD genes. They identified homozygous missense mutations that segregate with the disease in two families (p. Gln91His and p.Ala90Val) and homozygous protein-truncating mutations in two other unrelated consanguineous pedigrees (p.Gln155* and p.Glu354Alafs*39). The data suggest that *DZIP1L* mutations are not a common cause of the disease, but despite this fact, ARPKD NGS diagnostic multigene panels should target this gene for two reasons: mutations in *DZIP1L* may interact with other PKD or ciliopathy loci and would help to broaden the genetic complex understanding of the disease [37,53].

Genotype-Phenotype Correlation

Establishing a possible genotype/phenotype correlation is complicated by compound heterozygotes. The most common mutation, c.107C>T (T36M), only explains about 20% of mutant alleles, but no others are described as a cluster; in fact, many variants are unique to single lineages. There are plenty of pathogenic variants in *PKHD1*, including truncating, missense, and intronic/splice mutation. Typically genotype-phenotype studies are done regarding the type of mutation more than the specific site of the mutation [54,55].

Patients with two truncating mutations show a severe phenotype with high peri- or neonatal mortality and the presence of one or two missense mutations generally exhibits a moderate phenotype. However, there are exceptions, a patient described by Ebner and colleagues with two truncating mutations in *PKHD1* survived the first 30 months of life without renal replacement therapy or, on the contrary, several cases with two missense mutations that can be as severe as truncating variants [55–58].

Furthermore, up to 20% of siblings show a marked variation in phenotype, which means that the genotype is insufficient to explain the phenotypic variability in ARPKD, where complex transcriptional profiles may play an important role [54,59]. Moreover, there is evidence in ARPKD mouse models with variants in other genes that the disease is modified; for example, co-inheritance of mutations in *Pkhd1* and *Pkd1* worsens the phenotype; this correlates in humans where a mutation in the other ADPKD gene (*PKD2*) also worsen the kidney manifestation [60,61]. It is also thought that the epigenetics and environmental factors, as well as genetic variants in other genes associated with PKD, could explain the inter-and intrafamilial variability.

6. ARPKD Proteins: Structure and Function

The protein product of *PKHD1* is fibrocystin/polyductin/FPC (Figure 2B) [34,48], a membrane protein with a long extracellular N-terminus, a single transmembrane domain and a short cytoplasmic C-terminus tail. The extracellular domain contains twelve TIG/IPT domains (Ig-like domains) that have been described in cell surface receptors [62]. In addition, three potential protein kinase A (PKA) phosphorylation sites were identified in the cytoplasmic tail that may be relevant for its function [63]. A *PKHD1* homologue was reported, *PKHD1L*, with an identity of 25% and similarity of 41.5%, which encodes fibrocystin-L, a receptor with inducible T lymphocyte expression, and has not been implicated in PKD [64]. The longest open reading frame (ORF) of FPC is predicted with a length of 4074 amino acids [65]. However, the *PKHD1* gene undergoes a complicated splicing pattern and can encode several additional gene products. In the same way, FPC exhibits a highly complex pattern of Notch-like proteolytic processing validated at the in vitro level [66] and in the vivo level using mouse models [67], which make the investigation of *PKHD1*/FPC particularly difficult.

FPC is a 440 kDa membrane-bound protein that is expressed mainly in the kidney (cortical and medullary ducts), the liver (intra- and extra-hepatic biliary ducts) and the pancreas (pancreatic ducts) [65,68,69]. Two alternative FPC products of ~230 and ~140 kDa were detected and, more importantly, the ~140 kDa product was found in cellular fractions of secreted FPC products [65]. At the subcellular level, FPC is expressed in the primary apical cilia [65,68,70] and the basal body area of cilia [69] in renal epithelial cells and cholangiocytes [71]. Furthermore, FPC is also expressed in the apical membrane and cytoplasm of collecting duct cells [65]. It is controversial whether ARPKD tissues lack FPC expression, some studies support this idea [48,70], but other evidence suggests otherwise [72], suggesting a temporal and spatial expression complexity of FPC splicing variants.

The structure of FPC suggests a possible function of the cell surface receptor, which interacts with extracellular ligand through the N-terminus or transduces intracellular signals to the nucleus through its C-terminus [73]. The cytoplasmic tail can translocate to the nucleus after full-length cleavage [66,74]. However, the intrinsic mechanism of the C-terminus remains unclear, as its deletion in mouse models did not result in renal or hepatic cystic phenotype, suggesting that it is not essential for cyst formation in ARPKD [67].

DZIP1L encodes the DAZ (Deleted in AZoospermia) interacting protein 1-like, a zinc-finger protein with several coiled-coil domains and one C2H2-type zinc finger domain near its N-terminus [37]. The zinc finger protein DZIP1L is involved in primary cilium formation [75], and Lu and colleagues suggest a possible function in the polycystins/PCs (the ADPKD proteins) trafficking [37]. The results highlighted the transition zone of cilia as a new possible vital point to study ARPKD pathogenesis [53].

7. Pathogenesis of ARPKD/Molecular Basis/Disease Mechanism

7.1. ARPKD Rodent Models: Lessons from Animal Models

To date, several animal models have been developed in which they closely resemble human ARPKD (Table 1). Early models of PKD resulted from spontaneous mutations in non-orthologous genes that mimic the recessive trait and phenotype of the disease [76]. The first mouse model reported was the congenital polycystic kidney mouse, or *cpk*, in 1985 [77]. The development and expression/penetrance of disease and the genetics in this model were extensively studied [76,78]. The *cpk* model results in a spontaneous mutation in the C57BL/6J (B6) strain, which corresponds with the *Cys1* gene [62]. Later, during the 1990s, other models appeared with spontaneous mutations in other loci, including the well-studied *pcy* mouse [79] that has a mutation in the locus for *Nphp3* [80]. Furthermore, BALB/c polycystic kidney (*bpk*) [81] and the juvenile cystic kidney model (*jck*) [82] were described and characterized, that had spontaneous mutations in *Bicc1* [83] and *Nek8* [84] respectively. This was followed by animal models designed by chemical induction, as the juvenile congenital polycystic kidney (*jcpk*), which was obtained using a chlorambucil mutagenesis program [85]. Interestingly, later studies showed that *bpk* and *jcpk* models refined the mutated loci in *Bicc1* gene [83,86]. Furthermore, by insertional mutagenesis, the Oak Ridge polycystic kidney or *orpk* mouse was uncovered from a large-scale insertional mutagenesis program [87,88].

In the 2000s, and with the discovery of *PKHD1* as the main ARPKD gene [47,48], the first ARPKD animal models appeared. The polycystic kidney rat or PCK rat was initially proposed as ADPKD model due to its slow progressive kidney and liver disease [89], but later was confirmed that the *Pkhd1* gene was disrupted in the PCK rat [48]. The first *Pkhd1*-based transgenic mouse model was $Pkhd1^{ex40}$ [90], and later, many others appeared, as well as a *Dzip1l*-based model (Table 1). Remarkably, the hepatic ARPKD-like phenotype was always present in all *Pkhd1*-based models, but the renal phenotype was often absent. Interestingly, pancreatic cysts were often present in these models, unlike in patients with ARPKD [6].

The key or main molecular mechanism of cystogenesis in ARPKD remains unknown. Nevertheless, animal models have allowed us to expand our understanding of the different stages of the disease, from cyst formation to cyst progression. Throughout this complex process, several altered molecular pathways such as fluid secretion, abnormal cellular proliferation (such as mTOR, RAS-RAF-ERK and AKT), cAMP pathway regulated by PKA kinase and AC6, alterations in extracellular matrix (ECM), among others have been described [91,92]. Therefore, the understanding of the pathophysiology of ARPKD has improved in recent decades thanks to the existence of a good variety of animal models. However, the key intrinsic molecular mechanism of cystogenesis in ARPKD remains unknown, leading to increased interest in understanding the mechanism of the disease and developing new therapeutic strategies. Next, we review the main molecular pathways characterized in ARPKD.

Table 1. List of current animal models of ARPKD, or that mimics it phenotype. Animal models were ordered from least to most recent, according to published data.

Model Name	Specie	Gene	Allele Type (Mutation Type)	Liver Phenotype	Kidney Phenotype	Other Phenotypes	Ref.
			Animal models that mimics ARPKD				
cpk ($Cys1^{cpk}$) *	Mouse	*Cys1*	Spontaneous	·Liver cysts ·Dilated bile duct ·Hepatic fibrosis	·Kidney cysts ·Enlarged kidney	·Pancreas cysts	[77,93–96]
pcy (DBA/2-pcy/pcy) ($Nphp3^{pcy}$) *	Mouse	*Nphp3*	Spontaneous	None	·Kidney cysts ·Enlarged kidney ·Renal fibrosis	·Intracranial aneurysm	[79,80,97]
bpk ($Bicc1^{jcpk\text{-}bpk}$) *	Mouse	*Bicc1*	Spontaneous	·Enlarged bile duct	·Kidney cysts ·Enlarged kidney	·Premature death ·Postnatal lethality	[81]
jck ($Nek8^{jck}$) *	Mouse	*Nek8*	Spontaneous	·None	·Kidney cysts ·Enlarged kidney	·Premature death	[82]
orpk ($Ift88^{Tg737Rpw}$) *	Mouse	*Ift88*	Transgenic (insertion)	·Abnormal bile duct morphology ·Liver fibrosis	·Kidney cysts ·Enlarged kidney	·Pancreas cysts ·Polydactyly	[87,88]
jcpk	Mouse	*Bicc1*	Chemical induction	·Dilated bile duct	·Kidney cysts ·Dilated renal tubules	·Dilated pancreatic ducts	[85]
			ARPKD models				
PCK	Rat	*Pkhd1*	Spontaneous (splicing mutation)	·Liver cysts ·Dilated bile duct ·Hepatic fibrosis	·Kidney cysts ·Enlarged kidney ·Renal fibrosis	·Pancreas cysts	[89,98]
Pkhd1ex40	Mouse	*Pkhd1*	Targeted (KO by insertion)	·Liver cysts ·Hepatic fibrosis	·None	·Portal hypertension	[90]
$Pkhd1^{del2/del2}$ ($Pkhd1^{tm1Cjwa}$) *	Mouse	*Pkhd1*	Targeted (KO by deletion)	·Liver cysts ·Dilated bile duct ·Hepatic fibrosis	·Kidney cysts ·Dilated renal tubules	·Pancreas cysts ·Pancreatic duct abnormalities	[99]
$Pkhd1^{del3-4}$ ($Pkhd1^{tm1.1Gg8}$) *	Mouse	*Pkhd1*	Targeted (KO by deletion)	·Liver cysts ·Hepatic fibrosis	·Kidney cysts ·Renal fibrosis	·Pancreas cysts ·Choledochal cyst ·Ascending cholangitis	[60]
$Pkhd1^{del4/del4}$ ($Pkhd1^{tm1Som}$)*	Mouse	*Pkhd1*	Targeted (KO by deletion)	·Liver cysts ·Biliary cysts ·Hepatic fibrosis	·None	·Pancreas cysts ·Pancreas fibrosis ·Enlarged spleen	[100]
$Pkhd1^{e15GFP\Delta16}$ ($Pkhd1^{tm1Gwu}$) *	Mouse	*Pkhd1*	Targeted (KO by deletion)	·Liver cysts ·Hepatic fibrosis	·Kidney cysts ·Dilated renal tubules ·Renal fibrosis	·Pancreas cysts ·Dilated pancreatic duct ·Gastrointestinal ulcer	[101]
$Pkhd1^{lacZ}$ ($Pkhd1^{tm1Sswi}$) *	Mouse	*Pkhd1*	Targeted (KO by deletion)	·Dilated bile duct ·Hepatic fibrosis	·Kidney cysts ·Dilated renal tubules ·Renal fibrosis	·Pancreas cysts	[102]
$Pkhd1^{LSL(-)}$ ($Pkhd1^{tm2Cjwa}$) *	Mouse	*Pkhd1*	Targeted (KO by insertion)	·Liver cysts ·Hepatic fibrosis	·Kidney cysts ·Dilated renal tubules	·Unknown	[103]

Table 1. *Cont.*

Model Name	Specie	Gene	Allele Type (Mutation Type)	Liver Phenotype	Kidney Phenotype	Other Phenotypes	Ref.
Pkhd1$^{\Delta 67}$	Mouse	*Pkhd1*	Targeted (KO by deletion)	·None	·None	·None	[67]
Dzip1l$^{wpy/wpy}$ (*Dzip1l*^{warpy}) *	Mouse	*Dzip1l*	Targeted (KO by single point mutation)	·Abnormal bile duct morphology	·Kidney cysts ·Dilated renal tubules	·Polydactyly ·Abnormal eye morphology ·Cleft upper lip ·Cleft palate	[37]
Pkhd1^{C642*} (*Pkhd1*^{em1Mrug}) *	Mouse	*Pkhd1*	Targeted (KO by deletion)	·Liver cysts ·Biliary cysts ·Hepatic fibrosis	·Dilated renal tubules ·Proximal tubule ectasia	·Unknown	[67]

* Model name according to MGI (Mouse Genome Informatics) [104]. Ref. = reference. KO = knockout.

7.2. Abnormalities of EGFR-Axis Expression and Fluid Secretion

The first evidence that the epidermal growth factor receptor (EGFR) axis was altered in PKD was in 1992, by demonstrating that cells from primary cultures of PKD patients increased cyst expansion [105]. Subsequently, in primary cells isolated from ADPKD patients, epidermal growth factor (EGF) stimulated cyst formation [106]. In ARPKD, the first data were obtained from *cpk* mouse model renal extracts, which showed upregulation of EGF expression [107]. Progressively, other evidence has shown a significant role for EGFR in vitro [108] and murine models [88,109,110], and patients with ARPKD [111,112], where EGFR upregulation was located on the surface of the cystic epithelium. In the same way, abnormal expression of EGF [113,114] and transforming growth factor-alpha (TGFα) [115] have been demonstrated in ARPKD, and several members of EGFR family of receptors (EGFR1, ErbB2, and ErbB4) were found overexpressed in ARPKD rodent models [72,116] (Figure 3A). This overexpression includes increased mRNA, protein, and receptor activity or phosphorylation [92]. Furthermore, evidence from animal models suggests similar abnormalities in hepatic cystogenesis of the EGFR axis [117].

EGFR signaling has relevance in cystogenesis, a correlation between upregulation of EGFR tyrosine kinase activity and cystic growth. A modified *orpk* model with a point mutation that decreased EGF tyrosine kinase activity (wa2 mouse) [118], showed a significant decrease in cyst formation and improvement of renal function [110]. Furthermore, pharmacological inhibition with EGFR-specific tyrosine kinase inhibitors caused a decreased EGFR activity leading to a significant reduction of cyst progression [119–121]. Controversially, the EGFR tyrosine kinase inhibitor in the PCK rat did not slow the progression of renal cysts [122].

Epithelial secretion is a key physiopathology component of cyst formation. Reduced sodium uptake in ARPKD net fluid secretion has been proposed as being related to decreased EGF in the alpha subunit of the epithelial Na channel [123,124]. Nevertheless, conflicting data have been published on this topic [116]. In cells derived from cysts from patients with ARPKD, it was suggested that sodium absorption was mediated by the epithelial sodium channel (ENaC) [125]. Furthermore, this mechanism has been proposed as a regulator of hypertension in ARPKD [125–127].

On the other hand, in ADPKD, the data show that Cl^- secretion occurs through the cystic fibrosis transmembrane conductance regulator (CFTR) [128,129]. However, in ARPKD this mechanism has no apparent relevance. In the *bpk* murine model, the absence of CFTR did not slow the progression of the renal and hepatic cyst [130]. These data suggest that the mechanisms of Cl^- and fluid secretion are different in ADPKD and ARPKD.

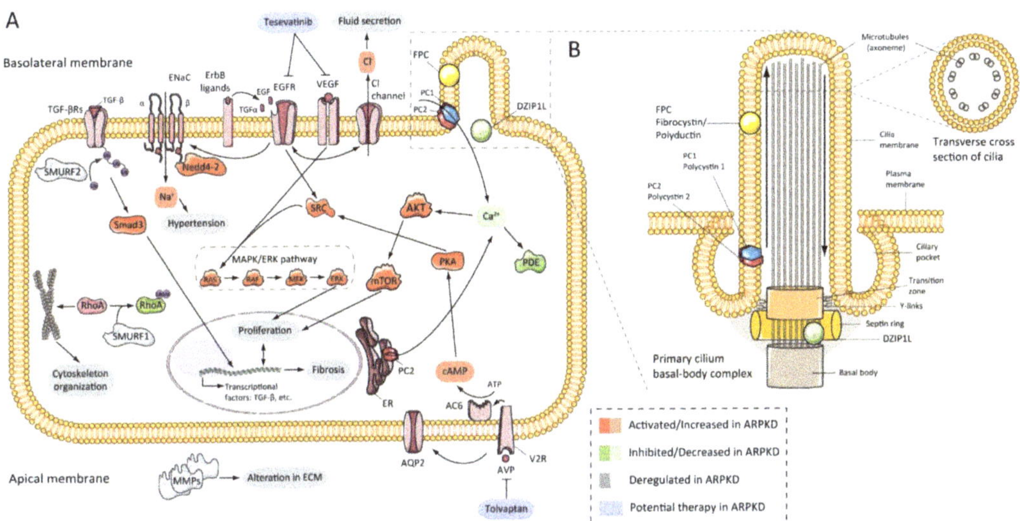

Figure 3. ARPKD molecular pathogenesis: (**A**) diagram representing the proposed up-, down-, or deregulated pathways in ARPKD renal epithelial cell and proposed potential therapies; (**B**) cartoon representing the localization of ARPKD protein in the cilium of a renal epithelial cell. FPC is located in the primary apical cilia and the basal body area of the cilia, whereas DZIP1L is located in the transition zone of the basal body of the cilium. (FPC—fibrocystin/polyductin; DZIP1L—DAZ interacting zinc finger protein 1 like; PC1—polycystin 1; PC2—polycystin 2; TGF—transforming growth factor; ENaC—epithelial sodium channels; Na^+—sodium cation; EGFR—epidermal growth factor receptor; VEGF—vascular endothelial growth factor; Cl^-—chlorine anion; SMURF—SMAD specific E3 ubiquitin-protein ligase; Nedd4-2—E3 ubiquitin-protein ligase Nedd4-2; SRC—proto-oncogene tyrosine-protein kinase Src; AKT—RAC-alpha serine/threonine-protein kinase; Ca^+—calcium cation; PDE—phosphodiesterase; mTOR—mammalian target of rapamycin; MAPK—mitogen-activated protein kinases; RAF—rapidly accelerated fibrosarcoma; MEK—mitogen-activated protein kinase kinase; ERK—extracellular-signal-regulated kinase; Rhoa—Ras homolog family member A; ER—endoplasmic reticulum; PKA—protein kinase A; cAMP—cyclic adenosine monophosphate (cyclic AMP); ATP—adenosine triphosphate; AC6—adenylate cyclase type 6; AQP2—aquaporin 2; V2R—vasopressin receptor 2; AVP—arginine vasopressin; MMPs—matrix metalloproteinases).

7.3. cAMP and Proliferation

Several studies have shown that adenylyl cyclase adenosine 3′,5′-cyclic monophosphate (cAMP) pathway stimulates cell proliferation in the renal epithelium of ARPKD and ADPKD. Production of cAMP is aberrant in the cyst epithelium, resulting in a large amount of this nucleotide in the cyst fluid [122,131–133]. cAMP activates the B-Raf, MEK, and ERK pathways in the cyst epithelium of the kidneys with ADPKD [134–136], and ADPKD and ARPKD cells in culture [133]. In the same way, these results were complemented with data showing upregulation at the protein level of MAPK and AKT/mTOR pathways in several rodent models of ARPKD [137–141]. These facts correlated with the reduction of intracellular Ca^{2+} and the phosphorylation of the SCR protein [142–144]. In particular, blocked intracellular Ca^{2+} elevated AKT and proliferative activity in ARPKD cells in culture [143]. This study opened the opportunity to use the level of intracellular calcium restoration as a therapeutic approach in PKD.

Dysregulation of calcium in PKD causes upregulation of the vasopressin V2 receptor [136,138], which activates the cAMP/PKA cascade [145,146]. In a preclinical trial, the V2 receptor antagonist has demonstrated its efficacy in ARPKD rat model, reducing the renal cAMP levels and improving cystic renal disease [137] (Figure 3A). This fact has been further studied, reviewed, and understood in ADPKD, at preclinical [147,148] and clinical levels [149], to the point that Tolvaptan (an vasopressin V2 receptor inhibitor) was approved for use in human patients [150] and an ongoing trial for children with ADPKD (Tolvaptan (NCT02964273)) [151]. ARPKD-related clinical trials will be reviewed later.

7.4. Other Pathways Involved in ARPKD Physiopathology

Other pathogenic features have been identified in ARPKD, as well as alterations in extracellular matrix (ECM) and metalloproteinase expression (MMPs) [152,153], upregulation of vascular endothelial growth factor (VEGF) and hypoxia-inducible factor-1 alpha (HIF-1α) in *Pkhd1* deficient cells [139], upregulation of peroxisome-proliferator-activated receptor-γ (PPAR-γ) in animal models [154,155], or metabolic alterations [156]. In a large and interesting study, Kaimori and colleagues published information about novel functional relationships between FPC and members of the C2-WWW-HECT domain E3 family of ubiquitin ligases. The authors localized FPC in vesicles where Ndfip2 was also present, a ubiquitin ligase interacting protein implicated in trafficking and regulating the Nedd4-2 ubiquitin ligase family and SMURF1 and SMURF2. These data may explain different universal phenotypes in ARPKD and renal and hepatic fibrosis through TGF-β signaling pathways, hypertension through to ENaC mediated sodium reabsorption, and cystogenesis through to RhoA ubiquitination and cytoskeleton organization [127] (Figure 3A). In other studies, tubular morphogenesis in PKD was associated with an abnormality planar cell polarity (PCP) [157]. However, later studies have shown contrary results [158].

7.5. Role of Cilia

Figure 3 shows the ARPKD proteins (zinc finger protein DZIP1L and FPC) are located in the cilia. The cilia are long and microtubular structures emanating from the surface of mammalian cells. The axoneme of primary cilia contains nine peripheral bundles of microtubules (9 + 0 pattern). Pathologies related to a loss of proper cilia function are called ciliopathies, including ARPKD [159]. PC2, also called TRPP2, is a member of the transient receptor channel (TRP) family and is a calcium-permeable non-selective channel [160]. PC2 and PC1 form a receptor-channel complex, that is involved in the calcium pathway and cilia response [161–163] (Figure 3B). FPC has been shown to interact with PC2 in primary cilia and regulates PC2 channel activity [101,164,165]. In addition, it has been reported that the C-terminus of FPC physically interacts with the N-terminus of PC2 in vivo and in vitro, and that *Pkhd1*-deficient cells exhibit dysregulation of PC2 channel activity [101]. However, Wang and colleagues found no differences in PC2 levels in cells with reduced FPC levels [165]. Other data using a novel *Pkhd1* mouse model have shown that deletion of the last exon of *Pkhd1*, the PC2 binding site, and the nuclear localization signal, had no apparent pathologic effects in mice [67]. In addition, the researchers were unable to co-precipitate FPC-PC2 in kidney samples from the transgenic mouse model. These results suggest that the PC2 binding domain of FPC is not essential for the fibrocystin function [67,166].

We have described a genetic interaction between *Pkd1* and *Pkhd1*, linking ADPKD and ARPKD [60]. Consequently, additional data has reported this genetic interaction between *Pkd1* and *Pkhd1* in other rodent models, describing mild cystic disease phenotypes of *Pkd1* and/or *Pkhd1* enhanced their severity in combination [38,167]. These studies and others have expanded the relationship between FPC and PC1. Using in-vitro models, the loss of FPC did not affect the biogenesis and location of PC1, suggesting that the genetic interaction between *Pkd1* and *Pkhd1* is indirect [30,167].

The results of these studies appear consistent with the idea that the PKD proteins form a functional complex in cilia with common downstream signaling pathways [168]. Interestingly, cilia loss suppresses renal cyst growth in murine models of ADPKD and autosomal dominant polycystic liver disease (ADPLD) murine models [169]. However, in a recent study, Gallagher and Somlo reported that this loss of cilia does not slow the progression of liver disease in ARPKD [170]. These data suggest that ADPKD and ARPKD, at least in the hepatic cystic phenotype, do not share a common cilia-related pathway.

On the other hand, the DAZ interacting protein 1-like protein (DZIP1L), based on several cell culture studies, zebrafish, and mice, localizes in centrioles and in the ciliary transition zone the primary cilia [37]. Lu and colleagues have demonstrated the interaction between DZIP1L and septin 2 (SEPT2), a protein involved in the maintenance of the periciliary diffusion barrier at the transition zone [171] and the co-localization with tectonic 1 (TCTN1), a ciliary transition zone protein. In *DZIP1L* mutant cells, the transport of polycystin-1 and -2 from the basal body to the axoneme of the cilia was altered; both were retained in the basal body when their normal distribution was in the ciliary axoneme. However, *DZIP1L* deficiency did not alter the localization or expression of FPC [37]. These findings suggest a role of *DZIP1L* in the trafficking of polycystins and new evidence that links ARPKD with ADPKD.

8. Clinical Trials

As we have noted, the central or key mechanism of cystogenesis in PKD remains unclear. There is evidence of several pathways involved in the pathogenesis of PKD from cellular and animal studies. These facts have allowed several drugs to reach the clinical phase (Table 2).

Table 2. Currently active or complete clinical trials for ARPKD.

Identifier	Intervention	Study Design and Characteristics	Study Description	Sponsor
NCT04782258	Tolvaptan	·Study type: interventional ·Primary purpose: treatment ·Period: April 2021–June 2025 (estimated) ·Patients: 20 (estimated, not yet recruitment) ·Allocation: non-randomized ·Intervention model: parallel assignment ·Masking: none (open label)	·The primary objective of this phase 3 trial is to evaluate the safety of Tolvaptan (OPC-41061) in infants and children, 8 days to less than 18 years old of age, with ARPKD. ·Participants in this study will be assigned to Tolvaptan for 18 months and closely monitored over the course of the study.	Otsuka Pharmaceutical Development & Commercialization, Inc. Princeton, New Jersey, USA.
NCT04786574	Tolvaptan	·Study type: interventional ·Primary purpose: treatment ·Period: April 2021–July 2025 (estimated) ·Patients: 20 (estimated, not yet recruitment) ·Allocation: N/A ·Intervention model: single group assignment ·Masking: none (open label)	·In this Phase 3 trial, the primary objective is to evaluate safety, tolerability, and efficacy of Tolvaptan (OPC-41061) in pediatric subjects, 28 days to less than 12 weeks of age, with ARPKD. ·Participants in this trial will be assigned to Tolvaptan for 24 months and closely monitored over the course of the study.	Otsuka Pharmaceutical Development & Commercialization, Inc. Princeton, New Jersey, USA.
NCT03096080	Tesevatinib	·Study type: interventional ·Primary purpose: treatment ·Period: March 2017–October 2019 (completed) ·Patients: 10 ·Allocation: Non-randomized ·Intervention model: sequential assignment ·Masking: none (open label)	·This trial in Phase 1 evaluates safety and tolerability of a single ascending dose of a Tesevatinib (KD019, XL647) liquid formulation administered to pediatric subjects (child with age 5–12 years) with ARPKD. ·To determine safety of the Tesevatinib liquid formulation in pediatric subjects with ARPKD, all participants receive active study drug on Day 1 of the study enrollment.	Kadmon Corporation, LLC Philadelphia, Pennsylvania, USA. Milwaukee, Wisconsin, USA.

Table 2. *Cont.*

Identifier	Intervention	Study Design and Characteristics	Study Description	Sponsor
NCT01401998	Observational	·Study type: observational ·Period: July 2011–December 2022 (recruitment) ·Patients: 200 (estimated) ·Observational model: cohort ·Time perspective: retrospective assignment	·This study captures clinical and genetic information of ARPKD patients to expand the knowledge of disease. ·The primary goal of this trial is create a clinical and mutational databases including clinical information and identifying genetic mutations from all patients enrolled in the study. ·Mutational database will be useful to facilitate genetic research as genotype-phenotype correlations, new disease gene studies, or modifier gene studies. ·Create a tissue resource with human tissue from both affected and controls individuals.	Lisa M. Guay-Woodford (Collaborator: National Institute of Diabetes and Digestive and Kidney Diseases (NIDDK)). Washington, District of Columbia, USA.
NCT00068224	Observational	·Study type: observational ·Period: September 2003–February 2021 (completed) ·Patients: 374 ·Observational model: cohort ·Time perspective: prospective	·This study evaluated patients with ciliopathies, including ARPKD. · The goal of the study is to better understand the medical complications of these disorders and identify characteristics that can help in the design of new treatments.	National Human Genome Research Institute (NHGRI). Bethesda, Maryland, USA.

Status according to https://clinicaltrials.gov/, accessed on 6 April 2021.

Based on the results of phase 3 and 4 clinical trials [149,172,173], two clinical trials are currently underway using the pharmacological intervention of Tolvaptan. The primary objective of these trials is to evaluate the safety of Tolvaptan in infants (8 days or less) and children (less than 18 years) (NCT04782258) and in pediatric patients (from 28 days to 12 weeks of age) (NCT04786574). On the other hand, PKD exhibits an abnormal c-Src activity, links the cAMP and EGFR molecular ways [174], and its inhibition ameliorates renal cystogenesis [144]. These data led Sweeney et al. to study the effect of a multi-kinase inhibitor of EGFR axis, c-Src and VEGFR, called tesevatinib (TSV), as a possible therapy in preclinical studies for ARPKD obtaining favorable results [175]. The positive and promising results led to the approval of phase I and II of TSV clinical trials for ARPKD (NCT03096080). Finally, two observational trials are being carried out to expand the knowledge of the disease (genotype-phenotype correlations, clinical aspects) and to create more precise mutational and clinical databases (NCT01401998 and NCT00068224).

9. Conclusions

The ARPKD field has experienced significant progress in the areas of genetics, diagnostics, and molecular biology. First, the identification of the *PKHD1* gene, for several years considered the only ARPKD gene, and more recently the *DZIP1L* gene, and its implementation in genetic diagnosis based on next-generation sequencing (NGS). Second, advances in the understanding of the function and localization of the fibrocystin/polyductin protein. Finally, different studies, especially in animal models, begin to elucidate the pathways involved in the pathogenesis of the disease identifying possible therapeutic approaches.

However, many unanswered questions should be answered in the future. The new finding of *DZIP1L* as a second genetic locus for ARPKD leaves open the possibility of the appearance of new genes that cause ARPKD in this regard it is necessary to apply genetically unresolved whole exome sequencing (WES) families (GUR) with phenotype ARPKD. The exact function of the FPC remains unknown, as well as a correct characterization of all its isoforms. Importantly, the key factor(s) driving cyst formation in PKD are not clear.

For these reasons, the pathogenicity of ARPKD is poorly understood and, even today, there is no approved therapy for existing replacement therapy. To explain the genetic and cellular basis of ARPKD, research on the subject emerges as the only way out of this situation.

Author Contributions: Writing—original draft preparation, A.C. and M.V.-G.; writing—review and editing, M.A.G.-G. A.C. and M.V.-G. contributed equally in this work. All authors have read and agreed to the published version of the manuscript.

Funding: This research was funded by Instituto de Salud Carlos III under FIS/FEDER funds PI15/01467 and PI18/00378 (to MA Garcia-Gonzalez) and by Xunta de Galicia award IN607B 2016/020 (to MA Garcia-Gonzalez).

Institutional Review Board Statement: Not applicable.

Informed Consent Statement: Not applicable.

Acknowledgments: The authors acknowledge Siddig Khallafalla and Anna-Rachel Gallagher for the useful English orthographic review.

Conflicts of Interest: The authors declare no conflict of interest.

References

1. Guay-Woodford, L.M.; Muecher, G.; Hopkins, S.D.; Avner, E.D.; Germino, G.G.; Guillot, A.P.; Herrin, J.; Holleman, R.; Irons, D.A.; Primack, W.; et al. The severe perinatal form of autosomal recessive polycystic kidney disease maps to chromosome 6p21.1-p12: Implications for genetic counseling. *Am. J. Hum. Genet.* **1995**, *56*, 1101–1107. [PubMed]
2. Kaariainen, H. Polycystic kidney disease in children: A genetic and epidemiological study of 82 Finnish patients. *J. Med. Genet.* **1987**, *24*, 474–481. [CrossRef] [PubMed]
3. Alzarka, B.; Morizono, H.; Bollman, J.W.; Kim, D.; Guay-Woodford, L.M. Design and implementation of the hepatorenal fibrocystic disease core center clinical database: A centralized resource for characterizing autosomal recessive polycystic kidney disease and other hepatorenal fibrocystic diseases. *Front. Pediatr.* **2017**, *5*. [CrossRef]
4. Bergmann, C. Genetics of Autosomal Recessive Polycystic Kidney Disease and Its Differential Diagnoses. *Front. Pediatr.* **2018**, *5*, 1–13. [CrossRef]
5. Cordido, A.; Besada-Cerecedo, L.; García-González, M.A. The Genetic and Cellular Basis of Autosomal Dominant Polycystic Kidney Disease—A Primer for Clinicians. *Front. Pediatr.* **2017**, *5*, 279. [CrossRef]
6. Bergmann, C.; Guay-Woodford, L.M.; Harris, P.C.; Horie, S.; Peters, D.J.M.; Torres, V.E. Polycystic kidney disease. *Nat. Rev. Dis. Prim.* **2018**, *4*, 50. [CrossRef] [PubMed]
7. Bergmann, C.; Senderek, J.; Schneider, F.; Dornia, C.; Küpper, F.; Eggermann, T.; Rudnik-Schöneborn, S.; Kirfel, J.; Moser, M.; Büttner, R.; et al. PKHD1 Mutations in Families Requesting Prenatal Diagnosis for Autosomal Recessive Polycystic Kidney Disease (ARPKD). *Hum. Mutat.* **2004**, *23*, 487–495. [CrossRef]
8. Guay-Woodford, L.M.; Desmond, R.A. Autosomal recessive polycystic kidney disease: The clinical experience in North America. *Pediatrics* **2003**, *111*, 1072–1080. [CrossRef]
9. Adeva, M.; El-Youssef, M.; Rossetti, S.; Kamath, P.S.; Kubly, V.; Consugar, M.B.; Milliner, D.M.; King, B.F.; Torres, V.E.; Harris, P.C. Clinical and molecular characterization defines a broadened spectrum of autosomal recessive polycystic kidney disease (ARPKD). *Medicine* **2006**, *85*, 1–21. [CrossRef]
10. Gunay-Aygun, M.; Avner, E.D.; Bacallao, R.L.; Choyke, P.L.; Flynn, J.T.; Germino, G.G.; Guay-Woodford, L.; Harris, P.; Heller, T.; Ingelfinger, J.; et al. Autosomal recessive polycystic kidney disease and congenital hepatic fibrosis: Summary statement of a First National Institutes of Health/Office of Rare Diseases conference. *J. Pediatr.* **2006**, *149*, 159–164. [CrossRef]
11. Liebau, M.C. Early clinical management of autosomal recessive polycystic kidney disease. *Pediatr. Nephrol.* **2021**. [CrossRef] [PubMed]
12. Bergmann, C.; Senderek, J.; Windelen, E.; Küpper, F.; Middeldorf, I.; Schneider, F.; Dornia, C.; Rudnik-Schöneborn, S.; Konrad, M.; Schmitt, C.P.; et al. Clinical consequences of PKHD1 mutations in 164 patients with autosomal-recessive polycystic kidney disease (ARPKD). *Kidney Int.* **2005**, *67*, 829–848. [CrossRef]
13. Erger, F.; Brüchle, N.O.; Gembruch, U.; Zerres, K. Prenatal ultrasound, genotype, and outcome in a large cohort of prenatally affected patients with autosomal-recessive polycystic kidney disease and other hereditary cystic kidney diseases. *Arch. Gynecol. Obstet.* **2017**, *295*, 897–906. [CrossRef]
14. Burgmaier, K.; Kilian, S.; Bammens, B.; Benzing, T.; Billing, H.; Büscher, A.; Galiano, M.; Grundmann, F.; Klaus, G.; Mekahli, D.; et al. Clinical courses and complications of young adults with Autosomal Recessive Polycystic Kidney Disease (ARPKD). *Sci. Rep.* **2019**, *9*. [CrossRef] [PubMed]
15. Belin, S.; Delco, C.; Parvex, P.; Hanquinet, S.; Fokstuen, S.; De Tejada, B.M.; Eperon, I. Management of delivery of a fetus with autosomal recessive polycystic kidney disease: A case report of abdominal dystocia and review of the literature. *J. Med. Case Rep.* **2019**, *13*. [CrossRef]

16. Fonck, C.; Chauveau, D.; Gagnadoux, M.F.; Pirson, Y.; Grünfeld, J.P. Autosomal recessive polycystic kidney disease. *Nephrol. Dial. Transplant.* **2001**, *16*, 1648–1652. [CrossRef] [PubMed]
17. Rubio San Simón, A.; Carbayo Jiménez, T.; Vara Martín, J.; Alonso Díaz, C.; Espino Hernández, M. Autosomal recessive polycystic kidney disease in the 21st century: Long-term follow up and outcomes. *An. Pediatr.* **2018**, *91*, 120–122. [CrossRef] [PubMed]
18. Burgmaier, K.; Brandt, J.; Shroff, R.; Witters, P.; Weber, L.T.; Dötsch, J.; Schaefer, F.; Mekahli, D.; Liebau, M.C. Gastrostomy tube insertion in pediatric patients with autosomal recessive polycystic kidney disease (ARPKD): Current practice. *Front. Pediatr.* **2018**, *6*. [CrossRef]
19. Cole, B.R.; Conley, S.B.; Stapleton, F.B. Polycystic kidney disease in the first year of life. *J. Pediatr.* **1987**, *111*, 693–699. [CrossRef]
20. Avni, F.E.; Guissard, G.; Hall, M.; Janssen, F.; DeMaertelaer, V.; Rypens, F. Hereditary polycystic kidney diseases in children: Changing sonographic patterns through childhood. *Pediatr. Radiol.* **2002**, *32*, 169–174. [CrossRef]
21. Bergmann, C. Early and Severe Polycystic Kidney Disease and Related Ciliopathies: An Emerging Field of Interest. *Nephron* **2019**, *141*, 50–60. [CrossRef]
22. Rivero, P.C.; Ecuador, U.R.D.C.L.; Andrade, R.E.C.; Baquero, S.M.; Espirel, V.M.; Ecuador, U.T.D.N. Polycystic kidney disease. *Annu. Rev. Med.* **2009**, *60*, 321–337. [CrossRef]
23. Gunay-Aygun, M.; Font-Montgomery, E.; Lukose, L.; Tuchman Gerstein, M.; Piwnica-Worms, K.; Choyke, P.; Daryanani, K.T.; Turkbey, B.; Fischer, R.; Bernardini, I.; et al. Characteristics of congenital hepatic fibrosis in a large cohort of patients with autosomal recessive polycystic kidney disease. *Gastroenterology* **2013**, *144*. [CrossRef]
24. Turkbey, B.; Ocak, I.; Daryanani, K.; Font-Montgomery, E.; Lukose, L.; Bryant, J.; Tuchman, M.; Mohan, P.; Heller, T.; Gahl, W.A.; et al. Autosomal recessive polycystic kidney disease and congenital hepatic fibrosis. *Pediatr. Radiol.* **2009**, *39*, 100–111. [CrossRef]
25. Hartung, E.A.; Wen, J.; Poznick, L.; Furth, S.L.; Darge, K. Ultrasound Elastography to Quantify Liver Disease Severity in Autosomal Recessive Polycystic Kidney Disease. *J. Pediatr.* **2019**, *209*, 107–115.e5. [CrossRef] [PubMed]
26. Guay-Woodford, L.M.; Bissler, J.J.; Braun, M.C.; Bockenhauer, D.; Cadnapaphornchai, M.A.; Dell, K.M.; Kerecuk, L.; Liebau, M.C.; Alonso-Peclet, M.H.; Shneider, B.; et al. Consensus expert recommendations for the diagnosis and management of autosomal recessive polycystic kidney disease: Report of an international conference. *J. Pediatr.* **2014**, *165*, 611–617. [CrossRef]
27. Gately, R.; Lock, G.; Patel, C.; Clouston, J.; Hawley, C.; Mallett, A. Multiple Cerebral Aneurysms in an Adult With Autosomal Recessive Polycystic Kidney Disease. *Kidney Int. Rep.* **2021**, *6*, 219–223. [CrossRef]
28. Perez, J.L.; McDowell, M.M.; Zussman, B.; Jadhav, A.P.; Miyashita, Y.; McKiernan, P.; Greene, S. Ruptured intracranial aneurysm in a patient with autosomal recessive polycystic kidney disease. *J. Neurosurg. Pediatr.* **2019**, *23*, 75–79. [CrossRef]
29. Blyth, H.; Ockenden, B.G. Polycystic disease of kidney and liver presenting in childhood. *J. Med. Genet.* **1971**, *8*, 257–284. [CrossRef] [PubMed]
30. Besse, W.; Dong, K.; Choi, J.; Punia, S.; Fedeles, S.V.; Choi, M.; Gallagher, A.-R.R.; Huang, E.B.; Gulati, A.; Knight, J.; et al. Isolated polycystic liver disease genes define effectors of polycystin-1 function. *J. Clin. Investig.* **2017**. [CrossRef] [PubMed]
31. Gunay-aygun, M.; Turkbey, B.I.; Bryant, J.; Daryanani, K.T.; Tuchman, M.; Piwnica-worms, K.; Choyke, P.; Heller, T.; Gahl, W.A. Hepatorenal findings in obligate heterozygotes for autosomal recessive polycystic kidney disease. *Mol. Genet. Metab.* **2011**, *104*, 677–681. [CrossRef]
32. Guay-Woodford, L.M. Autosomal recessive polycystic kidney disease: The prototype of the hepato-renal fibrocystic diseases. *J. Pediatr. Genet.* **2014**, *3*, 89–101. [CrossRef]
33. Raina, R.; Chakraborty, R.; Sethi, S.K.; Kumar, D.; Gibson, K.; Bergmann, C. Diagnosis and Management of Renal Cystic Disease of the Newborn: Core Curriculum 2021. *Am. J. Kidney Dis.* **2021**. [CrossRef]
34. Onuchic, L.F.; Furu, L.; Nagasawa, Y.; Hou, X.; Eggermann, T.; Ren, Z.; Bergmann, C.; Senderek, J.; Esquivel, E.; Zeltner, R.; et al. PKHD1, the polycystic kidney and hepatic disease 1 gene, encodes a novel large protein containing multiple immunoglobulin-like plexin-transcription-factor domains and parallel beta-helix 1 repeats. *Am. J. Hum. Genet.* **2002**, *70*, 1305–1317. [CrossRef]
35. Obeidova, L.; Seeman, T.; Fencl, F.; Blahova, K.; Hojny, J.; Elisakova, V.; Reiterova, J.; Stekrova, J. Results of targeted next-generation sequencing in children with cystic kidney diseases often change the clinical diagnosis. *PLoS ONE* **2020**, *15*. [CrossRef]
36. Bergmann, C. ARPKD and early manifestations of ADPKD: The original polycystic kidney disease and phenocopies. *Pediatr. Nephrol.* **2015**, *30*, 15–30. [CrossRef]
37. Lu, H.; Galeano, M.C.R.; Ott, E.; Kaeslin, G.; Kausalya, P.J.; Kramer, C.; Ortiz-Brüchle, N.; Hilger, N.; Metzis, V.; Hiersche, M.; et al. Mutations in DZIP1L, which encodes a ciliary-transition-zone protein, cause autosomal recessive polycystic kidney disease. *Nat. Genet.* **2017**, *49*, 1025–1034. [CrossRef]
38. Fedeles, S.V.; Tian, X.; Gallagher, A.R.; Mitobe, M.; Nishio, S.; Lee, S.H.; Cai, Y.; Geng, L.; Crews, C.M.; Somlo, S. A genetic interaction network of five genes for human polycystic kidney and liver diseases defines polycystin-1 as the central determinant of cyst formation. *Nat. Genet.* **2011**, *43*, 639–647. [CrossRef] [PubMed]
39. Torres, V.E.; Harris, P.C.; Pirson, Y. Autosomal dominant polycystic kidney disease. *Lancet* **2007**, *369*, 1287–1301. [CrossRef]
40. Schepis, C. The tuberous sclerosis complex. *Dermatol. Cryosurg. Cryother.* **2016**, 615–617. [CrossRef]
41. Shepherd, C.W.; Gomez, M.R.; Lie, J.; Crowson, C.S. Causes of Death in Patients With Tuberous Sclerosis. *Mayo Clin. Proc.* **1991**, *66*, 792–796. [CrossRef]

42. Decramer, S.; Parant, O.; Beaufils, S.; Clauin, S.; Guillou, C.; Kessler, S.; Aziza, J.; Bandin, F.; Schanstra, J.P.; Bellanné-Chantelot, C. Anomalies of the TCF2 Gene Are the Main Cause of Fetal Bilateral Hyperechogenic Kidneys. *J. Am. Soc. Nephrol.* **2007**, *18*, 923–933. [CrossRef]
43. Verhave, J.C.; Bech, A.P.; Wetzels, J.F.M.; Nijenhuis, T. Hepatocyte nuclear factor 1β-associated kidney disease: More than renal cysts and diabetes. *J. Am. Soc. Nephrol.* **2016**, *27*, 345–353. [CrossRef]
44. Hoff, S.; Halbritter, J.; Epting, D.; Frank, V.; Nguyen, T.M.T.; Van Reeuwijk, J.; Boehlke, C.; Schell, C.; Yasunaga, T.; Helmstädter, M.; et al. ANKS6 is a central component of a nephronophthisis module linking NEK8 to INVS and NPHP3. *Nat. Genet.* **2013**, *45*, 951–956. [CrossRef]
45. Bergmann, C.; Fliegauf, M.; Brüchle, N.O.; Frank, V.; Olbrich, H.; Kirschner, J.; Schermer, B.; Schmedding, I.; Kispert, A.; Kränzlin, B.; et al. Loss of Nephrocystin-3 Function Can Cause Embryonic Lethality, Meckel-Gruber-like Syndrome, Situs Inversus, and Renal-Hepatic-Pancreatic Dysplasia. *Am. J. Hum. Genet.* **2008**, *82*, 959–970. [CrossRef] [PubMed]
46. Huber, T.B.; Walz, G.; Kuehn, E.W. MTOR and rapamycin in the kidney: Signaling and therapeutic implications beyond immunosuppression. *Kidney Int.* **2011**, *79*, 502–511. [CrossRef]
47. Zerres, K.; Mücher, G.; Becker, J.; Steinkamm, C.; Rudnik-Schöneborn, S.; Heikkilä, P.; Rapola, J.; Salonen, R.; Germino, G.G.; Onuchic, L.; et al. Prenatal diagnosis of autosomal recessive polycystic kidney disease (ARPKD): Molecular genetics, clinical experience, and fetal morphology. *Am. J. Med. Genet.* **1998**, *76*, 137–144. [CrossRef]
48. Ward, C.J.; Hogan, M.C.; Rossetti, S.; Walker, D.; Sneddon, T.; Wang, X.; Kubly, V.; Cunningham, J.M.; Bacallao, R.; Ishibashi, M.; et al. The gene mutated in autosomal recessive polycystic kidney disease encodes a large, receptor-like protein. *Nat. Genet.* **2002**, *30*, 259–269. [CrossRef]
49. Zerres, K.; Rudnik-Schöneborn, S.; Deget, F.; Holtkamp, U.; Brodehl, J.; Geisert, J.; Schärer, K. Autosomal recessive polycystic kidney disease in 115 children: Clinical presentation, course and influence of gender. Arbeitsgemeinschaft für Pädiatrische, Nephrologie. *Acta Paediatr.* **1996**, *85*, 437–445. [CrossRef]
50. Deget, F.; Rudnik-Schöneborn, S.; Zerres, K. Course of autosomal recessive polycystic kidney disease (ARPKD) in siblings: A clinical comparison of 20 sibships. *Clin. Genet.* **1995**, *47*, 248–253. [CrossRef]
51. Consugar, M.B.; Anderson, S.A.; Rossetti, S.; Pankratz, V.S.; Ward, C.J.; Torra, R.; Coto, E.; El-youssef, M.; Kantarci, S.; Utsch, B.; et al. Haplotype Analysis Improves Molecular Diagnostics of Autosomal Recessive Polycystic Kidney Disease. *Am. J. Kidney Dis.* **2005**, *45*, 77–87. [CrossRef]
52. Perugorria, M.J.; Banales, J.M. Genetics: Novel causative genes for polycystic liver disease. *Nat. Rev. Gastroenterol. Hepatol.* **2017**, *14*, 391–392. [CrossRef]
53. Hartung, E.A.; Guay-Woodford, L.M. Polycystic kidney disease: DZIP1L defines a new functional zip code for autosomal recessive PKD. *Nat. Rev. Nephrol.* **2017**, *13*, 519–520. [CrossRef]
54. Bergmann, C.; Senderek, J.; Sedlacek, B.; Pegiazoglou, I.; Puglia, P.; Eggermann, T.; Rudnik-Schöneborn, S.; Furu, L.; Onuchic, L.F.; De Baca, M.; et al. Spectrum of mutations in the gene for autosomal recessive polycystic kidney disease (ARPKD/PKHD1). *J. Am. Soc. Nephrol.* **2003**, *14*, 76–89. [CrossRef] [PubMed]
55. Benz, E.G.; Hartung, E.A. Predictors of progression in autosomal dominant and autosomal recessive polycystic kidney disease. *Pediatr. Nephrol.* **2021**. [CrossRef]
56. Ebner, K.; Dafinger, C.; Ortiz-Bruechle, N.; Koerber, F.; Schermer, B.; Benzing, T.; Dötsch, J.; Zerres, K.; Weber, L.T.; Beck, B.B.; et al. Challenges in establishing genotype–phenotype correlations in ARPKD: Case report on a toddler with two severe PKHD1 mutations. *Pediatr. Nephrol.* **2017**, *32*, 1269–1273. [CrossRef] [PubMed]
57. Furu, L.; Onuchic, L.F.; Gharavi, A.; Hou, X.; Esquivel, E.L.; Nagasawa, Y.; Bergmann, C.; Senderek, J.; Avner, E.; Zerres, K.; et al. Milder Presentation of Recessive Polycystic Kidney Disease Requires Presence of Amino Acid Substitution Mutations. *J. Am. Soc. Nephrol.* **2003**. [CrossRef]
58. Burgmaier, K.; Kunzmann, K.; Ariceta, G.; Bergmann, C.; Buescher, A.K.; Burgmaier, M.; Dursun, I.; Duzova, A.; Eid, L.; Erger, F.; et al. Risk Factors for Early Dialysis Dependency in Autosomal Recessive Polycystic Kidney Disease. *J. Pediatr.* **2018**, *199*, 22–28.e6. [CrossRef]
59. Stevanovic, R.; Glumac, S.; Trifunovic, J.; Medjo, B.; Nastasovic, T.; Markovic-Lipkovski, J. Autosomal recessive polycystic kidney disease: Case report. *Clin. Genet.* **1995**, *47*, 248–253. [CrossRef]
60. Garcia-Gonzalez, M.A.; Menezes, L.F.; Piontek, K.B.; Kaimori, J.; Huso, D.L.; Watnick, T.; Onuchic, L.F.; Guay-Woodford, L.M.; Germino, G.G. Genetic interaction studies link autosomal dominant and recessive polycystic kidney disease in a common pathway. *Hum. Mol. Genet.* **2007**, *16*, 1940–1950. [CrossRef]
61. Bergmann, C.; von Bothmer, J.; Ortiz Brü chle, N.; Venghaus, A.; Frank, V.; Fehrenbach, H.; Hampel, T.; Pape, L.; Buske, A.; Jonsson, J.; et al. Mutations in Multiple PKD Genes May Explain Early and Severe Polycystic Kidney Disease. *J. Am. Soc. Nephrol.* **2011**, *22*, 2047–2056. [CrossRef] [PubMed]
62. Nagasawa, Y.; Matthiesen, S.; Onuchic, L.F.; Hou, X.; Bergmann, C.; Esquivel, E.; Senderek, J.; Ren, Z.; Zeltner, R.; Furu, L.; et al. Identification and Characterization of Pkhd1, the Mouse Orthologue of the Human ARPKD Gene. *J. Am. Soc. Nephrol.* **2002**, *13*, 2246–2258. [CrossRef]
63. Igarashi, P.; Somlo, S. Genetics and Pathogenesis of Polycystic Kidney Disease. *J. Am. Soc. Nephrol.* **2002**, *13*, 2384–2398. [CrossRef] [PubMed]

64. Hogan, M.C.; Griffin, M.D.; Rossetti, S.; Torres, V.E.; Ward, C.J.; Harris, P.C. PKHDL1, a homolog of the autosomal recessive polycystic kidney disease gene, encodes a receptor with inducible T lymphocyte expression. *Hum. Mol. Genet.* **2003**, *12*, 685–698. [CrossRef]
65. Menezes, L.F.C.; Cai, Y.; Nagasawa, Y.; Silva, A.M.G.; Watkins, M.L.; Da Silva, A.M.; Somlo, S.; Guay-Woodford, L.M.; Germino, G.G.; Onuchic, L.F. Polyductin, the PKHD1 gene product, comprises isoforms expressed in plasma membrane, primary cilium, and cytoplasm. *Kidney Int.* **2004**, *66*, 1345–1355. [CrossRef]
66. Kaimori, J.-Y.; Nagasawa, Y.; Menezes, L.F.; Garcia-Gonzalez, M.A.; Deng, J.; Imai, E.; Onuchic, L.F.; Guay-Woodford, L.M.; Germino, G.G. Polyductin undergoes notch-like processing and regulated release from primary cilia. *Hum. Mol. Genet.* **2007**, *16*, 942–956. [CrossRef]
67. Outeda, P.; Menezes, L.; Hartung, E.A.; Bridges, S.; Zhou, F.; Zhu, X.; Xu, H.; Huang, Q.; Yao, Q.; Qian, F.; et al. A novel model of autosomal recessive polycystic kidney questions the role of the fibrocystin C-terminus in disease mechanism. *Kidney Int.* **2017**, *92*, 1130–1144. [CrossRef]
68. Ward, C.J.; Yuan, D.; Masyuk, T.V.; Wang, X.; Punyashthiti, R.; Whelan, S.; Bacallao, R.; Torra, R.; LaRusso, N.F.; Torres, V.E.; et al. Cellular and subcellular localization of the ARPKD protein; fibrocystin is expressed on primary cilia. *Hum. Mol. Genet.* **2003**, *12*, 2703–2710. [CrossRef]
69. Wang, S.; Luo, Y.; Wilson, P.D.; Witman, G.B.; Zhou, J. The Autosomal Recessive Polycystic Kidney Disease Protein Is Localized to Primary Cilia, with Concentration in the Basal Body Area. *J. Am. Soc. Nephrol.* **2004**, *15*, 592–602. [CrossRef] [PubMed]
70. Zhang, M.-Z.; Mai, W.; Li, C.; Cho, S.; Hao, C.; Moeckel, G.; Zhao, R.; Kim, I.; Wang, J.; Xiong, H.; et al. PKHD1 protein encoded by the gene for autosomal recessive polycystic kidney disease associates with basal bodies and primary cilia in renal epithelial cells. *Proc. Natl. Acad. Sci. USA* **2004**, *101*, 2311–2316. [CrossRef]
71. Masyuk, T.V.; Huang, B.Q.; Ward, C.J.; Masyuk, A.I.; Yuan, D.; Splinter, P.L.; Punyashthiti, R.; Ritman, E.L.; Torres, V.E.; Harris, P.C.; et al. Defects in Cholangiocyte Fibrocystin Expression and Ciliary Structure in the PCK Rat. *Gastroenterology* **2003**, *125*, 1303–1310. [CrossRef] [PubMed]
72. Sweeney, W.E.; Avner, E.D. Molecular and cellular pathophysiology of autosomal recessive polycystic kidney disease (ARPKD). *Cell Tissue Res.* **2006**, *326*, 671–685. [CrossRef] [PubMed]
73. Wilson, P.D. Polycystic Kidney Disease. *N. Engl. J. Med.* **2004**, *350*, 151–164. [CrossRef] [PubMed]
74. Follit, J.A.; Li, X.; Vucica, Y.; Pazour, G.J. The cytoplasmic tail of fibrocystin contains a ciliary targeting sequence. *J. Cell Biol.* **2010**, *188*, 21–28. [CrossRef]
75. Glazer, A.M.; Wilkinson, A.W.; Backer, C.B.; Lapan, S.W.; Gutzman, J.H.; Cheeseman, I.M.; Reddien, P.W. The Zn Finger protein Iguana impacts Hedgehog signaling by promoting ciliogenesis. *Dev. Biol.* **2010**, *337*, 148–156. [CrossRef]
76. Guay-Woodford, L.M. Murine models of polycystic kidney disease: Molecular and therapeutic insights. *Am. J. Physiol. Ren. Physiol.* **2003**, *285*, 1034–1049. [CrossRef]
77. Fry, J.L.; Koch, W.E.; Jennette, J.C.; McFarland, E.; Fried, F.A.; Mandell, J. A genetically determined murine model of infantile polycystic kidney disease. *J. Urol.* **1985**, *134*, 828–833. [CrossRef]
78. Schieren, G.; Pey, R.; Bach, J.; Hafner, M.; Gretz, N. Murine models of polycystic kidney disease. *Nephrol. Dial. Transplant.* **1996**, *11*, 38–45. [CrossRef]
79. Takahashi, H.; Calvet, J.P.; Dittemore-Hoover, D.; Yoshida, K.; Grantham, J.J.; Gattone, V.H. A hereditary model of slowly progressive polycystic kidney disease in the mouse. *J. Am. Soc. Nephrol.* **1991**, *1*, 980–989. [CrossRef]
80. Woo, D.D.L.; Nguyen, D.K.P.; Khatibi, N.; Olsen, P. Genetic identification of two major modifier loci of polycystic kidney disease progression in pcy mice. *J. Clin. Investig.* **1997**, *100*, 1934–1940. [CrossRef]
81. Nauta, J.; Ozawa, Y.; Sweeney, W.E.; Rutledge, J.C.; Avner, E.D. Renal and biliary abnormalities in a new murine model of autosomal recessive polycystic kidney disease. *Pediatr. Nephrol.* **1993**, *7*, 163–172. [CrossRef] [PubMed]
82. Atala, A.; Freeman, M.R.; Mandell, J.; Beier, D.R. Juvenile cystic kidneys (jck): A new mouse mutation which causes polycystic kidneys. *Kidney Int.* **1993**, *43*, 1081–1085. [CrossRef]
83. Cogswell, C.; Price, S.J.; Hou, X.; Guay-Woodford, L.M.; Flaherty, L.; Bryda, E.C. Positional cloning of jcpk/bpk locus of the mouse. *Mamm. Genome* **2003**, *14*, 242–249. [CrossRef]
84. Liu, S.; Lu, W.; Obara, T.; Kuida, S.; Lehoczky, J.; Dewar, K.; Drummond, I.A.; Beier, D.R. A defect in a novel Nek-family kinase causes cystic kidney disease in the mouse and in zebrafish. *Development* **2002**, *129*, 5839–5846. [CrossRef]
85. Flaherty, L.; Bryda, E.C.; Collins, D.; Rudofsky, U.; Montgomery, J.C. New mouse model for polycystic kidney disease with both recessive and dominant gene effects. *Kidney Int.* **1995**, *47*, 552–558. [CrossRef] [PubMed]
86. Guay-Woodford, L.M.; Bryda, E.C.; Christine, B.; Lindsey, J.R.; Collier, W.R.; Avner, E.D.; D'eustachio, P.; Flaherty, L. Evidence that two phenotypically distinct mouse PKD mutations, bpk and jcpk, are allelic. *Kidney Int.* **1996**, *50*, 1158–1165. [CrossRef]
87. Moyer, J.H.; Lee-Tischler, M.J.; Kwon, H.Y.; Schrick, J.J.; Avner, E.D.; Sweeney, W.E.; Godfrey, V.L.; Cacheiro, N.L.A.; Wilkinson, J.E.; Woychik, R.P. Candidate gene associated with a mutation causing recessive polycystic kidney disease in mice. *Science* **1994**, *264*, 1329–1333. [CrossRef] [PubMed]
88. Sweeney, W.E.; Avner, E.D. Functional activity of epidermal growth factor receptors in autosomal recessive polycystic kidney disease. *Am. J. Physiol. Ren. Physiol.* **1998**, *275*, F387–F394. [CrossRef]
89. Lager, D.J.; Qian, Q.; Bengal, R.J.; Ishibashi, M.; Torres, V.E. The pck rat: A new model that resembles human autosomal dominant polycystic kidney and liver disease. *Kidney Int.* **2001**, *59*, 126–136. [CrossRef]

90. Moser, M.; Matthiesen, S.; Kirfel, J.; Schorle, H.; Bergmann, G.; Senderek, J.; Rudnik-Schöneborn, S.; Zerres, K.; Buettner, R. A mouse model for cystic biliary dysgenesis in autosomal recessive polycystic kidney disease (ARPKD). *Hepatology* **2005**, *41*, 1113–1121. [CrossRef]
91. Al-Bhalal, L.; Akhtar, M. Molecular basis of autosomal recessive polycystic kidney disease (ARPKD). *Adv. Anat. Pathol.* **2008**, *15*, 54–58. [CrossRef] [PubMed]
92. Sweeney, W.E.; Avner, E.D. Pathophysiology of childhood polycystic kidney diseases: New insights into disease-specific therapy. *Pediatr. Res.* **2014**, *75*, 148–157. [CrossRef]
93. Guay-Woodford, L.M.; D'eustachio, P.; Bruns, G.A.P. Identification of the syntenic human linkage group for the mouse congenital polycystic kidney (cpk) locus (Abstract). *J. Am. Soc. Nephrol.* **1993**, *4*, 814.
94. Orellana, S.A.; Sweeney, W.E.; Neff, C.D.; Avner, E.D. Epidermal growth factor receptor expression in abnormal in murine polycystic kidney. *Kidney Int.* **1995**, *47*, 490–499. [CrossRef] [PubMed]
95. Gattone, V.H.; MacNaughton, K.A.; Kraybill, A.L. Murine autosomal recessive polycystic kidney disease with multiorgan involvement induced by the cpk gene. *Anat. Rec.* **1996**, *245*, 488–499. [CrossRef]
96. Ricker, J.; Gattone, V.H.; Calvet, J.P.; Rankin, C.A. Development of Autosomal Recessive Polycystic Kidney Disease in BALB/c-cpk/cpk Mice | American Society of Nephrology. *J. Am. Soc. Nephrol.* **2000**, *10*, 1837–1847. [CrossRef]
97. Olbrich, H.; Fliegauf, M.; Hoefele, J.; Kispert, A.; Otto, E.; Volz, A.; Wolf, M.T.; Sasmaz, G.; Trauer, U.; Reinhardt, R.; et al. Mutations in a novel gene, NPHP3, cause adolescent nephronophthisis, tapeto-retinal degeneration and hepatic fibrosis. *Nat. Genet.* **2003**, *34*, 455–459. [CrossRef]
98. Muff, M.A.; Masyuk, T.V.; Stroope, A.J.; Huang, B.Q.; Splinter, P.L.; Lee, S.O.; LaRusso, N.F. Development and characterization of a cholangiocyte cell line from the PCK rat, an animal model of Autosomal Recessive Polycystic Kidney Disease. *Lab. Investig.* **2006**, *86*, 940–950. [CrossRef]
99. Woollard, J.R.; Punyashtiti, R.; Richardson, S.; Masyuk, T.V.; Whelan, S.; Huang, B.Q.; Lager, D.J.; Vandeursen, J.; Torres, V.E.; Gattone, V.H.; et al. A mouse model of autosomal recessive polycystic kidney disease with biliary duct and proximal tubule dilatation. *Kidney Int.* **2007**, *72*, 328–336. [CrossRef]
100. Gallagher, A.R.; Esquivel, E.L.; Briere, T.S.; Tian, X.; Mitobe, M.; Menezes, L.F.; Markowitz, G.S.; Jain, D.; Onuchic, L.F.; Somlo, S. Biliary and pancreatic dysgenesis in mice harboring a mutation in Pkhd1. *Am. J. Pathol.* **2008**, *172*, 417–429. [CrossRef]
101. Kim, I.; Fu, Y.; Hui, K.; Moeckel, G.; Mai, W.; Li, C.; Liang, D.; Zhao, P.; Ma, J.; Chen, X.-Z.; et al. Fibrocystin/Polyductin Modulates Renal Tubular Formation by Regulating Polycystin-2 Expression and Function. *J. Am. Soc. Nephrol.* **2008**, *19*, 455–468. [CrossRef]
102. Williams, S.S.; Cobo-Stark, P.; James, L.R.; Somlo, S.; Igarashi, P. Kidney cysts, pancreatic cysts, and biliary disease in a mouse model of autosomal recessive polycystic kidney disease. *Pediatr. Nephrol.* **2008**, *23*, 733–741. [CrossRef] [PubMed]
103. Bakeberg, J.L.; Tammachote, R.; Woollard, J.R.; Hogan, M.C.; Tuan, H.F.; Li, M.; Van Deursen, J.M.; Wu, Y.; Huang, B.Q.; Torres, V.E.; et al. Epitope-tagged Pkhd1 tracks the processing, secretion, and localization of fibrocystin. *J. Am. Soc. Nephrol.* **2011**, *22*, 2266–2277. [CrossRef]
104. Bult, C.J.; Kadin, J.A.; Richardson, J.E.; Blake, J.A.; Eppig, J.T. The mouse genome database: Enhancements and updates. *Nucleic Acids Res.* **2009**, *38*. [CrossRef]
105. Neufeld, T.K.; Douglass, D.; Grant, M.; Ye, M.; Silva, F.; Nadasdy, T.; Grantham, J.J. In vitro formation and expansion of cysts derived from human renal cortex epithelial cells. *Kidney Int.* **1992**, *41*, 1222–1236. [CrossRef] [PubMed]
106. Ye, M.; Grant, M.; Sharma, M.; Elzinga, L.; Swan, S.; Torres, V.E.; Grantham, J.J. Cyst fluid from human autosomal dominant polycystic kidneys promotes cyst formation and expansion by renal epithelial cells in vitro. *J. Am. Soc. Nephrol.* **1992**, *3*, 984–994. [CrossRef]
107. Lakshmanan, J.; Fisher, D.A. An inborn error in epidermal growth factor prohormone metabolism in a mouse model of autosomal recessive polycystic kidney disease. *Biochem. Biophys. Res. Commun.* **1993**, *196*, 892–901. [CrossRef]
108. Pugh, J.L.; Sweeney, W.E.; Avner, E.D. Tyrosine kinase activity of the EGF receptor in murine metanephric organ culture. *Kidney Int.* **1995**, *47*, 774–781. [CrossRef] [PubMed]
109. Dell, K.M.R.; Nemo, R.; Sweeney, W.E.; Avner, E.D. EGF-related growth factors in the pathogenesis of murine ARPKD. *Kidney Int.* **2004**, *65*, 2018–2029. [CrossRef] [PubMed]
110. Richards, W.G.; Sweeney, W.E.; Yoder, B.K.; Wilkinson, J.E.; Woychik, R.P.; Avner, E.D. Epidermal growth factor receptor activity mediates renal cyst formation in polycystic kidney disease. *J. Clin. Investig.* **1998**, *101*, 935–939. [CrossRef]
111. Arbeiter, A.; Büscher, R.; Bonzel, K.E.; Wingen, A.M.; Vester, U.; Wohlschläger, J.; Zerres, K.; Nürnberger, J.; Bergmann, C.; Hoyer, P.F. Nephrectomy in an autosomal recessive polycystic kidney disease (ARPKD) patient with rapid kidney enlargement and increased expression of EGFR. *Nephrol. Dial. Transplant.* **2008**, *23*, 3026–3029. [CrossRef] [PubMed]
112. Rohatgi, R.; Zavilowitz, B.; Vergara, M.; Woda, C.; Kim, P.; Satlin, L.M. Cyst fluid composition in human autosomal recessive polycystic kidney disease. *Pediatr. Nephrol.* **2005**, *20*, 552–553. [CrossRef] [PubMed]
113. Gattone, V.H.; Calvet, J.P. Murine Infantile Polycystic Kidney Disease: A Role for Reduced Renal Epidermal Growth Factor. *Am. J. Kidney Dis.* **1991**, *17*, 606–607. [CrossRef]
114. Nakanishi, K.; Gattone, V.H.; Sweeney, W.E.; Avner, E.D. Renal dysfunction but not cystic change is ameliorated by neonatal epidermal growth factor in bpk mice. *Pediatr. Nephrol.* **2001**, *16*, 45–50. [CrossRef]

115. Lowden, D.A.; Lindemann, G.W.; Merlino, G.; Barash, B.D.; Calvet, J.P.; Gattone, V.H. Renal cysts in transgenic mice expressing transforming gorwth factor-alpha. *J. Lab. Clin. Med.* **1994**, *124*, 386–394. [PubMed]
116. Zheleznova, N.N.; Wilson, P.D.; Staruschenko, A. Epidermal growth factor-mediated proliferation and sodium transport in normal and PKD epithelial cells. *Biochim. Biophys. Acta Mol. Basis Dis.* **2011**, *1812*, 1301–1313. [CrossRef] [PubMed]
117. Nauta, J.; Sweeney, W.E.; Rutledge, J.C.; Avner, E.D. Biliary epithelial cells from mice with congenital polycystic kidney disease are hyperresponsive to epidermal growth factor. *Pediatr. Res.* **1995**, *37*, 755–763. [CrossRef] [PubMed]
118. Luetteke, N.C.; Phillips, H.K.; Qiu, T.H.; Copeland, N.G.; Shelton Earp, H.; Jenkins, N.A.; Lee, D.C. The mouse waved-2 phenotype results from a point mutation in the EGF receptor tyrosine kinase. *Genes Dev.* **1994**, *8*, 399–413. [CrossRef]
119. Sweeney, W.E.; Chen, Y.; Nakanishi, K.; Frost, P.; Avner, E.D. Treatment of polycystic kidney disease with a novel tyrosine kinase inhibitor. *Kidney Int.* **2000**, *57*, 33–40. [CrossRef]
120. Sweeney, W.E.; Hamahira, K.; Sweeney, J.; Garcia-Gatrell, M.; Frost, P.; Avner, E.D. Combination treatment of PKD utilizing dual inhibition of EGF-receptor activity and ligand bioavailability. *Kidney Int.* **2003**, *64*, 1310–1319. [CrossRef]
121. Nemo, R.; Murcia, N.; Dell, K.M. Transforming Growth Factor Alpha (TGF-) and Other Targets of Tumor Necrosis Factor-Alpha Converting Enzyme (TACE) in Murine Polycystic Kidney Disease. *Pediatr. Res.* **2005**, *57*, 732–737. [CrossRef]
122. Torres, V.E.; Sweeney, W.E.; Wang, X.; Qian, Q.; Harris, P.C.; Frost, P.; Avner, E.D. Epidermal growth factor receptor tyrosine kinase inhibition is not protective in PCK rats. *Kidney Int.* **2004**, *66*, 1766–1773. [CrossRef] [PubMed]
123. Veizis, E.I.; Carlin, C.R.; Cotton, C.U. Decreased amiloride-sensitive Na+ absorption in collecting duct principal cells isolated from BPK ARPKD mice. *Am. J. Physiol. Ren. Physiol.* **2004**, *286*. [CrossRef]
124. Veizis, I.E.; Cotton, C.U. Abnormal EGF-dependent regulation of sodium absorption in ARPKD collecting duct cells. *Am. J. Physiol. Ren. Physiol.* **2005**, *288*. [CrossRef] [PubMed]
125. Rohatgi, R.; Greenberg, A.; Burrow, C.R.; Wilson, P.D.; Satlin, L.M. Na transport in autosomal recessive polycystic kidney disease (ARPKD) cyst lining epithelial cells. *J. Am. Soc. Nephrol.* **2003**, *14*, 827–836. [CrossRef] [PubMed]
126. Olteanu, D.; Yoder, B.K.; Liu, W.; Croyle, M.J.; Welty, E.A.; Rosborough, K.; Wyss, J.M.; Bell, P.D.; Guay-Woodford, L.M.; Bevensee, M.O.; et al. Heightened epithelial Na+ channel-mediated Na+ absorption in a murine polycystic kidney disease model epithelium lacking apical monocilia. *Am. J. Physiol. Cell Physiol.* **2006**, *290*. [CrossRef]
127. Kaimori, J.; Lin, C.-C.; Outeda, P.; Garcia-Gonzalez, M.A.; Menezes, L.F.; Hartung, E.A.; Li, A.; Wu, G.; Fujita, H.; Sato, Y.; et al. NEDD4-family E3 ligase dysfunction due to PKHD1/Pkhd1 defects suggests a mechanistic model for ARPKD pathobiology. *Sci. Rep.* **2017**, *7*, 7733. [CrossRef] [PubMed]
128. Davidow, C.J.; Maser, R.L.; Rome, L.A.; Calvet, J.P.; Grantham, J.J. The cystic fibrosis transmembrane conductance regulator mediates transepithelial fluid secretion by human autosomal dominant polycystic kidney disease epithelium in vitro. *Kidney Int.* **1996**, *50*, 208–218. [CrossRef]
129. Jouret, F.; Devuyst, O. Targeting chloride transport in autosomal dominant polycystic kidney disease. *Cell. Signal.* **2020**, *73*, 109703. [CrossRef]
130. Nakanishi, K.; Sweeney, W.E.; Dell, K.M.; Cotton, C.U.; Avner, E.D. Role of CFTR in Autosomal Recessive Polycystic Kidney Disease. *J. Am. Soc. Nephrol.* **2001**, *12*, 719–725. [CrossRef]
131. Yamaguchi, T.; Nagao, S.; Kasahara, M.; Takahashi, H.; Grantham, J.J. Renal accumulation and excretion of cyclic adenosine monophosphate in a murine model of slowly progressive polycystic kidney disease. *Am. J. Kidney Dis.* **1997**, *30*, 703–709. [CrossRef]
132. Smith, L.A.; Bukanov, N.O.; Husson, H.; Russo, R.J.; Barry, T.C.; Taylor, A.L.; Beier, D.R.; Ibraghimov-Beskrovnaya, O. Development of polycystic kidney disease in juvenile cystic kidney mice: Insights into pathogenesis, ciliary abnormalities, and common features with human disease. *J. Am. Soc. Nephrol.* **2006**, *17*, 2821–2831. [CrossRef]
133. Belibi, F.A.; Reif, G.; Wallace, D.P.; Yamaguchi, T.; Olsen, L.; Li, H.; Helmkamp, G.M.; Grantham, J.J. Cyclic AMP promotes growth and secretion in human polycystic kidney epithelial cells. *Kidney Int.* **2004**, *66*, 964–973. [CrossRef] [PubMed]
134. Yamaguchi, T.; Pelling, J.C.; Ramaswamy, N.T.; Eppler, J.W.; Wallace, D.P.; Nagao, S.; Rome, L.A.; Sullivan, L.P.; Grantham, J.J. cAMP stimulates the in vitro proliferation of renal cyst epithelial cells by activating the extracellular signal-regulated kinase pathway. *Kidney Int.* **2000**, *57*, 1460–1471. [CrossRef]
135. Yamaguchi, T.; Nagao, S.; Wallace, D.P.; Belibi, F.A.; Cowley, B.D.; Pelling, J.C.; Grantham, J.J. Cyclic AMP activates B-Raf and ERK in cyst epithelial cells from autosomal-dominant polycystic kidneys. *Kidney Int.* **2003**, *63*, 1983–1994. [CrossRef] [PubMed]
136. Yamaguchi, T.; Wallace, D.P.; Magenheimer, B.S.; Hempson, S.J.; Grantham, J.J.; Calvet, J.P. Calcium restriction allows cAMP activation of the B-Raf/ERK pathway, switching cells to a cAMP-dependent growth-stimulated phenotype. *J. Biol. Chem.* **2004**, *279*, 40419–40430. [CrossRef]
137. Wang, X. Effectiveness of Vasopressin V2 Receptor Antagonists OPC-31260 and OPC-41061 on Polycystic Kidney Disease Development in the PCK Rat. *J. Am. Soc. Nephrol.* **2005**, *16*, 846–851. [CrossRef]
138. Wang, X.; Wu, Y.; Ward, C.J.; Harris, P.C.; Torres, V.E. Vasopressin directly regulates cyst growth in polycystic kidney disease. *J. Am. Soc. Nephrol.* **2008**, *19*, 102–108. [CrossRef] [PubMed]
139. Zheng, R.; Wang, L.; Fan, J.; Zhou, Q. Inhibition of PKHD1 may cause S-phase entry via mTOR signaling pathway. *Cell Biol. Int.* **2009**, *33*, 926–933. [CrossRef]

140. Fischer, D.C.; Jacoby, U.; Pape, L.; Ward, C.J.; Kuwertz-Broeking, E.; Renken, C.; Nizze, H.; Querfeld, U.; Rudolph, B.; Mueller-Wiefel, D.E.; et al. Activation of the AKTmTOR pathway in autosomal recessive polycystic kidney disease (ARPKD). *Nephrol. Dial. Transplant.* **2009**, *24*, 1819–1827. [CrossRef]
141. Ren, X.S.; Sato, Y.; Harada, K.; Sasaki, M.; Furubo, S.; Song, J.Y.; Nakanuma, Y. Activation of the PI3K/mTOR Pathway Is Involved in Cystic Proliferation of Cholangiocytes of the PCK Rat. *PLoS ONE* **2014**, *9*, e87660. [CrossRef] [PubMed]
142. Hovater, M.B.; Olteanu, D.; Hanson, E.L.; Cheng, N.-L.; Siroky, B.; Fintha, A.; Komlosi, P.; Liu, W.; Satlin, L.M.; Darwin Bell, P.; et al. Loss of apical monocilia on collecting duct principal cells impairs ATP secretion across the apical cell surface and ATP-dependent and flow-induced calcium signals. *Purinergic Signal.* **2007**. [CrossRef]
143. Yamaguchi, T.; Hempson, S.J.; Reif, G.A.; Hedge, A.M.; Wallace, D.P. Calcium restores a normal proliferation phenotype in human polycystic kidney disease epithelial cells. *J. Am. Soc. Nephrol.* **2006**, *17*, 178–187. [CrossRef]
144. Sweeney, W.E.; von Vigier, R.O.; Frost, P.; Avner, E.D. Src inhibition ameliorates polycystic kidney disease. *J. Am. Soc. Nephrol.* **2008**, *19*, 1331–1341. [CrossRef]
145. Goel, M.; Zuo, C.-D.; Schilling, W.P. Role of cAMP/PKA signaling cascade in vasopressin-induced trafficking of TRPC3 channels in principal cells of the collecting duct. *Am. J. Physiol. Ren. Physiol.* **2010**, *298*. [CrossRef] [PubMed]
146. Chebib, F.T.; Sussman, C.R.; Wang, X.; Harris, P.C.; Torres, V.E. Vasopressin and disruption of calcium signalling in polycystic kidney disease. *Nat. Rev. Nephrol.* **2015**, *11*, 451–464. [CrossRef] [PubMed]
147. Gattone, V.H.; Wang, X.; Harris, P.C.; Torres, V.E. Inhibition of renal cystic disease development and progression by a vasopressin V2 receptor antagonist. *Nat. Med.* **2003**, *9*, 1323–1326. [CrossRef] [PubMed]
148. Torres, V.E.; Wang, X.; Qian, Q.; Somlo, S.; Harris, P.C.; Gattone, V.H. Effective treatment of an orthologous model of autosomal dominant polycystic kidney disease. *Nat. Med.* **2004**, *10*, 363–364. [CrossRef]
149. Torres, V.E.; Chapman, A.B.; Devuyst, O.; Gansevoort, R.T.; Grantham, J.J.; Higashihara, E.; Perrone, R.D.; Krasa, H.B.; Ouyang, J.; Czerwiec, F.S. Tolvaptan in patients with autosomal dominant polycystic kidney disease. *N. Engl. J. Med.* **2012**, *367*, 2407–2418. [CrossRef]
150. Gansevoort, R.T.; Arici, M.; Benzing, T.; Birn, H.; Capasso, G.; Covic, A.; Devuyst, O.; Drechsler, C.; Eckardt, K.U.; Emma, F.; et al. Recommendations for the use of tolvaptan in autosomal dominant polycystic kidney disease: A position statement on behalf of the ERA-EDTA Working Groups on Inherited Kidney Disorders and European Renal Best Practice. *Nephrol. Dial. Transplant.* **2016**, *31*, 337–348. [CrossRef]
151. Sweeney, W.E.; Avner, E.D. Emerging Therapies for Childhood Polycystic Kidney Disease. *Front. Pediatr.* **2017**, *5*, 1–7. [CrossRef]
152. Liu, B.; Li, C.; Liu, Z.; Dai, Z.; Tao, Y. Increasing extracellular matrix collagen level and MMP activity induces cyst development in polycystic kidney disease. *BMC Nephrol.* **2012**, *13*, 1–8. [CrossRef]
153. Urribarri, A.D.; Munoz-Garrido, P.; Perugorria, M.J.; Erice, O.; Merino-Azpitarte, M.; Arbelaiz, A.; Lozano, E.; Hijona, E.; Jiménez-Agüero, R.; Fernandez-Barrena, M.G.; et al. Inhibition of metalloprotease hyperactivity in cystic cholangiocytes halts the development of polycystic liver diseases. *Gut* **2014**, *63*, 1658–1667. [CrossRef]
154. Dai, B.; Liu, Y.; Mei, C.; Fu, L.; Xiong, X.; Zhang, Y.; Shen, X.; Hua, Z. Rosiglitazone attenuates development of polycystic kidney disease and prolongs survival in Han:SPRD rats. *Clin. Sci.* **2010**, *119*, 323–333. [CrossRef] [PubMed]
155. Yoshihara, D.; Kugita, M.; Yamaguchi, T.; Aukema, H.M.; Kurahashi, H.; Morita, M.; Hiki, Y.; Calvet, J.P.; Wallace, D.P.; Toyohara, T.; et al. Global gene expression profiling in PPAR-γ agonist-treated kidneys in an orthologous rat model of human autosomal recessive polycystic kidney disease. *PPAR Res.* **2012**. [CrossRef] [PubMed]
156. Riwanto, M.; Kapoor, S.; Rodriguez, D.; Edenhofer, I.; Segerer, S.; Wüthrich, R.P. Inhibition of aerobic glycolysis attenuates disease progression in polycystic kidney disease. *PLoS ONE* **2016**, *11*. [CrossRef] [PubMed]
157. Fischer, E.; Legue, E.; Doyen, A.; Nato, F.; Nicolas, J.F.; Torres, V.; Yaniv, M.; Pontoglio, M. Defective planar cell polarity in polycystic kidney disease. *Nat. Genet.* **2006**, *38*, 21–23. [CrossRef] [PubMed]
158. Nishio, S.; Tian, X.; Gallagher, A.R.; Yu, Z.; Patel, V.; Igarashi, P.; Somlo, S. Loss of Oriented Cell Division Does not Initiate Cyst Formation. *J. Am. Soc. Nephrol.* **2010**, *21*, 295–302. [CrossRef] [PubMed]
159. Hildebrandt, F.; Benzing, T.; Katsanis, N. Ciliopathies. *N. Engl. J. Med.* **2011**, *364*, 1533–1543. [CrossRef] [PubMed]
160. Busch, T.; Köttgen, M.; Hofherr, A. TRPP2 ion channels: Critical regulators of organ morphogenesis in health and disease. *Cell Calcium* **2017**, *66*, 25–32. [CrossRef]
161. Hanaoka, K.; Qian, F.; Boletta, A.; Bhunia, A.K.; Piontek, K.; Tsiokas, L.; Sukhatme, V.P.; Guggino, W.B.; Germino, G.G. Co-assembly of polycystin-1 and -2 produces unique cation-permeable currents. *Nature* **2000**, *408*, 990–994. [CrossRef]
162. Nauli, S.M.; Alenghat, F.J.; Luo, Y.; Williams, E.; Vassilev, P.; Li, X.; Elia, A.E.H.; Lu, W.; Brown, E.M.; Quinn, S.J.; et al. Polycystins 1 and 2 mediate mechanosensation in the primary cilium of kidney cells. *Nat. Genet.* **2003**, *33*, 129–137. [CrossRef]
163. Wang, Z.; Ng, C.; Liu, X.; Wang, Y.; Li, B.; Kashyap, P.; Chaudhry, H.A.; Castro, A.; Kalontar, E.M.; Ilyayev, L.; et al. The ion channel function of polycystin-1 in the polycystin-1/polycystin-2 complex. *EMBO Rep.* **2019**, 1–18. [CrossRef] [PubMed]
164. Wu, Y.; Dai, X.-Q.Q.; Li, Q.; Chen, C.X.; Mai, W.; Hussain, Z.; Long, W.; Montalbetti, N.; Li, G.; Glynne, R.; et al. Kinesin-2 mediates physical and functional interactions between polycystin-2 and fibrocystin. *Hum. Mol. Genet.* **2006**, *15*, 3280–3292. [CrossRef]
165. Wang, S.; Zhang, J.; Nauli, S.M.; Li, X.; Starremans, P.G.; Luo, Y.; Roberts, K.A.; Zhou, J. Fibrocystin/Polyductin, Found in the Same Protein Complex with Polycystin-2, Regulates Calcium Responses in Kidney Epithelia Fibrocystin/Polyductin, Found in the Same Protein Complex with Polycystin-2, Regulates Calcium Responses in Kidney Epithel. *Mol. Cell. Biol.* **2007**, *27*, 3241–3252. [CrossRef]

166. Allison, S.J. Polycystic kidney disease: FPC in ARPKD. *Nat. Rev. Nephrol.* **2017**, *13*, 597. [CrossRef] [PubMed]
167. Olson, R.J.; Hopp, K.; Wells, H.; Smith, J.M.; Furtado, J.; Constans, M.M.; Escobar, D.L.; Geurts, A.M.; Torres, V.E.; Harris, P.C. Synergistic Genetic Interactions between Pkhd1 and Pkd1 Result in an ARPKD-Like Phenotype in Murine Models. *J. Am. Soc. Nephrol.* **2019**. [CrossRef] [PubMed]
168. Kaimori, J.; Germino, G.G. ARPKD and ADPKD: First cousins or more distant relatives? *J. Am. Soc. Nephrol.* **2008**, *19*, 416–418. [CrossRef] [PubMed]
169. Ma, M.; Tian, X.; Igarashi, P.; Pazour, G.J.; Somlo, S. Loss of cilia suppresses cyst growth in genetic models of autosomal dominant polycystic kidney disease. *Nat. Genet.* **2013**, *45*, 1004–1012. [CrossRef]
170. Gallagher, A.R.; Somlo, S. Loss of cilia does not slow liver disease progression in mouse models of autosomal recessive polycystic kidney disease. *Kidney360* **2020**, 1–14. [CrossRef] [PubMed]
171. Palander, O.; El-Zeiry, M.; Trimble, W.S. Uncovering the Roles of Septins in Cilia. *Front. Cell Dev. Biol.* **2017**, *5*, 36. [CrossRef]
172. Torres, V.E.; Chapman, A.B.; Devuyst, O.; Gansevoort, R.T.; Perrone, R.D.; Koch, G.; Ouyang, J.; McQuade, R.D.; Blais, J.D.; Czerwiec, F.S.; et al. Tolvaptan in Later-Stage Autosomal Dominant Polycystic Kidney Disease. *N. Engl. J. Med.* **2017**, *377*, 1930–1942. [CrossRef]
173. Edwards, M.E.; Chebib, F.T.; Irazabal, M.V.; Ofstie, T.G.; Bungum, L.A.; Metzger, A.J.; Senum, S.R.; Hogan, M.C.; El-Zoghby, Z.M.; Kline, T.L.; et al. Long-term administration of tolvaptan in autosomal dominant polycystic kidney disease. *Clin. J. Am. Soc. Nephrol.* **2018**, *13*, 1153–1161. [CrossRef] [PubMed]
174. Talbot, J.J.; Song, X.; Wang, X.; Rinschen, M.M.; Doerr, N.; LaRiviere, W.B.; Schermer, B.; Pei, Y.P.; Torres, V.E.; Weimbs, T. The cleaved cytoplasmic tail of polycystin-1 regulates Src-dependent STAT3 activation. *J. Am. Soc. Nephrol.* **2014**, *25*, 1737–1748. [CrossRef] [PubMed]
175. Sweeney, W.E.; Frost, P.; Avner, E.D. Tesevatinib ameliorates progression of polycystic kidney disease in rodent models of autosomal recessive polycystic kidney disease. *World J. Nephrol.* **2017**, *6*, 188–200. [CrossRef] [PubMed]

International Journal of
Molecular Sciences

MDPI

Review

Molecular Basis, Diagnostic Challenges and Therapeutic Approaches of Bartter and Gitelman Syndromes: A Primer for Clinicians

Laura Nuñez-Gonzalez [1,2], Noa Carrera [1,2,3,*] and Miguel A. Garcia-Gonzalez [1,2,3,4,*]

1 Grupo de Xenetica e Bioloxia do Desenvolvemento das Enfermidades Renais, Laboratorio de Nefroloxia (No. 11), Instituto de Investigacion Sanitaria de Santiago (IDIS), Complexo Hospitalario de Santiago de Compostela (CHUS), 15706 Santiago de Compostela, Spain; laura.nunez.gonzalez@rai.usc.es
2 Grupo de Medicina Xenomica, Complexo Hospitalario de Santiago de Compostela (CHUS), 15706 Santiago de Compostela, Spain
3 RedInRen (Red en Investigación Renal) RETIC (Redes Temáticas de Investigación Cooperativa en Salud), ISCIII (Instituto de Salud Carlos III), 28029 Madrid, Spain
4 Fundación Pública Galega de Medicina Xenomica—SERGAS, Complexo Hospitalario de Santiago de Compotela (CHUS), 15706 Santiago de Compostela, Spain
* Correspondence: noa.carrera.cachaza@sergas.es (N.C.); miguel.garcia.gonzalez@sergas.es (M.A.G.-G.)

Citation: Nuñez-Gonzalez, L.; Carrera, N.; Garcia-Gonzalez, M.A. Molecular Basis, Diagnostic Challenges and Therapeutic Approaches of Bartter and Gitelman Syndromes: A Primer for Clinicians. *Int. J. Mol. Sci.* **2021**, *22*, 11414. https://doi.org/10.3390/ijms222111414

Academic Editor: Anastasios Lymperopoulos

Received: 15 September 2021
Accepted: 14 October 2021
Published: 22 October 2021

Abstract: Gitelman and Bartter syndromes are rare inherited diseases that belong to the category of renal tubulopathies. The genes associated with these pathologies encode electrolyte transport proteins located in the nephron, particularly in the Distal Convoluted Tubule and Ascending Loop of Henle. Therefore, both syndromes are characterized by alterations in the secretion and reabsorption processes that occur in these regions. Patients suffer from deficiencies in the concentration of electrolytes in the blood and urine, which leads to different systemic consequences related to these salt-wasting processes. The main clinical features of both syndromes are hypokalemia, hypochloremia, metabolic alkalosis, hyperreninemia and hyperaldosteronism. Despite having a different molecular etiology, Gitelman and Bartter syndromes share a relevant number of clinical symptoms, and they have similar therapeutic approaches. The main basis of their treatment consists of electrolytes supplements accompanied by dietary changes. Specifically for Bartter syndrome, the use of non-steroidal anti-inflammatory drugs is also strongly supported. This review aims to address the latest diagnostic challenges and therapeutic approaches, as well as relevant recent research on the biology of the proteins involved in disease. Finally, we highlight several objectives to continue advancing in the characterization of both etiologies.

Keywords: Bartter syndrome; Gitelman syndrome; genetics; genetic diagnosis; therapeutic targets; hyponatremia; hypokalemia; hypercalciuria; hypomagnesemia

1. Introduction

The main function of renal tubules is the control of reabsorption and secretion of electrolytes, in order to maintain correct homeostasis [1]. There are numerous proteins involved directly or indirectly in this function. Any impairment that compromises their correct function will cause, to a greater or lesser degree, a dysregulation of homeostasis [2], thus, giving rise to different clinical manifestations (kidney tubulopathy).

GS and BS are monogenic diseases belonging the group of inherited renal tubulopathies [2]. Gitelman (GS) and Bartter syndromes (BS) pathognomonic symptoms include hypokalemia, hypochloremic metabolic alkalosis, hyperreninism and secondary hyperaldostheronism due to volume-contraction consequences of the activation of the renin-angiotensin-aldosterone system (RAAS) [3,4]. The overall prevalence of inherited tubulopathies remains unknown [5]. For GS, the estimated prevalence varies from 1 to 10 per 40,000 individuals worldwide [6], whereas, for BS, the data is not well-defined but an annual incidence of 1 per 1,000,000 is estimated [7].

In 1962, Dr. C. Frederic described the clinical symptoms of a new disease named Bartter syndrome [8]. Since then, several studies have revealed the great genetic heterogeneity related to this syndrome and, to date, six genes have been linked to this pathology [9–15], subclassifying the disease according to the underlying genes (Table 1). On the other hand, Gitelman syndrome (GS) was first named as an independent clinical entity in 1969 [16]. Despite the fact that the causative gene (*SLC12A3*) was discovered in 1996 [17], Gitelman syndrome has been confused with a subtype of Bartter syndrome, leading to misdiagnoses for many years [18].

Table 1. Bartter syndrome classification, associated genes and encoded proteins. All the UniProtein codes are referred to Homo sapiens. AR: Autosomal Recessive; DR: Digenic Recessive; and XLR: X-Linked Recessive.

Bartter Subtype	OMIM	Inheritance	Causative Gene	Related Protein	UniProt Code
Type 1	601,678	AR	*SLC12A1*	NKCC2 (Solute carrier family 12 member 1)	Q13621
Type 2	600,359	AR	*KCNJ1*	ROMK (ATP-sensitive inward rectifier potassium channel 1)	P48048
Type 3	607,364	AR	*CLCNKB*	CLCNKB (Chloride channel protein ClC-Kb)	P51801
Type 4a	602,522	AR	*BSND*	BSND (Barttin)	Q8WZ55
Type 4b	613,090	AR DR	*CLCNKB* + *CLCNKA*	CLCNKB (Chloride channel protein ClC-Kb) + CLCNKA (Chloride channel protein ClC-Ka)	P51801 + P51800
Type 5	300,470	XLR	*MAGED2*	MAGED2 (Melanoma-associated antigen D2)	Q9UNF1

2. Renal Physiology of Electrolytes

One of the functions of the kidney is the maintenance and regulation of acid-base homeostasis [19]. This balance is maintained by the interregulation between glomerular filtration and tubular reabsorption [20,21], which makes the urine not only a way of eliminating toxic metabolites but also a complex milieu adapted to these processes [2]. In different segments of the nephron, 99% of the total glomerular filtration volume is reabsorbed via transcellular (active or passive transporters) or paracellular mechanisms [21]. The force generated by the reabsorption of the solutes is what drives the reabsorption of water into the interstitium [22]. Urine's ability to concentrate resides in the loop of Henle, which creates a hypertonic medullary interstitium [19,23,24]. It is in the thick ascending loop of Henle (TAL) where active transport of solutes takes place through different co-transporting channels (Figure 1).

In this process, K^+ is the limiting substrate for NKCC2 channels and, in turn, for total electrolyte reabsorption [19]. Due to the correct K^+ concentration gradient between tubules and the interstitium, a positive electrostatic charge is present in the tubular lumen. This charge allows the paracellular reabsorption of Mg^{2+} and Ca^{2+} [19,25]. The distal convoluted tubule (DCT) mediates the correct regulation of electrolytes and their secretion [26]. The DCT plays an important role in K^+ secretion [27] and the transcellular reabsorption of calcium, via the transient receptor potential cation channel subfamily V member 5 (TRPV5) [25], and magnesium, via the transient receptor potential cation channel subfamily M member 6 (TRPM6) [28].

The reabsorption of Mg^{2+} in this segment depends on the proper function of a sodium-chloride cotransporter, the NCC channel (encoded by the *SLC12A3* gene), which is exclusively located in the cells of the DCT (see Figure 1) [26,29]. NCC, together with uromodulin and calcineurin, have been described as important regulators of magnesium homeostasis in the DCT [28]. In "salt-losing tubulopathies", such as BS and GS, any alteration in this transport mechanism causes a loss of electrolytes that interrupts the global reabsorption process, resulting in a wide range of progressive alterations in renal physiology [20,30,31].

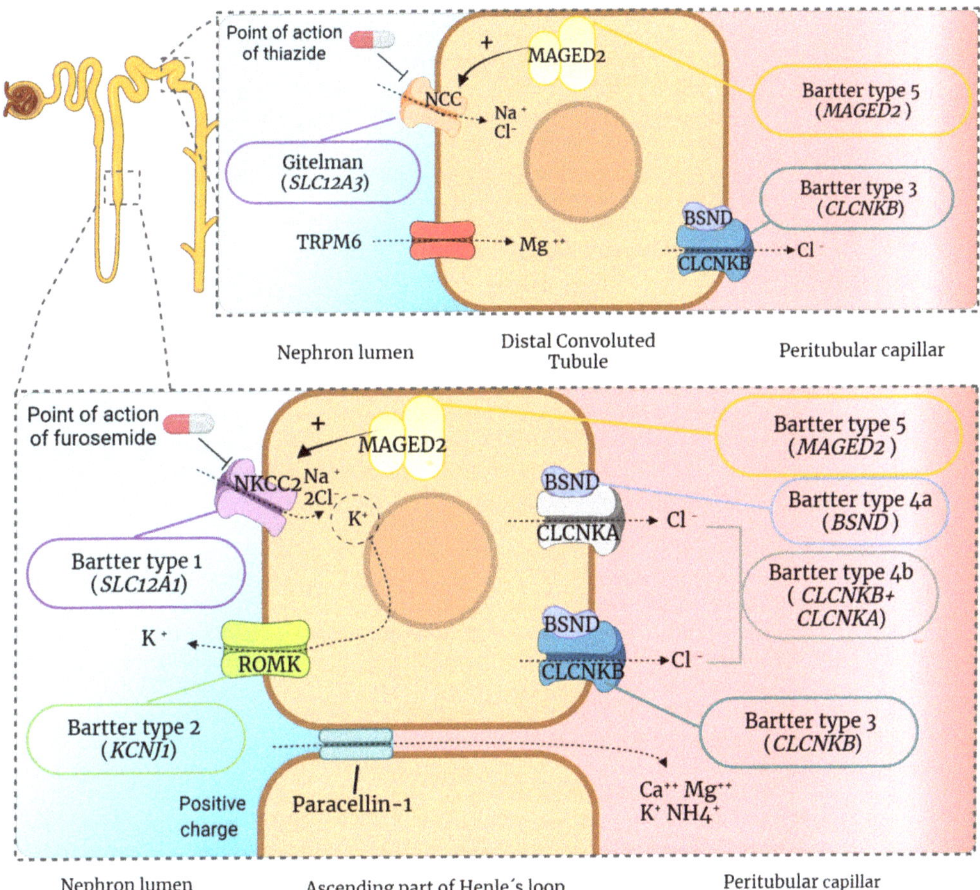

Figure 1. Proteins and channels implicated in the pathogenesis of Gitelman and Bartter syndromes. The electrolyte transports of the most important channels for the diseases are represented as well as the channels related to the inhibition by thiazide (NCC) and furosemide diuretics (NKCC2). Each disease is accompanied by the causative gene (in capital letters, brackets and italics), whereas the corresponding protein is indicated above the channel (only in capital letters). NCC: Solute carrier family 12 member 3; MAGED2: Melanoma-associated antigen D2; TRPM6: Transient receptor potential cation channel subfamily M member 6; CLCNKB: Chloride channel protein ClC-Kb; NKCC2: Solute carrier family 12 member 1; BSND: Barttin; CLCNKA: Chloride channel protein ClC-Ka; and ROMK: ATP-sensitive inward rectifier potassium channel 1. The positive charge of the DCT makes an electrochemical gradient from the luminal tubule from the interstitium possible.

3. Molecular Basis and Clinical Features of the Diseases

Although GS and the different types of BS share most of their clinical symptoms, they have different etiologies. Their associated proteins reside in different parts along the nephron resulting in different pathophysiological mechanisms and compensatory phenomena regarding their loss of function. For example, poor TAL functionality associated with BS causes a decrease in electrolyte bioavailability, which leads to activation of the juxtaglomerular apparatus with secondary hyperaldosteronism and glomerular hyperfiltration due to afferent arteriole dilation [3].

This compensatory effect results in the hypertrophy of the juxtaglomerular apparatus and a constant elevation of renin and aldosterone in plasma [20]. In contrast, the typical salt loss of the DCT associated with GS causes volume contraction, leading to hyperaldosteronism with elevated plasma renin levels [25]. In GS, there is no direct involvement of the juxtaglomerular apparatus [20], which marks a relevant molecular difference from BS.

3.1. Molecular Basis of Gitelman Syndrome and Clinical Consequences

As mentioned previously, GS and BS share pathological conditions, including hypokalemia, hypochloremic metabolic alkalosis, hyperreninemia and secondary hyperaldosteronism [3,4]. However, the main difference is that GS also presents with hypocalciuria. This is due to an increase in Ca^{2+} reabsorption in an attempt to compensate for the loss of salts [25]. This compensatory process does not occur in patients with BS, since electrolyte dysregulation in the TAL causes a lack of the positive light gradient necessary for paracellular Ca^{2+} reabsorption [25,32]. Active transport of Ca^{2+} remains unaltered in GS [25,32].

Another relevant molecular characteristic is the presence of hypomagnesemia, mainly in cases of GS. However, this anomaly is not a complete differentiation between GS and BS [33] since it can also be found in certain cases of BS [34]. Although this phenomenon has been studied for more than 20 years [32,35,36], the reason for this magnesium loss is not completely understood. One of the most accepted hypotheses points to the alteration in the expression of the TRPM6 channel [28], the principal channel by which Mg^{2+} is reabsorbed in DCT (Figure 1). Moreover, there is a strict correlation between hypomagnesemia and chondrocalcinosis [37].

Mg^{2+} is a cofactor of the group of pyrophosphatases, particularly for alkaline phosphatase [38]. A decrease in the concentration of magnesium causes a dysfunction of these proteins, which increases the levels of pyrophosphate. Inorganic pyrophosphate binds to Ca^{2+} ions by ionic interaction, resulting in crystal formation. These crystals are deposited over time, eventually causing chondrocalcinosis [39]. In fact, magnesium is a crucial factor in the prognosis of GS, due to the possible development of such pathological conditions [37].

Historically, GS has been considered a benign disease, which was generally diagnosed incidentally by the presence of cramps, extreme fatigue, tetany or muscle weakness [5]. However, there is a wide phenotypic variability among patients, ranging from asymptomatic individuals to individuals with a severe phenotype [33]. Furthermore, it was considered as a "desirable disease" in the sense that patients typically have normal or low blood pressure [33,40]. For instance, a recent case report described a kidney transplant from a donor with GS, in which the possibility of lowering the blood pressure in the recipient was considered beneficial [41]. However, clinical surgeries like this can be seen as controversial.

Similarly, GS (as well as BS) was considered as a human model of hypotension, since it was seen that carriers of pathogenic heterozygous mutations in their respective genes were associated with a lower blood pressure than that of the control population [42]. Interestingly, a study performed in 2000 had already anticipated that some pathogenic mutations in *SLC12A3* could protect against hypertension, which coincides with the literature [41–46]. Nevertheless, the presence of some homozygous variants is associated with primary hypertension [47]. A relevant study confirmed the possibility that Gitelman's patients may develop hypertension due to continuous activation of the RAAS axis [48]. The role of zygosity in this process has yet to be elucidated.

In addition to hypertension, several phenotypes have also been related to this disease. For example, it has been stated that patients with GS have a greater predisposition to viral infections [49,50], and they are more prone to develop type II diabetes mellitus [28,51,52]. Moreover, it was recently postulated that they also have an abnormal glycosylation pattern in angiotensin converting enzyme 2 (ACE2), which leads to the activation of RAAS (also seen in BS patients) [45].

Long-term population studies might provide a better understanding and anticipation of the comorbidities that Gitelman's patients might face. It is essential to make a great effort in the study of genotype-phenotype correlations, emphasizing the correlations with different levels of blood pressure values.

3.2. Molecular Basis of Bartter Syndrome and Clinical Consequences

In terms of presentation, Bartter Syndrome has been traditionally grouped into neonatal or classic Bartter Syndrome. The neonatal refers to a severe form with an antenatal presentation that leads to serious polyuria. Consequently, polyhydramnios, premature delivery and severe cases of electrolyte and water loss occur [53]. The classic type refers to a more subtle presentation that can occur at any time, but typically in early childhood, with polyuria, polydipsia, volume contraction and muscle weakness [34]. Currently, the different forms of Bartter Syndrome are classified into six subtypes based on the underlying gene. Moreover, the different subtypes can be further grouped into three categories, based on the similarity between the main molecular mechanisms in which the encoded products participate and their associated pathophysiology: BS type 1 and 2; BS types 3 and 4; and BS type 5.

- Bartter syndrome types 1 and 2

The proteins NKCC2 (BS type 1) and ROMK (BS type 2) are the main players in the reabsorption of solutes in the TAL [19]. When they are disrupted, the physiological abnormalities lead to an early phenotype, usually with manifestations appearing during the prenatal stage [54]. The main consequences of its dysfunction in the embryonic stage includes electrolyte imbalances that can cause polyuria due to isosthenuria, polyhydramnios, preterm birth [55] as well as subsequent growth retardation, serious episodes of salt loss, hypercalciuria and metabolic alkalosis with hypokalemia and hypochloremia [54]. Although type 1 and type 2 share most of these symptoms, the appearance of episodes of early transient hyperkalemia is seen mostly in type 2 [56].

Newborns with mutations in the *KCNJ1* gene may not be able to excrete potassium via ROMK and other non-canonical mechanisms restore potassium levels later [54]. Due to the strict correlation and functional of the transporters in the TAL, the loss of function of the ROMK channels (BS type 2) could lead to the inactivation of NKCC2 (BS type 1) [57], which could justify the overlapping phenotype in both types. Furthermore, an interaction of both channels has also been demonstrated in the secretion of uromodulin protein [58,59], which is decreased in patients with type 2 Bartter [59]. Thus, these results point to a possible role for uromodulin in tubular disorders.

In addition, new studies on the function of NKCC2 channels have shown that stoichiometry can change due to mutations in the *SLC12A1* gene [60]. The type of mutation can determine differences in electrolyte disorders, which would explain the high phenotypic variability between individuals. For example, particular genetic alterations could mean that the Na^+ K^+ $2Cl^-$ cotransporter changes to the unique Na^+ Cl^- transport. Genotype-phenotype correlation studies of large cohorts could help determine the relevance of the type of mutations in relation to the phenotype.

- Bartter syndrome types 3 and 4

BS type 3 is caused by genetic abnormalities in *CLCNKB* gene. The transporter encoded by this gene mediates the reabsorption of chloride from tubular cells to peritubular capillaries in TAL and DCT [2,34]. This type of BS is one of the most investigated, due to the clinical similarity with GS and the need to differentiate them for an accurate diagnosis. It is characterized by enormous clinical variability. Thus, the first manifestations can appear at any time, from the antenatal to the adult stage, and truncating variants are mainly associated with early onset of the disease [34].

The central feature of BS type 3 is severe hypochloremia [54]. CLCNKB is one of the channels necessary for the reabsorption of NaCl [34] in TAL, and its function has to be intact for the reabsorption of chlorine in DCT [61,62]. In fact, its total genetic inactivation is incompatible with life [63]. An incorrect reabsorption can affect the functionality in the $HCO3^-/Cl^-$ exchanger [54], and thus chloride homeostasis is totally damaged in this segment of the nephron. CLCNKA cannot compensate for the loss of function in CLCNKB, on the basis that *CLCNKA* expression is decreased in orthologous *CLCNKB* null mice [61].

BS type 4 can be caused by mutations in *BSND* (type 4a) or digenic recessive mutations in CLCK– channels (*CLCNKB* and *CLCNKA*, Type 4b). In both cases, congenital deafness is a differential symptom [64,65], due to the loss of potential load in the inner ear and the incorrect function of chloride channels.

Furthermore, the BSND is not properly y localized when CLCNK channels are absent [63]. BSND is a mandatory subunit for the normal function of all CLCNK channels [12,65], so the phenotype of BS type 4 is more severe than BS type 3. Interestingly, although BSND is also present in DCT, BS type 4 is generally not confused with GS, as is the case with type 3 BS. In contrast to the other types of BS, renal failure is mainly associated with patients with type 4 BS [12].

Since the first discovery in 2004 of recessive digenic inheritance of the *CLCNKB* and *CLCNKA* genes [14,15], only a few cases of BS type 4b have been reported since then. This is because these types of cases, in addition to being rare, are not usually included in population-based studies of BS. This complexity can add another intriguing aspect. Given that large heterozygous deletions of the physically contiguous genes, *CLCNKA* to *CLCNKB* have been reported, the question remains, is CLCNKA actually associated with the disease? Until now, no homozygous or heterozygous patients with *CLCNKA* have been phenotypically reported.

The homozygous inactivation of the orthologous *CLCNKA* gene in animal models resembles the phenotype of diabetes insipidus, although the mice do not lose salt [66]. As in the case of Alport syndrome associated with the contiguous deletion of the *COL4A5* and *COL4A6* genes [67], excluding the association of the *COL4A6* gene with said syn-drome [68], it would be interesting to identify the clinical or molecular relevance of the inactivation of the *CLCNKA* gene alone in BS. Recently, the p.R83G variant in *CLCNKA* has been postulated as the putative gene loci for a major incidence of heart failure in dilated cardiopathy [69]. Except for this association, thus far, no possible phenotypes directly related to the functional deficiency of the *CLCNKA* gene have been identified.Bartter syndrome type 5.

The clinical characteristics of patients with mutations in the *MAGED2* gene are very similar to classic Bartter, highlighting polyuria, hyperreninism and hyperaldosteronism. However, the most relevant clinical findings consist of severe polyhydramnios, premature birth and perinatal complications. Despite starting as a severe form of Bartter syndrome [13], phenotypic restoration occurs spontaneously, without the need for any specific treatment.

The explanation for this temporal manifestation lies in the fact that the apical localization of NCC (*SLC12A3*) and NKCC2 (*SLC12A1*) depends on MAGED2 during the developmental stages in humans [13]. It is logical that the diminished ubieties for both cotransporters bring about a severe phenotype [70]. Two theories have been put forward on the pathophysiology related to MAGED deficiency. The first is that MAGED2 binds to Hsp40 to regulate endoplasmic reticulum-associated degradation (ERAD), and therefore mutations in *MAGED2* cause alterations in this process [71] and NCC and NKCC2 are retained intracellularly.

The second postulates that the mutations in *MAGED2* prevent sufficient concentrations of cyclic adenosine monophosphate (cAMP) for the correct function of the antidiuretic hormone (ADH), causing the mislocalization of the channels [70]. It is essential to bear in mind that ERAD, lysosomal degradation and the specific ubiquitination of unfolded and immature NCC and NKCC2 have also been widely described as key mechanisms of protein expression and localization [72–75], for which further study of Bartter's disease type 5 will contribute to a greater understanding of these fundamental processes. Similarly, the presence of mutations in the *MAGED2* gene could explain the cases of idiopathic polyhydramnios and unresolved prenatal tubulopathies [76].

3.3. *Long-Term Outcomes in GS and BS*

Investigation of tubular channels as well as micropuncture and patch clamp studies have significantly benefited our understanding of GS and BS [27,77–80]. Despite this, it is still necessary to continue contributing to basic research on these pathologies. The different studies that focused on the characterization of these diseases in various stages of the disease had small sample sizes. Therefore, many unknowns about the progression of these diseases and their possible complications remain to be clarified.

For example, it is necessary to investigate the systemic consequences of these tubulopathies. Recent studies suggested that *SLC12A1* (NKCC2) could be involved in the pathogenesis of hyperparathyroidism [81,82]. The current evidence is not sufficient to establish the link, since the sample sizes only reached a dozen patients, and the biochemical relationship is unproven. It would be interesting to expand these clinical case studies, paying particular attention to the interactions between NKCC2 and CASR (extracellular calcium-sensing receptor) (the molecular target of drugs for hyperparathyroidism) [82]. Recent studies have suggested that GS [83–85] and BS [34,85,86] may be associated with glomerular damage, despite being classic tubulopathies. In both cases, moderate proteinuria associated with focal lesions of segmental glomerular sclerosis and thickening of the glomerular basement membrane have been reported [84]. These alterations could be due to possible glomerular damage due to sustained tubular insufficiency, as occurs, for example, in Dent's disease [87]. This can lead to an erroneous clinical diagnosis, and for this reason it is very important to consider certain diagnostic tools for a correct clinical diagnosis.

4. Diagnostic Approaches

Figure 2 summarizes the main diagnostic methods for Gitelman and Bartter syndromes, as well as the relevant aspects of each to take into consideration.

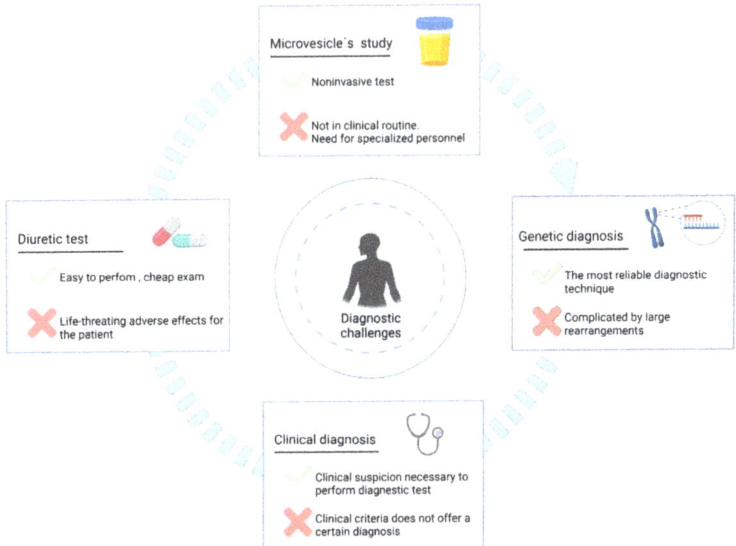

Figure 2. Main methods for the diagnosis of Gitelman and Bartter syndromes. Each method is accompanied by its principal benefit and the major drawback.

4.1. Clinical Diagnosis

A differential clinical diagnosis based solely on symptomatology is currently considered difficult and inaccurate. Since there is phenotypic overlap between GS, and the different types of BS, as well as great clinical variability between individuals with the same syndrome. Generally, a prenatal presentation of BS can automatically rule out BS type 3 however, BS type 3 can, in some cases, lead to an early or prenatal manifestation of the disorder [54].

Adding a further degree of complexity, there are other clinical entities that resemble BS and GS that can be confused and thus lead to inaccurate diagnoses. Among others, autosomal dominant familial hypocalcemia stands out, which before the discovery of *MAGED2*, it was considered to be the fifth type of Bartter syndrome [25]. Familial hypocalcemia is caused by pathogenic mutations in the gene *CASR*, these include activating mutations (OMIM #601199). This gene encodes a plasma membrane G protein-coupled receptor, CASR, which in the kidney tubule negatively regulates calcium resorption. The inappropriate activation of the CASR receptor by these activating mutations leads to hypochloremic metabolic alkalosis, hypokalemia and hypocalcemia, a phenotype that mimics BS [88].

Moreover, the inadequate activation of this receptor entails a reduction in the activity of other channels involved in BS, including ROMK [88], NKCC2 [89] and Na^+/K^+ ATPase [90,91] therefore, this resembles the loss of function of NKCC2 and NCC accomplished by mutations in *MAGED2* (previously discussed) [88]. Of note, the dominant inheritance pattern of familial hypocalcemia differs from the characteristic pattern of BS which is recessive [92].

Given the degree of complexity associated with the physiology of these syndromes, it is important to prioritize clinical peculiarities that, in some cases, allow a differential diagnosis, as well as to identify symptoms that characterize other conditions similar to GS and BS that may be confused with each other [93]. Some of the most relevant indistinguishable conditions are summarized in Table 2 (for GS) and in Table 3 (for BS). Furthermore, in Figure 3 we summarize the best differential symptoms that could help to discern between GS and the different types of BS.

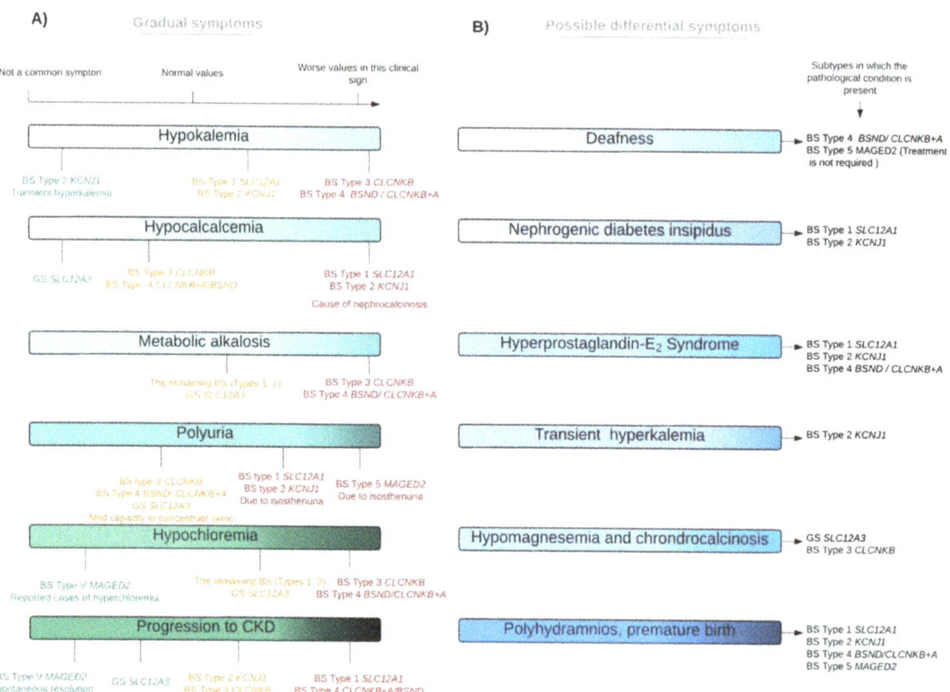

Figure 3. The phenotypic spectrum of GS and BS; principal signs. This figure reflects the high variability that exists among the different subtypes of BS and GS. (**A**) Clinical parameters present in several subtypes, with different levels of severity. Red colour is indicative of a worse severity, whilst orange and green correspond to a lesser disease severity (in this descending order). (**B**) Characteristic (but not necessary) clinical parameters of specific subtypes, which can be used for differential diagnosis (if present). For these cases, there are not differential grades in each subtype.

Generally, a pattern of electrolyte alterations in the biochemical studies support the first proof for the clinical diagnosis [94]. Additionally, an elevation in blood of renin, aldosterone and prostaglandin E_2 (PGE_2) are specific parameters that can shed light on the precise clinical diagnosis for GS and BS, disregarding other tubulopathies [95]. One possible parameter that can provide a differential diagnosis of Bartter and Gitelman over other diseases that exhibit metabolic alkalosis, includes the assessment on Cl^- concentration in urine, which is persistently high [96]. Moreover, chloride status is considered as an important biomarker before the start of any pharmacological treatment [85].

Table 2. Clinical entities that mimic Gitelman syndrome.

Gitelman-Like Diseases	Clinical Similarities to Gitelman	Clinical Differences to Gitelman	References
Abuse of thiazide diuretics	Hypochloremic metabolic alkalosis Hyperaldostheronism Hypokalemia Hypomagnesemia	The symptoms will be ruled out after the withdrawal. Hyperchloruric response as indicative parameter	[3,96,97]
ADTKD-HNF1-β	Hypomagnesemia Hypokalaemia Hyperparathiroidism Young adult presentation (but not mandatory at this ages)	Dominant inheritance pattern Extrarenal manifestations: MODY type 3 Renal cysts	[98–100]
Cystic fibrosis	Hypochloremic metabolic alkalosis Hypokalemia Hypochloremia	Levels of chloride in response to treatment Similarities are present with hot weather	[101,102]
Autoinmune diseases; SLE [1], Sjögren syndrome, autoimmune thyroiditis	Muscular weakness Hypokalemia Hypomagenesemia Hypocalciuria Cramping	Presence of auto-antibodies against NCC [2]	[103,104]
Congenital Chloride Diarrhea	Hyponatremia Hypochloremia	Low chloride in urine but high in stool Watery stool, similar to urine	[105–107]

[1] SLE: Systemic Lupus Erythematosus; [2] NCC: Solute carrier family 12 member 3.

Table 3. Clinical entities that mimic Bartter syndrome.

Bartter-Like Diseases	Clinical Similarities to Bartter	Clinical Differences to Bartter	References
Abuse of loop diuretics	Hypochloremic metabolic alkalosis Hyperaldostheronism Hypokalemia Hypercalciuria Ototoxicity Hyperuricemiaç	The symptoms will be ruled out after the withdrawal. Hyperchloruric response as indicative parameter	[3,96,97]
Autosomal dominant hypocalcemia due to *CASR*	Hypokalemia Metabolic alkalosis Hyperreninemia, hyperaldosteronism Hypocalcemia with hypercalciuria Nephrocalcinosis	Dominant inheritance pattern Low concentration of PTH [3]	[88,108,109]
ADTKD-*REN*	Early presentation in life (childhood) Hyperuricemia Hypokalemia	Dominant inheritance pattern Hypoaldosteronism Hyporeninism	[110,111]
Nephrogenic diabetes insipidus	Polyuria Failure to thrive Hypokalemia, hypercalciuria Nephrocalcinosis	Hypernatremia, hyperchloremia No response to desmopressin	[112–114]
Cystic fibrosis	Hypochloremic metabolic alkalosis Failure to thrive Hypokalemia Hypochloremia	Levels of chloride in response to treatment Similarities are present in hot weather	[101,102,115]
Dent's disease	Hypokalemic metabolic alkalosis Hypercalciuria Hyperreninism Hyperaldostheronism	Proximal characteristic Proteinuria with low molecular weights	[87,116]
Congenital Chloride Diarrhea	Polyhydramnios Premature delivery Hypokalemia Loss of Na^+ and Cl^- in urine	Low chloride in urine but high in stool Watery stool, similar to urine	[105–107]

Table 3. *Cont.*

Bartter-Like Diseases	Clinical Similarities to Bartter	Clinical Differences to Bartter	References
Aminoglycosides antibiotics: gentamicin and amikacin	Metabolic alkalosis Hypocalcemia Hypomagensemia Polyuria Hearing loss (amikacin)	Slow recovery after the finish of the treatment	[96,117–119]

[3] PTH: Parathyroid hormone.

4.2. Diuretics Tests

Diuretics are active natriuretic principles that modify electrolyte transport and are widely used for the treatment of hypertension, heart failure, and fluid overload [120,121]. The main idea of conducting a diuretic test is to demonstrate the lack of effect in patients with GS and BS [122,123]. Specifically, the lack of effect is observed for the diuretics that block the channels related to the diseases; thiazide diuretics for GS and loop diuretics for BS (Figure 1). Briefly, thiazide diuretics cannot manifest its effects in GS because NCC channel has an altered function. In BS, loop diuretics (furosemide as a classic example) will have no effect because the NKCC2 channel is not at its optimal activity. To analyze this effect, the natriuretic and chloruretic effect of the drugs are assessed in spot urine samples, approximately 6 hours after administration [122,123]. The main objective of the clinical trial NCT00822107 was to evaluate the chloruretic response to thiazide in Gitelman syndrome. The results were published in April 2019 and revealed that the thiazide response in GS was null, but only five patients were analyzed (all of them women). This should be considered a clue, but it does not reflect the usefulness of taking these types of exams. Likewise, clinical information rescued by these tests is not totally enlightening.

On the one hand, it is difficult to obtain an absolute answer (yes or no) to a drug treatment. This will only be useful for GS (*SLC12A3*) or BS type 1 patients (*SLC12A1*), being the remaining types of BS only partially interpreted. On the other hand, it is controversial to withdraw daily treatments to patients (K^+ sparing diuretics) to perform the diagnostic test [124]. In addition, other treatments cannot be withdrawn (such as antiprostaglandins or antialdosterone) and this could introduce a relevant bias [123]. Finally, in the case of BS, diuretic tests also carry possible serious side effects, such as severe volume depletion [125]. For GS, this consequence is not very serious but there is a wide variability in the response for diuretic drugs in GS [123], that can lead to misinterpretation.

4.3. Analysis of Urinary Microvesicles

Another possible recent clinical approach has been the study of urinary microvesicles in order to diagnose and predict possible biomarkers of GS and BS [126,127]. All the channels mainly related to these diseases are apical proteins that can be manifestly secreted into multivesicular bodies (MVBs). Therefore, the vesicular proteomic profile of each patient can be studied using different proteomic techniques, in order to identify possible biomarkers of disease. Despite being a non-invasive method, this approach is not yet a reality in current clinical routine due to its cost and time [128].

4.4. Genetic Diagnosis

Given the high phenotypic inter- and intra-familial variability of these syndromes and the large number of phenotypes that often lead to confusing them, genetic studies for GS and BS may be the most conclusive tool for diagnosis. The identification of causal mutations in any of the associated genes can not only determine the diagnosis, but also allow for establishing a better prognosis of the disease [37,54] and a better treatment strategy for the patient [33,125].

In addition, information related to mutations could help to inform about possible comorbidities. From a public health point of view, genetic studies are also the basis for prenatal and preimplantation diagnosis. Prenatal tests in BS are very relevant in prenatal forms of types 1, 2, 4 and 5. In fact, any clinical case with polyhydramnios without morphological anomalies must raise suspicions of Bartter syndrome, and the case is candidate for genetic testing [13,34,71].

Additionally, in GS, prenatal tests would be of great value [129] since some cases have been reported and diagnosed during pregnancy [130,131], with episodes of hypokalemia and salt craving. Nevertheless, as we will discuss later, genetic testing is not without problems and a conclusive diagnosis is not always achieved.

The development and constant improvement of different techniques for the study of the genome, such as massive sequencing technologies (ultra-sequencing or Next Generation Sequencing-NGS) and the continuous reduction in cost allows more health systems to include this type of test as part of the diagnostic routine. The process of searching for pathogenic mutations generally begins with the sequencing of the known associated genes by NGS technologies. The GS and BS genes are often part of predesigned panels of known genes involved in tubulopathies [2], chronic hypokalemia or salt-acid disorders [5].

Another complementary method is the search for CNVs (copy number variations) in candidate genes, through specific NGS products and software, arrays of SNPs (single nucleotide polymorphism) or MLPA® (Multiplex Ligation-dependent Probe Amplification), among others [12,132,133]. This method is of particular interest for *CLCNKB* gene, where a large deletion was found to be the most common mutation, particularly among the Chinese population [34,54,132,134,135]. This strategy will normally end up with the identification of candidate diagnostic mutations in one of the BS or GS associated genes.

A recent study demonstrated an efficacy greater than 85% in finding the causal BS mutations [76]. For GS, the diagnostic efficacy is lower, around 62% [5,136–139], and a considerable number of clinical-diagnosed patients carries only one mutation (around 25%) [136–139]. In both scenarios, it is recommended to carry out a study of carriers of the variants identified in the parents, in order to determine that each variant is located in different copies of the gene. In a subset of the cases studied, an inconclusive or negative result can be obtained.

An inconclusive result refers to studies in which only one candidate variant is found, or where the variant or variants found are classified as "variants of uncertain significance" (VUS). VUS are variants for which there is not enough knowledge to predict whether they could cause a pathogenic effect on the organism, and thus it is not possible to unequivocally establish a causal relationship between the variant and the patient´s disease. Therefore, they cannot be classified as pathogenic or neutral [140].

The lack of identification of a second mutation or the absence of any variant may be due to different situations: (a) the nature of the mutation it is not detected by the technology used; (b) the pathogenic mutations are in regions of the genome, with an unknown, or little-known function. These regions are generally excluded from analysis due to the difficulty of predicting their impact on the phenotype [141]. This may be the case of variants that affect promoters, poly (A) regions, non-translated regions (UTR) or deep intronic regions, among others; (c) the causative gene has not yet been discovered and is therefore not included in diagnostic gene panels. Investigation of unresolved cases and their parents using whole genome strategies could lead to the discovery of new genes associated with GS or BS; or (d) the identification of a second mutation is not described since it does not really exist.

As with other disorders, it could be the case that the presence of a single heterozygous pathogenic variant located in a gene associated with recessive inheritance is associated with mild forms of the disease. Advances in basic research are providing new insights into the function of hitherto poorly understood genomic region, which may lead to improved interpretation of the effect of variants in these regions [142]. In the same way, the constant improvement in the technology for studying the genome will probably increase the rate of positive studies.

To date, no large populations of GS have been carried out that allow establishing a representative genotype-phenotype correlation of the disease [50]. However, peculiarities of different populations have been identified, such as, such as hotspot mutations in the Japanese population associated with higher levels of magnesium in the blood than others [50]. Another example is the exclusivity of two mutations in the Roma ethnic group: *SLC12A3* c.1180+1G>T and *SLC12A3* c.1939G>A [143].

For BS there is a defined phenotype for each subtype; however, these manifestations may overlap with others, especially early in the disease [54]. In any case, there is not a strict correlation for the type of mutation and the disease severity, as in the case of other inherited diseases such as Autosomal Dominant Polycystic Kidney Disease [144,145]. Perhaps, this is due to the lower prevalence of GS and BS, which makes it more challenging to study.

Heterozygous Carriers for Gitelman Mutations

Individuals carrying only one heterozygous pathogenic mutation are not considered as patients of this disease. However, as with other pathologies, it has been described that these patients may present a mild phenotype [51,146–149]. It is arguable that these individuals might have a second mutation, but several independent studies have ruled this out. In a study carried out in 147 carriers in the Amish population [146], lower blood potassium values were associated with heterozygous carriers.

However, a recent cross-sectional study (with 81 heterozygous individuals) did not reflect this difference. The lower K^+ value shown in this article [146] could be related to the type of mutation (*SLC12A3* p.R642G, rs200697179). However, correla-tion of genotype-phenotype has not been established in GS to date, and only recent studies are beginning to determine these variabilities [50]. Another possibility is that either genetic background in this population or environmental-acquired factor might also play a key role.

As in previous studies, recent publications [51,146,147] highlighted higher resting heart rates as the main characteristic of heterozygous carriers. In contrast to previous studies, they did not identify lower values in blood pressure for heterozygotes [148,149]. In one of them, the lower blood pressure was only referenced in children [149]. Although there is a current debate on this topic, the general conclusions thus far do not relate lower blood pressure in heterozygotes. Therefore, clinical peculiarities ascribed to *SLC12A3* in heterozygotes may not result in a pathological condition but will be paired with a different clinical profile. In any case, this situation supports a genetic complexity in GS, rather than it being a classical recessive inherited disease.

5. Therapeutic Approaches

5.1. Current Pharmacological Treatments

GS and BS are determined by the lack of clinical interventions and therapeutic possibilities. There is not a definitive treatment and therapeutic interventions are based on supportive treatments. Furthermore, information on the efficacy of each therapy is scarce that there is not a well-established algorithm for the treatment of these patients [125]. Thus, each patient receives a treatment based on the individuals' effectiveness [20]. The first goal of any treatment for these syndromes is the restoration of the electrolyte profile. Due to the genetic nature of the diseases, these disorders are chronic, and the treatment is for the entire life of the patient.

5.1.1. Oral Salt Supplementation

Water and salt supplements are the first-line approach since their goal is to restore electrolyte loss. Although the commercially available supplements have improved since ab libitum salt treatment [33], patients face several problems with such formulations. For example, the number of tablets and pills per day is often high, which creates an uncomfortable daily situation. An examination of the electrolyte abnormality profile of each patient is necessary to prescribe adequate supplementation. KCl is the preferred form of salt supplementation because it replaces the loss of potassium and chloride (both of them are affected) [85].

In patients with hypomagnesemia, magnesium supplementation becomes very important to prevent associated chondrocalcinosis [150]. Magnesium values are not easily restored and the bioavailability is quite variable depending of the quelant for Mg^{2+}, causing adverse effects and ongoing inconveniences such as diarrhoea [151]. Within salt supplementation, it is cruicial to assist the diet that patients follow carefully. The reason is that patients often have an appetite for salty foods, and continued and uncontrolled intake of salts can worsen the disease state [95]. In addition, traditional herbal medical preparations and dietary supplements can be enriched in salts and must therefore be administered with care or prohibited [24]. Pharmacokinetic studies would be very useful for these patients to establish the bioavailability achieved by each supplement in each individual.

5.1.2. Non-Steroidal Anti-Inflammatory Drugs (NSAIDs)

Non-steroidal anti-inflammatory drugs (NSAIDs) are medicines widely used to relieve pain, reduce inflammation and lower the temperature [152]. Its mechanism of action is based on the non-selective inhibition of the cyclooxygenase (COX), the enzyme that is involved in the synthesis of PGE_2 [153]. Therefore, they are useful to prevent the elevation of PGE_2 that typically occurs in BS patients, as the ones that benefit the most from this treatment compared with GS patients. However, these drugs are used in both diseases, as they can alleviate hyperprostaglandinuria, secondary hyperaldosteronism, hypochloremic hypokalemic metabolic alkalosis, polyuria, hypercalciuria and growth retardation [154].

Indomethacin is the most prescribed NSAID for the treatment of BS [125]. Its use has showed strong benefits and is crucial at an early start in the treatment. The most relevant effect is the restoration of electrolyte abnormalities, which improve the development of the disease. Another important point is the reduction of nephrocalcinosis since indomethacin is able to reduce the hypercalciuria [155,156].

Although the symptomatology of GS usually appears during later in childhood or adulthood, there is wide variability in the spectrum of GS and symptoms can appear early in childhood [157]. In these cases it is common to suffer failure to thrive. Hence, the use of indomethacin could result as beneficial as for Bartter´s patients. This drug was also effective in the correction of hypokalemia during adulthood in a study that included 22 Gitelman´s patients [158]. Indomethacin showed better results than eplenerone or amiloride (discussed later).

The paradox of these treatments is the unavoidable adverse effects of NSAIDs on the kidney [159,160] and gastrointestinal tract [161], even when administered with proton pump inhibitors, such as omeprazole [125,152]. Gasongo et al. [156] suggested the monitoring of renin concentration to achieve the minimal dose with which it is possible to see a beneficial effect. This proposal points towards a more personalized medicine for each patient and could be strongly advantageous. This proposal is particularly important for affected children with different needs for indomethacin or ibuprofen treatments to avoid developmental delays. In this scenario, hospital pharmacies are very useful to prepare the adequate oral suspensions, in relation to the weight and dose required by these patients [162].

The toxicity of NSAIDs is a logical reason to try alternative medications such as COXIBs (ciclooxygenase-2 inhibitors) which, in contrast to NSAIDs, can selectively inhibit COX-2 [163]. However, COXIBs can also cause adverse effects, such as an increased risk of cardiac problems and, in particular, myocardial infarction [164]. To date, there are no well-documented records of secondary effects under the treatment of COXIBs in BS or GS patients. Thus, there is a strong preference to NSAIDs over COXIBs [165], and COXIBs only supports a better therapy when the use of NSAIDs plus proton pumps inhibitors causes an uncontrolled severe hypomagnesemia [125].

In summary, NSAIDs constitute one of the most reliable treatments for the symp-toms of BS and GS, despite not being a definitive pharmacological approach.

5.1.3. Potassium Sparing Diuretics

Potassium sparing diuretics include a series of diuretics without a strong diuretic effect. However, they are capable of restoring hypokalemia caused by diuretics furosemide and thiazide, this being the main reason for their use [166,167]. They act in the distal part of the nephron and are classically subdivided as aldosterone antagonists or Na^+ channel blockers [166]. Considering the suppressive effect of these drugs on aldosterone and secondary hyperkalemia, they can be considered as promising pharmacologic agents in GS and BS. However, if the distal region is blocked, partial reabsorption of Na^+ and water cannot occur, leading to worsened salt loss. Therefore, the patient could be at risk of life-threating hypovolemia [20,168–170].

The first investigations that deepened the usefulness of these drugs found good efficacy in the use of amiloride in patients with BS [171–173]. Notably, amiloride was reported during pregnancy in a few cases [174,175] (B category for pregnancy [176]). However, a recent study reiterated the dangerous consequences of amiloride treatment, including hypovolemia and fibrosis in the kidney and heart in long-term treatments [158]. Another complication of its discontinuation is hypotension [169,170]. On the contrary, a beneficial point of its use is a minor loss of magnesium with this treatment, although only slight modifications were observed [158].

Essentially, sparing diuretics are only accepted clinically when the other therapeutic options, such as NSAIDs or supplemental drugs are have no impact on phenotype improvement [170,171]. Moreover, cases with a severe phenotype do not improve even with these treatments [177] and the risk of hypovolemia must be carefully assessed [20,168–170].

5.1.4. Renin-Angiotensin-Aldosterone System Inhibitors

GS and BS are characterized by a constant activation of the RAAS axis in an attempt to recover lost salt and water [25]. In fact, this alteration is the cause of both hyperreninism and hypertrophy of the juxtaglomerular apparatus (specific for BS) [20]. Hence, the use of inhibitors of this axis was proposed as a therapeutic strategy. This clinical approach is similar to sparing diuretics, because the potential risk of hypovolemia and hypotension caused by these medications causes restrictions in the prescription of these drugs [20,178]. ACE inhibitors cannot resolve the clinical abnormalities of GS and BS patients. In fact, they do not improve electrolyte disturbances [179]. Despite this, they can be a therapeutic possibility for cases of refractory NSAIDs, nephrotoxicity with NSAIDs [180] or even superior to sparing diuretics [181]. In addition, a possible benefit is the partial correction of hypomagnesemia, which was described in one case with enalapril [179].

5.1.5. Growth Hormone (GH)

Growth hormone (GH) is intended to prevent growth abnormalities [182]. Im-portantly, GH is considered only after the failure of NSAID treatment [125] for the same purpose. GS is a disease in which growth retardation is less frequent than in BS [33] but growth hormone is considered as beneficial [33,50]. Perhaps, GH shows a more beneficial effect in GS than in BS due to the worse phenotype in BS in terms of growth retardation. In fact, Fujimura´s studies [50] stated that GH treatment was not accurate in all patients. The duration of treatment may need to be longer to achieve visible efficacy even into adolescence or young adulthood.

This strongly correlates with clinical guidance, in which treatment should be prolonged until a coherent mineral density is achieved [183,184]. To date, GH has been proven in cases of kidney damage and does not impact further damage [185]. However, the major drawback is the controversy among endocrinologists over the side-effects of the long-term use of the drug. The most recent guidelines [186,187] have pointed out two essential points: The relevance of genetic testing to ensure the causes and the possibility of discontinuation and the improvements in the adherence with lower frequencies in the administration of the recombinant human growth hormone.

5.1.6. Overview of the Effectiveness of These Treatments

It is important to note that all the current treatments are supportive, and none of them can solve the main problems later in life. Therefore, more research essential to provide better solutions for the patients. Another key point is that BS and GS therapeutics overlap, sharing the vast majority of possible treatments as reflected in Table 4.

Treatment in GS and BS is based on the administration of a wide range of drugs, which depends both on the criteria of the physician and on the severity of each patient [33,125]. Therefore, clinicians pay special attention to the manifestations of hypomagnesemia and its consequences in GS, while in the case of BS, indomethacin is the main approach [125]. In this scenario, there is no comprehensive understanding of the disease or greater improvements in the phenotype of the patients. In fact, knowledge is acquired through description of cases from independent reports, which is why large population studies are required.

Table 4. Currently used pharmacological treatments for Bartter and Gitelman syndromes.

Therapeutic Approaches	Gitelman Syndrome	Bartter Syndrome
Supplemental electrolyte drugs	Mandatory, especially with magnesium loss	Mandatory
NSAIDs	Possible	Indomethacin as principal treatment in BS
Potassium Sparing Diuretics	Possible, but not recommendable	Possible, but not recommendable
Inhibitors of RAAS axis	Poorly described, but possible	Possible, especially with nephrotic damage from NSAIDs
Growth Hormone	Possible, poor evidence of efficacy	Possible, poor evidence of efficacy

On the other hand, currently there is still debate on different topics. There is no consensus on the treatment of practically asymptomatic patients with GS [95] or the definition of the threshold that symptomatic patients with salt-wasting tubulopathy must reach to ensure a good prognosis [20]. For example, one question is what level should be reached to have a correct intracellular magnesium concentration in GS [188]. In this sense, the basic concepts in a treatment, such as the "treatment objective" are not yet defined. The "goal of treatment" is to define the point at which treatment can be interrupted or discontinued [189]. A relevant study of A. Blanchar et al. [158] argues that a later period of treatments may improve efficacy.

Perhaps, the correction of the electrolytes in the blood is a gratifying achievement, as it is the most reliable surrogate endpoint, however, another specific point to consider should be the prevention of renal and extrarrenal manifestations such as end stage renal disease (ESRD), chondrocalcinosis or mineral bone fractures. In conclusion, therapeutic possibilities in GS and BS are dominated by a paradigm with fairly variable prognosis.

Currently, despite the lack of knowledge regarding prognosis, this is usually acceptable with correct adherence to the respective treatments [85,190]. Gitelman´s patients have an acceptable prognosis [85], although asthenia and muscle weakness are common daily manifestations which impair quality of life [33]. Bartter´s patients with early presentations can start treatment with indomethacin in the 4th–6th week of life, showing a good response and adequate prognosis [190]. However, BS type 4 shows the worst phenotype, being possible the appearance of nephrocalcinosis and progression to ESRD [191,192]. There is no information on the prognosis in the case of lack or suspension of treatments or the influence of a delayed diagnosis and treatment. Furthermore, there is no information on the differences among different therapeutic combinations in terms of prognosis.

5.2. Future Therapeutic Perspectives

The clinical trials for GS and BS do not offer new drugs for their treatment. In fact, the trials inlvolve different combinations of commonly used drugs, and there is no drug with a newly transitional mechanism of action [193,194].

One of these studies uses acetazolamide (a diuretic that inhibits carbonic anhydrase [195]) to alleviate the metabolic alkalosis of BS (NCT03847571). Although the results have not been yet published, a recent article from the same group suggests the use of acetazolamide as another possible adjuvant of the classical therapies, never as monotherapy [196].

Another study has the intervetion of progesterone to alleviate the hypokalemia (NCT02297048). In line with this hypothesis, Blanchard et al. [197] have reported that adrenal glands adapt their response to levels of hypokalemia and hyponatremia. This could be the first step towards a better understanding of the pathophysiology and, perhaps, towards a new therapeutic group.

6. Conclusions

GS and BS are inherited tubulopathies with challenges in their precise diagnosis since both pathologies have overlapping clinical symptoms. Regarding GS, it used to be considered as a mild disease, but patients frequently report extreme fatigue, showing an impaired quality of life. Regarding BS, these patients tend to have a worse clinical outcome. In fact, symptoms can appear before birth, resulting in delayed growth and propensity to ESRD.

In terms of diagnosis, genetic tests, regardless of the difficulties they may have, are the optimal tool for an accurate and definitive diagnosis. On the other hand, prognosis is poorly known currently, and therefore, population studies that investigate following up with patients are a huge necessity. Lastly, basic research enables the discovery of the molecular basis of the pathologies as well as the identification of novel therapeutic targets.

Author Contributions: L.N.-G., N.C. wrote the manuscript, which was critically edited and reviewed by L.N.-G., N.C. and M.A.G.-G. All authors have read and agreed to the published version of the manuscript.

Funding: This research was funded by Instituto de Salud Carlos III under FIS/FEDER fund PI15/01467 and PI18/00378 (to MA Garcia-Gonzalez) and by Xunta de Galicia award IN607B 2016/020 (to MA Garcia-Gonzalez).

Institutional Review Board Statement: Ethical review and approval were waived for this study, due to the nature of the study (review study).

Informed Consent Statement: Not applicable.

Data Availability Statement: Not applicable.

Acknowledgments: The authors acknowledge Anna-Rachel Gallagher for the useful English orthographic review. Figures 1 and 2 were created with Biorender (www.biorender.com, accessed on 30 May 2021).

Conflicts of Interest: The authors declare no conflict of interest.

Abbreviations

ACE2	angiotensin converting enzyme 2
ADH	antidiuretic hormone
BS	Bartter syndrome
BSND	barttin
cAMP	cyclic adenosine monophosphate
CASR	extracellular calcium-sensing receptor
CLCNKA	chloride channel protein ClC-Ka
CLCNKB	chloride channel protein ClC-Kb
CNVs	copy number variants
DCT	distal convoluted tubule
ERAD	endoplasmic reticulum-associated degradation
GS	Gitelman syndrome
MLPA®	Multiplex ligation-dependent amplification probe
MVBs	multivesicular bodies
NGS	next generation sequencing
PGE_2	prostaglandin E_2
RAAS	renin-angiotensin-aldosterone system
NKCC2	solute carrier family 12 member 1
NCC	solute carrier family 12 member 3
ROMK	ATP-sensitive inward rectifier potassium channel 1

SNPs	single nucleotide polymorfism
TAL	ascending loop of Henle
TRPM6	transient receptor potential cation channel subfamily M member 6
TRPV5	transient receptor potential cation channel subfamily V
UTR	non-translated regions
VUS	variants of uncertain significance

References

1. Arakawa, H.; Kubo, H.; Washio, I.; Staub, A.Y.; Nedachi, S.; Ishiguro, N.; Nakanishi, T.; Tamai, I. Rat Kidney Slices for Evaluation of Apical Membrane Transporters in Proximal Tubular Cells. *J. Pharm. Sci.* **2019**, *108*, 2798–2804. [CrossRef] [PubMed]
2. Downie, M.L.; Lopez Garcia, S.C.; Kleta, R.; Bockenhauer, D. Inherited Tubulopathies of the Kidney. *Clin. J. Am. Soc. Nephrol.* **2020**, *16*, CJN14481119. [CrossRef] [PubMed]
3. Mitchell, J.E.; Pomeroy, C.; Seppala, M.; Huber, M. Pseudo-Bartter's syndrome, diuretic abuse, idiopathic edema, and eating disorders. *Int. J. Eat. Disord.* **1988**, *7*, 225–237. [CrossRef]
4. Bartter and Gitelman Syndromes—UpToDate. Available online: https://www.uptodate.com/contents/inherited-hypokalemic-salt-losing-tubulopathies-pathophysiology-and-overview-of-clinical-manifestations (accessed on 14 May 2021).
5. Hureaux, M.; Ashton, E.; Dahan, K.; Houillier, P.; Blanchard, A.; Cormier, C.; Koumakis, E.; Iancu, D.; Belge, H.; Hilbert, P.; et al. High-throughput sequencing contributes to the diagnosis of tubulopathies and familial hypercalcemia hypocalciuria in adults. *Kidney Int.* **2019**, *96*, 1408–1416. [CrossRef]
6. Bao, M.; Cai, J.; Yang, X.; Ma, W. Genetic screening for Bartter syndrome and Gitelman syndrome pathogenic genes among individuals with hypertension and hypokalemia. *Clin. Exp. Hypertens.* **2019**, *41*, 381–388. [CrossRef]
7. Orphanet: Bartter Syndrome. Available online: https://www.orpha.net/consor/cgi-bin/OC_Exp.php?Lng=GB&Expert=112 (accessed on 26 May 2021).
8. Frederic, C. Hyperplasia of the juxtaglomerular complex with hyperaldosteronism and hypokalemic alkalosis. *A new syndrome. Am. J. Med.* **1962**, *33*, 811–828. [CrossRef]
9. Simon, D.B.; Karet, F.E.; Hamdan, J.M.; Di Pietro, A.; Sanjad, S.A.; Lifton, R.P. Bartter's syndrome, hypokalaemic alkalosis with hypercalciuria, is caused by mutations in the Na-K-2CI cotransporter NKCC2. *Nat. Genet.* **1996**, *13*, 183–188. [CrossRef]
10. Simon, D.B.; Karet, F.E.; Rodriguez-Soriano, J.; Hamdan, J.H.; DiPietro, A.; Trachtman, H.; Sanjad, S.A.; Lifton, R.P. Genetic heterogeneity of Bartter's syndrome revealed by mutations in the K^+ channel, ROMK. *Nat. Genet.* **1996**, *14*, 152–156. [CrossRef]
11. Simon, D.B.; Bindra, R.S.; Mansfield, T.A.; Nelson-Williams, C.; Mendonca, E.; Stone, R.; Schurman, S.; Nayir, A.; Alpay, H.; Bakkaloglu, A.; et al. Mutations in the chloride channel gene, *CLCNKB*, cause Bartter's syndrome type III. *Nat. Genet.* **1997**, *17*, 171–178. [CrossRef]
12. Birkenhäger, R.; Otto, E.; Schürmann, M.J.; Vollmer, M.; Ruf, E.M.; Maier-Lutz, I.; Beekmann, F.; Fekete, A.; Omran, H.; Feldmann, D.; et al. Mutation of *BSND* causes Bartter syndrome with sensorineural deafness and kidney failure. *Nat. Genet.* **2001**, *29*, 310–314. [CrossRef]
13. Laghmani, K.; Beck, B.B.; Yang, S.-S.; Seaayfan, E.; Wenzel, A.; Reusch, B.; Vitzthum, H.; Priem, D.; Demaretz, S.; Bergmann, K.; et al. Polyhydramnios, Transient Antenatal Bartter's Syndrome, and *MAGED2* Mutations. *N. Engl. J. Med.* **2016**, *374*, 1853–1863. [CrossRef]
14. Schlingmann, K.P.; Konrad, M.; Jeck, N.; Waldegger, P.; Reinalter, S.C.; Holder, M.; Seyberth, H.W.; Waldegger, S. Salt Wasting and Deafness Resulting from Mutations in Two Chloride Channels. *N. Engl. J. Med.* **2004**, *350*, 1314–1319. [CrossRef]
15. Bichet, D.G.; Fujiwara, T.M. Reabsorption of Sodium Chloride—Lessons from the Chloride Channels. *N. Engl. J. Med.* **2004**, *350*, 1281–1283. [CrossRef]
16. Gitelman, H.J.; Graham, J.B.; Welt, L.G. A familial disorder characterized by hypokalemia and hypomagnesemia. *Ann. N. Y. Acad. Sci.* **1969**, *162*, 856–864. [CrossRef]
17. Simon, D.B.; Nelson-Williams, C.; Bia, M.J.; Ellison, D.; Karet, F.E.; Molina, A.M.; Vaara, I.; Iwata, F.; Cushner, H.M.; Koolen, M.; et al. Gitelman's variant of Bartter's syndrome, inherited hypokalaemic alkalosis, is caused by mutations in the thiazide-sensitive Na-Cl cotransporter. *Nat. Genet.* **1996**, *12*, 24–30. [CrossRef]
18. Punzi, L.; Calo, L.; Schiavon, F.; Pianon, M.; Rosada, M.; Todesco, S. Chondrocalcinosis is a feature of Gitelman's variant of Bartter's syndrome: A new look at the hypomagnesemia associated with calcium pyrophosphate dihydrate crystal deposition disease. *Rev. Rhum.* **1998**, *65*, 571–574.
19. George, A.L.; Neilson, E.G. Biología celular y molecular de los riñones. In *Harrison. Principios de Medicina Interna*, 19th ed.; Kasper, D., Fauci, A., Hauser, S., Longo, D., Jameson, J.L., Loscalzo, J., Eds.; McGraw-Hill Education: New York, NY, USA, 2019.
20. Kleta, R.; Bockenhauer, D. Salt-losing tubulopathies in children: What's new, what's controversial? *J. Am. Soc. Nephrol.* **2018**, *29*, 727–739. [CrossRef]
21. Fox, S.I. Fisiología de los riñones. In *Fisiología Humana*, 14th ed.; McGraw-Hill Education: New York, NY, USA, 2017.
22. Eaton, D.C.; Pooler, J.P. Mecanismos de transporte tubular. In *Fisiología Médica. Un Enfoque por Aparatos y Sistemas*; Raff, H., Levitzky, M., Eds.; McGraw-Hill Education: New York, NY, USA, 2015.
23. Seyberth, H.W.; Schlingmann, K.P. Bartter- and Gitelman-like syndromes: Salt-losing tubulopathies with loop or DCT defects. *Pediatr. Nephrol.* **2011**, *26*, 1789–1802. [CrossRef]

24. Beck Laurence, H.J.; Salant, D.J. Enfermedades tubulointersticiales del riñón. In *Harrison. Principios de Medicina Interna*, 20th ed.; Jameson, J.L., Fauci, A.S., Kasper, D.L., Hauser, S.L., Longo, D.L., Loscalzo, J., Eds.; McGraw-Hill Education: New York, NY, USA, 2018.
25. Besouw, M.T.P.; Kleta, R.; Bockenhauer, D. Bartter and Gitelman syndromes: Questions of class. *Pediatr. Nephrol.* **2020**, *35*, 1815–1824. [CrossRef]
26. Velázquez, H.; Náray-Fejes-Tóth, A.; Silva, T.; Andújar, E.; Reilly, R.F.; Desir, G.V.; Ellison, D.H. Rabbit distal convoluted tubule coexpresses NaCl cotransporter and 11β- hydroxysteroid dehydrogenase II mRNA. *Kidney Int.* **1998**, *54*, 464–472. [CrossRef]
27. Ellison, D.H.; Terker, A.S.; Gamba, G. Potassium and its discontents: New insight, new treatments. *J. Am. Soc. Nephrol.* **2016**, *27*, 981–989. [CrossRef]
28. Maeoka, Y.; McCormick, J.A. NaCl cotransporter activity and Mg^{2+} handling by the distal convoluted tubule. *Am. J. Physiol.—Ren. Physiol.* **2020**, *319*, F1043–F1053. [CrossRef]
29. Kunchaparty, S.; Palcso, M.; Berkman, J.; Velázquez, H.; Desir, G.V.; Bernstein, P.; Reilly, R.F.; Ellison, D.H. Defective processing and expression of thiazide-sensitive Na-Cl cotransporter as a cause of Gitelman's syndrome. *Am. J. Physiol.—Ren. Physiol.* **1999**, *277*. [CrossRef]
30. Kurtz, I. Molecular pathogenesis of Bartter's and Gitelman's syndromes. *Kidney Int.* **1998**, *54*, 1396–1410. [CrossRef]
31. Peters, M.; Jeck, N.; Reinalter, S.; Leonhardt, A.; Tönshoff, B.; Klaus, G.; Konrad, M.; Seyberth, H.W. Clinical presentation of genetically defined patients with hypokalemic salt-losing tubulopathies. *Am. J. Med.* **2002**, *112*, 183–190. [CrossRef]
32. Nijenhuis, T.; Vallon, V.; Van Der Kemp, A.W.C.M.; Loffing, J.; Hoenderop, J.G.J.; Bindels, R.J.M. Enhanced passive Ca^{2+} reabsorption and reduced Mg^{2+} channel abundance explains thiazide-induced hypocalciuria and hypomagnesemia. *J. Clin. Investig.* **2005**, *115*, 1651–1658. [CrossRef]
33. Blanchard, A.; Bockenhauer, D.; Bolignano, D.; Calò, L.A.; Cosyns, E.; Devuyst, O.; Ellison, D.H.; Karet Frankl, F.E.; Knoers, N.V.A.M.; Konrad, M.; et al. Gitelman syndrome: Consensus and guidance from a Kidney Disease: Improving Global Outcomes (KDIGO) Controversies Conference. In *Proceedings of the Kidney International*; Elsevier B.V.: Amsterdam, The Netherlands, 2017; Volume 91, pp. 24–33.
34. Seys, E.; Andrini, O.; Keck, M.; Mansour-Hendili, L.; Courand, P.Y.; Simian, C.; Deschenes, G.; Kwon, T.; Bertholet-Thomas, A.; Bobrie, G.; et al. Clinical and genetic spectrum of bartter syndrome type 3. *J. Am. Soc. Nephrol.* **2017**, *28*, 2540–2552. [CrossRef]
35. Dai, L.J.; Ritchie, G.; Kerstan, D.; Kang, H.S.; Cole, D.E.C.; Quamme, G.A. Magnesium transport in the renal distal convoluted tubule. *Physiol. Rev.* **2001**, *81*, 51–84. [CrossRef]
36. Franken, G.A.C.; Adella, A.; Bindels, R.J.M.; de Baaij, J.H.F. Mechanisms coupling sodium and magnesium reabsorption in the distal convoluted tubule of the kidney. *Acta Physiol.* **2021**, *231*, e13528. [CrossRef]
37. Ea, H.K.; Blanchard, A.; Dougados, M.; Roux, C. Chondrocalcinosis secondary to hypomagnesemia in Gitelman's syndrome. *J. Rheumatol.* **2005**, *32*, 1840–1842. [PubMed]
38. Leone, F.A.; Rezende, L.A.; Ciancaglini, P.; Pizauro, J.M. Allosteric modulation of pyrophosphatase activity of rat osseous plate alkaline phosphatase by magnesium ions. *Int. J. Biochem. Cell Biol.* **1998**, *30*, 89–97. [CrossRef]
39. Calò, L.; Punzi, L.; Semplicini, A. Hypomagnesemia and chondrocalcinosis in Bartter's and Gitelman's syndrome: Review of the pathogenetic mechanisms. *Am. J. Nephrol.* **2000**, *20*, 347–350. [CrossRef] [PubMed]
40. Pollak, M.R.; Friedman, D.J. The genetic architecture of kidney disease. *Clin. J. Am. Soc. Nephrol.* **2020**, *15*, 268–275. [CrossRef] [PubMed]
41. Stewart, D.; Iancu, D.; Ashton, E.; Courtney, A.E.; Connor, A.; Walsh, S.B. Transplantation of a Gitelman Syndrome Kidney Ameliorates Hypertension: A Case Report. *Am. J. Kidney Dis.* **2019**, *73*, 421–424. [CrossRef]
42. Ji, W.; Foo, J.N.; O'Roak, B.J.; Zhao, H.; Larson, M.G.; Simon, D.B.; Newton-Cheh, C.; State, M.W.; Levy, D.; Lifton, R.P. Rare independent mutations in renal salt handling genes contribute to blood pressure variation. *Nat. Genet.* **2008**, *40*, 592–599. [CrossRef]
43. Monette, M.Y.; Rinehart, J.; Lifton, R.P.; Forbush, B. Rare mutations in the human NA-K-CL cotransporter (NKCC2) associated with lower blood pressure exhibit impaired processing and transport function. *Am. J. Physiol.—Ren. Physiol.* **2011**, *300*, 840–847. [CrossRef]
44. Balavoine, A.S.; Bataille, P.; Vanhille, P.; Azar, R.; Noël, C.; Asseman, P.; Soudan, B.; Wémeau, J.L.; Vantyghem, M.C. Phenotype-genotype correlation and follow-up in adult patients with hypokalaemia of renal origin suggesting Gitelman syndrome. *Eur. J. Endocrinol.* **2011**, *165*, 665–673. [CrossRef]
45. Calò, L.A.; Davis, P.A. Are the clinical presentations (Phenotypes) of gitelman's and bartter's syndromes gene mutations driven by their effects on intracellular ph, their "ph" enotype? *Int. J. Mol. Sci.* **2020**, *21*, 5660. [CrossRef]
46. Calò, L.A.; Davis, P.A.; Rossi, G.P. Understanding themechanisms of angiotensin II signaling involved in hypertension and its long-term sequelae: Insights from Bartter's and Gitelman's syndromes, humanmodels of endogenous angiotensin II signaling antagonism. *J. Hypertens.* **2014**, *32*, 2109–2119. [CrossRef]
47. Melander, O.; Orho-Melander, M.; Bengtsson, K.; Lindblad, U.; Råstam, L.; Groop, L.; Hulthén, U.L. Genetic variants of thiazide-sensitive NaCl-cotransporter in Gitelman's syndrome and primary hypertension. *Hypertension* **2000**, *36*, 389–394. [CrossRef]
48. Berry, M.R.; Robinson, C.; Karet Frankl, F.E. Unexpected clinical sequelae of Gitelman syndrome: Hypertension in adulthood is common and females have higher potassium requirements. *Nephrol. Dial. Transplant.* **2013**, *28*, 1533–1542. [CrossRef]

49. Evans, R.D.R.; Antonelou, M.; Sathiananthamoorthy, S.; Rega, M.; Henderson, S.; Ceron-Gutierrez, L.; Barcenas-Morales, G.; Müller, C.A.; Doffinger, R.; Walsh, S.B.; et al. Inherited salt-losing tubulopathies are associated with immunodeficiency due to impaired IL-17 responses. *Nat. Commun.* **2020**, *11*, 1–19. [CrossRef]
50. Fujimura, J.; Nozu, K.; Yamamura, T.; Minamikawa, S.; Nakanishi, K.; Horinouchi, T.; Nagano, C.; Sakakibara, N.; Nakanishi, K.; Shima, Y.; et al. Clinical and Genetic Characteristics in Patients With Gitelman Syndrome. *Kidney Int. Rep.* **2019**, *4*, 119–125. [CrossRef]
51. Blanchard, A.; Vallet, M.; Dubourg, L.; Hureaux, M.; Allard, J.; Haymann, J.P.; de la Faille, R.; Arnoux, A.; Dinut, A.; Bergerot, D.; et al. Resistance to insulin in patients with gitelman syndrome and a subtle intermediate phenotype in heterozygous carriers: A cross-sectional study. *J. Am. Soc. Nephrol.* **2019**, *30*, 1534–1545. [CrossRef]
52. Ren, H.; Qin, L.; Wang, W.; Ma, J.; Zhang, W.; Shen, P.Y.; Shi, H.; Li, X.; Chen, N. Abnormal glucose metabolism and insulin sensitivity in Chinese patients with Gitelman syndrome. *Am. J. Nephrol.* **2013**, *37*, 152–157. [CrossRef]
53. Han, Y.; Zhao, X.; Wang, S.; Wang, C.; Tian, D.; Lang, Y.; Bottillo, I.; Wang, X.; Shao, L. Eleven novel SLC12A1 variants and an exonic mutation cause exon skipping in Bartter syndrome type I. *Endocrine* **2020**, *64*, 708–718. [CrossRef]
54. Brochard, K.; Boyer, O.; Blanchard, A.; Loirat, C.; Niaudet, P.; MacHer, M.A.; Deschenes, G.; Bensman, A.; Decramer, S.; Cochat, P.; et al. Phenotype-genotype correlation in antenatal and neonatal variants of Bartter syndrome. *Nephrol. Dial. Transplant.* **2009**, *24*, 1455–1464. [CrossRef]
55. Mourani, C.C.; Sanjad, S.A.; Akatcherian, C.Y. Bartter syndrome in a neonate: Early treatment with indomethacin. *Pediatr. Nephrol.* **2000**, *14*, 143–145. [CrossRef]
56. Finer, G.; Shalev, H.; Birk, O.S.; Galron, D.; Jeck, N.; Sinai-Treiman, L.; Landau, D. Transient neonatal hyperkalemia in the antenatal (ROMK defective) Bartter syndrome. *J. Pediatr.* **2003**, *142*, 318–323. [CrossRef]
57. Gamba, G.; Friedman, P.A. Thick ascending limb: The Na^{+}:K^{+}:$2Cl^{-}$ co-transporter, NKCC2, and the calcium-sensing receptor, CaSR. *Pflugers Arch. Eur. J. Physiol.* **2009**, *458*, 61–76. [CrossRef]
58. Mutig, K.; Kahl, T.; Saritas, T.; Godes, M.; Persson, P.; Bates, J.; Raffi, H.; Rampoldi, L.; Uchida, S.; Hille, C.; et al. Activation of the bumetanide-sensitive Na +,K +, 2Cl -Cotransporter (NKCC2) is facilitated by Tamm-Horsfall protein in a chloride-sensitive manner. *J. Biol. Chem.* **2011**, *286*, 30200–30210. [CrossRef]
59. Schiano, G.; Glaudemans, B.; Olinger, E.; Goelz, N.; Müller, M.; Loffing-Cueni, D.; Deschenes, G.; Loffing, J.; Devuyst, O. The Urinary Excretion of Uromodulin is Regulated by the Potassium Channel ROMK. *Sci. Rep.* **2019**, *9*, 1–12. [CrossRef]
60. Marcoux, A.A.; Tremblay, L.E.; Slimani, S.; Fiola, M.J.; Mac-Way, F.; Garneau, A.P.; Isenring, P. Molecular characteristics and physiological roles of Na^{+}–K^{+}–Cl^{-} cotransporter 2. *J. Cell. Physiol.* **2021**, *236*, 1712–1729. [CrossRef]
61. Grill, A.; Schießl, I.M.; Gess, B.; Fremter, K.; Hammer, A.; Castrop, H. Salt-losing nephropathy in mice with a null mutation of the Clcnk2 gene. *Acta Physiol.* **2016**, *218*, 198–211. [CrossRef]
62. Hennings, J.C.; Andrini, O.; Picard, N.; Paulais, M.; Huebner, A.K.; Cayuqueo, I.K.L.; Bignon, Y.; Keck, M.; Cornière, N.; Böhm, D.; et al. The ClC-K2 chloride channel is critical for salt handling in the distal nephron. *J. Am. Soc. Nephrol.* **2017**, *28*, 209–217. [CrossRef]
63. Pérez-Rius, C.; Castellanos, A.; Gaitán-Peñas, H.; Navarro, A.; Artuch, R.; Barrallo-Gimeno, A.; Estévez, R. Role of zebrafish ClC-K/barttin channels in apical kidney chloride reabsorption. *J. Physiol.* **2019**, *597*, 3969–3983. [CrossRef]
64. Nozu, K.; Inagaki, T.; Fu, X.J.; Nozu, Y.; Kaito, H.; Kanda, K.; Sekine, T.; Igarashi, T.; Nakanishi, K.; Yoshikawa, N.; et al. Molecular analysis of digenic inheritance in Bartter syndrome with sensorineural deafness. *J. Med. Genet.* **2008**, *45*, 182–186. [CrossRef]
65. Estévez, R.; Boettger, T.; Stein, V.; Birkenhäger, R.; Otto, E.; Hildebrandt, F.; Jentsch, T.J. Barttin is a Cl^{-} channel β-subunit crucial for renal Cl^{-} reabsorption and inner ear K^{+} secretion. *Nature* **2001**, *414*, 558–561. [CrossRef] [PubMed]
66. Matsumura, Y.; Uchida, S.; Kondo, Y.; Miyazaki, H.; Ko, S.B.H.; Hayama, A.; Morimoto, T.; Liu, W.; Arisawa, M.; Sasaki, S.; et al. Overt nephrogenic diabetes insipidus in mice lacking the CLC-K1 chloride channel. *Nat. Genet.* **1999**, *21*, 95–98. [CrossRef] [PubMed]
67. Kruegel, J.; Rubel, D.; Gross, O. Alport syndrome—Insights from basic and clinical research. *Nat. Rev. Nephrol.* **2013**, *9*, 170–178. [CrossRef] [PubMed]
68. Rost, S.; Bach, E.; Neuner, C.; Nanda, I.; Dysek, S.; Bittner, R.E.; Keller, A.; Bartsch, O.; Mlynski, R.; Haaf, T.; et al. Novel form of X-linked nonsyndromic hearing loss with cochlear malformation caused by a mutation in the type IV collagen gene COL4A6. *Eur. J. Hum. Genet.* **2014**, *22*, 208–215. [CrossRef]
69. Villard, E.; Perret, C.; Gary, F.; Proust, C.; Dilanian, G.; Hengstenberg, C.; Ruppert, V.; Arbustini, E.; Wichter, T.; Germain, M.; et al. A genome-wide association study identifies two loci associated with heart failure due to dilated cardiomyopathy. *Eur. Heart J.* **2011**, *32*, 1065–1076. [CrossRef]
70. Quigley, R.; Saland, J.M. Transient antenatal Bartter's Syndrome and X-linked polyhydramnios: Insights from the genetics of a rare condition. *Kidney Int.* **2016**, *90*, 721–723. [CrossRef]
71. Allison, S.J. Renal physiology: *MAGED2* mutations in transient antenatal Bartter syndrome. *Nat. Rev. Nephrol.* **2016**, *12*, 377.
72. Bakhos-douaihy, D.; Seaayfan, E.; Demaretz, S.; Komhoff, M.; Laghmani, K. Differential effects of stch and stress—Inducible hsp70 on the stability and maturation of nkcc2. *Int. J. Mol. Sci.* **2021**, *7*, 2207. [CrossRef]
73. Donnelly, B.F.; Needham, P.G.; Snyder, A.C.; Roy, A.; Khadem, S.; Brodsky, J.L.; Subramanya, A.R. Hsp70 and Hsp90 multichaperone complexes sequentially regulate thiazide-sensitive cotransporter endoplasmic reticulum-associated degradation and biogenesis. *J. Biol. Chem.* **2013**, *288*, 13124–13135. [CrossRef]

74. Needham, P.G.; Mikoluk, K.; Dhakarwal, P.; Khadem, S.; Snyder, A.C.; Subramanya, A.R.; Brodsky, J.L. The thiazide-sensitive NaCl cotransporter is targeted for chaperone-dependent endoplasmic reticulum-associated degradation. *J. Biol. Chem.* **2011**, *286*, 43611–43621. [CrossRef]
75. Rosenbaek, L.L.; Rizzo, F.; Wu, Q.; Rojas-Vega, L.; Gamba, G.; MacAulay, N.; Staub, O.; Fenton, R.A. The thiazide sensitive sodium chloride co-transporter NCC is modulated by site-specific ubiquitylation. *Sci. Rep.* **2017**, *7*. [CrossRef]
76. Legrand, A.; Treard, C.; Roncelin, I.; Dreux, S.; Bertholet-Thomas, A.; Broux, F.; Bruno, D.; Decramer, S.; Deschenes, G.; Djeddi, D.; et al. Prevalence of novel *MAGED2* mutations in antenatal bartter syndrome. *Clin. J. Am. Soc. Nephrol.* **2018**, *13*, 242–250. [CrossRef]
77. Loffing, J.; Vallon, V.; Loffing-Cueni, D.; Aregger, F.; Richter, K.; Pietri, L.; Bloch-Faure, M.; Hoenderop, J.G.J.; Shull, G.E.; Meneton, P.; et al. Altered renal distal tubule structure and renal Na+ and Ca 2+ handling in a mouse model for Gitelman's syndrome. *J. Am. Soc. Nephrol.* **2004**, *15*, 2276–2288. [CrossRef]
78. Wang, L.; Zhang, C.; Su, X.; Lin, D.H.; Wang, W. Caveolin-1 deficiency inhibits the basolateral K^+ channels in the distal convoluted tubule and impairs renal K^+ and Mg^{2+} transport. *J. Am. Soc. Nephrol.* **2015**, *26*, 2678–2690. [CrossRef]
79. Giebisch, G. Renal Potassium Channels: Function, Regulation, and Structure. In *Proceedings of the Kidney International*; Blackwell Publishing Inc.: Hoboken, NJ, USA, 2001; Volume 60, pp. 436–445.
80. Waldegger, S.; Jentsch, T.J. Functional and structural analysis of ClC-K chloride channels involved in renal disease. *J. Biol. Chem.* **2000**, *275*, 24527–24533. [CrossRef]
81. Li, D.; Tian, L.; Hou, C.; Kim, C.E.; Hakonarson, H.; Levine, M.A. Association of mutations in *SLC12A1* encoding the NKCC2 cotransporter with neonatal primary hyperparathyroidism. *J. Clin. Endocrinol. Metab.* **2016**, *101*, 2196–2200. [CrossRef]
82. Wongsaengsak, S.; Vidmar, A.P.; Addala, A.; Kamil, E.S.; Sequeira, P.; Fass, B.; Pitukcheewanont, P. A novel *SLC12A1* gene mutation associated with hyperparathyroidism, hypercalcemia, nephrogenic diabetes insipidus, and nephrocalcinosis in four patients. *Bone* **2017**, *97*, 121–125. [CrossRef]
83. Chen, Q.; Wang, X.; Min, J.; Wang, L.; Mou, L. Kidney stones and moderate proteinuria as the rare manifestations of Gitelman syndrome. *BMC Nephrol.* **2021**, *22*, 12. [CrossRef]
84. Demoulin, N.; Aydin, S.; Cosyns, J.P.; Dahan, K.; Cornet, G.; Auberger, I.; Loffing, J.; Devuyst, O. Gitelman syndrome and glomerular proteinuria: A link between loss of sodium-chloride cotransporter and podocyte dysfunction? *Nephrol. Dial. Transplant.* **2014**, *29*, iv117–iv120. [CrossRef]
85. Walsh, P.R.; Tse, Y.; Ashton, E.; Iancu, D.; Jenkins, L.; Bienias, M.; Kleta, R.; Van't Hoff, W.; Bockenhauer, D. Clinical and diagnostic features of Bartter and Gitelman syndromes. *Clin. Kidney J.* **2018**, *11*, 302–309. [CrossRef]
86. Bettinelli, A.; Borsa, N.; Bellantuono, R.; Syrèn, M.L.; Calabrese, R.; Edefonti, A.; Komninos, J.; Santostefano, M.; Beccaria, L.; Pela, I.; et al. Patients With Biallelic Mutations in the Chloride Channel Gene CLCNKB: Long-Term Management and Outcome. *Am. J. Kidney Dis.* **2007**, *49*, 91–98. [CrossRef]
87. Stokman, M.F.; Renkema, K.Y.; Giles, R.H.; Schaefer, F.; Knoers, N.V.A.M.; van Eerde, A.M. The expanding phenotypic spectra of kidney diseases: Insights from genetic studies. *Nat. Rev. Nephrol.* **2016**, *12*, 472–483. [CrossRef]
88. Watanabe, S.; Fukumoto, S.; Chang, H.; Takeuchi, Y.; Hasegawa, Y.; Okazaki, R.; Chikatsu, N.; Fujita, T. Association between activating mutations of calcium-sensing receptor and Bartter's syndrome. *Lancet* **2002**, *360*, 692–694. [CrossRef]
89. Carmosino, M.; Gerbino, A.; Hendy, G.N.; Torretta, S.; Rizzo, F.; Debellis, L.; Procino, G.; Svelto, M. NKCC2 activity is inhibited by the Bartter's syndrome type 5 gain-of-function CaR-A843E mutant in renal cells. *Biol. Cell* **2015**, *107*, 98–110. [CrossRef] [PubMed]
90. Vargas-Poussou, R.; Huang, C.; Hulin, P.; Houillier, P.; Jeunemaître, X.; Paillard, M.; Planelles, G.; Déchaux, M.; Miller, R.T.; Antignac, C. Functional characterization of a calcium-sensing receptor mutation in severe autosomal dominant hypocalcemia with a Bartter-like syndrome. *J. Am. Soc. Nephrol.* **2002**, *13*, 2259–2266. [CrossRef] [PubMed]
91. Huang, C.; Sindic, A.; Hill, C.E.; Hujer, K.M.; Chan, K.W.; Sassen, M.; Wu, Z.; Kurachi, Y.; Nielsen, S.; Romero, M.F.; et al. Interaction of the Ca^{2+}-sensing receptor with the inwardly rectifying potassium channels Kir4.1 and Kir4.2 results in inhibition of channel function. *Am. J. Physiol.—Ren. Physiol.* **2007**, *292*, F1073–F1081. [CrossRef]
92. Vezzoli, G.; Arcidiacono, T.; Paloschi, V.; Terranegra, A.; Biasion, R.; Weber, G.; Mora, S.; Syren, M.L.; Coviello, D.; Cusi, D.; et al. Autosomal dominant hypocalcemia with mild type 5 Bartter syndrome. *J. Nephrol.* **2006**, *19*, 525–528. [PubMed]
93. Matsunoshita, N.; Nozu, K.; Shono, A.; Nozu, Y.; Fu, X.J.; Morisada, N.; Kamiyoshi, N.; Ohtsubo, H.; Ninchoji, T.; Minamikawa, S.; et al. Differential diagnosis of Bartter syndrome, Gitelman syndrome, and pseudo-Bartter/Gitelman syndrome based on clinical characteristics. *Genet. Med.* **2016**, *18*, 180–188. [CrossRef]
94. Alfandary, H.; Landau, D. Future considerations based on the information from Barrter's and Gitelman's syndromes. *Curr. Opin. Nephrol. Hypertens.* **2017**, *26*, 9–13. [CrossRef]
95. Jain, G.; Ong, S.; Warnock, D.G. Genetic disorders of potassium homeostasis. *Semin. Nephrol.* **2013**, *33*, 300–309. [CrossRef]
96. Kamel, K.S.; Halperin, M.L. Use of Urine Electrolytes and Urine Osmolalityin Clinical Diagnosis of Fluid, Electrolytes, and Acid-Base Disorders. *Kidney Int. Rep.* **2021**, *6*, 1211–1224. [CrossRef]
97. Persu, A.; Lafontaine, J.J.; Devuyst, O. Chronic hypokalaemia in young women—It is not always abuse of diuretics. *Nephrol. Dial. Transplant.* **1999**, *14*, 1021–1025. [CrossRef]
98. Adalat, S.; Woolf, A.S.; Johnstone, K.A.; Wirsing, A.; Harries, L.W.; Long, D.A.; Hennekam, R.C.; Ledermann, S.E.; Rees, L.; Van't Hoff, W.; et al. HNF1B mutations associate with hypomagnesemia and renal magnesium wasting. *J. Am. Soc. Nephrol.* **2009**, *20*, 1123–1131. [CrossRef]

99. Verhave, J.C.; Bech, A.P.; Wetzels, J.F.M.; Nijenhuis, T. Hepatocyte nuclear factor 1β-associated kidney disease: More than renal cysts and diabetes. *J. Am. Soc. Nephrol.* **2016**, *27*, 345–353. [CrossRef]
100. Viering, D.H.H.M.; de Baaij, J.H.F.; Walsh, S.B.; Kleta, R.; Bockenhauer, D. Genetic causes of hypomagnesemia, a clinical overview. *Pediatr. Nephrol.* **2017**, *32*, 1123–1135. [CrossRef]
101. Bamgbola, O.F.; Ahmed, Y. Differential diagnosis of perinatal Bartter, Bartter and Gitelman syndromes. *Clin. Kidney J.* **2021**, *14*, 36–48. [CrossRef]
102. Kintu, B.; Brightwell, A. Episodic seasonal pseudo-bartter syndrome in cystic fibrosis. *Paediatr. Respir. Rev.* **2014**, *15*, 19–21. [CrossRef]
103. Kim, Y.K.; Song, H.C.; Kim, Y.-S.; Choi, E.J. Acquired Gitelman Syndrome. *Electrolytes Blood Press.* **2009**, *7*, 5. [CrossRef]
104. Barathidasan, G.S.; Krishnamurthy, S.; Karunakar, P.; Rajendran, R.; Ramya, K.; Dhandapany, G.; Ramamoorthy, J.G.; Ganesh, R.N. Systemic lupus erythematosus complicated by a Gitelman-like syndrome in an 8-year-old girl. *CEN Case Rep.* **2020**, *9*, 129–132. [CrossRef]
105. Matsunoshita, N.; Nozu, K.; Yoshikane, M.; Kawaguchi, A.; Fujita, N.; Morisada, N.; Ishimori, S.; Yamamura, T.; Minamikawa, S.; Horinouchi, T.; et al. Congenital chloride diarrhea needs to be distinguished from Bartter and Gitelman syndrome. *J. Hum. Genet.* **2018**, *63*, 887–892. [CrossRef]
106. Kurteva, E.; Lindley, K.J.; Hill, S.M.; Köglmeier, J. Mucosal Abnormalities in Children With Congenital Chloride Diarrhea—An Underestimated Phenotypic Feature? *Front. Pediatr.* **2020**, *8*. [CrossRef]
107. Nozu, K.; Yamamura, T.; Horinouchi, T.; Nagano, C.; Sakakibara, N.; Ishikura, K.; Hamada, R.; Morisada, N.; Iijima, K. Inherited salt-losing tubulopathy: An old condition but a new category of tubulopathy. *Pediatr. Int.* **2020**, *62*, 428–437. [CrossRef]
108. Nesbit, M.A.; Hannan, F.M.; Howles, S.A.; Babinsky, V.N.; Head, R.A.; Cranston, T.; Rust, N.; Hobbs, M.R.; Heath, H.; Thakker, R.V. Mutations Affecting G-Protein Subunit α 11 in Hypercalcemia and Hypocalcemia. *N. Engl. J. Med.* **2013**, *368*, 2476–2486. [CrossRef]
109. Okazaki, R.; Chikatsu, N.; Nakatsu, M.; Takeuchi, Y.; Ajima, M.; Miki, J.; Fujita, T.; Arai, M.; Totsuka, Y.; Tanaka, K.; et al. A Novel Activating Mutation in Calcium-Sensing Receptor Gene Associated with a Family of Autosomal Dominant Hypocalcemia 1. *J. Clin. Endocrinol. Metab.* **1999**, *84*, 363–366. [CrossRef] [PubMed]
110. Devuyst, O.; Olinger, E.; Weber, S.; Eckardt, K.U.; Kmoch, S.; Rampoldi, L.; Bleyer, A.J. Autosomal dominant tubulointerstitial kidney disease. *Nat. Rev. Dis. Prim.* **2019**, *5*, 1–20. [CrossRef] [PubMed]
111. Schaeffer, C.; Olinger, E. Clinical and genetic spectra of kidney disease caused by REN mutations. *Kidney Int.* **2020**, *98*, 1397–1400. [CrossRef] [PubMed]
112. Bockenhauer, D.; Bichet, D.G. Inherited secondary nephrogenic diabetes insipidus: Concentrating on humans. *Am. J. Physiol.—Ren. Physiol.* **2013**, *304*, 1037–1042. [CrossRef] [PubMed]
113. Bockenhauer, D.; Van'T Hoff, W.; Dattani, M.; Lehnhardt, A.; Subtirelu, M.; Hildebrandt, F.; Bichet, D.G. Secondary nephrogenic diabetes insipidus as a complication of inherited renal diseases. *Nephron—Physiol.* **2010**, *116*. [CrossRef] [PubMed]
114. D'Alessandri-Silva, C.; Carpenter, M.; Ayoob, R.; Barcia, J.; Chishti, A.; Constantinescu, A.; Dell, K.M.; Goodwin, J.; Hashmat, S.; Iragorri, S.; et al. Diagnosis, Treatment, and Outcomes in Children With Congenital Nephrogenic Diabetes Insipidus: A Pediatric Nephrology Research Consortium Study. *Front. Pediatr.* **2020**, *7*, 550. [CrossRef]
115. Kennedy, J.D.; Dinwiddie, R.; Daman-Willems, C.; Dillon, M.J.; Matthew, D.J. Pseudo-Bartter's syndrome in cystic fibrosis. *Arch. Dis. Child.* **1990**, *65*, 786–787. [CrossRef]
116. Okamoto, T.; Tajima, T.; Hirayama, T.; Sasaki, S. A patient with Dent disease and features of Bartter syndrome caused by a novel mutation of CLCN5. *Eur. J. Pediatr.* **2012**, *171*, 401–404. [CrossRef]
117. Chen, Y.S.; Fang, H.C.; Chou, K.J.; Lee, P.T.; Hsu, C.Y.; Huang, W.C.; Chung, H.M.; Chen, C.L. Gentamicin-Induced Bartter-like Syndrome. *Am. J. Kidney Dis.* **2009**, *54*, 1158–1161. [CrossRef]
118. Chrispal, A.; Boorugu, H.; Prabhakar, A.; Moses, V. Amikacin-induced type 5 Bartter-like syndrome with severe hypocalcemia. *J. Postgrad. Med.* **2009**, *55*, 208. [CrossRef]
119. Medication Guides. Arikayce. Available online: https://www.accessdata.fda.gov/scripts/cder/daf/index.cfm?event=medguide.page (accessed on 13 May 2021).
120. Cinotti, R.; Lascarrou, J.B.; Azais, M.A.; Colin, G.; Quenot, J.P.; Mahé, P.J.; Roquilly, A.; Gaultier, A.; Asehnoune, K.; Reignier, J. Diuretics decrease fluid balance in patients on invasive mechanical ventilation: The randomized-controlled single blind, IRIHS study. *Crit. Care* **2021**, *25*, 1–9. [CrossRef]
121. Blowey, D.L. Diuretics in the treatment of hypertension. *Pediatr. Nephrol.* **2016**, *31*, 2223–2233. [CrossRef]
122. Nozu, K.; Iijima, K.; Kanda, K.; Nakanishi, K.; Yoshikawa, N.; Satomura, K.; Kaito, H.; Hashimura, Y.; Ninchoji, T.; Komatsu, H.; et al. The pharmacological characteristics of molecular-based inherited salt-losing tubulopathies. *J. Clin. Endocrinol. Metab.* **2010**, *95*. [CrossRef]
123. Colussi, G.; Bettinelli, A.; Tedeschi, S.; De Ferrari, M.E.; Syrén, M.L.; Borsa, N.; Mattiello, C.; Casari, G.; Bianchetti, M.G. A thiazide test for the diagnosis of renal tubular hypokalemic disorders. *Clin. J. Am. Soc. Nephrol.* **2007**, *2*, 454–460. [CrossRef]
124. A Translational Approach to Gitelman Syndrome—Full Text View—ClinicalTrials.gov. Available online: https://clinicaltrials.gov/ct2/show/NCT00822107 (accessed on 26 May 2021).

125. Konrad, M.; Nijenhuis, T.; Ariceta, G.; Bertholet-Thomas, A.; Calo, L.A.; Capasso, G.; Emma, F.; Schlingmann, K.P.; Singh, M.; Trepiccione, F.; et al. Diagnosis and management of Bartter syndrome: Executive summary of the consensus and recommendations from the European Rare Kidney Disease Reference Network Working Group for Tubular Disorders. *Kidney Int.* **2021**, *99*, 324–335. [CrossRef]
126. Corbetta, S.; Raimondo, F.; Tedeschi, S.; Syrèn, M.L.; Rebora, P.; Savoia, A.; Baldi, L.; Bettinelli, A.; Pitto, M. Urinary exosomes in the diagnosis of Gitelman and Bartter syndromes. *Nephrol. Dial. Transplant.* **2015**, *30*, 621–630. [CrossRef]
127. Pisitkun, T.; Shen, R.F.; Knepper, M.A. Identification and proteomic profiling of exosomes in human urine. *Proc. Natl. Acad. Sci. USA* **2004**, *101*, 13368–13373. [CrossRef]
128. Williams, T.L.; Bastos, C.; Faria, N.; Karet Frankl, F.E. Making urinary extracellular vesicles a clinically tractable source of biomarkers for inherited tubulopathies using a small volume precipitation method: Proof of concept. *J. Nephrol.* **2020**, *33*, 383–386. [CrossRef]
129. Snoek, R.; Stokman, M.F.; Lichtenbelt, K.D.; van Tilborg, T.C.; Simcox, C.E.; Paulussen, A.D.C.; Dreesen, J.C.M.F.; van Reekum, F.; Lely, A.T.; Knoers, N.V.A.M.; et al. Preimplantation Genetic Testing for Monogenic Kidney Disease. *Clin. J. Am. Soc. Nephrol.* **2020**, CJN.03550320. [CrossRef]
130. Lim, M.; Gannon, D. Diagnosis and outpatient management of Gitelman syndrome from the first trimester of pregnancy. *BMJ Case Rep.* **2021**, *14*, e241756. [CrossRef]
131. Wu, W.F.; Pan, M. The outcome of two pregnancies in a patient with Gitelman syndrome: Case report and review of the literature. *J. Matern. Neonatal Med.* **2020**, *33*, 4171–4173. [CrossRef]
132. Han, Y.; Lin, Y.; Sun, Q.; Wang, S.; Gao, Y.; Shao, L. Mutation spectrum of Chinese patients with bartter syndrome. *Oncotarget* **2017**, *8*, 101614–101622. [CrossRef]
133. Feldmann, D.; Alessandri, J.L.; Deschênes, G. Large deletion of the 5′ end of the *ROMK1* gene causes antenatal Bartter syndrome. *J. Am. Soc. Nephrol.* **1998**, *9*, 2357–2359. [CrossRef]
134. García Castaño, A.; Pérez de Nanclares, G.; Madariaga, L.; Aguirre, M.; Madrid, A.; Nadal, I.; Navarro, M.; Lucas, E.; Fijo, J.; Espino, M.; et al. Genetics of Type III Bartter Syndrome in Spain, Proposed Diagnostic Algorithm. *PLoS ONE* **2013**, *8*, e74673. [CrossRef]
135. Castaño, A.G.; De Nanclares, G.P.; Madariaga, L.; Aguirre, M.; Madrid, Á.; Chocrón, S.; Nadal, I.; Navarro, M.; Lucas, E.; Fijo, J.; et al. Poor phenotype-genotype association in a large series of patients with Type III Bartter syndrome. *PLoS ONE* **2017**, *12*, e173581. [CrossRef]
136. Ma, J.; Ren, H.; Lin, L.; Zhang, C.; Wang, Z.; Xie, J.; Shen, P.Y.; Zhang, W.; Wang, W.; Chen, X.N.; et al. Genetic Features of Chinese Patients with Gitelman Syndrome: Sixteen Novel *SLC12A3* Mutations Identified in a New Cohort. *Am. J. Nephrol.* **2016**, *44*, 113–121. [CrossRef]
137. Vargas-Poussou, R.; Dahan, K.; Kahila, D.; Venisse, A.; Riveira-Munoz, E.; Debaix, H.; Grisart, B.; Bridoux, F.; Unwin, R.; Moulin, B.; et al. Spectrum of mutations in Gitelman syndrome. *J. Am. Soc. Nephrol.* **2011**, *22*, 693–703. [CrossRef]
138. Shen, Q.; Chen, J.; Yu, M.; Lin, Z.; Nan, X.; Dong, B.; Fang, X.; Chen, J.; Ding, G.; Zhang, A.; et al. Multi-centre study of the clinical features and gene variant spectrum of Gitelman syndrome in Chinese children. *Clin. Genet.* **2021**, *99*, 558–564. [CrossRef]
139. Zhang, L.; Peng, X.; Zhao, B.; Zhu, Z.; Wang, Y.; Tian, D.; Yan, Z.; Yao, L.; Liu, J.; Qiu, L.; et al. Clinical and laboratory features of female Gitelman syndrome and the pregnancy outcomes in a Chinese cohort. *Nephrology* **2020**, *25*, 749–757. [CrossRef]
140. Mighton, C.; Shickh, S.; Uleryk, E.; Pechlivanoglou, P.; Bombard, Y. Clinical and psychological outcomes of receiving a variant of uncertain significance from multigene panel testing or genomic sequencing: A systematic review and meta-analysis. *Genet. Med.* **2021**, *23*, 22–33. [CrossRef]
141. Lo, Y.F.; Nozu, K.; Iijima, K.; Morishita, T.; Huang, C.C.; Yang, S.S.; Sytwu, H.K.; Fang, Y.W.; Tseng, M.H.; Lin, S.H. Recurrent deep intronic mutations in the *SLC12A3* gene responsible for Gitelman's syndrome. *Clin. J. Am. Soc. Nephrol.* **2011**, *6*, 630–639. [CrossRef]
142. Zhang, L.; Huang, K.; Wang, S.; Fu, H.; Wang, J.; Shen, H.; Lu, Z.; Chen, J.; Bao, Y.; Feng, C.; et al. Clinical and Genetic Features in 31 Serial Chinese Children With Gitelman Syndrome. *Front. Pediatr.* **2021**, *9*, 299. [CrossRef]
143. García-Nieto, V.M.; Claverie-Martín, F.; Perdomo-Ramírez, A.; Cárdoba-Lanus, E.; Ramos-Trujillo, E.; Mura-Escorche, G.; Tejera-Carreño, P.; Luis-Yanes, M.I. Considerations about the molecular basis of some kidney tubule disorders in relation to inbreeding and population displacement. *Nefrologia* **2020**, *40*, 126–132. [CrossRef] [PubMed]
144. Le Gall, E.C.; Audrézet, M.-P.P.; Rousseau, A.; Hourmant, M.; Renaudineau, E.; Charasse, C.; Morin, M.-P.P.; Moal, M.-C.C.; Dantal, J.; Wehbe, B.; et al. No Title. *J. Am. Soc. Nephrol.* **2016**, *27*, 942–951. [CrossRef]
145. Le Gall, E.C.; Audrézet, M.P.; Chen, J.M.; Hourmant, M.; Morin, M.P.; Perrichot, R.; Charasse, C.; Whebe, B.; Renaudineau, E.; Jousset, P.; et al. Type of PKD1 mutation influences renal outcome in ADPKD. *J. Am. Soc. Nephrol.* **2013**, *24*, 1006–1013. [CrossRef] [PubMed]
146. Wan, X.; Perry, J.; Zhang, H.; Jin, F.; Ryan, K.A.; Van Hout, C.; Reid, J.; Overton, J.; Baras, A.; Han, Z.; et al. Heterozygosity for a Pathogenic Variant in *SLC12A3* That Causes Autosomal Recessive Gitelman Syndrome Is Associated with Lower Serum Potassium. *J. Am. Soc. Nephrol.* **2021**, *32*, ASN2020071030. [CrossRef]
147. Hsu, Y.J.; Yang, S.S.; Chu, N.F.; Sytwu, H.K.; Cheng, C.J.; Lin, S.H. Heterozygous mutations of the sodium chloride cotransporter in Chinese children: Prevalence and association with blood pressure. *Nephrol. Dial. Transplant.* **2009**, *24*, 1170–1175. [CrossRef]

148. Fava, C.; Montagnana, M.; Rosberg, L.; Burri, P.; Almgren, P.; Jönsson, A.; Wanby, P.; Lippi, G.; Minuz, P.; Hulthèn, L.U.; et al. Subjects heterozygous for genetic loss of function of the thiazide-sensitive cotransporter have reduced blood pressure. *Hum. Mol. Genet.* **2008**, *17*, 413–418. [CrossRef]
149. Cruz, D.N.; Simon, D.B.; Nelson-Williams, C.; Farhi, A.; Finberg, K.; Burleson, L.; Gill, J.R.; Lifton, R.P. Mutations in the Na-Cl cotransporter reduce blood pressure in humans. *Hypertension* **2001**, *37*, 1458–1464. [CrossRef]
150. Knoers, N.V.A.M. Gitelman syndrome. *Adv. Chronic Kidney Dis.* **2006**, *13*, 148–154. [CrossRef]
151. Ranade, V.V.; Somberg, J.C. Bioavailability and pharmacokinetics of magnesium after administration of magnesium salts to humans. *Am. J. Ther.* **2001**, *8*, 345–357. [CrossRef]
152. Tarnawski, A.S.; Jones, M.K. Inhibition of angiogenesis by NSAIDs: Molecular mechanisms and clinical implications. *J. Mol. Med.* **2003**, *81*, 627–636. [CrossRef]
153. Suleyman, H.; Cadirci, E.; Albayrak, A.; Halici, Z. Nimesulide is a Selective COX-2 Inhibitory, Atypical Non-Steroidal Anti-Inflammatory Drug. *Curr. Med. Chem.* **2008**, *15*, 278–283. [CrossRef]
154. Fulchiero, R.; Seo-Mayer, P. Bartter Syndrome and Gitelman Syndrome. *Pediatr. Clin. N. Am.* **2019**, *66*, 121–134. [CrossRef]
155. Mackie, F.E.; Hodson, E.M.; Roy, L.P.; Knight, J.F. Neonatal Bartter syndrome—Use of indomethacin in the newborn period and prevention of growth failure. *Pediatr. Nephrol.* **1996**, *10*, 756–758. [CrossRef]
156. Gasongo, G.; Greenbaum, L.A.; Niel, O.; Kwon, T.; Macher, M.A.; Maisin, A.; Baudouin, V.; Dossier, C.; Deschênes, G.; Hogan, J. Effect of nonsteroidal anti-inflammatory drugs in children with Bartter syndrome. *Pediatr. Nephrol.* **2019**, *34*, 679–684. [CrossRef]
157. Larkins, N.; Wallis, M.; McGillivray, B.; Mammen, C. A severe phenotype of Gitelman syndrome with increased prostaglandin excretion and favorable response to indomethacin. *Clin. Kidney J.* **2014**, *7*, 306–310. [CrossRef]
158. Blanchard, A.; Vargas-Poussou, R.; Vallet, M.; Caumont-Prim, A.; Allard, J.; Desport, E.; Dubourg, L.; Monge, M.; Bergerot, D.; Baron, S.; et al. Indomethacin, amiloride, or eplerenone for treating hypokalemia in Gitelman syndrome. *J. Am. Soc. Nephrol.* **2015**, *26*, 468–475. [CrossRef]
159. Brater, D.C. Anti-inflammatory agents and renal function. *Semin. Arthritis Rheum.* **2002**, *32*, 33–42. [CrossRef]
160. Kleinknecht, D. Interstitial nephritis, the nephrotic syndrome, and chronic renal failure secondary to nonsteroidal anti-inflammatory drugs. *Semin. Nephrol.* **1995**, *15*, 228–235.
161. Marlicz, W.; Łoniewski, I.; Grimes, D.S.; Quigley, E.M. Nonsteroidal anti-inflammatory drugs, proton pump inhibitors, and gastrointestinal injury: Contrasting interactions in the stomach and small Intestine. *Mayo Clin. Proc.* **2014**, *89*, 1699–1709. [CrossRef]
162. The Hospital for Sick Children: Indomethacin, Oral Suspension. Available online: https://www.sickkids.ca/en/care-services/for-health-care-providers/pharmacy/ (accessed on 18 October 2021).
163. Dembo, G.; Park, S.B.; Kharasch, E.D. Central nervous system concentrations of cyclooxygenase-2 inhibitors in humans. *Anesthesiology* **2005**, *102*, 409–415. [CrossRef]
164. Solomon, D.H.; Schneeweiss, S.; Glynn, R.J.; Kiyota, Y.; Levin, R.; Mogun, H.; Avorn, J. Relationship between Selective Cyclooxygenase-2 Inhibitors and Acute Myocardial Infarction in Older Adults. *Circulation* **2004**, *109*, 2068–2073. [CrossRef]
165. Baigent, C.; Bhala, N.; Emberson, J.; Merhi, A.; Abramson, S.; Arber, N.; Baron, J.A.; Bombardier, C.; Cannon, C.; Farkouh, M.E.; et al. Vascular and upper gastrointestinal effects of non-steroidal anti-inflammatory drugs: Meta-analyses of individual participant data from randomised trials. *Lancet* **2013**, *382*, 769–779. [CrossRef]
166. Jackson, E.K. Fármacos que afectan la función excretora renal. In *Goodman & Gilman: Las Bases Farmacológicas, De La Terapéutic*, 13th ed.; Brunton, L.L., Chabner, B.A., Knollmann, B.C., Eds.; McGraw-Hill Education: New York, NY, USA, 2019.
167. Perazella, M.A. Trimethoprim is a potassium-sparing diuretic like amiloride and causes hyperkalemia in high-risk patients. *Am. J. Ther.* **1997**, *4*, 343–348. [CrossRef]
168. Plumb, L.A.; van't Hoff, W.; Kleta, R.; Reid, C.; Ashton, E.; Samuels, M.; Bockenhauer, D. Renal apnoea: Extreme disturbance of homoeostasis in a child with Bartter syndrome type IV. *Lancet* **2016**, *388*, 631–632. [CrossRef]
169. Schepkens, H.; Lameire, N. Gitelman's syndrome: An overlooked cause of chronic hypokalemia and hypomagnesemia in adults. *Acta Clin. Belg.* **2001**, *56*, 248–254. [CrossRef]
170. Koudsi, L.; Nikolova, S.; Mishra, V. Management of a severe case of Gitelman syndrome with poor response to standard treatment. *BMJ Case Rep.* **2016**, 1–3. [CrossRef]
171. Griffing, G.T.; Komanicky, P.; Aurecchis, S.A.; Sindler, B.H.; Melby, J.C. Amiloride in Bartter's syndrome. *Clin. Pharmacol. Ther.* **1982**, *31*, 713–718. [CrossRef]
172. Griffing, G.T.; Melby, J.C. The therapeutic use of a new potassium-sparing diuretic, amiloride, and a converting enzyme inhibitor, mk-421, in preventing hypokalemia associated with primary and secondary hyperaldosteronism. *Clin. Exp. Hypertens.* **1983**, *A5*, 779–801. [CrossRef]
173. Griffing, G.T.; Aurecchia, S.A.; Sindler, B.H.; Melby, J.C. The Effect of Amiloride on the Renin—Aldosterone System in Primary Hyperaldosteronism and Bartter's Syndrome. *J. Clin. Pharmacol.* **1982**, *22*, 505–512. [CrossRef]
174. Luqman, A.; Kazmi, A.; Wall, B.M. Bartter's syndrome in pregnancy: Review of potassium homeostasis in gestation. *Am. J. Med. Sci.* **2009**, *338*, 500–504. [CrossRef] [PubMed]
175. Deruelle, P.; Dufour, P.; Magnenant, E.; Courouble, N.; Puech, F. Maternal Bartter's syndrome in pregnancy treated by amiloride. *Eur. J. Obstet. Gynecol. Reprod. Biol.* **2004**, *115*, 106–107. [CrossRef] [PubMed]
176. Federal Register: Content and Format of Labeling for Human Prescription Drug and Biological Products. Requirements for Pregnancy and Lactation Labeling. Available online: https://www.federalregister.gov/documents/2008/05/29/E8-1

1806/content-and-format-of-labeling-for-human-prescription-drug-and-biological-products-requirements-for (accessed on 24 May 2021).
177. Santos, F.; Gil-Peña, H.; Blázquez, C.; Coto, E. Gitelman syndrome: A review of clinical features, genetic diagnosis and therapeutic management. *Expert Opin. Orphan Drugs* **2016**, *4*, 1005–1009. [CrossRef]
178. Morales, J.M.; Ruilope, L.M.; Praga, M.; Coto, A.; Alcazar, J.M.; Prieto, C.; Nieto, J.; Rodicio, J.L. Long-term enalapril therapy in Bartter's syndrome. *Nephron* **1988**, *48*, 327. [CrossRef]
179. Hené, R.J.; Koomans, H.A.; Mees, E.J.D.; Stolpe, A.; Verhoef, G.E.G.; Boer, P. Correction of Hypokalemia in Bartter's Syndrome by Enalapril. *Am. J. Kidney Dis.* **1987**, *9*, 200–205. [CrossRef]
180. Nascimento, C.L.P.; Garcia, C.L.; Schvartsman, B.G.S.; Vaisbich, M.H. Treatment of Bartter syndrome. Unsolved issue. *J. Pediatr.* **2014**, *90*, 512–517. [CrossRef]
181. Jest, P.; Pedersen, K.E.; Klitgaard, N.A.; Thomsen, N.; Kjaer, K.; Simonsen, E. Angiotensin-converting enzyme inhibition as a therapeutic principle in Bartter's syndrome. *Eur. J. Clin. Pharmacol.* **1991**, *41*, 303–305. [CrossRef]
182. Álvarez-Nava, F.; Lanes, R. GH/IGF-1 signaling and current knowledge of epigenetics; A review and considerations on possible therapeutic options. *Int. J. Mol. Sci.* **2017**, *18*, 1624. [CrossRef]
183. Ficha Tecnica Genotonorm Kabipen 12 mg Polvo Y Disolvente Para Solucion Inyectable. Available online: https://cima.aemps.es/cima/dochtml/ft/60117/FT_60117.html (accessed on 4 May 2021).
184. Cook, D.M.; Rose, S.R. A review of guidelines for use of growth hormone in pediatric and transition patients. *Pituitary* **2012**, *15*, 301–310. [CrossRef]
185. Drube, J.; Wan, M.; Bonthuis, M.; Wühl, E.; Bacchetta, J.; Santos, F.; Grenda, R.; Edefonti, A.; Harambat, J.; Shroff, R.; et al. Clinical practice recommendations for growth hormone treatment in children with chronic kidney disease. *Nat. Rev. Nephrol.* **2019**, *15*, 577–589. [CrossRef]
186. Collett-Solberg, P.F.; Ambler, G.; Backeljauw, P.F.; Bidlingmaier, M.; Biller, B.M.K.; Boguszewski, M.C.S.; Cheung, P.T.; Choong, C.S.Y.; Cohen, L.E.; Cohen, P.; et al. Diagnosis, Genetics, and Therapy of Short Stature in Children: A Growth Hormone Research Society International Perspective. *Horm. Res. Paediatr.* **2019**, *92*, 1–14. [CrossRef]
187. Yuen, K.C.J.; Biller, B.M.K.; Radovick, S.; Carmichael, J.D.; Jasim, S.; Pantalone, K.M.; Hoffman, A.R. American Association of Clinical endocrinologists and American College of Endocrinology guidelines for management of growth hormone deficiency in adults and patients transitioning from pediatric to adult care. *Endocr. Pract.* **2019**, *25*, 1191–1232. [CrossRef]
188. Touyz, R.M. Magnesium in clinical medicine. *Front. Biosci.* **2004**, *9*, 1278–1293. [CrossRef]
189. Sackett, D.L.; Haynes, R.B.; Tugwell, P. *Clinical Epidemiology: A Basic Science for Clinical Medicine*; Little, Brown & Co.: Boston, MA, USA, 1985; p. 370, ISBN 0-316-76595-3.
190. Ariceta, G.; Rodríguez-Soriano, J. Inherited Renal Tubulopathies Associated With Metabolic Alkalosis: Effects on Blood Pressure. *Semin. Nephrol.* **2006**, *26*, 422–433. [CrossRef]
191. Shalev, H.; Ohali, M.; Kachko, L.; Landau, D. The neonatal variant of Bartter syndrome and deafness: Preservation of renal function. *Pediatrics* **2003**, *112*, 628–633. [CrossRef]
192. Jeck, N.; Reinalter, S.C.; Henne, T.; Marg, W.; Mallmann, R.; Pasel, K.; Vollmer, M.; Klaus, G.; Leonhardt, A.; Seyberth, H.W.; et al. Hypokalemic salt-losing tubulopathy with chronic renal failure and sensorineural deafness. *Pediatrics* **2001**, *108*, e5. [CrossRef]
193. ClinicalTrials.gov. Bartter Syndrome. Available online: https://clinicaltrials.gov/ct2/results?cond=bartter&term=&cntry=&state=&city=&dist= (accessed on 27 May 2021).
194. ClinicalTrials.gov. Gitelman Syndrome. Available online: https://clinicaltrials.gov/ct2/results?cond=gitelman&draw=2&rank=3#rowId2 (accessed on 27 May 2021).
195. Diamox® (Acetazolamide Extended-Release Capsules). Available online: https://www.accessdata.fda.gov/drugsatfda_docs/label/2005/12945s037,038lbl.pdf (accessed on 27 May 2021).
196. Mazaheri, M.; Assadi, F.; Sadeghi-Bojd, S. Adjunctive acetazolamide therapy for the treatment of Bartter syndrome. *Int. Urol. Nephrol.* **2020**, *52*, 121–128. [CrossRef]
197. Blanchard, A.; Tabard, S.B.; Lamaziere, A.; Bergerot, D.; Zhygalina, V.; Lorthioir, A.; Jacques, A.; Hourton, D.; Azizi, M.; Crambert, G. Adrenal adaptation in potassium-depleted men: Role of progesterone? *Nephrol. Dial. Transplant.* **2020**, *35*, 1901–1908. [CrossRef]

International Journal of
Molecular Sciences

MDPI

Review

Role of miR-24 in Multiple Endocrine Neoplasia Type 1: A Potential Target for Molecular Therapy

Francesca Marini [1,2] and Maria Luisa Brandi [2,*]

1 Department of Experimental and Clinical Biomedical Sciences, University of Florence, Viale Pieraccini 6, 50139 Florence, Italy; francesca.marini@unifi.it
2 F.I.R.M.O., Italian Foundation for the Research on Bone Diseases, Via Reginaldo Giuliani 195/A, 50141 Florence, Italy
* Correspondence: marialuisa.brandi@unifi.it or marialuisa@marialuisabrandi.it; Tel.: +39-055-23-36-663

Abstract: Multiple endocrine neoplasia type 1 (MEN1) is a rare autosomal dominant inherited multiple cancer syndrome of neuroendocrine tissues. Tumors are caused by an inherited germinal heterozygote inactivating mutation of the *MEN1* tumor suppressor gene, followed by a somatic loss of heterozygosity (LOH) of the *MEN1* gene in target neuroendocrine cells, mainly at parathyroids, pancreas islets, and anterior pituitary. Over 1500 different germline and somatic mutations of the *MEN1* gene have been identified, but the syndrome is completely missing a direct genotype-phenotype correlation, thus supporting the hypothesis that exogenous and endogenous factors, other than *MEN1* specific mutation, are involved in MEN1 tumorigenesis and definition of individual clinical phenotype. Epigenetic factors, such as microRNAs (miRNAs), are strongly suspected to have a role in MEN1 tumor initiation and development. Recently, a direct autoregulatory network between miR-24, *MEN1* mRNA, and menin was demonstrated in parathyroids and endocrine pancreas, showing a miR-24-induced silencing of menin expression that could have a key role in initiation of tumors in MEN1-target neuroendocrine cells. Here, we review the current knowledge on the post-transcriptional regulation of *MEN1* and menin expression by miR-24, and its possible direct role in MEN1 syndrome, describing the possibility and the potential approaches to target and silence this miRNA, to permit the correct expression of the wild type menin, and thereby prevent the development of cancers in the target tissues.

Keywords: multiple endocrine neoplasia type 1 (MEN1); *MEN1* gene; loss of heterozygosity (LOH); microRNA (miRNAs); miR-24

Citation: Marini, F.; Brandi, M.L. Role of miR-24 in Multiple Endocrine Neoplasia Type 1: A Potential Target for Molecular Therapy. *Int. J. Mol. Sci.* **2021**, *22*, 7352. https://doi.org/10.3390/ijms22147352

Academic Editor: Ivano Condò

Received: 16 June 2021
Accepted: 7 July 2021
Published: 8 July 2021

1. Introduction

Multiple endocrine neoplasia type 1 (MEN1) is a rare autosomal dominant inherited cancer syndrome that causes the development of multiple endocrine and non-endocrine tumors in a single patient [1,2]. The main affected organs are parathyroid glands, anterior pituitary, and the neuroendocrine cells of the gastro-entero-pancreatic tract. Morbidity and mortality of the disease are related to hormone over-secretion by endocrine functioning tumors, leading to the development of specific syndromes, and/or to the malignant progression of silent tumors, such as non-functioning neuroendocrine neoplasms of the pancreas and the thymus.

Medical therapies of MEN1 aim to control hormone over-secretion and tumor growth. Surgery is the main treatment employed for parathyroid adenomas and gastro-entero-pancreatic neuroendocrine tumors (GEP-NETs) [3]. No therapeutic intervention is definitively resolutive; given the genetic nature of the syndrome and the asynchronous development of tumors, MEN1 patients have a high prevalence of post-operative tumor recurrences, both in the parathyroids and the gastro-entero-pancreatic tract [4]. Therefore, there is a strong need for novel therapies acting at the molecular level and able to prevent tumors in the target neuroendocrine cells. The comprehension of molecular mechanisms underlying MEN1 tumorigenesis is fundamental to identify possible targets for the design of novel therapies [2].

In 1997, the causative gene, *MEN1*, was identified at the 11q13 locus [5]. The *MEN1* gene is a classic tumor suppressor gene: The first inactivating heterozygote mutation is inherited by the affected parent (first hit), while the second copy of the gene is somatically lost in target neuroendocrine cells (second hit), mainly by a large deletion at the 11q13 locus or, more rarely, by a second intragenic loss-of-function mutation (loss of heterozygosity; LOH) [6,7]. The *MEN1* gene encodes menin, a nuclear protein which exerts a wide spectrum of key activities, such as control of cell cycle and apoptosis, regulation of gene transcription and chromatin structure, and DNA repair [8]. Loss of both wild type *MEN1* copies, resulting in loss of menin functions, appears to be the trigger of tumor initiation in MEN1 target neuroendocrine cells. However, the absence of a complete genotype-phenotype correlation and the different tumor manifestations between carriers of the same *MEN1* mutation (even homozygote twins) suggest that other factors concur to cause MEN1 individual tumorigenesis. Epigenetic factors are the main suspected co-actors in driving tumor development and progression in MEN1 target neuroendocrine cells [9]. Alterations in the normal epigenetic regulation of gene transcription (histone modification and/or DNA methylation), following the loss of wild type menin activity, have been demonstrated to play an important role in the progression of MEN1 pancreatic neuroendocrine tumors [10].

Among epigenetic regulators of gene expression, microRNAs (miRNAs) have recently been shown to be involved in the development of various human malignancies, either acting directly as oncogenes (oncomiRs) or inhibiting the expression of tumor suppressor genes [11]. These molecules are non-coding small RNAs that normally negatively regulate gene expression by directly binding the 3′UTR of their target mRNAs [12–14]. Through the activity of tissue- and cell-specific miRNAs, the organism regulates the expression of numerous genes, in a spatial and temporal way, granting the correct functionality of various and important biological processes [15,16]. Alterations of expression and/or activity of one or more miRNAs can lead to disease development, including cancer. A role of miRNAs has been demonstrated in the initiation of various human malignancies [17–19] and in development of metastases [20,21].

In the last two decades, tissue-specific altered activity and/or expression of miRNAs have been suggested as possible modulators of MEN1 tumorigenesis [22–25], acting synergically with the *MEN1* mutation, indicating the miR-24 as a possible effector of tumor development.

Here, we review results from recent studies that demonstrate the existence of an autoregulatory network between miR-24, *MEN1* mRNA, and menin, suggesting possible roles of this miRNA in MEN1 tumorigenesis, and we discuss the possibility to silence this molecule in *MEN1* mutation carriers to prevent/reduce cancer development and/or progression.

2. The Autoregulatory Network between miR-24, *MEN1*, and Menin: A Possible Effector of MEN1 Tumorigenesis

miR-24 is encoded by two separated chromosomal locations: one gene cluster located on chromosome 9q22, which includes miR-23b, miR-27b, and miR-24-1; and a second gene cluster located on chromosome 19p13, which includes miR-23a, miR-27a, and miR-24-2 [26]. The two miR-24 mature isoforms are identical in nucleotide sequencing, regardless of the different chromosomal origin.

Using Miranda, Target Scan, and Pictar Vert prediction software, Luzi et al. predicted the 599-605 nt position of the 832 nt-3′UTR of *MEN1* mRNA as a direct target of miR-24-1 [27]. The seed site of miR-24-1, which binds to *MEN1* mRNA 3′UTR, is highly conserved in humans, rats, mice, chickens, and dogs. The direct targeting of *MEN1* mRNA was demonstrated in vitro for the first time in 2012 by Luzi et al. [27] in the BON1 cells, a human cell line derived from a lymph node metastasis of a serotonin-secreting pancreatic neuroendocrine tumor [28], via luciferase report assays. Authors showed that the induced over-expression of miR-24-1, through the transfection of pri-miR-24, inhibited the expression of menin protein, while the silencing of the endogenous miR-24-1, through the transfection of a specific 2′-O-methyl-RNA miR-24-1 antisense, resulted in an increased expression of menin, suggesting the existence of a direct negative feedback loop between miR-24-1 and menin, which could have a role in MEN1 tumorigenesis.

Later, in 2016, the same research group [29] demonstrated in BON1 cells that, in addition to the negative feedback loop between miR-24-1, *MEN1* mRNA, and menin, there was also a feedforward loop in which menin protein directly binds to the pri-miR-24-1 (the primary mRNA precursor of miR-24-1), promoting the DROSHA-mediated processing to pre-miR-24-1 and, thus, the biogenesis of mature miR-24-1. Moreover, the specific siRNA-induced silencing of menin expression in BON1 cells resulted in a complete suppression of pri-miR-24-1 expression, indicating an essential and direct role of menin in miR-24-1 synthesis, that is exerted by acting both at the transcriptional and at the post-transcriptional modification levels of the miR-24-1 synthesis. Conversely, when an overexpression of menin is induced in BON1 cells, this leads to an increased expression of mature miR-24-1. Authors showed that menin specifically binds to pri-miR-24-1, but not to pri-miR-24-2, and that the silencing of menin expression had no effect on the pri-miR-24-2 expression and the processing to pre-miR-24-2.

The proposed autoregulatory network between miR-24-1, *MEN1* mRNA, and menin is shown in Figure 1.

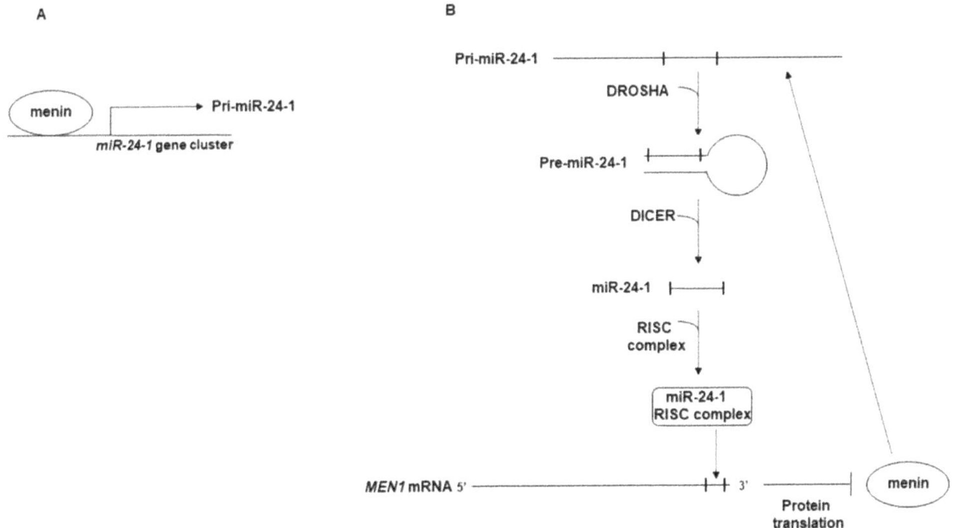

Figure 1. Schematic representation of the autoregulatory network between miR-24-1, *MEN1* mRNA, and menin. (**A**) menin binds the upstream region of the miR-24-1-encoding gene cluster on chromosome 9, promoting the transcription of pri-miR-24-1; (**B**) menin directly interacts with the primary transcript of miR-24-1, pri-miR-24-1, facilitating the DROSHA-mediated processing to pre-miR-24-1, in a positive feedforward loop in which menin promotes the maturation of the repressor of its own expression, miR-24-1. Mature miR-24-1, together with the RISC complex, binds the 3′ UTR of *MEN1* mRNA blocking translation of menin.

The regulatory network between miR-24, *MEN1*, and menin was demonstrated also in other species, than humans, in vitro and in vivo [30–32]. Interestingly, Gao et al. [30] demonstrated, both by Target Scan algorithm-based prediction and by an in vitro study in neonatal rat cardiac myocytes, that miR-24 targets the *CDNK1B* mRNA, blocking translation of the encoded $p27^{kip1}$ protein, a negative regulator of cell cycle, resulting in a decreased G0/G1 arrest and in cell hypertrophy. Inactivating mutations of the *CDNK1B* tumor suppressor gene are responsible for the development of multiple endocrine neoplasia type 4 (MEN4), a clinical phenocopy of MEN1 [33]. Therefore, miR-24 could also have a role in MEN4 tumorigenesis.

2.1. miR-24, MEN1 mRNA, and Menin in Parathyroid Glands

Luzi et al. first demonstrated a direct regulation of menin expression by miR-24-1 in parathyroid tissues from MEN1 patients [27], suggesting an epigenetic oncogenic role of this miRNA in the tissue-specific tumorigenesis of these glands in *MEN1* mutation carriers. The analysis of miR-24-1 expression profiles in parathyroid adenomas from *MEN1* mutation carriers, in sporadic non-MEN1 tumor counterparts and in healthy parathyroid tissue, showed an inverse correlation in the expression profiles of miR-24-1 and menin, indicating a direct role of miR-24-1 in the post-transcriptional negative regulation of menin expression itself. This inverse correlation was present only when the wild type copy of the *MEN1* gene was still retained in the tumoral tissue, while in MEN1 adenomas presenting LOH at the 11q13 locus, menin resulted to be unexpressed and miR-24-1 was expressed at a very low level. The chromatin immunoprecipitation (ChIP) analysis, with an anti-menin antibody, in parathyroid tissues from MEN1 patients, showed the occupancy of miR-24-1 promoter region by menin, only in parathyroid adenomas conserving one wild type copy of the *MEN1* gene, but not in tumors presenting *MEN1* LOH, and it confirmed menin as a positive regulator of miR-24-1 expression, as previously demonstrated in BON1 cells [27].

Results from this study suggested that, in MEN1-associated parathyroid tumors, after the first inherited germinal "hit", the somatic onset and progression of neoplasia could be under the control of a negative feedback loop between menin protein and miR-24-1. Authors hypothesized that this regulatory network could mimic and substitute the second somatic "hit" of tumor suppressor inactivation in tissues in which *MEN1* LOH has not yet occurred, probably representing an intermediate step before the irreversible genetic *MEN1* LOH. The pathway leading to MEN1 tumor development and progression could be explained by the proposed negative feedback loop between menin and miR-24-1 acting as a "homeostatic regulatory network" that needs to be "broken" to induce the somatic LOH "hit" and, consequently, the progression to neoplasia. This mechanism could explain the proliferative changes in the neuroendocrine cells (hyperplasia) that precede neoplasia, and this could also be hypothesized for pancreatic, duodenal, and other MEN1-associated tumors.

Some years later, the same Research Group [34] failed to find both miR-24-1 and miR-24-2 as differentially expressed between MEN1 parathyroid adenomas with somatic *MEN1* LOH at 11q13 and MEN1 parathyroid adenomas still retaining one wild type copy of the *MEN1* gene when they performed a microarray expression profiling covering about 1890 human miRNAs and using a p-value < 0.05 and a $\log_2$ fold change > 1.5 as parameters of statistical significance.

Grolmusz et al. [23] compared the expression of six potential *MEN1*-targeting miRNAs, including miR-24, in MEN1-associated parathyroid adenomas/hyperplasia with a germinal *MEN1* mutation (16), in sporadic counterparts bearing a somatic *MEN1* mutation (10) and in sporadic parathyroid lesions wild type for the *MEN1* gene (40). Expression levels of miR-24 and miR-28 were found to be significantly higher in sporadic primary hyperparathyroidism (PHPT) tissues with respect to MEN1-associated adenomas/hyperplasia, both when all the sporadic samples were considered and when considering the two sporadic PHPT subgroups of samples positive for nuclear menin staining or of samples negative for nuclear menin staining separately. No significant expression differences were found between the *MEN1* mutated vs. the *MEN1* wild type nuclear menin-negative sporadic lesions. All these data seemed to identify the higher expression of miR-24 and miR-28 as two universal signatures of sporadic non-syndromic parathyroid hyperplasia/adenoma, independent of their somatic genetic profile.

2.2. miR-24, MEN1 mRNA, and Menin in the Endocrine Pancreas

In the duodenum and the pancreas, the *MEN1* gene-associated heterozygote germline mutation causes hyperplasia of insulin- gastrin-, somatostatin-, and glucagon-secreting cells, resulting in a subsequent multifocal development of tumors. The great majority of tumors analyzed showed allelic deletion/inactivation of the second copy of the *MEN1* gene, whereas the precursor lesions retained their *MEN1* heterozygosity [35]. Pancreatic endocrine tumors develop from beta islets, through a stage of islet hyperplasia in which the wild-type *MEN1* allele is still retained. LOH at *MEN1* locus has never been found in these enlarged islets and, therefore, these lesions are considered to be a precursor stage of the tumors. It is possible that these pancreatic lesions present the same mechanism evidenced in MEN1 parathyroid glands: The proposed negative feedback loop between miR-24, *MEN1* mRNA and menin.

In 2014, Vijayaraghavan et al. [36] confirmed the presence of a feedback loop between miR-24 and menin, similar to that identified in parathyroids, in cell lines of endocrine pancreas (the MIN6 cells derived from a mouse insulinoma and the Blox5 cells, an immortalized cell line produced from a purified population of human pancreas beta cells by infection with retroviral vectors). Authors confirmed the direct interaction of miR-24 with the 3′UTR of *MEN1* mRNA. Cell transfection with pre-miR-24 increased miR-24 levels and resulted in a reduction of *MEN1* and menin expression, of 35% and 23% in MIN6 cells and 61% and 38% in Blox5 cells, respectively. Conversely, cell transfection with an anti-miR-24 antisense decreased endogenous miR-24 levels, in association with an increased *MEN1* expression in MIN6 cells but not in Blox5 cells, and with no significant difference of menin expression in both MIN6 and Blox5 cells. At the same time, an induced overexpression of menin increased the expression of mature miR-24 both in MIN6 (2.5-fold increase) and Blox5 (0.6-fold increase) cells, confirming that miR-24 and menin reciprocally regulate their expression. The negative effect of menin depletion on miR-24 expression was also confirmed in vivo in a conditional knockout mouse model with a selective deletion of the *MEN1* gene in pancreas beta cells, in which reductions of about 60% and 90% of miR-24 levels were found in heterozygote and homozygote knocked-out animals, respectively [36].

In the same study, since in the pancreas menin is known to regulate gene expression in a transcription complex with mixed-lineage leukemia (MLL) factor, the Authors [36] hypothesized the possibility that the miR-24-1- and miR-24-2-encoding genes could be targets of menin/MLL complex. The ChiP analysis in MIN6 and Blox5 cells showed that both menin and MLL proteins were present in the region upstream of miR-24, both on chromosomes 9 and 19, confirming the importance of the menin/MLL complex in miR-24 expression in pancreas islets.

Finally, since menin is known to negatively regulate cell proliferation by transcriptionally regulating expression of two main cell cycle inhibitors, $p27^{kip1}$ and $p18^{ink4c}$, the Authors also investigated the impact of miR-24 expression on these two proteins. Cell transfection with pre-miR-24 induced a 42% reduction of both p27 mRNA and protein in MIN6 cells, and of 57% p27 mRNA, but not protein, in Blox5 cells, and a reduction of 48% of p18 mRNA, but not protein, in Blox5 cells. No effect on both expressions of p18 mRNA and protein was found in MIN6 cells. Inhibition of miR-24 expression by transfection with an anti-miR-24 antisense did not significantly affect p18 mRNA nor protein expression in MIN6 and Blox5 cells [36]. The fact that an increased expression of miR-24 reduced the expression of p27 and p18 mRNAs suggested this miRNA as negatively regulating the expression of these two cell cycle inhibitors in an indirect manner, via the suppression of menin translation. In this way, the increased expression of miR-24, and the subsequent silencing of menin, lead to a significantly increased cell proliferation in Blox5 cells, but not in insulinoma-derived MIN6 cells, in which cell growth is already at such a high level as to not be further influenced by menin, p27, and p18 inhibition. Considering this, it can be assumed that an increase in miR-24 expression could be responsible for enhanced proliferation of beta-cells and hyperplasia of pancreas islets in the first stage of MEN1 tumorigenesis.

Molecular effects of miR-24 in parathyroid glands and endocrine pancreas and possible roles in MEN1 tumorigenesis, reported in the currently available studies, are summarized in Table 1.

Table 1. Molecular effects of miR-24 parathyroid glands and endocrine pancreas, and possible roles in MEN1 tumorigenesis.

miR-24 Target	Effect	Possible Role in MEN1 Tumorigenesis	Reference
	Parathyroid glands		
MEN1	No effect on *MEN1* mRNA expression. Loss of menin protein expression.	Uncontrolled cell proliferation	[27]
	Endocrine pancreas		
MEN1	Reduction of both *MEN1* mRNA and menin expression.	Uncontrolled cell proliferation	[36]
CDKN1B	Reduction of expression of both *CDKN1B* mRNA and of $p27^{kip1}$ protein in a mouse insulinoma cell line (MIN6).	Enhanced proliferation of beta-cells and hyperplasia of pancreas islets	[36]
CDKN1B	Reduction of expression of *CDKN1B* mRNA in an immortalized human pancreas beta cell line (Blox5). No data on expression of $p27^{kip1}$ protein.	Enhanced proliferation of beta-cells and hyperplasia of pancreas islets	[36]
CDKN2C	No effect on expression of both *CDKN2C* mRNA and of $p18^{Ink4c}$ protein in a mouse insulinoma cell line (MIN6).	Non applicable	[36]
CDKN2C	Reduction of expression of *CDKN2C* mRNA in an immortalized human pancreas beta cell line (Blox5). No data on expression of $p18^{Ink4c}$ protein.	Enhanced proliferation of beta-cells and hyperplasia of pancreas islets	[36]

2.3. *miR-24, MEN1 mRNA, and Menin in Non-MEN1 Tumors*

Menin has been found to exert its tumor suppression activity by blocking cell hyperplasia in tissues other than those commonly affected by MEN1 syndrome, such as prostate, breast, lung, liver, and stomach [37].

Using the A549 cells (a continuous cell line derived from a human adenocarcinoma of the alveolar basal epithelium of the lungs) and the NCI-H446 cells (a cell line derived from a human small-cell lung tumor), Pan et al. [38] also confirmed the interaction of miR-24 with the 3′UTR of *MEN1* mRNA and the miR-24-mediated significant silencing of menin expression in lung cancer.

The induced overexpression of miR-24 in A549 cells inhibited menin expression and upregulated SMAD3 and cyclin D1, stimulating cell proliferation and, at the same time, increased expression of Bcl-2 and inhibited expression of Bax, resulting in the inhibition of cell apoptosis [38]. Conversely, the induced silencing of endogenous miR-24 increased menin expression and downregulated SMAD3 and cyclin D1, inhibiting cell growth, and, at the same time, inhibited Bcl-2 expression and promoted Bax expression, allowing cell apoptosis [38]. In addition, the induced overexpression of miR-24, in both A549 and NCI-H446 cell lines, upregulated SMAD3 and MMP2 and significantly enhanced cell migration and invasion, while downregulation of miR-24 decreased the expression of both SMAD3 and MMP2, preventing cell migration and invasion. Taken together, these data indicate miR-24 as an oncogenic effector in lung cancer cells, acting as promoter of both cell proliferation and metastatic spread.

In the same study, the analysis of surgical specimens from human lung cancer showed a higher expression of miR-24, and a mirrored lower expression of menin, in tumors with respect to the adjacent healthy tissues. miR-24 expression inversely correlated with tumor size and patient overall survival, while patients with a higher menin expression showed a longer survival rate [38].

Bronchopulmonary carcinoids of the neuroendocrine cells of the bronchial epithelium arise in approximately 5–8% of MEN1 patients [39]. Despite the different cytological origin with respect to lung cancers studied by Pan et al. [38], an involvement of the miR-24-*MEN1* mRNA-menin network in the tumorigenesis of these carcinoids in MEN1 syndrome cannot be excluded.

Although menin has been found to regulate cell proliferation in hepatocellular carcinoma [40], the exact role of menin in liver carcinogenesis has not yet been widely studied, and the liver is not a site of primary tumor development in MEN1 syndrome, but only metastases from pancreas and duodenal MEN1-related neuroendocrine tumors. Ehrlich et al. [37] studied the miR-24-*MEN1* mRNA-menin network in six human cell lines of intrahepatic and extrahepatic cholangiocarcinoma (CCA), a biliary epithelial adenocarcinoma whose cells adopt a neuroendocrine-like phenotype during their proliferation. Both *MEN1* mRNA and menin expression resulted significantly decreased in all six CCA-derived cell lines compared to a human immortalized, non-malignant, cholangiocyte cell line (H69). A further-induced silencing of menin in tumoral cells increased cell growth, while the induced menin overexpression reduced the cell proliferation rate together with a decreased cell migration and invasion.

All the six tumoral cell lines showed a basal overexpression of miR-24 with respect to the H69 control, as well as human CCA tumor samples compared with their matched healthy tissues. The induced knock-down of miR-24 was associated with an increase of menin expression, a decreased expression of pro-angiogenic and pro-proliferative factors, and a reduction of cell proliferation and migration.

3. Targeting miR-24: A Potential Therapeutic Tool for MEN1 Tumorigenesis

The miR-24-driven post-transcriptional inhibition of menin synthesis, in parathyroid cells and pancreas beta islets, could represent an intermediate epigenetic step guiding the MEN1 tumorigenesis. menin has been demonstrated to have a role in the maintenance of genome stability and DNA repair, protecting cells from DNA damage [41,42]. At the same time, the epigenetic silencing of wild type menin by miR-24 promotes cell proliferation and exposes cells to DNA damage, which could result in deletion at the 11q13 locus or mutation of the second copy of the *MEN1* gene, favoring the somatic *MEN1* LOH, which represents a common hallmark of MEN1 cancers in parathyroids and endocrine pancreas.

In light of this, it is conceivable that targeting and silencing the expression of miR-24 at this intermediate, still reversible, step, before the occurrence of the genetic, irreversible, *MEN1* LOH, could restore expression and activity of wild type menin and block neoplastic initiation and progression, representing a promising innovative tool for preventing tumor development and progression in MEN1 patients [43].

Silencing of a miRNA that has an oncogenic activity (oncomiR), such as miR-24 in MEN1, can be obtained by miRNA inhibition target therapies, that include antisense anti-miRNAs (synthetic anti-miRNA oligonucleotides (AMOs), modified synthetic AMOs, locked nucleic acids (LNA)-based AMOs, and antagomirs), small molecules miRNA inhibitors, miRNA sponges, and miRNA masks. These anti-oncomiR tools, and their characteristics, are depicted in Table 2.

Table 2. miRNA inhibition target therapies for oncomiRs.

Therapeutic Tool	Description	Mechanism of Action on the Target oncomiR	Reference
Anti-miRNA oligonucleotides (AMOs)	Synthetic single-stranded RNA molecules complementary to the target miRNA	Competitive inhibition of the target mature miRNA by base pair	[44]
Modified AMOs	AMOs with a chemical modification of the 2′-OH into 2′-O′methyl- or 2′-O′methoxyethyl- groups, to increase intracellular stability	Competitive inhibition of the target mature miRNA by base pair	[44]
Antagomirs	AMOs with the 2′-O chemical modification and phosphorothioate bonds to increase intracellular stability and with a conjugated cholesterol tail at the 3′-end to favor cell membrane permeation	Competitive inhibition of the target mature miRNA by base pair	[45]
Locked nucleic acids (LNA)-based AMOs	AMOs containing an additional methylene link between the 2′-O atom and the 4′-C atom, that locks ribose into a more thermodynamically stable conformation	High-affinity base pair with their target mature miRNA. Inhibition of miRNA activity	[46]
Small molecules miRNA inhibitors	Small molecules (chemical compounds) that interfere with miRNA biogenesis and/or activity	Inhibition of a specific miRNA biogenesis and/or activity by chemical structure-based docking to miRNA precursor or to mature miRNA	[47]
miRNA sponges	RNA transcripts presenting multiple tandem repeats of the binding site (seed sequence) of the miRNA to be targeted	They stably interact with the endogenous target miRNA, preventing its interaction with its target mRNAs	[48]
miRNA masks	Single-stranded 2′-O′methyl-modified antisense oligonucleotides totally complementary to the miRNA binding sites in the 3′-UTR of the target mRNA	They "mask" the target mRNA from the endogenous miRNA, preventing the miRNA-driven suppression of protein translation	[49]

Among the above mentioned oncomiR inhibitor target therapies, transfection of target cells with RNA antagomirs [45] appears to be a promising tool for miR-24 silencing in MEN1. Synthetic RNA antagomirs for the in vivo silencing of endogenous miRNAs were first created in 2005; intravenous injection of these molecules in mice has been shown to considerably reduce the expression of target miRNA, with a silencing effect lasting up to three weeks after systemic administration [50]. Currently, this technology appears to be a promising tool for the treatment of different types of human cancer and other diseases [51–53], also thanks to the fact that antagomirs do not activate any immune response.

RNA antagomirs consist in single strand RNAs, complementary to the target miRNAs to be inhibited, presenting specific chemical modifications (introduction of a 2′-O-methyl residue to nucleotides and substitution of an oxygen atom with an atom of sulfur within the phosphodiester bonds; phosphorothioate bond), to reduce RNA degradation by endogenous ribonucleases and conjugated with a cholesterol tail at the 3′-end to enhance permeation of cell membrane and intracellular distribution [54,55].

To further improve the efficacy of this exogenous delivery system, it is possible to conjugate, through electrostatic interactions, the anionic RNA antagomir with polymeric cations possessing extremely low toxicity and immunogenicity (i.e., polyethylenimine; PEI), able to grant high efficiency of cell transfection, increase the bond to the target miRNA, and, at the same time, protect the transfected molecule from intracellular lysosomal degradation [56,57].

In addition, it is possible to conjugate the RNA antagomir-PEI complex with paramagnetic nanoparticles (PMNP), which give the RNA antagomir-PEI-PMNP complex the potential to be specifically conveyed in vivo, after a systemic intravenous injection, to target cells and tissues, in which we want to silence the target miRNA, through the application of external magnetic fields [58–60].

Luzi et al. [29] tested a 2′-O-methyl-modified antisense RNA complementary to hsa-miR-24-1 in the BON1 cells, measuring the effect of transfection of this molecule on both miR-24-1 and menin expression. Transfection of this antisense RNA caused a specific inhibition of activity of the endogenous miR-24-1, with a subsequent over-expression of menin, with respect to BON1 cells transfected with a scrambled sequence RNA as control.

These preliminary and promising results, obtained in basic research, show the ability of the anti-miR-24-1 antagomir to target and inhibit the human endogenous miR-24-1 and restore the normal expression of wild type menin, and prompt the possibility to develop and design RNA antagomir-PEI-PMNP complexes against hsa-miR-24, for systemic administration to MEN1 patients and prevention of tumor development.

A first attempt to evaluate in vivo the therapeutic effect of the miR-24 silencing on tumor growth was performed by Ehrlich et al. [37] in an athymic nude mouse model in which the CCA were induced by injection of Mz-ChA-1, a human cell line of extrahepatic CCA. After the development of tumors, treated mice were injected, three times a week, with a commercial miR-24 hairpin inhibitor, while the control group received an injection of a scramble (inactive) hairpin inhibitor. Ten weeks after the first injection, treated animals showed a significant reduction in tumor size compared to untreated controls, in association with a decreased nuclear Ki-67 staining in tumor specimens. Quantitative expression analysis, performed on tumor samples from treated and control mice, showed a decreased expression of both pro-angiogenic and pro-proliferative factors in treated animals. Results from this study clearly show that targeting miR-24 represents a promising therapeutic tool for reducing biliary tumor growth and suggest the possibility of using the same approach for other tumors, the tumorigenesis of which involves the oncomiR miR-24 and its inhibitory action on the *MEN1* tumor suppressor gene.

4. Future Research Needed in the Field of MEN1 Syndrome and miRNAs

In addition to miR-24, other miRNAs showed to be possibly involved in the MEN1 tumorigenesis or to be deregulated in the sporadic tumor counterparts of the neuroendocrine tissues commonly affected in MEN1 syndrome [25], and they should be further investigated by functional studies, as was done for miR-24, to assess their importance in MEN1-related tumor development and progression.

In this paragraph we briefly summarized the most promising miRNAs, that could be involved in the tumorigenesis of the three MEN1 main affected endocrine tissues.

Luzi et al. [34] profiled the expression of 1890 human miRNAs between MEN1 parathyroid adenomas with somatic *MEN1* LOH or still retain one wild type copy of the *MEN1* gene, and with respect to non-MEN1 sporadic parathyroid adenomas. After both the microarray analysis and the quantitative RT-PCR validation, three miRNAs were identified as differentially expressed between the two groups of MEN1 tumors. Authors analyzed, by ComiR tool, the putative gene targets of these three differentially expressed miRNAs, with a focus on genes known to be involved in parathyroid tumorigenesis. miR-4258 resulted remarkably downregulated in MEN1 parathyroid adenomas with *MEN1* LOH, with respect to those without *MEN1* LOH. miR-4258 was predicted to target the gene encoding the cyclin D1 (*CCND1*), a positive regulator of cell cycle and cell growth, suggesting that, in the absence of wild type menin, cells lose the miR-4258-driven negative control of *CCND1* expression, presumably leading to an uncontrolled cell growth. Conversely, miR-664 resulted to be upregulated in *MEN1* non-LOH adenomas and down-regulated in *MEN1* LOH adenomas, both with respect to non-MEN1 controls. miR-664 was predicted to target the *CDC73* tumor suppressor gene, whose germinal inactivating mutations are responsible for the hyperparathyroidism-jaw tumor syndrome [61] and somatic mutations have been associated with the development of sporadic parathyroid carcinoma [62], and the *CDKN2C* tumor suppressor gene encoding the $p18^{ink4c}$, an inhibitor of the cyclin kinases CDK4 and CDK6, that negatively controls the cell cycle G1 progression. When a copy of the wild type *MEN1* gene is conserved, it can be speculated that the over-expression of miR-664 leads to silence these tumor suppressor genes, resulting in promoting cell growth and favoring tumor initiation. miR-1301 resulted upregulated in *MEN1* LOH parathyroid adenomas, both with respect to the *MEN1* non-LOH parathyroid adenomas and the non-MEN1 controls. miR-1301 was predicted to target the *CDKN1B* tumor suppressor gene, whose inactivating mutations are responsible for MEN4 syndrome, the *RET* oncogene, whose activating mutations are responsible for the multiple endocrine neoplasia type 2 (MEN2) syndrome, the *AP2S1* gene, whose loss-of-function mutations have been associated with familial hypocalciuric hypercalcemia type 3, the *CCND2* gene encoding cyclin D2 a positive regulator of cell cycle, and the *CTNNB1* gene encoding the catenin beta 1, involved in the regulation of cell adhesion. These data on expression and in silico analyses suggested these three miRNAs as potentially involved in MEN1 parathyroid tumorigenesis. The interaction of these three miRNAs with *MEN1* and menin protein should be functionally investigated, to better understand the exact role of these epigenetic regulators in the development and progression of parathyroid tumors, including their possible role in parathyroid carcinoma, which is extremely rare in MEN1 patients and appeared not to be directly correlated with the *MEN1* mutation. In addition, expression and functional role of these three miRNAs should be also studied in other MEN1 target tissues, to verify if their possible oncogenic role is tissue-specific only to parathyroid glands or, as for miR-24, they act in other MEN1-associated malignancies. Indeed, the *CDKN2C* gene, which could be a target of miR-664, has been demonstrated to be frequently lost in pituitary adenomas [63].

Regarding the endocrine pancreas, Lu et al. [64] showed that the miR-17, whose expression is induced by glucose and hyperglycemia, directly downregulated menin expression, through targeting the 3′UTR of the *MEN1* mRNA, promoting proliferation of the pancreas beta cells. On the other hand, *MEN1* is not only a direct target of miRNAs, but it has been showed also to regulate the biogenesis of some miRNAs that are involved in insulin-signaling and insulin-induced proliferation of pancreas beta cells. Through its direct physical interaction with the arsenite-resistance protein 2 (ARS2), a component of the nuclear RNA cap-binding complex that stabilizes pri-miRNA, menin facilitates the processing of pri-let-7a and pri-miR-155 to pre-let-7a and pre-miR-155 [65]. IRS2, a protein involved in the insulin-signaling pathway, is a target of mature let-7a, and resulted to be over-expressed when *MEN1* expression is lost. miR-155, whose biogenesis is positively regulated by menin, had been previously showed to be s among the top downregulated miRNAs in pancreatic neuroendocrine tumors [66], and its downregulation could have a role also in MEN1 tumorigenesis of the endocrine pancreas.

Two miRNAs, miR-15a and miR-16-1, that had been previously shown as having a significantly reduced expression in human sporadic pituitary adenomas [67], were reported to be downregulated also in MEN1 pituitary tumors from $Men1^{+/-}$ mice, with respect to the healthy pituitary tissue [68]. The expression of these two miRNAs inversely correlated with the expression of cyclin D1 [68], suggesting their possible tumor suppressor gene-like role in pituitary tumorigenesis. The *Men1* knockdown in the AtT20 mouse pituitary cell lines resulted in a significantly decreased expression of miR-15a [68], suggesting that the decrease of miR-15a expression in the pituitary cells of MEN1 patients may be a direct result of menin expression lost, that may concur to MEN1 pituitary tumorigenesis. Study on human specimens and functional studies are needed to confirm these preliminary findings.

5. Conclusions

- The absence of a genotype-phenotype correlation in MEN1 syndrome suggested a possible role of epigenetic factors in the development of the individual clinical phenotype in any single patient, even in presence of the same *MEN1* mutation.
- miRNAs have shown increasing evidence of a direct role in human malignancies, both for sporadic and hereditary cancers. Several miRNAs resulted to be deregulated in the sporadic tumor counterparts of the neuroendocrine tissues commonly affected in

MEN1 syndrome. A role of specific mi-RNA deregulation also in MEN1 tumorigenesis can be suspected.
- miR-24 has been demonstrated to negatively regulate menin expression in the parathyroids and the endocrine pancreas in MEN1 syndrome, and in other non-MEN1 sporadic tumors, suggesting it as initiator of menin loss-derived tumorigenesis.
- Targeting/silencing miR-24, during the hyperplastic phase of parathyroid and endocrine pancreas tumorigenesis and before the occurrence of the somatic *MEN1* LOH, could by a promising tissue-specific RNA-based anti-cancer therapy, aimed to restore the correct expression of menin in pre-cancerous cells.

Author Contributions: F.M. and M.L.B. contributed equally to write the manuscript. Both authors have read and agreed to the published version of the manuscript.

Funding: This review received no external funding.

Conflicts of Interest: The authors declare no conflict of interest.

References

1. Thakker, R.V.; Newey, P.J.; Walls, G.V.; Bilezikian, J.; Dralle, H.; Ebeling, P.R.; Melmed, S.; Sakurai, A.; Tonelli, F.; Brandi, M.L. Endocrine Society. Clinical practice guidelines for multiple endocrine neoplasia type 1 (MEN1). *J. Clin. Endocrinol. Metab.* **2012**, *97*, 2990–3011. [CrossRef]
2. Brandi, M.L.; Agarwal, S.K.; Perrier, N.D.; Lines, K.E.; Valk, G.D.; Thakker, R.V. Multiple Endocrine Neoplasia Type 1: Latest Insights. *Endocr. Rev.* **2021**, *42*, 133–170. [CrossRef]
3. Geslot, A.; Vialon, M.; Caron, P.; Grunenwald, S.; Vezzosi, D. New therapies for patients with multiple endocrine neoplasia type 1. *Ann. Endocrinol.* **2021**, *82*, 112–120. [CrossRef]
4. Al-Salameh, A.; Cadiot, G.; Calender, A.; Goudet, P.; Chanson, P. Clinical aspects of multiple endocrine neoplasia type 1. *Nat. Rev. Endocrinol.* **2021**, *17*, 207–224. [CrossRef]
5. Chandrasekharappa, S.C.; Guru, S.C.; Manickam, P.; Olufemi, S.E.; Collins, F.S.; Emmert-Buck, M.R.; Debelenko, L.V.; Zhuang, Z.; Lubensky, I.A.; Liotta, L.A.; et al. Positional cloning of the gene for multiple endocrine neoplasia-type 1. *Science* **1997**, *276*, 404–407. [CrossRef] [PubMed]
6. Pannett, A.A.; Thakker, R.V. Somatic mutations in MEN type 1 tumors, consistent with the Knudson "two-hit" hypothesis. *J. Clin. Endocrinol. Metab.* **2001**, *86*, 4371–4374. [CrossRef]
7. Valdes, N.; Alvarez, V.; Diaz-Cadorniga, F.; Aller, J.; Villazon, F.; Garcia, I.; Herrero, A.; Coto, E. Multiple endocrine neoplasia type 1 (MEN1): LOH studies in an affected family and in sporadic cases. *Anticancer Res.* **1998**, *18*, 2685–2689. [PubMed]
8. Dreijerink, K.M.A.; Timmers, H.T.M.; Brown, M. Twenty years of menin: Emerging opportunities for restoration of transcriptional regulation in MEN1. *Endocr. Relat. Cancer* **2017**, *24*, T135–T145. [CrossRef] [PubMed]
9. Feng, Z.; Ma, J.; Hua, X. Epigenetic regulation by the menin pathway. *Endocr. Relat. Cancer* **2017**, *24*, T147–T159. [CrossRef] [PubMed]
10. Marini, F.; Giusti, F.; Tonelli, F.; Brandi, M.L. Pancreatic Neuroendocrine Neoplasms in Multiple Endocrine Neoplasia Type 1. *Int. J. Mol. Sci.* **2021**, *22*, 4041. [CrossRef]
11. Svoronos, A.A.; Engelman, D.M.; Slack, F.J. OncomiR or Tumor Suppressor? The Duplicity of MicroRNAs in Cancer. *Cancer Res.* **2016**, *76*, 3666–3670. [CrossRef]
12. O'Brien, J.; Hayder, H.; Zayed, Y.; Peng, C. Overview of MicroRNA Biogenesis, Mechanisms of Actions, and Circulation. *Front. Endocrinol.* **2018**, *9*, 402. [CrossRef]
13. Bartel, D.P. MicroRNAs: Target recognition and regulatory functions. *Cell* **2009**, *136*, 215–233. [CrossRef]
14. Shukla, G.C.; Singh, J.; Barik, S. MicroRNAs: Processing, Maturation, Target Recognition and Regulatory Functions. *Mol. Cell Pharmacol.* **2011**, *3*, 83–92.
15. Tüfekci, K.U.; Meuwissen, R.L.J.; Genç, S. The role of microRNAs in biological processes. *Methods Mol. Biol.* **2014**, *1107*, 15–31. [CrossRef] [PubMed]
16. Vidigal, J.A.; Ventura, A. The biological functions of miRNAs: Lessons from in vivo studies. *Trends Cell Biol.* **2015**, *25*, 137–147. [CrossRef] [PubMed]
17. Peng, Y.; Croce, C.M. The role of MicroRNAs in human cancer. *Signal. Transduct. Target. Ther.* **2016**, *1*, 15004. [CrossRef]
18. Lin, Y.-C.; Tso-Hsiao Chen, T.-H.; Huang, Y.-M.; Wei, P.-L.; Lin, J.-C. Involvement of microRNA in Solid Cancer: Role and Regulatory Mechanisms. *Biomedicines* **2021**, *9*, 343. [CrossRef] [PubMed]
19. Tan, W.; Liu, B.; Qu, S.; Liang, G.; Luo, W.; Gong, C. MicroRNAs and cancer: Key paradigms in molecular therapy. *Oncol. Lett.* **2018**, *15*, 2735–2742. [CrossRef]
20. Bouyssou, J.M.; Manier, S.; Huynh, D.; Issa, S.; Roccaro, A.M.; Ghobrial, I.M. Regulation of microRNAs in cancer metastasis. *Biochim. Biophys. Acta* **2014**, *1845*, 255–265. [CrossRef]

21. Jafri, M.A.; Al-Qahtani, M.H.; Shay, J.W. Role of miRNAs in human cancer metastasis: Implications for therapeutic intervention. *Semin. Cancer Biol.* **2017**, *44*, 117–131. [CrossRef] [PubMed]
22. Nagy, Z.; Szabó, P.M.; Grolmusz, V.K.; Perge, P.; Igaz, I.; Patócs, A.; Igaz, P. MEN1 and microRNAs: The link between sporadic pituitary, parathyroid and adrenocortical tumors? *Med. Hypotheses* **2017**, *99*, 40–44. [CrossRef] [PubMed]
23. Grolmusz, V.K.; Borka, K.; Kövesdi, A.; Németh, K.; Balogh, K.; Dékány, C.; Kiss, A.; Szentpéteri, A.; Sármán, B.; Somogyi, A.; et al. MEN1 mutations and potentially MEN1-targeting miRNAs are responsible for menin deficiency in sporadic and MEN1 syndrome-associated primary hyperparathyroidism. *Virchows Arch.* **2017**, *471*, 401–411. [CrossRef]
24. Luzi, E.; Pandolfini, L.; Ciuffi, S.; Marini, F.; Cremisi, F.; Nesi, G.; Brandi, M.L. MicroRNAs regulatory networks governing the epigenetic landscape of MEN1 gastro-entero-pancreatic neuroendocrine tumor: A case report. *Clin. Transl. Med.* **2021**, *11*, e351. [CrossRef]
25. Donati, S.; Ciuffi, S.; Marini, F.; Palmini, G.; Miglietta, F.; Aurilia, C.; Brandi, M.L. Multiple Endocrine Neoplasia Type 1: The Potential Role of microRNAs in the Management of the Syndrome. *Int. J. Mol. Sci.* **2020**, *21*, 7592. [CrossRef]
26. Lal, A.; Navarro, F.; Maher, C.A.; Maliszewski, L.E.; Yan, N.; O'Day, E.; Chowdhury, D.; Dykxhoorn, D.M.; Tsai, P.; Hofmann, O.; et al. miR-24 Inhibits cell proliferation by targeting E2F2, MYC, and other cell-cycle genes via binding to "seedless" 3'UTR microRNA recognition elements. *Mol. Cell* **2009**, *35*, 610–625. [CrossRef]
27. Luzi, E.; Marini, F.; Giusti, F.; Galli, G.; Cavalli, L.; Brandi, M.L. The negative feedback-loop between the oncomir Mir-24-1 and menin modulates the Men1 tumorigenesis by mimicking the "Knudson's second hit". *PLoS ONE* **2012**, *7*, e39767. [CrossRef] [PubMed]
28. Evers, B.M.; Ishizuka, J.; Townsend, C.M., Jr.; Thompson, J.C. The human carcinoid cell line, BON. A model system for the study of carcinoid tumors. *Ann. N. Y. Acad. Sci.* **1994**, *733*, 393–406. [CrossRef] [PubMed]
29. Luzi, E.; Marini, F.; Ciuffi, S.; Galli, G.; Brandi, M.L. An autoregulatory network between menin and pri-miR-24-1 is required for the processing of its specific modulator miR-24-1 in BON1 cells. *Mol. Biosyst.* **2016**, *12*, 1922–1928. [CrossRef]
30. Gao, J.; Zhu, M.; Liu, R.-F.; Zhang, J.-S.; Xu, M. Cardiac Hypertrophy is Positively Regulated by MicroRNA-24 in Rats. *Chin. Med. J.* **2018**, *131*, 1333–1341. [CrossRef]
31. Hall, C.; Ehrlich, L.; Meng, F.; Invernizzi, P.; Bernuzzi, F.; Lairmore, T.C.; Alpini, G.; Glaser, S. Inhibition of microRNA-24 increases liver fibrosis by enhanced menin expression in Mdr2(−/−) mice. *J. Surg. Res.* **2017**, *217*, 160–169. [CrossRef] [PubMed]
32. Qiaoqiao, C.; Li, H.; Liu, X.; Yan, Z.; Zhao, M.; Xu, Z.; Wang, Z.; Shi, K. MiR-24-3p regulates cell proliferation and milk protein synthesis of mammary epithelial cells through menin in dairy cows. *J. Cell Physiol.* **2019**, *234*, 1522–1533. [CrossRef] [PubMed]
33. Alrezk, R.; Hannah-Shmouni, F.; Stratakis, C.A. MEN4 and CDKN1B mutations: The latest of the MEN syndromes. *Endocr. Relat. Cancer.* **2017**, *24*, T195–T208. [CrossRef]
34. Luzi, E.; Ciuffi, S.; Marini, F.; Mavilia, C.; Galli, G.; Brandi, M.L. Analysis of differentially expressed microRNAs in MEN1 parathyroid adenomas. *Am. J. Transl Res.* **2017**, *9*, 1743–1753. [PubMed]
35. Falchetti, A.; Marin, F.; Luzi, E.; Giusti, F.; Cavalli, L.; Cavalli, T.; Brandi, M.L. Multiple endocrine neoplasia type 1 (MEN1): Not only inherited endocrine tumors. *Genet. Med.* **2009**, *11*, 825–835. [CrossRef] [PubMed]
36. Vijayaraghavan, J.; Maggi, E.C.; Crabtree, J.S. miR-24 regulates menin in the endocrine pancreas. *Am. J. Physiol Endocrinol. Metab.* **2014**, *307*, E84–E92. [CrossRef]
37. Ehrlich, L.; Hall, C.; Venter, J.; Dostal, D.; Bernuzzi, F.; Invernizzi, P.; Meng, F.; Trzeciakowski, J.P.; Zhou, T.; Standeford, H.; et al. miR-24 Inhibition Increases Menin Expression and Decreases Cholangiocarcinoma Proliferation. *Am. J. Pathol.* **2017**, *187*, 570–580. [CrossRef]
38. Pan, Y.; Wang, H.; Ma, D.; Ji, Z.; Luo, L.; Cao, F.; Huang, F.; Liu, Y.; Dong, Y.; Chen, Y. miR-24 may be a negative regulator of menin in lung cancer. *Oncol. Rep.* **2018**, *39*, 2342–2350. [CrossRef]
39. Montero, C.; Sanjuán, P.; del Mar Fernández, M.; Vidal, I.; Verea, H.; Cordido, F. Bronchial carcinoid and type 1 multiple endocrine neoplasia syndrome. A case report. *Arch. Bronconeumol.* **2010**, *46*, 559–561. (In Spanish) [CrossRef]
40. Gang, D.; Hongwei, H.; Hedai, L.; Ming, Z.; Qian, H.; Zhijun, L. The tumor suppressor protein menin inhibits NF-κB-mediated transactivation through recruitment of Sirt1 in hepatocellular carcinoma. *Mol. Biol. Rep.* **2013**, *40*, 2461–2466. [CrossRef]
41. Gallo, A.; Agnese, S.; Esposito, I.; Galgani, M.; Avvedimento, V.E. Menin stimulates homology-directed DNA repair. *FEBS Lett.* **2010**, *584*, 4531–4536. [CrossRef]
42. Jin, S.; Mao, H.; Schnepp, R.W.; Sykes, S.M.; Silva, A.C.; D'Andrea, A.D.; Hua, X. Menin associates with FANCD2, a protein involved in repair of DNA damage. *Cancer Res.* **2003**, *63*, 4204–4210.
43. Gambari, R.; Brognara, E.; Spandidos, D.A.; Fabbri, E. Targeting oncomiRNAs and mimicking tumor suppressor miRNAs: New trends in the development of miRNA therapeutic strategies in oncology (Review). *Int. J. Oncol.* **2016**, *49*, 5–32. [CrossRef]
44. Dias, N.; Stein, C.A. Antisense oligonucleotides: Basic concepts and mechanisms. *Mol. Cancer Ther.* **2002**, *1*, 347–355.
45. Mattes, J.; Yang, M.; Foster, P.S. Regulation of microRNA by antagomirs: A new class of pharmacological antagonists for the specific regulation of gene function? *Am. J. Respir. Cell Mol. Biol.* **2007**, *36*, 8–12. [CrossRef]
46. Vester, B.; Wengel, J. LNA (locked nucleic acid): High-affinity targeting of complementary RNA and DNA. *Biochemistry* **2004**, *43*, 13233–13241. [CrossRef] [PubMed]
47. Wen, D.; Danquah, M.; Chaudhary, A.K.; Mahato, R.I. Small molecules targeting microRNA for cancer therapy: Promises and obstacles. *J. Control. Release* **2015**, *219*, 237–247. [CrossRef]
48. Ebert, M.S.; Sharp, P.A. MicroRNA sponges: Progress and possibilities. *RNA* **2010**, *16*, 2043–2050. [CrossRef] [PubMed]

49. Wang, Z. The principles of MiRNA-masking antisense oligonucleotides technology. *Methods Mol. Biol.* **2011**, *676*, 43–49. [CrossRef] [PubMed]
50. Krützfeldt, J.; Rajewsky, N.; Braich, R.; Rajeev, K.G.; Tuschl, T.; Manoharan, M.; Stoffel, M. Silencing of microRNAs in vivo with 'antagomirs'. *Nature* **2005**, *438*, 685–689. [CrossRef]
51. Preethi, K.A.; Lakshmanan, G.; Sekar, D. Antagomir technology in the treatment of different types of cancer. *Epigenomics* **2021**, *13*, 481–484. [CrossRef]
52. Murdaca, G.; Tonacci, A.; Negrini, S.; Greco, M.; Borro, M.; Puppo, F.; Gangemi, S. Effects of AntagomiRs on Different Lung Diseases in Human, Cellular, and Animal Models. *Int. J. Mol. Sci.* **2019**, *20*, 3938. [CrossRef]
53. Innao, V.; Allegra, A.; Pulvirenti, N.; Allegra, A.G.; Musolino, C. Therapeutic potential of antagomiRs in haematological and oncological neoplasms. *Eur. J. Cancer Care* **2020**, *29*, e13208. [CrossRef]
54. Krutzfeldt, J.; Kuwajima, S.; Braich, R.; Rajeev, K.G.; Pena, J.; Tuschl, T.; Manoharan, M.; Stoffel, M. Specificity, duplex degradation and subcellular localization of antagomirs. *Nucleic Acids Res.* **2007**, *35*, 2885–2892. [CrossRef] [PubMed]
55. Lennox, K.A.; Behlke, M.A. Chemical modification and design of anti-miRNA oli-gonucleotides. *Gene Ther.* **2011**, *18*, 1111–1120. [CrossRef] [PubMed]
56. Fu, Y.; Chen, J.; Huang, Z. Recent progress in microRNA-based delivery systems for the treatment of human disease. *ExRNA* **2019**, *1*, 24. [CrossRef]
57. Dasgupta, I.; Anushila Chatterjee, A. Recent Advances in miRNA Delivery Systems. *Methods Protoc.* **2021**, *4*, 10. [CrossRef] [PubMed]
58. Muthana, M.; Scott, S.D.; Farrow, N.; Morrow, F.; Murdoch, C.; Grubb, S.; Brown, N.; Dobson, J.; Lewis, C.E. A novel magnetic approach to enhance the efficacy of cell-based gene therapies. *Gene Ther.* **2008**, *15*, 902–910. [CrossRef]
59. Revia, R.A.; Stephen, Z.R.; Zhang, M. Theranostic Nanoparticles for RNA-Based Cancer Treatment. *Acc. Chem Res.* **2019**, *52*, 1496–1506. [CrossRef]
60. Hofmann, A.; Wenzel, D.; Becher, U.M.; Freitag, D.F.; Klein, A.M.; Eberbeck, D.; Schulte, M.; Zimmermann, K.; Bergemann, C.; Gleich, B.; et al. Combined targeting of lentiviral vectors and positioning of transduced cells by magnetic nanoparticles. *Proc. Natl. Acad. Sci. USA* **2009**, *106*, 44–49. [CrossRef]
61. Carpten, J.D.; Robbins, C.M.; Villablanca, A.; Forsberg, L.; Presciuttini, S.; Bailey-Wilson, J.; Simonds, W.F.; Gillanders, E.M.; Kennedy, A.M.; Chen, J.D.; et al. HRPT2, encoding parafibromin, is mutated in hyperparathyroidism-jaw tumor syndrome. *Nat. Genet.* **2002**, *32*, 676–680. [CrossRef]
62. Shattuck, T.M.; Välimäki, S.; Obara, T.; Gaz, R.D.; Clark, O.H.; Shoback, D.; Wierman, M.E.; Tojo, K.; Robbins, C.M.; Carpten, J.D.; et al. Somatic and germ-line mutations of the HRPT2 gene in sporadic parathyroid carcinoma. *N. Engl. J. Med.* **2003**, *349*, 1722–1729. [CrossRef] [PubMed]
63. Kirsch, M.; Mörz, M.; Pinzer, T.; Schackert, H.K.; Schackert, G. Frequent loss of the CDKN2C (p18INK4c) gene product in pituitary adenomas. *Genes Chromosomes Cancer* **2009**, *48*, 143–154. [CrossRef] [PubMed]
64. Lu, Y.; Fei, X.-Q.; Yang, S.-F.; Xu, B.-K.; Li, Y.-Y. Glucose-induced microRNA-17 promotes pancreatic beta cell proliferation through down-regulation of Menin. *Eur. Rev. Med. Pharmacol. Sci.* **2015**, *19*, 624–629. [PubMed]
65. Gurung, B.; Katona, B.W.; Hua, X. Menin-mediated regulation of miRNA biogenesis uncovers the IRS2 pathway as a target for regulating pancreatic beta cells. *Oncoscience* **2014**, *1*, 562–566. [CrossRef]
66. Roldo, C.; Missiaglia, E.; Hagan, J.P.; Falconi, M.; Capelli, P.; Bersani, S.; Calin, G.A.; Volinia, S.; Liu, C.G.; Scarpa, A.; et al. MicroRNA expression abnormalities in pancreatic endocrine and acinar tumors are associated with distinctive pathologic features and clinical behavior. *J. Clin. Oncol.* **2006**, *24*, 4677–4684. [CrossRef]
67. Bottoni, A.; Piccin, D.; Tagliati, F.; Luchin, A.; Zatelli, M.C.; degli Uberti, E.C. miR-15a and miR-16-1 down-regulation in pituitary adenomas. *J. Cell Physiol.* **2005**, *204*, 280–285. [CrossRef]
68. Lines, K.E.; Newey, P.J.; Yates, C.J.; Stevenson, M.; Dyar, R.; Walls, G.V.; Bowl, M.R.; Thakker, R.V. miR-15a/miR-16-1 expression inversely correlates with cyclin D1 levels in Men1 pituitary NETs. *J. Endocrinol.* **2018**, *240*, 41–50. [CrossRef]

MDPI
St. Alban-Anlage 66
4052 Basel
Switzerland
Tel. +41 61 683 77 34
Fax +41 61 302 89 18
www.mdpi.com

International Journal of Molecular Sciences Editorial Office
E-mail: ijms@mdpi.com
www.mdpi.com/journal/ijms

www.ingramcontent.com/pod-product-compliance
Lightning Source LLC
LaVergne TN
LVHW070114170726
843515LV00007B/1687

9783036574615